Grzimek's ANIMAL LIFE ENCYCLOPEDIA

Volume 1

LOWER ANIMALS

Volume 2

INSECTS

Volume 3

MOLLUSKS AND ECHINODERMS

Volume 4

FISHES I

Volume 5

FISHES II AND AMPHIBIA

Volume 6

REPTILES

Volume 7

BIRDS I

Volume 8

BIRDS II

Volume 9

BIRDS III

Volume 10

MAMMALS I

Volume 11

MAMMALS II

Volume 12

MAMMALS III

Volume 13

MAMMALS IV

Grzimek's
ANIMAL LIFE
ENCYCLOPEDIA

Editor-in-Chief

Dr. Dr. h.c. Bernhard Grzimek

Professor, Justus Liebig University of Giessen
Director, Frankfurt Zoological Garden, Germany
Trustee, Tanzania National Parks, Tanzania

 VAN NOSTRAND REINHOLD COMPANY

New York   Cincinnati   Toronto   London   Melbourne

First published in paperback in 1984

Copyright © 1968 Kindler Verlag A.G. Zurich

Library of Congress Catalog Card Number 79-183178

ISBN 0-442-23038-9

Printed in Federal Republic of Germany

Van Nostrand Reinhold Company Inc.
135 West 50th Street
New York, New York 10020

Van Nostrand Reinhold Company Limited
Molly Millars Lane
Wokingham, Berkshire RG11 2PY, England

Van Nostrand Reinhold
480 Latrobe Street
Melbourne, Victoria 3000, Australia

Macmillan of Canada
Division of Gage Publishing Limited
164 Commander Boulevard
Agincourt, Ontario M1S 3C7 Canada

16 15 14 13 12 11 10 9 8 7 6 5 4 3 2 1

EDITORS AND CONTRIBUTORS

Editor-in-Chief
DR. DR. H.C. BERNHARD GRZIMEK
Professor, Justus Liebig University of Giessen, Germany
Director, Frankfurt Zoological Garden, Germany
Trustee, Tanzanian National Parks, Tanzania

Volume 4

FISH I

Edited by:

BERNHARD GRZIMEK
WERNER LADIGES
ADOLF PORTMANN
ERICH THENIUS

ENGLISH EDITION

GENERAL EDITOR:
George M. Narita

SCIENTIFIC EDITOR:
Erich Klinghammer

TRANSLATOR:
David R. Martinez

ASSISTANT EDITORS:
John B. Brown
Peter W. Mehren

PRODUCTION DIRECTOR:
James V. Leone

EDITORIAL ASSISTANT:
Karen Boikess

ART DIRECTOR:
Lorraine K. Hohman

INDEX:
Suzanne C. Klinghammer

CONTENTS

For a more complete listing
of animal names, see systematic classification or the index.

1. **THE VERTEBRATES**   19

    Introduction by Adolf Portmann

2. **JAWLESS FISHES**   30

    by Heinrich Kühl
    Phylogeny by Erich Thenius

3. **FISHES**   45

    Introduction by Paul Kähsbauer
    Phylogeny by Erich Thenius
    Modern fishes by Paul Kähsbauer
    Fishery and fish breeding by Kurt Lillelund
    Water pollution by Bernhard Grzimek

4. **THE CARTILAGINOUS FISHES**   86

    by Wolfgang Klausewitz

5. **THE BONY FISHES**   130

    Introduction by Werner Ladiges

6. **POLYPTERIDS, STURGEON, AND RELATED FORMS**   132

    by Fritz Terofal

7. **GARS AND BOWFINS**   144

    by Fritz Terofal

8. **TARPONS**   152

    by Kurt Schubert

9. **EELS**   159

    Eels by Carl-Heinz Brandes
    Spiny eels by Werner Ladiges

10. **HERRING**   172

    by Kurt Schubert

11. **OSTEOGLOSSIDS AND MORMYRIDS**   202

    Osteoglossids by Werner Ladiges
    Mormyrids by Jacques Géry

12. **SALMON**   213

    Salmonids by Ludwig Karbe
    Galaxiids by Werner Ladiges
    Deepsea salmon by Wolfgang Villwock

13. **CETOMIMIFORMES, CTENOTHRISSIFORMES,
    AND GONORHYNCHIFORMES**   264

    Cetomimiformes by Wolfgang Villwock
    Ctenothrissiformes by Werner Ladiges
    Gonorynchiformes by Nicolaus Peters

14.	**CHARACINS AND ELECTRIC EELS**	276

Introduction to carp by Jacques Géry
Characins and electric eels by Jacques Géry

15.	**CARP**	305

Introduction and carp by Pedru Banarescu
The goldfish by Bernhard Grzimek and Werner Ladiges
Suckers, gyrinocheilids, psilorhynchids, and hillstream loaches
  by Pedru Banarescu
Loaches by T. Nalbant

16.	**CATFISHES**	363

by Dieter Vogt

17.	**TROUT-PERCHES, TOADFISHES, CLINGFISHES,** **AND ANGLERFISHES**	395

Trout-perches, toadfishes, and clingfishes by Werner Ladiges
Anglerfishes by Wolfgang Villwock

18.	**CODFISHES**	405

by Hans-Joachim Messtorff

19.	**FLYINGFISHES, TOOTHED CARP, AND SILVERSIDES**	426

Flyingfishes by Wolfgang Villwock
Toothed carp by Michael Dzwillo and Wolfgang Villwock
Silversides by Michael Dzwillo

Appendix	Systematic Classification	461

On the Zoological Classification and Names	485
Animal Dictionary:	487
English-German-French-Russian	487
German-English-French-Russian	492
French-German-English-Russian	502
Russian-German-English-French	507
Conversion Tables of Metric to U.S. and British Systems	512
Supplementary Readings	517
Picture Credits	520
Index	521
Abbreviations and Symbols	531

1   The Vertebrates

Subphylum:
vertebrates,
by A. Portmann

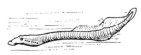

Fig. 1-1. Lamprey *(Lampetra)* fin edge.

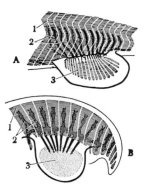

Fig. 1-2. Cross-sectional diagrams of limbs. A. Shark (Selachii) and skate (Rajiformes) embryos; B. Mammalian embryo; 1. Muscle (myotome); 2. Nerves enervating the muscle; 3. Musculature of the limbs.

The vertebrates (Vertebrata) are a subphylum of the chordates (Phylum Chordata; see Vol. III); thus they share essential structural features with other members of this phylum, for example with tunicates (Tunicata) and lancelets *(Amphioxus).* These include a supporting organ in the embryonic stage, a notochord, a central nervous system above the spinal chord and an alimentary canal beneath the spinal chord. The development of the coelom among all organisms with dorsal spinal cords, including vertebrates, has elements in common.

Geological investigations have lent support to the theory of evolution by attesting to one of the greatest developments in the animal kingdom. Fishlike forms predominated in the oldest time periods (recognized by deposits in stratified rock) in the earth's history. Amphibious forms have been present since Devonian times, and reptiles appeared in the Carboniferous Age. The first mammals occurred in the Triassic and the first birds in the Jurassic Period. Primates (see Vol. X) developed in the transition period to the Tertiary and manlike forms (Hominoidea; see Vol. XI) developed in the Miocene (approximately ten to twenty-five million years ago). These findings show us not only how new forms developed but clearly demonstrate the adaptations of these basic forms to their various environmental conditions.

In considering the earliest vertebrate forms we take as a basis two sets of findings: the fossil record in stratified rock and embryological investigations. In both, conditions occur which in all probability repeat certain stages of evolution—although not everything that we find in the development of present-day animals should be taken as such a recapitulation of evolution. Vertebrate embryos show early developmental stages in which paired extremities are missing. This is comparable to a situation in the Silurian Period (approximately 440 million years ago) in which vertebrates without paired fins appeared along with finned fish. Thus the former type is considered to be closer to the original vertebrate form, and older than the latter. In considering present animal forms, it is evident that these early Silurian vertebrates without paired fins have very much in common with the contemporary hagfish (Myxini) and lampreys (Petromyzones). Therefore these two contemporary groups are considered to be living examples of much earlier vertebrate forms.

The earliest form of vertebrate locomotion is an undulating movement. This is activated by a large lateral muscle which extends on the left and right from the spinal cord and spinal column over the trunk and tail. It is divided segmentally, the segments being attached to partitions consisting of connective tissue. The nerve supply corresponds to a segmental organization via pairs of spinal nerves. The paired pectoral fins form behind the head; their muscles and nerves are derived embryonically from the lateral muscle and the spinal nerves.

Among fishes the pectoral fins are anchored firmly in the region of the strong pectoral girdle, which is connected to the skull. In contrast the pelvic fins are attached to the trunk musculature with only weak fibers. The effectiveness of undulating motion is heightened by fins formed in the vertical plane of the body. In numerous cases an associated fin seam is formed which runs along the back and on the lower side to the anterior region. In more highly developed forms we find specialized dorsal, caudal and anal fins. Even the aquatic forms of amphibians—and this includes larvae as well as adults of those species which spend prolonged periods or live permanently in water (i.e., salamanders)—develop such a complete fin system.

The development of land forms—that is, types of animals which correspond to amphibians of today—required a new type of locomotion, a type of "lifting mechanism" with which they could push themselves forward. Such legs and their precursors have developed by means of a modification of the paired fins. In the lobefinned fish group (Crossopterygii; see Vol. V) we find fossil forms with fin systems indicating the transition from swimming fins to the "lifting" type of extremities. The oldest amphibian forms known are from Devonian deposits (which are about 400 million years old); they have a trunk and tail structured similarly to that of a salamander which demonstrates that the lateral muscle was associated with undulating movements, especially in water. The forward push of the extremities originates from the hind legs, the bones of which are firmly connected with the vertebral column. One of the vertebrae becomes a sacral vertebra. The number of vertebrae involved in this sacral region increases with the development of terrestrial vertebrates. Among the earliest reptiles there are two; man has five and birds have as many as twenty-three such vertebrae.

Further changes evolved simultaneously with the new demands made on the pelvic region, such as the modification of the gills, the

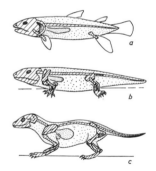

Fig. 1-3. Development of organs of motion. Release of pectoral girdle from skull, development of a neck and disposition of heart and respiratory organs in the thorax. Raising off the ground, extension of limbs, unification of the pelvic girdle with the vertebrae; 1. Lobefin fish (Crossopterygii); b. Amphibian (Stegocephalia); c. Higher mammals (Eutheria).

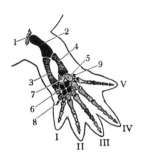

Fig. 1-4. Structure of limb of four-footed animals: I–V, original branching elements of the third segment (autopodium). Front limb: 1. Pectoral girdle; 2. Upper arm (humerus); 3-4. Lower arm; 3. Radius; 4. Ulna; 5-9. Carpals; 5. Ulnare; 6. Radiale; 7. Intermedium; 8. Pisiform; 9. Trapezium; Rear limb: 1. Pelvic girdle; 2. Upper leg (femur); 3-4. Lower leg; 3. Tibia; 4. Fibula; 5-9. Tarsals; 5. Fibulare; 6. Tibiale; 7. Intermedium; 8. Entocuneiform; 9. Calcaneous.

separation of the pectoral girdle from the head region and the development of a new neck segment. The limbs themselves are arranged in a very similar manner among all vertebrates, and thus give striking evidence of their common development. Of the three parts of each arm or leg, the one that is attached to the body bears a single skeletal element (the humerus or femur, respectively). The next part is a pair of supporting bones (radius and ulna in the arms, tibia and fibula in the legs), and finally the hand or foot has a complex structure consisting of several elements, the carpals and tarsals. The outward-branching elements of the hand and foot are attached to these. Originally there were five elements which were divided into a longer element, the mid-hand or foot, which led into the several segments of the fingers or toes.

The changes in this basic plan are of astonishing variety. In bats (see Vol. XI) and flying saurians (see Vol. VI) the hand bones become large wing support structures, which in the latter forms are an elongation of a finger. In the front limbs of birds (see Vol. VII) the outward-branching elements are less developed, consisting of three elements; they support the wings, whose surface is largely built up by the plumage. In hoofed animals the limbs can also have a reduced number of branching elements, as with two hooves in even-toed ungulates (see Vol. XIII) or reduced to the single mid-member in horses (see Vol. XII).

Such a regression of paired limbs could only occur in the early stages of vertebrate evolution—that is, when the side musculature is still well developed. In the course of evolution this occurred in many groups of fishes, amphibians and reptiles; and in about a dozen different lines of development it culminated in apody, a complete lack of feet. In lizards the relationship of trunk and tail has remained in such cases, and in snakes the tail segment is very small.

As in all highly developed animal groups, the vertebrate sensory organs, the central nervous organ and access to the digestive tract are united at the front end of the body to form a head. Thus in the head region a number of complex structures lie very close together, and that has led to a correspondingly complex skeletal configuration: the skull. This skull arises from three structures which during embryonic development are initially separate: brain case, or neurocranium; the skull roof, or splanchnocranium; and the palatal complex, or dermatocranium. The brain case protects the brain and the large sensory organs (nose, eyes and ears). The skull roof is a portion associated with the foregut and thus originally with the gills. The palatal complex is formed from skeletal elements in the dermis without a previous cartilaginous stage. These so-called "dermal bones" rest on top of the cartilaginous parts already formed.

The primitive structures, as early fish forms show, could become modified to such a degree that one major skeletal part (i.e., the lower jawbone of mammals) is formed from just one primitive element (in this case from one of the dermal bones). Thus very little of the original structure still remains. In contrast to that the lower jaw of pre-mammalian forms—from the fishes to the birds—is made up of several dermal bone elements, in which two bones of the old skull roof have remained in the jaw joint.

A single bone pair from the arch of the skull roof is modified in the transition from an aquatic to a terrestrial life; it is the uppermost part of the hyoid bone arch. On each side of the head it forms a bar which conducts sound (the columella of the middle ear, but in mammals the stapes; compare Vols. VI and X).

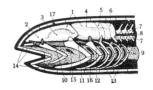

Fig. 1–5. Structure of the early skull: Neuro-cranium, splanchnocran-ium and dermatocranium. Arched elements are shown by hatched region; 3. Eye region; 4. Ear region; 5 and 6. Occipital region; 7. Arcualia; 8. Spi-nal cord; 9. Foregut; 10. Premandibular arch; 11. Mandibular arch; 12. Hyoid arch; 13. Bran-chial arch; 14. Bone plates of the skull; 15. Preman-dibular slit; 16. Spiracu-lum; 17. Pinealforamen.

Already in the oldest known forms the vertebrate brain is highly developed; however among the lowest vertebrates the interrelation-ship of the parts of the brain is quite unlike that of the highest forms. Five relatively distinct brain segments are arranged in the following order along the longitudinal axis: 1. Forebrain (Telencephalon). This consists of two hemispheres, and in birds and mammals is the most substantial brain component. 2. Diencephalon. Here the pineal gland is located on its upper side and on its underside we find the pituitary, a particularly important hormonal gland. Furthermore in many fishes, amphibians and a few reptiles there is a light sensitive organ in the upper part of this region, the parietal organ. 3. Midbrain (Mesence-phalon). Generally this consists of two symmetrical formations which in the early brain are the endpoints of the optic nerves. This function is less developed in mammals and both projections form only two forward protrusions of the corpora quadrigemina. 4. Cerebellum. This is very poorly developed in the earliest vertebrates, but in many fishes is the most massive part of the brain and it is pushed over the other parts of the brain towards the front and back. This part is the center for equilibrium and movement coordination. 5. Brain stem (Myelen-cephalon). Of all brain segments this one is most like the spinal cord. Important cranial nerves radiate from the myelencephalon. Of the twelve cranial nerves, long known to anatomists, the so-called bran-chial nerves lie in the myelencephalon. These are nerves which in the original vertebrate head were each associated with a gill slit.

In the early aquatic vertebrate forms, the olfactory organ is found towards the front; it consists of a cavity with sensory cells, which in most fish is separated from the mouth, but in terrestrial vertebrates is connected with it. The flowing water which conducts olfactory sub-stances to this organ is in fishes diverted by a particular opening, a canal which in all probability has been modified into the tear ducts of terrestrial vertebrate eyes. In all early forms the olfactory function is well developed; these animals are called macrosmates, in contrast to the forms with a less-developed olfactory sense, among which the eyes and ears play a greater role and which are known as microsmates. For example the shark, skate and eel are olfactory-oriented organisms

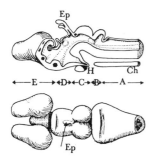

Fig. 1-6. Early vertebrate brain. Upper sketch: longitudinal cross-section: A. Brain stem; B. Cerebellum; C. Midbrain; D. Diencephalon; E. Forebrain; Ep. Pineal gland; Ch. Spinal cord; H. Pituitary.

unlike the pike. Birds, utilizing eyes and ears to a great degree, have only a poor sense of smell. In mammals it is again the earlier forms, such as marsupials, insectivores, rodents and small predators, which are more oriented to olfactory sensations than are the more highly developed forms.

At the front end of the body there is often a special organ which also has an olfactory function: the Jacobson's or vomeronasal organ. In amphibians it is an appendage to the nasal cavity, and in reptiles it is an independent cavity with its own opening into the mouth cavity (see Vol. VI).

Vertebrates have lens eyes and represent the highest developmental stage of these organs; only the visual organ of the cuttlefish (see Vol. III) matches its capability. The way the eyes develop—as protrusions of the brain up to the outer surface of the body—results in the light-sensitive cells of the retina lying well away from the light, and the optic nerve must enter the retina at the so-called "blind spot", so that its fibers can pass back to the central brain region. The cuttlefish eye does not utilize this indirect method because it originates in a different way. The adaptation of the eye to various distances occurs differently in the original aquatic vertebrates than among terrestrial forms. When at rest the fish eye is focused on nearby objects; changing to distant focus is achieved by moving the relatively rigid spherical lens on the optical axis. In contrast the relaxed eye of terrestrial forms is focused at a distance and changing to near focus is done by altering the shape of the lens. The terrestrial vertebrate eye has a complex gland and protective eyelid system which are found in aquatic forms only rarely.

The auditory system develops in invertebrates as a vesicle under the skin, as does the vestibular apparatus, which in vertebrates has an upper element (forming in lampreys just two, but in all other vertebrates three, arched passageways) which is sensitive to three-dimensional changes in spatial orientation. The lower part of the labyrinth contains a surface of sensory cells in the first enclosure; these are divided into three parts which are associated with the arched processes. This portion of the labyrinth also has two sensitive spots which in bony fishes rest on specially developed bodies (statoliths, also called otoliths), and otherwise are covered by a gelatinous mass in which calcium crystals are embedded. Another auditory sensory area in the lower labyrinth takes the shape of a snail in birds and mammals; in mammals this structure has as many as four and one-half winding convolutions.

The labyrinthine auditory organ contains sound transmitting organs which may exist in highly different forms. The modification of an embryonic gill slit into the middle ear is a widespread phenomenon. The middle ear is closed to the outside by an eardrum, the vibrations of which are transmitted into the labyrinth by small bones. In bony fishes sound transmission can derive from the swim bladder and be accomplished by modified vertebrae. Many different sound transmission systems exist in amphibians. Localizing the source of a sound is refined in mammals with the outer ear which often has a highly movable ear muscle.

The conversion of matter necessary for maintaining the vertebrae is accomplished by highly developed organs of digestion, respiration and excretion; they are all connected by a circulatory system, in which blood flow is maintained by a pulsing center, the heart. All these organs are enclosed together with those of reproduction in the body cavity. In all vertebrates the heart is separated from the body cavity by the pericardium. Mammals also have a muscular wall, the diaphragm, which separates the chest cavity from the visceral cavity. Association with the outer environment is maintained by the mouth opening and, when breathing, by the nasal openings. Elimination of body wastes generally takes place at the end of the body cavity, where specialized openings are located. One is the urogenital tract, with which a cloaca for the intestinal tract may be associated.

The mouth region is structured in a great variety of ways to accomplish the task of food intake. Bony tooth structures occur in general; this, however, is not the case for lampreys (which have horny teeth), tadpoles (with horny jaws), and the larvae of fishes and toads. Tooth structure is highly differentiated. Teeth contain bonelike dentine and a coating of a harder substance, enamel, which is excreted by specialized cells but in very different amounts influenced by the tooth-forming layer. Hard formations corresponding to teeth are found in early vertebrates even in the external skin (e.g., in sharks). In some cases many bones bear teeth in the mouth cavity; even the esophagus can have teeth. The simplest forms are rootless teeth anchored to the bones by connective tissue. Tooth structure becomes more complex in the evolutionary development of the various vertebrate groups; the forms of the single teeth as well as the order of the rows of teeth vary. Changing of teeth is a continuous process in primitive vertebrate species, but ultimately a single change occurs, and some mammals do not change certain teeth at all. A complete retrogressive development of the teeth can be found in all large vertebrate groups; in these there are occasional horny beak formations (such as in turtles and birds).

Types of nutrients utilized have varied considerably in the course of vertebrate development. The earliest forms took only the simplest nourishment or fed by hunting. With the coming of land plants in later periods, highly specialized herbivores become widespread. Large ungulates have developed in connection with this new feeding style. Only the oceans have made possible the existence of large animal forms utilizing small organisms as food (such as the whalebone whales) and the primeval predation methods (such as used by the giant sperm whales).

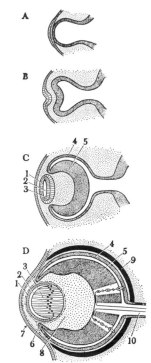

Fig. 1-7. Vertebrate eyes originate from the lower side of the diencephalon; they can be recognized as optic vesicles, the forward wall of which folds inward and leads into the double-layered optic cup. The light-sensitive cells are formed from the inner layer and are derived from the brain. The lens system is disassociated at the point of contact between the optic cup and the epidermis. A. Optic vesicle; B. Beginning of inward folding; C. Developing eye; D. Eye: 1. Epithelial layer; 2. Lens vesicle; 3. Lens fibers; 4. Outer wall; 5. Inner wall; 6. Optic cup edge; 7. Corneal epithelium; 8. Vitreous body; 9. Scleral cartilage; 10. Optic shaft.

The front of the alimentary canal is the site of important respiratory organs. The gills of marine vertebrates develop in this region; they consist of paired slits in the front end of the alimentary tract. The large respiratory organs of terrestrial animals—the lungs—develop from the same embryonic structures in the rear gill region. In fish these structures form large bladders, paired or unpaired swim bladders, with or without a connecting passage to the alimentary tract (this passage can disappear during embryonic development). Beginning with amphibians the lungs develop from an increased growth of the inner upper part of the tract; they serve to exchange gases with the blood. Lungs have become adapted to the newly appearing neck segment by an elongation of the trachea. Their connection with the mouth cavity, the larynx, can become a vocal organ, which is highly differentiated in mammals. In birds the branching of the trachea into both lung bronchules takes over this task.

All vertebrates have a closed circulatory system. Muscular arteries become capillaries at their fine endings; from the capillaries the blood flow is transferred to the venous system which leads back to the heart. Blood circulation requires free blood cells, of which the red blood cells (erythrocytes), containing hemoglobin, primarily serve gaseous exchange. In all non-mammalian vertebrates these erythrocytes have nuclei. White blood cells, or leucocytes, have several functions. The blood of higher vertebrates (reptiles, birds and mammals) also contains platelets (thrombocytes) which participate in clotting. In adult vertebrates red and white blood cells develop in the spleen and bone marrow. Homeiotherms ("warm-blooded animals") have lymph nodes which are also involved in white blood cell formation.

The circulatory pump, the heart, lies within a pericardium or sac which helps prevent distension of the softer parts of the heart wall. In the early marine vertebrates the venous (oxygen-poor) blood gathers in a specialized vena cava; it then flows through valves into a forechamber (auricle) with relatively weak musculature and from there into the muscular ventricles, from which it is directed to the gills. In the gills gas exchange takes place. With the formation of swim bladders and lungs, other pathways develop so that blood destined for gas exchange is directed into the heart itself.

Metabolic wastes pass into the kidneys for excretion. To a large extent these wastes are the result of complex processes which take place in the liver; ammonia, urea, uric acid and related substances are formed, the excretion of which takes place in the kidneys. The kidneys are formed at the gastric wall on both sides of the spine. In embryonic stages they consist of small, segmentally arranged canals which originate as siphons on the gastric cavity. Then blood glomeruli form within specialized kidney capsules (Bowman's Capsules), which filter urine from the blood into the kidney canals. Convoluted tubules of the canals, richly supplied with venules, take additional materials from

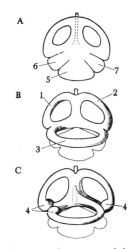

Fig. 1-8. Structure of the ear labyrinth: A. Development of the two vertical arched passages; B. Development of the horizontal archways; C. Development of the three ampullae; 1, 2, and 3. Archways; 4. Ampullae; 5. Sacculus; 6. Descensus utriculi; 7. Lagena, which in higher vertebrates develops into the cochlea.

the blood and also replace vital materials from the original wastes back into the blood. In mammals, venules in the renal glomeruli attend to this supply, so that complete urine exchange takes place in the glomeruli associated with the capsules.

Particular nerves whose action is largely involuntary coordinate the mutually dependent tasks of metabolism. They may be contrasted to the so-called somatic nerves whose function is dependent on the will. This autonomic nervous system (autonom = independent) is organized into sympathetic and parasympathetic nerves.

In addition to these nerves there are particular glands which participate in metabolic control. They excrete hormones into the circulatory system and are collectively known as glands with internal secretion. Among them the pituitary plays a leading role. This gland develops in part from the brain, in part from the roof of the mouth of the embryo. Other such glands are the thyroid, thymus and parathyroid glands, all of which develop from the front portion of the alimentary tract. The pineal gland has already been mentioned as a brain organ. But there are also cell groups within the brain which excrete hormones, and in the gastric region special cells of the pancreas form hormones. In lower vertebrates there are cell concentrations in the region of the kidneys, which in higher forms exist as a single organ, the adrenal gland. It secretes several substances; in mammals the outer part is thereby functionally differentiated from the inner part. In birds and reptiles these two parts of the adrenal are not so greatly different. The male and female genitals are also important hormone-producing glands. They produce male and female sex hormones which act in various ways.

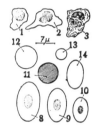

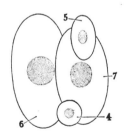

The "cold-blooded" (poikilothermic) vertebrates include fishes, amphibians and reptiles, while "warm-blooded" (homeiothermic) animals are the higher groups, the birds and mammals. Actually the terms cold- and warm-blooded are misleading, if well-known, ones. A reptile sunning itself on a rock is not "cold-blooded", and neither is a tuna, the body temperature of which has climbed by about 10° C. above that of the water by rapid swimming (which in the Mediterranean can mean body temperature approaching that of our own).

Poikilothermic vertebrates are to a great extent dependent on their surroundings. Therefore they have developed behavioral adaptations in order to withstand environmental extremes, such as extreme summer heat or winter temperatures, during the latter of which the metabolic rate could fall dangerously low.

Fig. 1-9. Blood cells: 1–3. Leucocytes (white blood cells); 4–14. Erythrocytes (red blood cells); 4. Lamprey; 5. Skates; 6. Salamander (Amphiumidae); 7. Proteid; 8. Frog; 9. Lizard; 10. Hummingbird; 11. Man; 12. Elephant; 13. Dwarf musk deer; 14. Camel, llama.

In contrast homeiothermic vertebrates have mechanisms which largely prevent great variations in body temperature, controlled by the higher nervous organization (see Vol VI, Chapter 1 and Vol. III).

As in all highly developed animal groups, adult vertebrates are bisexual, but in some species, e.g., frogs, intermediate forms can be

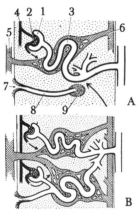

Fig. 1-10. Vertebrate
kidney systems.
A. Sharks and skates
(Selachii and Rajoidei);
B. Salamanders (Urodela);
above without, below
with ciliated funnels;
C. Frogs (Anura);
1. Bowman's Capsule;
2. Glomerulus; 3. Kidney
tubule; 4. Artery; 5. Vein
leading to kidney; 6. Vein
leading away from
kidney; 7. Ciliated funnel;
8. Early kidney canal;
9. Lymphatic organ.

found (juvenile hermaphrodites). Such forms are found in the adult stage only in scattered bony fish species.

The gonads—whether testes or ovaries—develop in the kidney region and are associated with these organs. In the simplest forms the germ cells and gonadal secretions are passed out without specialized tracts. In lampreys the paired gonads develop into a single unit at an early stage of growth; their contents are emptied into the gastric cavity and reach the outside via two short canals near the opening of the urinary passage. In all higher groups the tracts from the gonads are more strongly associated with the urinary tracts in females than in males. The descent of the male gonads and their deposition in an external scrotum is found in numerous mammals. In both sexes the germ duct is embryonically formed from the urinary tract; this duct becomes an independent tract early in development and becomes established for germ cell production. The ovaries are always separated from the kidneys so that mature egg cells are directed through the gastric cavity to the opening of the Fallopian tubes.

In the lowest vertebrates (cyclostomes, many fishes and amphibians) fertilization of the eggs occurs in open water; complex behavioral patterns, often associated with nestbuilding, assure that the germ cells meet. Many cartilaginous fishes, a few bony fishes and the highest vertebrate groups (reptiles, birds and mammals) have reproductive organs; as paired organs they are formed from the ventral fins, and they are also paired in snakes and lizards, in which they are in the cloacal region. They are unpaired organs in crocodiles, turtles and mammals. Only a few bird species (e.g., ostriches and geese) have unpaired reproductive organs; in other bird species a courtship ceremony, which is often highly developed, precedes copulation.

The primitive vertebrate egg is rather rich in yolk. In the first developmental stage the entire egg divides into cells and undergoes embryonic growth (it is known as a holoblast). The holoblastic germ cells of some early fishes (i.e., of sturgeon and lung fish) and those of some salamanders, develop in a few days to a juvenile form which does not become greatly modified in future development. This is made possible by the rich yolk supply in all germ cells. Other species such as lampreys, many fishes and the majority of salamanders, frogs and toads, form specialized transition organs early in embryonic development. These are sucking organs, adhering cilia, external gills and others. The embryos become larvae, which must often undergo a metamorphosis, an extreme structural modification, before achieving the mature species form. During metamorphosis the transitory organs are redeveloped in a process often associated with a change in feeding. Thus the larvae of frogs and toads—tadpoles—feed on plant substances while the adult forms are carnivorous.

In numerous vertebrate species another basic egg cell plan has been developed; only a part of the living egg mass develops into an embryo. The rest remains a nutritional source in the yolk sac. Furthermore additional organs appear on the embryo, which function in the embryonic development. In the simplest cases this consists of a visceral network surrounding the yolk sac, creating a circulating yolk supply. The yolk sac stays in connection with the embryonic gut by a yolk shaft; this shaft and the yolk sac blood vessels are joined by an umbilical cord. Such eggs, which develop only in part into embryos, are designated "meroblastic". In vertebrates these meroblastic eggs exist in their simplest form among classes of bony fishes, and are more developed in sharks, rays and their relatives.

Meroblastic eggs do not necessarily require a rich yolk supply. There are bony fish embryos with only one-half millimeter diameters, in which only a germ disc nourishes the embryo, in contrast to the amphibian egg cell of the same size which develops completely into an embryo. But very large yolk supplies occur only in meroblastic eggs; ostriches and large ray species have eggs with diameters of ten to twenty centimeters. Shark and ray eggs are similar, in terms of possessing egg white, to the most well-known eggs in the animal kingdom: those of reptiles and birds.

Reptile eggs are covered by a thick outer coat in which minerals in various quantities are stored. As in birds, the eggs develop out of the water, which is actually the original environment of vertebrate eggs. But in spite of terrestrial development, the vital "watery surrounding" is assured for the embryo, for a skin-like layer (the amnion) grows around it and serves as an inner envelope sealing in water with the embryo. Thus the classes of reptiles, birds and mammals are classified as amnionic organisms (Amniota) and classes of other vertebrates are considered anamnionic (Anamnia; see Vol. VI).

Amnionic animals also develop a specialized urinary bladder from the embryonic cloacal region, which via the umbilical cord and together with the yolk passage and blood vessels leads from the embryo and extends outside the embryo. This "allantois" (see Vols. VI and X) can be less well developed in some mammal species, such as in rodents and primates, in which the mother takes over elimination of wastes through the placenta. Almost all reptilian eggs are dependent on an external, additional water supply. During the long, drawn-out embryological development period the total embryo weight is reduced. In birds the water supply is measured in advance such that unavoidable evaporation during the incubation period is equalized by the number of eggs laid.

It seems quite certain that mammalian species have undergone parallel development in each of their evolutionary lines; in the process many deviant lines have appeared, of which no traces remain. Many such lines have been indicated by the development of the Theriodontia

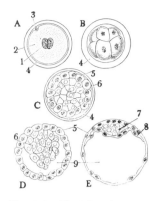

Fig. 1-11. First developmental stages in mammals: A. Mature, fertilized egg cell at the moment of union of the pronucleus; B. Eight-cell stage; C and D. Separation of the trophoblast and embryonic knot situated outside the embryo; E. Embryonic vesicle with the initiation of migration of entoderm cells; 1. Pronucleus; 2. Cytoplasm; 3. Pole body; 4. Yolk sac; 5. Nutritive layer outside the embryo; 6. Embryonic node; 7. Embryonic substance; 8. Inner cells; 9. Blastocyst.

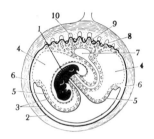

Fig. 1-12. Development of the placenta in the ground squirrel *(Citellus):*
1. Embryo; 2. Trophoblast; 3. Wall of the umbilicus; 4. Extra-embryonic coelom;
5. Yolk sac vesicles;
6. Edge sinus; 7. Allantois (bladder); 8. Allantoic vesicles; 9. Nutritive layer of the placenta outside the embryo, as far as it has developed from the allantois; 10. Uterus glands.

(the ancestral reptilian group from which the mammals developed), whose own origin is not understood. The cloacal animals (see Vol. X, Chapter 2) lay yolk-rich eggs which are strikingly similar to reptilian and bird eggs. Marsupials and the higher mammals characteristically have very small egg cells with little yolk content. Their embryonic development is described in Volume X (Chapters 1, 3 and 10).

The present-day species of subphylum Vertebrata (or Craniota) are divided into seven classes: 1. Cyclostomes (Cyclostomata); 2. Cartilaginous fishes (Chondrichthyes); 3. Bony fishes (Osteichthyes); 4. Amphibians (Amphibia; see Vol. V); 5. Reptiles (Reptilia; see Vol. VI; 6. Birds (Aves; Vols. VII–IX); 7. Mammals (Mammalia; Vols. X–XIII). Many zoologists classify the cyclostomes as jawless vertebrates (superclass Agnatha) and the other six classes as jawed vertebrates (Gnathostomata).

2 Jawless Fishes

The superclass of jawless fishes (Agnatha), with the sole extant class of cyclostomes (Cyclostomata), is placed at the beginning of the vertebrates. The cyclostomes have been well investigated with respect to their internal and external anatomy and many different opinions have been brought forth about the origin and relationships of these animals. But their way of life and behavior have not been described to the same extent as those of fishes or higher vertebrates.

Yet it is quite exciting to see the most well-known cyclostomes, the lampreys, engaged in spawning behavior in the aquarium or in open water. I still clearly recall that my institute colleagues could not be dragged away from the 100-liter aquarium when the lampreys dragged stones about, built nests, and then spawned, after previously having done nothing else for months other than hanging by their suckers on the walls. The jawless fishes are also among the most interesting animals in other respects. It is in this group that many phenomena may be observed which are significant for students of evolution, for example the species development and the regression and modification of organs. G. Sterba, who has done extensive work with lampreys, rightly noted that Darwin observed only those finch species prevalent on the Galapagos Islands in the Pacific Ocean during his visit and thus first came upon the idea of transformation of species (see Darwin finches, Vol. IX) and that Harms could then prove the existence of species evolution from expeditions made on the Malayan islands. But, as Sterba added, "has anyone ever noted lamprey evolution, which is a prime subject for scientific inquiry? Tomorrow perhaps only preserved lamprey will decorate German museums, and then we will undertake trips throughout the world in order to capture them somewhere in North America, Asia or Australia. Ask your grandfather; he'll know lampreys for sure, because they were prevalent in all our flowing waters fifty years ago".

The cyclostomes (Cyclostomata) are water inhabitants with elongated, wormlike or eel-shaped bodies which lack paired extremities.

Superclass: jawless fishes,
by H. Kühl

Speciation

Class: cyclostomes

Distinguishing characteristics

They also lack jaws having instead a toothed sucking mouth that is long or round. They have gills but no gill covers and their skin is naked and slimy. Instead of a skull lampreys have a few cartilaginous elements and a cartilaginous strip without a pectoral or pelvic girdle. Lampreys have an unpaired nasal opening. There are two subclasses: 1. Hagfishes (Myxini) with the family Myxinidae, and 2. Lampreys (Petromyzones) which includes the family Petromyzonidae.

Origin of the jawless fishes

It is fairly probable that the origins of jawless fishes can be found in brackish water, i.e., in the intermediate region between saltwater and freshwater with its high variations in water temperature, salinity, water flow, depth and other characteristics, which accelerate an evolutionary process. Some structural features of the cyclostomes were present primevally and others have been modified as a result of the cyclostome life habits.

Distribution

The distribution of jawless fishes is limited to the temperate to cold waters of the northern and southern hemispheres. The northern species are more highly differentiated from the southern ones than those in the same hemisphere. Water temperature is an important factor in distribution of the sea-living hagfishes. The ten degree (°C.) boundary is crucial for them. In the cold northern and southern seas hagfishes can penetrate to approximately thirty meters, while at the equator they are found at depths of over 1000 m.

Subclass: hagfishes

Hagfishes (subclass Myxini, family Myxinidae) are worm-shaped jawless fishes whose nasal opening lies at the fore-end of the body and is joined with the mouth opening. The species has four to six barbels on the head; the mouth has two rows of protruding teeth and a row of mucous glands are on each side of the stomach. A single fin seam is present and the cartilage gill skeleton is poorly developed. Hagfishes have underdeveloped eyes which are not visible externally. They are pure sea inhabitants, and they lay only a few large eggs which in development lack a larval stage. There are two hagfish subfamilies: 1. Myxininae, having 5–15 gill sacs with but one opening; 2. Bdellostomatinae, the members of which have gill sacs with separate orifices.

Distinguishing characteristics

The NORTH ATLANTIC HAGFISH or COMMON HAGFISH (*Myxine glutinosa*; see Color plate, p. 41) lives on soft ground. Boring into the floor is accomplished by powerful swimming motions of the rear body portion in a vertical position. As aquarium observations show, these motions subside when the front half of the fish has disappeared into the ground. Further boring in is then done with the head end. In this way disappearing into the ground can take place very rapidly, in about one minute. Sometimes an hour can pass before the head again appears on the surface. First the nose appears; a funnel then forms around the head from the force of respiring in the water, the funnel lying in a small hill. The bored-in passages are not coated with mucous; should someone step in them, they collapse immediately.

Hagfishes in aquariums usually lie on one side. Before swimming

Fig. 2–1. Distribution of hagfishes (Myxinidae).

they first lie on their stomach, and then make undulating movements whereby the head is greatly protruded—further than in eels, for example. The four nasal barbels radiate outward, but the mouth barbels are laid back so that the mouth remains closed. The barbels are tactile and chemical sensory organs. The common hagfish can also readily swim backwards when threatened. One special movement is that of "knotting up", whereby the fish can form a figure eight. In this manner it frees itself of clinging dirt and slime. When hagfishes are grabbed firmly, they can free themselves very successfully by this motion. The knotting also serves as a means of tearing out parts of larger food fragments; for example the hagfishes grab onto a dead fish, form a knot (a figure eight), push against the fish and draw their mouths back through the figure eight loop, so that a piece can be torn off.

Surfaces inhabited by hagfishes have the appearance of little volcanic regions. Underwater observations and photos taken at thirty meter depths in Hardangerfjord show that many little hills touch one another. These little hills have diameters of thirty three centimeters and are eight centimeters high. A hole with a two-centimeter diameter is on top, surrounded by a small crater. Common hagfishes can also find smaller silt hills and live more densely. As a marking experiment has shown, they are even able to find these spots from great distances. A common hagfish was taken several kilometers away from its place and successfully returned to it. Hagfishes always live near the sea floor and therefore they will not take bait suspended more than one meter above the floor.

The hagfish eye is covered by skin and has neither lens, iris nor muscles, and the corresponding connections to the brain are poorly developed. Experiments have shown that this eye has no particular visual significance; interestingly, light-sensitive organs have been found in the skin, especially in the head and rear regions. A similar condition exists in other boring and hole-inhabiting animals.

The equilibrium organ is also greatly simplified; it consists of a ring with two bubble-like thickenings. Structurally it is not as underdeveloped as the eye, however. Thus one can assume that this organ transmits some information about spatial orientation, although the common hagfish can swim either on its back or abdomen.

The nasal opening is directly above the mouth cavity; it has a saclike vesicle and a direct connection to the pharynx. The nose also has nervous connections with the brain lying above. Thus water flow is directed through the nose and the·olfactory substances therein can immediately alarm the animal. Aquarium and field experiments have shown that common hagfishes become active immediately after some bait has been laid and can find it within two minutes. Little is known about distances from which food can be detected, but on the basis of some observations this would amount to fifty to·sixty centimeters.

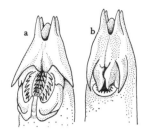

Fig. 2-2. Head of the North Atlantic hagfish (*Myxine glutinosa*); a. with opened mouth; b. with closed mouth.

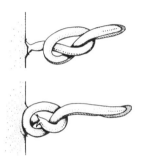

Fig. 2-3. Hagfish (*Myxine glutinosa*) knotting.

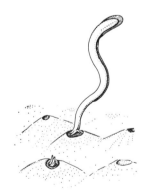

Fig. 2-4. Where hagfishes live in great concentrations the floor looks like a lunar landscape. In the foreground the head of a hagfish looks out.

Large numbers of hagfishes collect around a good-sized piece of food after a short time. In one case 123 hagfishes were found at a dead cod, and another time 100 had collected at a bait.

Hagfishes grab on with the teeth and tear out pieces utilizing the technique described above. They thereby reach the internal organs of the dead fishes, usually by way of the gills. When fish are caught with lines or baskets and are hindered in their motions, one often finds hagfishes in their bodies. Until recently this led to the widespread belief that hagfishes are parasites and that their structure evolved from the parasitic behavior. It was said that they penetrate a living fish through the skin and gills, kill the fish, and then so completely devour the fish's insides that after a short time nothing was left but a sack of skin and bones. Now all investigators agree that hagfishes feed only on dead or dying fishes. It is possible that they attack fishes or octopods (order Octobrachia; see Vol. III) which have been caught in nets and cannot move or have suffered injuries. Dean has reported this for the Japanese hagfish *(Bdellostoma burgeri),* which feeds on living fishes caught in nets.

Non-parasitic

Dead animals are quickly devoured by hagfishes. However there is not a sufficient number of dead fishes or other organisms lying on the ocean floor to serve as the prime food source for larger hagfish populations. Extensive investigations have shown that the main hagfish diet does not consist of carrion but organisms inhabiting the ocean floor such as annelids, echiurid worms *(Echiurus echiurus)* and others. Small snails and mussels less than a millimeter in size have also been found in hagfishes, as well as the remains of shrimp and hermit crabs. These very active crabs were almost certainly eaten in injured, dying or dead states. It is at any rate certain that hagfishes feed when burrowing through the mud, for mud masses with unicellular organisms and bacteria have been found in the guts of hagfishes.

Although hagfishes of the North Atlantic occur in great numbers and their prevalence has been investigated, much of their reproductive behavior is not understood, but hagfishes have some characteristic features in this respect. Sexual differentiation is incomplete in hagfishes, but the hermaphroditic condition is in a stage of regressive development. These organisms appear to be in a transition stage leading to bisexuality. Males and females cannot be distinguished even when mature.

The unpaired and unfolded gonads do not have their own genital tract. Germ cells pass into the body cavity and are sent through the kidneys into the cloaca and thus out of the body. The bean-shaped eggs are quite large, being fourteen to twenty-five millimeters long. They are equipped with anchoring fibers on their ends which harden after the eggs are laid into the water and with the help of which the eggs adhere to each other and the floor. Eggs often adhere to moss

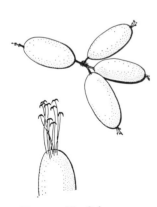

Fig. 2-5. Hagfish eggs. The large bean-shaped eggs have anchoring fibers with which they can adhere to each other and the floor. Left: anchoring fibers enlarged.

animals (Polyzoa or Bryozoa). A female can probably lay up to twenty eggs throughout the year, although whether this is done all at once or in groups is unknown. In general hardly anything is known about hagfish spawning behavior. Presumably the eggs are fertilized during or after laying. The young hagfish is about 4–5 cm long when leaving the egg, which breaks along a seam. Hagfishes undergo no larval stage.

As pure ocean organisms hagfishes are very sensitive to a lowered salinity, but still more to raised water temperature. Therefore they occur only in colder regions in shallow water and in warmer waters at great depths. They can be kept for long periods in aquariums which have a salinity between thirty-one and thirty-four percent and a maximum water temperature of 10° C.

In summary, the characteristic hagfish structures do not proceed from earlier assumptions of a parasitic life mode, but rather because these jawless fishes live like floor-dwelling worms which feed on small organisms and also fresh carrion. They can be designated as "vertebrate worms". Hagfishes have become quite distinct from their ancestors at a very early evolutionary stage.

LAMPREYS (subclass Petromyzones, family Petromyzonidae) are also eel-shaped, but as adults have well-developed eyes. The nasal orifice on top of the head has a blind end. There are seven gill openings and a circular sucking mouth well supplied with teeth is on the lower side of the head. Lampreys lack barbels. They have two dorsal fins and one caudal fin. Lampreys are free-living inhabitants of salt and fresh water. Development is indirect, with a distinct larval stage. There are two lamprey subfamilies which are distinguished by the structure of the sucking mouth: 1. Petromyzoninae with five genera (with but one horny plate and cilia on the edge of the mouth); 2. Mordacinae with three genera (these species have two horny plates and no cilia). The Australian lampreys *(Geotria)* have a throat sack.

Lampreys undergo metamorphosis similar to amphibians and some fishes: they have a juvenile stage, which as a larva is structured differently than the adult form and leads a completely different way of life. In many cases the larvae even inhabit another environment. Two groups can be differentiated: migrating forms and fresh water forms. Members of the former group live as adults in the ocean or in brackish water near the coast, but after a period of time migrate into rivers and streams, lay their eggs, and their larvae live there until they metamorphose. After metamorphosis the young larvae migrate again to the coast. Species of the second group spend their entire lives in fresh water and do not migrate.

All lampreys are born in fresh water and live there as larvae until metamorphosis. These conditions play a significant role in lamprey speciation. Lamprey larvae were once considered to belong to another species and were designated *Ammocoetes branchialis.* The larvae are

Subclass: lampreys

Distinguishing characteristics

Fig. 2-6. Distribution of lampreys (Petromyzonidae).

Lamprey larvae

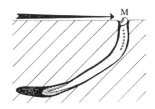

Fig. 2-7. Lamprey larva in mud tube. The mouth opening (M) is directed against the water current (arrow).

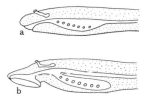

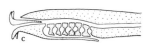

Fig. 2-8. Nasal passage and gill sac in the lamprey larva (a), adult lamprey (b), and hagfish (c).

blind, with a toothless mouth which is double-lobed and has brushy processes (the so-called cilia); the gill sack joins the alimentary tract in the immediate proximity. The larval body is ringed and has the appearance of a worm.

Lamprey larvae appear in streams where the water flows slowly and their fine slimy parts can easily find a firm sitting place. In soft muddy ground the larvae sit in tubes which develop from the larvae lifting their heads off the ground and boring in with powerful blows of the tail. Near the gills the larvae secrete a substance through which the front part of the mud tube is fixed. These conditions vary somewhat, often consisting of U-shaped tubes that are usually a few centimeters longer than the larvae themselves. At the slightest disturbance, for example when shaken, the larvae withdraw into their tubes.

The larval mouth always protrudes somewhat from the mud bottom and is oriented against the water current. The larvae are filter feeders; they hold back all larger particles with their cilia and from time to time expectorate them by drawing the cilia together. Their food consists of small organisms such as diatoms and detritus which are captured in the gills and pass into the gut. According to Sterba food is not captured by pumping of the gill pouches, but by ciliary movement. The larvae often change their location.

As a result of filter feeding, the growth of lamprey proceeds slowly. The larvae reach a length of 2 cm after one year, as Applegate has shown for the SEA LAMPREY (*Petromyzon marinus*) and Sterba for the BROOK LAMPREY (*Lampetra planeri*). After four to five years the larvae are 10–20 cm long. In sea lampreys the larval period, at the same rate of growth, is perhaps somewhat shorter; metamorphosed lampreys only eight centimeters long have been found. A larval period of seven years is given for the AMERICAN BROOK LAMPREY (*Ichthyomyzon fossor*). Sterba claims that the onset of metamorphosis is influenced by the nutritive development of the larvae. Metamorphosis itself is activated by the pituitary gland. During this time feeding ceases so that body size has decreased by one-ninth when metamorphosis is complete.

The structural modifications which lampreys undergo are considerable. The larval head takes on a new appearance: the circular mouth develops from the double-lipped mouth; the rasping tongue-like structure develops in the mouth cavity, and teeth erupt. The larval gut is modified and the connection to the gut is closed. The gill pouch develops, which serves only respiration. In the following life stage metabolism is increased and the oxygen demands are considerably higher. The most striking change, however, is that the eyes develop. The brown-whitish larva becomes an organism with a blue-green back and a silver-white belly. Two separate dorsal fins develop. After three to four months the entire metamorphosis is complete. At that point the migration drive sets in, which is of crucial importance in speciation.

In the FRESHWATER LAMPREY *(Lampetra fluviatilis)*, young which have just metamorphosed migrate to the coast into brackish and sea water. This occurs before sexual maturity in the early part of the year. The lampreys assume a predatory life, grow faster and in a short time can reach a L up to fifty centimeters. This period on the coast can last several years (however some sources list this period as ending in the same year). Lampreys at this stage feed on small organisms; according to Ladiges they also feed on carrion, but primarily on other fishes (herring, cod, smelt, salmon, etc.), as Bahr and others report.

The rasping teeth grab a hold on the prey and boring in is accomplished by rasping movements with the toothed head. The attack of a lamprey on a flounder has been observed in an aquarium. The flounder swam wildly about, turned on its back and attempted to shake off the lamprey. But the lamprey held fast and slid along the upper part of the flounder body without losing its grip. Although the struggle lasted some time, the flounder could not rid itself of the lamprey. Later a long muscle injury could be seen on the flounder which in places was quite deep. Bahr reports that lampreys primarily eat muscle and blood but not viscera.

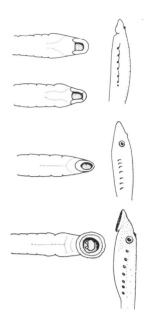

Fig. 2-9. From top to bottom: development of the head, sucking mouth and eyes during lamprey metamorphosis.

Spawning migration begins in the rivers in summer and can last until March; spawning itself occurs in the grayling and trout regions of streams. In extensive studies Bahr found that before spawning lampreys have a critical salinity limit, which is lower as the development of the gonads is lower. This means that lampreys can migrate from salt water to fresh water but not vice-versa. Young lampreys are less sensitive to higher salinity.

Before spawning, lampreys feed less; the gut regresses, and the teeth become blunt. This indicates that lampreys die after spawning. Significant structural modifications appear shortly before spawning. In females the dorsal fins become enlarged (especially the rear one, which becomes thickened in its front portion) and push together. A thickening also appears at the rear end. In males the dorsal fins also get larger, and the cloacal region thickens and elongates, forming a copulatory organ. Behavioral changes appear as well. Previously preferring dark places, lampreys now orient more toward light and actively seek it. This is understandable, since lampreys spawn in shallow streams and generally during sunshine.

Bahr reports that the male digs out spawning pits by lifting out stones with the sucking mouth and carrying them off. Later the female assists with this. The male, already present at the pit, becomes strongly stimulated by the approaching female. To copulate the male swims from behind, attaches itself to the gills of the female, and wraps itself around the female with the rear part of its body and in front of the forward dorsal fin of the female. The male then slides rearward and remains between the thickenings of the rear part and caudal fins of the female. The copulatory organ of the male, modified from the

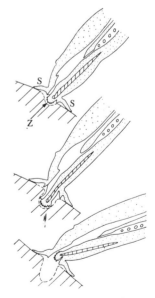

Fig. 2-10. Rasping and sucking motions of an adult lamprey. With the sucker (S) firmly attached to the prey the rasping tongue (Z) digs into the body.

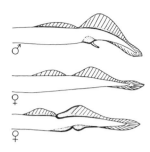

Fig. 2–11. Lamprey (Petromyzonidae) characteristics. In the male, enlargement of the dorsal fins; the genital papilla develops into the copulatory organ. In the female the dorsal fins grow together with thickenings on the second dorsal fin and in front of the anal fin, which hinders the male from sliding off during copulation. Center drawing: female before spawning.

cloaca, comes to rest thereby right before the cloaca of the female, and the eggs (which were pressed out from the action of the male wrapped around the female) are fertilized. Thus the secondary sexual structures which arose before the spawning insure successful copulation. If the male is considerably smaller than the female, it grabs hold somewhat more to the rear. Spawning only lasts a few seconds. Usually several males copulate with a single female, and at each copulation the female only releases a few eggs. The eggs, which are as big as millet seed, fall among the sand and stones of the stream bottom from the action of the mating lampreys. The number of eggs laid is very high (since there is but a single spawning season), and in lampreys of the Elbe River, Germany, they number between 9,000 and 26,000. After spawning, lampreys die from exhaustion and weakening, and Bahr reports that the outer skin degenerates.

The larvae hatch in nine to twenty days, depending on water temperature. As soon as they have eaten the yolk, which lies in the rear part of the body, they leave the sand and due to the water current or their own swimming pass into calm, muddy areas in the stream. During this short migration many larvae are eaten by fishes, and great losses are suffered. Meanwhile the organs have developed and the lampreys, now bored into the floor, begin the floor-living life which has been described.

Metamorphosis in brook lampreys (Lampetra planeri) is similar to that of the freshwater lamprey. But according to Sterba brook lampreys do not build nests, and the copulatory site is often exchanged in sunny places. Lohnisky reports that brook lampreys spawn in groups, whereby the female becomes wrapped around by several males. Pauses between individual spawnings last only a few minutes. In egg development the structure and life of the brook lamprey larvae is similar to freshwater lamprey species. But after metamorphosis is complete brook lampreys first rest and only thereafter have a weak migratory drive. The young brook lampreys do not migrate to the coast, but to the spawning area, where spawning behavior shortly begins. The metamorphosed brook lamprey does not feed any longer; an intestinal tract is present but is not utilized as it is closed at several points. The teeth are blunt even when formed and in the same condition as in freshwater lampreys just before spawning. Thus the brook lamprey does not grow any more and is only twelve to twenty centimeters long. The eggs (600 to 1500 are laid) are somewhat larger than those of freshwater lampreys and are likewise laid a few at a time. Fertilization is external; Sterba also found sperm in the cloacal cavity of a few females and assumes that this species possibly represents a transition from external to internal fertilization. The females die ten to fifteen days after spawning, and males survive about another twenty to forty days.

Very little is known about the SEA LAMPREY (Petromyzon marinus

marinus), and only their migration has been followed. These large lampreys probably spend several years on the coast feeding on cod, mackerel, and salmon. Sucking traces from these lampreys have even been found on whales. Before spawning they migrate into rivers to unknown spawning sites. Sea lampreys have been found in the Rhine River as far as Basel, Switzerland, and in the Elbe River in Bohemia (Czechoslovakia). Since they can maintain suction on boats and other floating objects it is believed that they can be "freighted" up rivers. This would be a matter of coincidence, however, for lampreys are excellent swimmers.

The AMERICAN SEA LAMPREY *(Petromyzon marinus dorsatus)* has been well investigated, in contrast to the European sea lamprey. During the last decades the sea lamprey has spread widely in the Great Lakes of North America and has become seriously damaging to fishes. These lampreys are presumed to have migrated long ago through the St. Lawrence seaway into Lake Ontario, but were prevented by Niagara Falls from further spreading. When the Welland Canal was built in the last century, and later the lake canal system was enlarged, a passageway into the Great Lakes was opened for sea lampreys. In 1921 the first sea lamprey was caught in Lake Erie, and since that time they have penetrated Lakes Huron, Michigan and Superior. As the number of lampreys increased, catches of economically important lake trout decreased. For example, in Lake Michigan the 3000-ton catch of 1944 had decreased to sixteen kilograms by 1955, and a similar situation was found in the other Great Lakes.

Biologically it is interesting that the American sea lampreys have modified their way of life and spend their entire lives in fresh water. Structural modifications have also appeared, so that a new form has been described. As Applegate's extensive investigations show, the American subspecies has much the same life as the species already depicted. For spawning they seek the numerous tributaries of the lakes and form groups on sandy, gravelly bottom, where they build their nests in shallow water up to one meter deep. Only the males engage in nestbuilding. The females release from 60,000 to 240,000 eggs in groups, the individual eggs being about one millimeter in size. The larvae hatch after eight to twenty days. Migration to the Great Lakes takes place after three to five years, where the lampreys carry on a predatory life for one and one-half to three and one-half years.

A great attempt has been made in North America to overcome the lamprey plague. Many traps and electrical barriers were set up, and even poisons specific for lampreys were developed. This program was not very successful, in spite of the fact that huge numbers of migrating lampreys were caught. In contrast a considerable reduction of lamprey populations is occurring in many parts of Europe. In Germany one can even speak of their extinction. In Germany the larval stage is the weak-

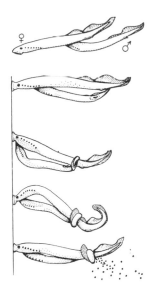

Fig. 2-12. Lamprey (Petromyzonidae) copulation. From top to bottom: the male, approaching from behind, grabs hold of the gill region of the female with its sucking mechanism, wraps around the female and slides to the rear where the eggs are fertilized as they are released.

Fig. 2-13. Recently the sea lamprey *(Petromyzon marinus)* has found its way into the Great Lakes of North America and now spends its entire life in fresh water.

est link in the chain of lamprey life. Since they are mud filter feeders, they are easily harmed by industrial poisons and deposits of such; since larvae of all European species go through this particular stage, all species have likewise suffered.

For the fishing industry the freshwater lamprey is—or in Europe was—of special significance, for the similar tasting sea lamprey is caught too infrequently and the brook lamprey is too small. In earlier times lampreys were considered a delicacy. In France it was forbidden by royal declaration to buy lampreys from dealers before the dealers reached Paris. The city of Gloucester gave the kings of England a lamprey paste at Christmas. On the upper Elbe River there were particular privileges associated with the lamprey catch. And whoever has eaten "Lamproie a la Bordelaise" (lampreys in red wine with leeks) knows how tasty they can be!

The lamprey catch played a large role in the Baltic Sea. In the Gdansk area alone there were 450–500 lamprey trapping sites. Large catches were also made in the western European rivers. Migrating lampreys are caught in large anchored nets, baskets, or from boats with draw nets. In earlier times the nightly catch on the upper Elbe could weigh as much as 200 kg. Lampreys are most desired in the fall and command the highest prices at that time. Since they are migrating then, and do not feed, their fat content is correspondingly reduced. Today lampreys are usually roasted and canned. In America they are not consumed by people but are used only for fertilizing.

Evolution of jawless fishes, by E. Thenius

Class: ostracoderms

Jawless fishes are the oldest vertebrate forms. Fossil remains, mostly small fishlike organisms, are distinguished from extant cyclostomes in that the former possess a bony external skeleton. They are therefore classified as ostracoderms (Ostracodermata). Earlier they were placed in the order with the jawed placoderms (Placodermi) as armored fishes. The oldest fossils are from the Ordovician period (450 million years ago). The ostracoderm head and chest were protected by an armored plate. The brain was developed as it is in lampreys. Being sluggish swimmers they usually were found on the floor and fed there. These ostracoderms flourished in the Silurian and Devonian periods (from 300 to 400 million years ago), disappearing sometime after the end of the Devonian. Today their fossils can only be found in the northern hemisphere (Europe, Asia, Spitzbergen and North America), although in earlier times they were more widespread.

In spite of the scant external similarity between the ostracoderms and present-day hagfishes and lampreys, thorough investigations have shown that these old forms belong to the jawless fishes. The Swedish paleontologist E. A. Stensiö made serial cuttings on cephalaspis forms embedded in rock; that is, he carefully cut away thin layers of rock and thereby could compare the structure of the organism and its organs in the various cut layers. In this way he gained a three-dimensional insight into the external and internal anatomy of the ostracoderms. The cephalaspis forms have an extensive bony internal skeleton. Originally the appearance of the armored covering about the head caused them to be interpreted as trilobites until the renowned paleontologist and ichthyologist Louis Agassiz recognized their vertebrate character in 1835. The fossil material on hand has increased considerably since Stensiö's work from the Devonian on Spitzbergen in 1927. Presently a great number of ostracoderms are known, which are divided as follows: 1. Cephalaspidomorphi: a) Osteostraci with the family Cephalaspidae and the principal genus *Cephalaspis;* b) Anaspida with the genera *Pterygolepis* and *Jamoytius;* 2. Pteraspidomorphi, with Heterostraci containing the major genera *Pteraspis* and *Palaeodus;* 3. Thelodonti with genus *Thelodus.*

The ostracoderm form varies from fishlike, armorless, scaly swimming forms (Anaspida) to armored forms with a snoutlike skull process (Heterostraci) to flattened bottom dwellers in which either the head has been modified to a single armored element (Osteostraci) or the upper body surface was only covered by small scales (Thelodonti). Thus these jawless fossil fishes already represented the most important fish forms. When the jawed fishes appeared later, they almost completely displaced the less-successful Agnatha, except for the hagfishes and lampreys of today.

In most ostracoderms a unit (Osteostraci) or composite (Heterostraci) head armor can be recognized as a distinctive characteristic. The rest of the body is mostly covered by high, narrow scales. Based on the geologically oldest forms *(Astraspis, Tesseraspis)* it is presumed that the unit armor is the result of fusion, and in geologically younger forms (e.g., *Drepanaspis*) a degeneration appears in which the scales appear as tiny toothlike structures. The heterocercal caudal fin, dorsal or ventral fins or fin seams are all unpaired fins. A movable rear spine is present in some forms (e.g., *Pteraspis*). Paired fins appear only in some *Cephalaspis* species. A mandibular arch is not present. A single nasal opening associated with the pituitary is located on the upper side of the head between the eyes. There are a great number of gill slit arches. The equilibrium organ consists only of two arched tubes (which together with the absence of paired body structures are all characteristics, at least in the older forms, which otherwise appear only in lampreys).

In contrast to extant cyclostomes, the fossil ancestors mainly ate decaying plant and animal matter, which they ingested by the gill gut with its filter apparatus. This is presumed from the lack of a jaw and teeth as well as the well-developed state of the gills. Only the pteraspids and cyathaspids had mouth plates permitting biting. Stensiö's assumption that cephalaspids had a rasping tongue is questionable.

As a result of serial cutting investigations the various sensory organs

Distinguishing characteristics

▷
Extant vertebrates: 1. Sea lamprey *(Petromyzon marinus)* on a cod *(Gadus morhua);* 1a. Sucking mouth; 2. Freshwater lamprey *(Lampetra fluviatilis);* 3. Hagfish *(Myxine glutinosa).*

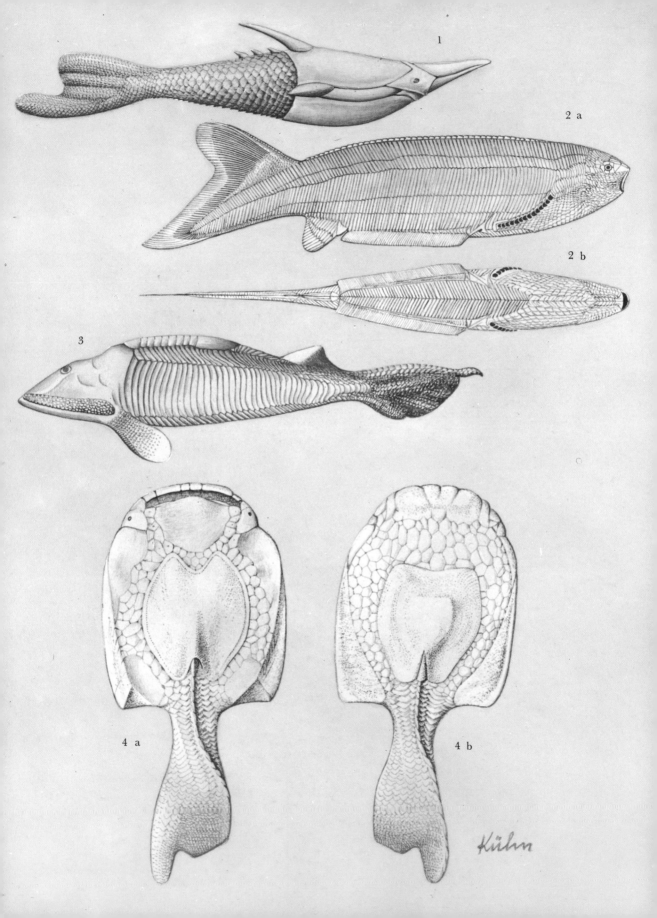

of these oldest vertebrates have been identified. The pineal organ, situated next to the fairly well developed eye, served the visual sense; a nasal depression, which also contained the pituitary organ, functioned for smell. The inner ear with its sensorial field appeared in bottom-dwelling species. As they are richly supplied with nerves, these sensorial fields were originally thought to be electrical regions, but instead were found to be receptors of an equilibrium and auditory organ. E. Jarvik has found the evolutionary remains of this organ in extant lampreys as simple projections from the auditory labyrinth. Lateral organs are also present in some present-day species which are similar to those of the primeval jawed fishes.

On the basis of ostracoderms from the early Ordovician *(Palaeodus)* compared with structural modifications found in those from the middle Ordovician, the origin of jawless fishes is considered to lie in the Cambrian period, about 500 million years ago. The fact that the oldest vertebrates were heavily armored and that present forms possess only a cartilaginous supporting skeleton has given rise to various explanations regarding the original condition. Before zoologists had fossils available it appeared that cartilage, and thus cartilaginous skeletons, preceded bony skeletons. But the fossil record has shown that bones are very old structural elements which in some fishes were later replaced by cartilage. Furthermore fossil finds have shown that the cartilaginous fishes were not the earliest forms. The armored fishes, the jawless ostracoderms, and the jawed plate-covered organisms appeared considerably earlier than cartilaginous fishes. And in armored fishes a degeneration and regression of the original unit skeleton appears. All of this supports the hypothesis that the bones were the original supporting mass.

How, then, have contemporary cyclostomes evolved from the fossil ostracoderms? Based on a few similarities, Stensiö felt that lampreys evolved from Cephalaspidomorphi species and the hagfishes from Pteraspidomorphi, and thereby assumes that this divergence took place long ago in evolutionary development. A recent investigation by Ritchie showed that the small anaspid from the Upper Silurian period in Scotland, *Jamoytius kerwoodi,* (originally classified as Acrania; see Vol. III), is very close to lamprey ancestral forms. It has a long trunk covered with narrow plates, a fin seam on the back and sides and a heterocercal caudal fin. The terminal mouth opening supported by a cartilage ring and a cartilaginous gill skeleton are particularly significant; they are similar to the present cyclostomes and their larvae. From this anaspid it can be deducted that the bone-like plates of cyclostomes are regressive structures and that they diverged from jawless fishes in the oldest geological period of the earth.

The ostracoderms have also answered the question dealing with the original environment of vertebrates. For a long time the opinion was

◁

Pteraspis (Pteraspidomorphi) from the lower Devonian in Europe; 2a, 2b. Lateral and ventral view of *Pterygolepis* (Anaspida); 3. *Hemicyclaspis* (Cephalaspidae) from the late Silurian and early Devonian; 4a, 4b. Dorsal and ventral view of *Drepanaspis* (Heterostraci) from the lower Devonian.

held that the earliest vertebrates inhabited fresh water. But extensive studies of all fossil forms from the Ordovician and Silurian have shown that the oldest forms are all from sea water deposits. The first primarily fresh-water forms were the ostracoderms from the Devonian period (310 to 350 million years ago). Thus, on the basis of these forms, the origin of vertebrates is to be found in the sea and not in fresh water.

3 Fishes

Superclass: fishes,
by P. Kähsbauer

What is a fish?

Fishes (superclass Pisces) are vertebrates adapted to living in water. They possess fins and their skin is usually covered with scales, but they can also have dermal teeth or bony plates or entirely lack scales. Respiration usually takes place through the gills. Fishes are cold-blooded animals with two-chambered hearts, the heart circulating only venous blood. Thus the defining concept of fishes has been greatly altered since the jawless fishes have been placed into a special vertebrate superclass.

In earlier times biologists did not make this conceptual distinction. Until the 16th Century some natural historians considered not only aquatic mammals such as the whales, dolphins, seals and hippopotamuses as fish, but also crocodiles, snails, crabs, sea urchins, echinoderms and tunicates. Later biologists classified only whales and amphibians together with fish. As late as 1858 the British zoologist Richard Owen put fishes, amphibians and reptiles together in one class.

Fish divisions

Aristotle (384–322 B.C.) attempted organizing fishes phylogenetically. He divided them into cartilaginous and bony fishes. Linné distinguished four major fish groups on the basis of the location of the pectoral fins. Louis Agassiz (1807–1873), using fossil forms, classified on the basis of scales, which Johannes Müller showed to be untenable. Today approximately 25,000 fish species are known, which are divided into two classes with four subclasses: 1. Cartilaginous fishes (Chondrichthyes) with subclasses sharks and rays (Elasmobranchii) and chimaeras (Holocephali); and 2. Bony fishes (Osteichthys) with subclasses higher bony fishes (Actinopterygii) and lungfishes and lobefins (Sarcopterygii; see Vol. V). These comprise a total of thirty-four orders and 418 families. Twenty-nine orders and 111 families of fossil cartilaginous and bony fishes are known.

Evolution,
by E. Thenius

Various forms can be distinguished among the oldest jawed fishes which are known. These show that differential development began in the Silurian period (450 million years ago) and that the actual ancestral group of fishes has not yet been identified. Since it has become known that the ostracoderms (Ostracodermata) belong with jawless fishes, the oldest jawed fish types are considered to be the spiny sharks (Acanthodii) and the placoderms (Placodermi).

The spiny sharks are generally small fishes only rarely reaching a L of 50 cm. The body is covered with diamond-shaped scales, and the head has numerous irregular dermal bones. The nose is short; the mouth opening is at the end of the body, and the eye openings very large. A long gill slit separates the mandibular arch from the hyoid arch. The spiny sharks existed from the Upper Silurian until the Permian (from 450 to 240 million years ago).

The designation of spiny sharks has come about from the distant relationship of these fishes with sharks. Thus, they were originally classed together with cartilaginous fishes. Later they were considered to be transitional forms between jawless and jawed species. But extensive studies have shown that this group has many characteristics of early bony (teleost) fish types. Although they are not considered to be the direct bony fish ancestral group, they are close relatives of the common ancestor and afford valuable insights into the organization of early fish forms.

The placoderms, earlier classed together with the jawless ostracoderms as armored fishes, were covered on the head, shoulder and underside with bony plates. Their jaws had teeth or cutting plates. Various groups can be distinguished which have only the armored covering (in some cases regressively developed) in common. As in the spiny sharks, the placoderm gill arch was modified to form a mandibular arch. In all likelihood this was the most crucial development in the formation of jawed fishes, for thereby the prerequisite for an effective cutting or chewing structure was met. In addition the development of teeth opened up food sources and environments which had been unobtainable for the jawless fishes. Since the structure of the various groups is highly differentiated, it is questionable whether the placoderms actually belong to one unified group.

The Arthrodira, from the Devonian period (310 to 350 million years ago), are among the best known placoderms. They were marine predators, of which some (e.g., *Titanichthys*) reached a L of over 8 m. A bony armor with many plates protected the head and the fore-end of the trunk. The head and breast armor were connected by a bone joint which enabled up and down movement of the head. The remainder of the body was naked and culminated in a heterocercal caudal fin. The Arthrodira had two unpaired dorsal fins, an anal fin, and a pair each of pectoral and pelvic fins. The internal skeleton was partially ossified. The terminal mouth opening had strong jaws, which were sometimes jagged and then acted as teeth. The eyes, protected by hard rings, were relatively large. The gills had a single external gill opening.

Spiny sharks

Fig. 3-1. *Climatius.* Spiny shark *(Acanthodii)* with large fins barbs and additional fins between the paired pectoral and ventral fins.

Placoderms

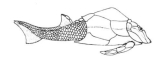

Fig. 3-2. *Pterichthys,* a genus of the armored fishes.

Fig. 3-3. *Dinichythys,* a North American placoderm form.

Antiarchs

In contrast the Antiarchi, Devonian small fishes, were chiefly freshwater forms which in their appearance are similar to extant callichthyid armored catfishes (Callichthyidae). The flattened head-breast-armor suggests a bottom-dwelling creature. This is further confirmed not only by the mouth opening but also by the structure of the pectoral appendages which diverge from typical fins completely and are known as "arthropterygia". Generally these appendages consist of two segments joined with each other, which are covered by many bony plates. The head and trunk are separated, but a neck joint has not been developed. The mouth opening is on the lower side of the body. The part of the body which is not covered by bony armor has scales or is naked and terminates in a heterocercal caudal fin. The fossil Antiarchi armor has been found throughout the world, but particularly in the freshwater deposits of Scotland and Canada, in which the remains of coelacanths and lungfishes (see Vol. V) also exist.

Among other placoderms only the Rhenanida and Ptyctodontida from the Devonian will be mentioned here. They belong to those forms in which the armor is more or less regressively developed. Both groups have been found in marine deposits. The ptyctodontids are of special evolutionary interest, for they have the same appearance as chimaeras (Holocephali), which flourished in the Cretaceous period (60 to 140 million years ago). Recent studies have described further similarities which imply that the chimaeras developed from the ptyctodontids. Finds on other placoderms also suggest a closer relationship to the cartilaginous fishes. In any case the placoderms are closer to the ancestral group, from which the cartilaginous fishes evolved, than the primeval bony fishes.

The ancestors of sharks appeared in the Devonian and Carboniferous periods (from 350 to 240 million years ago). Initially these were primeval sharks (Cladoselachii) which are distinguished from present-day forms by the terminal mouth opening, the large eyes, small teeth, symmetrical caudal fin and the number of gills. Thus the true cartilaginous fishes developed after bony fishes had evolved. Ptyctodontid finds make it probable that the chimaeras developed from a different line than sharks and rays, which are today classified as cartilaginous fishes. In the Upper Devonian more highly developed sharks (Hybodontoidea) appeared which later died out. Examples of genera still existing today (e.g., *Heterodontus* and *Hexanchus*) appeared in the Jurassic period (175 to 140 million years ago). These (as *Chlamydoselachus*, which appeared in the Miocene) belong to the most primitive sharks living today. "Modern" sharks appeared in the Carboniferous period. The rays and skates (Rajiformes), descendants of sharks, are considerably younger. With the exception of Pleuracanthodii sharks, which lived from the Devonian until the Triassic, the cartilaginous fishes are salt water inhabitants.

Fig. 3-4. *Cladoselache*, from the late Devonian.

Fig. 3-5. *Dipterus*, a lung fish, the oldest fossil form from the Devonian.

The oldest lungfishes stem from the Devonian of the Rhineland region and "old red sandstone" of Scotland. They are about 350 million years old. The predecessors of lobefin fishes also appeared in the old red sandstone. Today they are grouped together with lungfishes in subclass Sarcopterygii. These Rhipidistia lungfishes were elongated forms with a meat-eating mouth structure and flaplike pectoral and pelvic fins. They have been found since 1938 in the waters between South Africa and Malagasy (see genus *Latimeria*, in Vol. V).

Geologically the oldest bony fish is *Cheirolepsis* from the middle Devonian. Presumably primeval teleosts already existed in the Upper Silurian, for scattered fossil scales of them have been found. Teleosts are divided into several groups, as will be shown. Among them the bichirs (superorder Polypteri), sturgeons and paddlefishes (Chondrostei), and bowfins and gars (Holostei) have a firm external skeleton of ganoid scales, which is sometimes regressively developed, while the true bony fishes have cycloid or ctenoid scales, in association with which an ossified internal skeleton develops.

Fig. 3-6. *Cheirolepsis*, the oldest bone (teleost) fish.

The high point of Chondrostei was in the Permian Triassic period (240 to 175 million years ago), and they contained the fossil genera *Palaeoniscus*, *Platysomus*, *Dorypterus* and *Redfieldius*. The more highly developed Holostei flourished particularly in the Jurassic (175 to 140 million years ago) and included genera *Dapedius*, *Lepidotus*, *Gyrodus* and *Microdon*. The remains of these ganoid fishes have been well preserved in Solnhofen lithographic slate. They are particularly prominent for the genus *Protosphyraena* from Kansas, which resembles the bowfin (Amiiformes) of today, for *Protosphyraena* has two tusklike teeth in the upper jaw and one sharp tooth in the lower jaw.

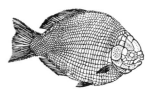

Fig. 3-7. *Dapedius*, a fossil ganoid fish.

The bony (teleost) fishes of today, found throughout the world, developed from the ganoid fishes 140 to 175 million years ago. The true teleosts appear for the first time with genus *Leptolepsis* in the Lower Jurassic. Only in modern times have the teleosts invaded all sea and fresh-water habitats. *Portheus molossus* from the Carboniferous in Kansas is a relative of the tarpons of today, but with its L of 4.5 m is almost three times as long as the tarpons.

In the recent period of the earth, which began 60 million years ago, the earth assumed its present appearance more and more. The continents and oceans developed, and mountain ranges such as the Alps, Himalayas and the American Rocky Mountains appeared. As the seas withdrew in the late Miocene (about 20 million years ago) the fishes increasingly invaded fresh water. The European continent was flat at that time, but fish forms which are known to us today existed in the lakes and rivers. Salmon were not in the warm Tertiary waters and did not appear until the Ice Age.

Based on investigations of the prominent marine biologist A. F. Thienemann we know that thirty-seven percent of German fish species existed there before the Ice Age, which would comprise a time period of more than 800,000 years. Among these species are the burbot, three-spined stickleback, perch, pike, dace, bleak, carp, silver bream, chub, rudd, and loach. During the Ice Age fish forms appeared which moved from the northern ice masses to the south, and those which were forced from the Alpine glacial formations to the north. These included all fishes preferring cold water and those that spawn in the winter (e.g., char, trout, whitefish)—representing about forty-five percent of German fish species. After the Ice Age, a period beginning 12,000 years ago, the northern glaciers withdrew from the European continent. With that, additional species appeared which comprise about eighteen percent of fish species found in Germany, and among these are the smelt, greylings, Danube salmon, European bitterling, pike, sheatfish, bowfin, sterlet, Zingel and about nineteen carplike forms, including carp itself.

Present-day fishes, by P. Kähsbauer

Size and age

The size of present-day fishes varies considerably. The basking shark (*Cetorhinus maximus*) can be fourteen meters long and the manta ray (*Manta birostris*) is up to seven meters wide and four meters long. One hundred years ago in the Caspian Sea there were sturgeons nine meters long weighing 1500 kg. The Brazilian arapaima (*Arapaima gigas*) is said to be four and one-half meters long and weighs up to 400 kg. Halibut (genus *Hippoglossus*; see Vol. V) two and one-half meters long weighing 200 kg are not rare.

In contrast the eleven millimeter long goby (*Mistichthys luzonensis*; see Vol. V) from the Philippines is the smallest fish in the world, and at that an important commercial fish caught in great quantities. A few tropical toothed carps two and one-half centimeters long are also among the world's smallest fishes, as well as the blennioid fish genus *Schindleria*, a close relative of the goby which with its twelve millimeter length is scarcely bigger than the goby. The goby fish *Hyrcanogobius bergi* of the Caspian Sea is twenty-one millimeters long as a sexually mature adult.

The age which fish can attain is just as variable. Some Salangidae fishes (in the salmon order Salmoniformes) and gobies (genera *Latrunculus, Benthophilus* and *Bubyr*) reach an age of just one year and die after spawning. Yet carp, sheatfish, and pike can live to eighty years of age. Eels begin migrating into the sea at an age of from six to ten years; and a sardine never surpasses two years of age. The belief is widespread that fishes reach a biblical age. Thus carps 150 years old and even a 267 year old pike 5.7 m long weighing 250 kg have been reported by the Mannheim, Germany, cathedral. Those, however, are only legends. Most commercial fishes do not even attain their natural life span, because they are caught either by predators or man before then. Age can be precisely determined by the scales or vestibular apparatus, which has growth rings just as trees do.

An aquatic life has different demands on the animal than terrestrial life. Water is heavier than air and the resulting pressure is much greater than air pressure. But as a result of this, water is much more buoyant than air, which makes it possible for so many fishes to survive in water. Other aquatic creatures, whose specific gravity exceeds that of water, sink to the floor and can reach the upper levels only by strenuous muscular exertion, and then only for a short time. These are bottom-dwelling fishes; but most fishes are able to swim freely. In fact the fish form has evolved as it is from swimming and the physical characteristics of water.

The characteristic fish form is a spindle-shaped body, pressed together and flattened at the sides (which is reduced at the front and rear ends). Fast long-distance swimmers have a torpedo-like body which offers minimum resistance to the water (examples are blue sharks, tuna, and trout). Fishes inhabiting standing water (e.g., carp) are flattened on the sides, a characteristic also shared by all poor and slow swimmers (among them the demoiselles; see Vol. V). Pure bottom dwellers such as rays and flatfishes are extremely flattened, but those bottom dwellers that move very rapidly have an eel-like shape (e.g., eels, conger eels) or a fiberlike body (e.g., snipe eels). Fishes that stand quietly and then rapidly attack their prey have arrow-shaped bodies (e.g., pike, gar fishes). Those that are not very mobile have a balloon shape (i.e., puffers); the motionless pipefishes found in eel grass have a needle-like form. The form of sea horses is distinctive, in which the front end of the body is enclosed in bony armor so that only the tail assists in propelling the fish. The slow-swimming short sunfish, the nimble butterfly fishes and squaretails *Nematobrycon palmeri,* and the slow characin *Crenuchus spilurus* have bodies flattened on the sides, great height and narrow width, as is enountered in some deep-sea fishes.

Swimming is an undulating movement carried out by the strong lateral muscles together with the caudal fin. In geologically early times fishes swam with rhythmic contractions of the trunk and tail muscles, which made the body bend. The fish was propelled through the water by the sequential force of various body parts. Most fishes have preserved this primeval muscular arrangement and are distinguished from terrestrial vertebrates, in which the principal muscles are located on the fore and rear extremities.

In order to move forward in the water, three actions are utilized: 1. Body movement through extension and contraction of the muscles; 2. Fin movement; and 3. Effect of the water forced out of the gills during respiration. Although the muscular action is the most significant, the majority of fishes use all three forms, alone or simultaneously. Trunk fishes (Ostraciontidae; see Vol. V) have a completely different swimming technique; the trunk fish body is in a small, hard immobile body housing. The tail sticks out in the rear and moves back and forth

Technique of propulsion

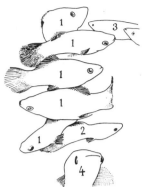

Lebendgebärende Zahnkärpflinge gehören zu den beliebtesten Aquarienfischen. Neben einer Anzahl (1) Roter Schwertträger (Zuchtrasse von *Xiphophorus helleri,* s. S. 466) sind hier (2) Guppys (*Poecilia reticulata,* s. S. 463) sowie (3) Gambusen (*Gambusia,* s. S. 467) und (4) ein Schwarzer Molly (Zuchtrasse des Spitzmaulkärpflings, *Poecilia sphenops,* s. S. 465) in einem Aquarium vereinigt.

ANATOMY OF A FISH (CARP)

A. EXTERNAL ANATOMY

1-12 Head: 1. Upper lip;
2. Barb; 3. Nasal opening;
4. Eye (the segment from
the upper lip to the front
edge of the eye is desig-
nated the snout);
5. Cheek; 6. Neck;
7. Lower lip; 8. Throat;
9. Front gill cover; 10. Intermediate gill cover;
11. Lower gill cover;
12. Gill cover; 13-20
Trunk: 13. Back;
14. Dorsal fin; 15. Chest;
16. Pectoral fin; 17. Belly;
18. Pelvic fin; 19. Lateral
line; 20. Anus; 21. Anal
fin; 22. Tail shaft;
23. Caudal fin.
B. Skeleton

B. SKELETON

1-14 Skull: 1. Premaxil-
lary; 2. Maxillary;
3. Dental; 4. Lacrimal;
5. Orbital; 6. Frontal;
7. Parietal; 8. Supraoccipi-
tal; 9. Quadratum;
10. Radii branchiostegi;
11. Preoperculum; 12. In-
teroperculum; 13. Subo-
perculum; 14. Opercu-
lum; 15-23 Paired
members; 15-20 Pectoral
girdle with associated
bones: 15. Supra-
cheithrum; 16. Clavicle;
17. Coracoid; 18. Scapula;
19. Basalia; 20. Pectoral
fin rays; 21-23 Pelvis and
associated bones:
21. Pelvis; 22. Basalia;
23. Pelvic fin rays; 24-30
Vetrebral column and
skeletal elements of un-
paired members: 24. First
to third vertebrae;
25. Modified third rib;
26. Radialia; 27. Ribs;
28. Neurapophysis;
29. Hemapophysis;
30. Hypural.
C. Internal Organs

C. INTERNAL ORGANS

1-2 Respiratory organs
and swim bladder: 1. Gill
elements; 2a. Anterior
swim bladder compart-
ment; 2b. Posterior swim
bladder compartment;
3-6 Digestive organs:
3. Esophagus; 4. Liver;
5. Intestinal tract; 6. Anus;
7-9 Excretory organs and
gonads: 7. Kidney;
8. Urinary bladder;
9. Testes; 10-12 Circula-
tory organs: 10. Atrium;
11. Ventriculum;
12. Bulbus arteriosus.

(caused by alternating contraction of the tail musculature), which moves the stiff trunk structure forward much as a boat with a simple single rudder can be moved forward. Between these two extremes— undulating movement of the whole body on one hand, and pure fin movement on the other—is a whole continuum of swimming techniques. In bottom dwellers such as bullheads or gobies with their springing movement, the tail is the chief propelling organ. Many species can only move forward slowly by wavelike motions of the caudal fin, in which the waves run perpendicularly along the longitudinal axis of the body. Fishes with large tails are relatively slow swimmers; their hind portion is rounded or cut straight off and they cannot swim rapidly for long periods of time. In contrast those species in which the tips of the caudal fin are long and pointed are among the swiftest swimmers. Thus the oceanic bonito (*Katsuwonus*; see Vol. V), with its almost sickle-shaped caudal fin, is one of the fastest fishes.

Species in which the body form deviates greatly from a streamlined shape can utilize the dorsal and anal fins for propulsion. The North American bowfishes move forward with the help of undulating motions of the long anal fin, and electric eels use their anal fin in a similar fashion. Sea horses and pipefishes swim with wavy movements emanating from the caudal fin, and in flatfishes the dorsal and anal fins undulate. Ventral fins help maintain balance and only rarely function in propulsion. Moderately fast fishes use the pectoral fins for movement. Thus wrasses swim by simultaneously beating the pectoral fins. In rays and skates the pectoral fins are the only means of propulsion. They also function as brakes by stopping their motion. Water passed through the gills in respiration can sometimes also play a role in forward movement.

The velocity of fishes is highly variable. Some carp can swim twelve kilometers per hour; barbels reach eighteen kilometers per hour, while pike attain twenty-five and trout thirty-five kilometers per hour. Among marine fishes tuna can swim twenty-two, sharks thirty-six, and swordfish up to ninety kilometers per hour. — Velocity

The skin of fishes is either naked (e.g., catfishes), or has scales, dermal teeth or barbs. The skin consists of a thin epidermis and subcutis of several fibrous layers. In addition to the numerous mucous glands in the epidermis, some bony fishes (e.g., the scorpion fish) have poison glands on the gill cover and dorsal fin which are activated by energetic fin strokes. Large stingrays also have poison glands on their tail stinging spine. Scales are round plates; the rear edge overlays that of the adjacent scale much like tiles on the roof of a house. They are made up of firm, fibrous connective tissue and each one rests in a scale cavity within the subcutis. The scales are covered by the transparent, soft epidermis. Scales are arranged in diagonal rows according to the muscles beneath them. — Skin

Four types of scales are distinguished: placoid, ganoid, cycloid and — Scale types ctenoid. Placoid (platelike) scales are common in sharks and trigger fishes (Balistidae; see Vol. V). These scales have an ectodermal cap underlaid by a body of dentine. The ganoid scales of lobefins, sturgeon and gar are bony diamond-shaped scales with a ganoine outer layer, a hard inorganic salt. This results in a high skin sheen. The ganoine layer is formed by the subcutis. Lungfishes and bowfins have cycloid scales, and true teleost fishes have either cycloid or ctenoid scales.

Cycloid scales are true scales; their rear edge projects from the scale cavity and is smooth and not toothed. The rear, free edge of ctenoid scales is toothed. Generally, cycloid scales are found in fishes with smooth fins such as herring, salmon and carp. Most fishes with bristled fins have ctenoid scales (e.g., perch). Yet exceptions do exist. Many members of genus *Epinephelus* (see Vol. V) have ctenoid scales above the lateral midline, but cycloid scales beneath it. The lemon dab (*Pleuronectes microcephalus;* see Vol. V) has ctenoid scales on its back and cycloid scales on the stomach. Scale formulas, which are of systematic importance, are determined by counting the rows lengthwise and diagonally. Sometimes scales develop into large plates, as in the armor of armored catfishes, trunk fishes and gurnards. Eel scales are microscopically small, in tarpons five centimeters wide, and in an Indian barb *(Barbus mosal)* they can be as wide as the palm of one's hand. Carp usually have typical scales, but there is one race with scattered giant scales and another completely lacking scales.

The epidermis which covers the scales contains innumerable mucous glands; their secretions protect the fish and its body fluids and cause the sliminess which allows the fish to swim with a minimum of resistance to the water. The mucus is normally transparent but can become colored when boiled, accounting for the blue color of boiled trout. This darkening can also occur under the influence of acids or lye, such as from industrial water pollution.

Coloration

Fish coloration stems from chromatophores located along the border of the epidermis and the subcutis, and also from fatty dyes (red, orange, or yellow lipophores, which are carotinoids dissolved in fat). Color cells are under central nervous system control and when stimulated either cause color to appear or eliminate coloration already present. Changing coloration enables fishes to adapt quickly to their surroundings. Skates and flatfishes not only assume the color of the floor but also the sandy or gravel texture by forming imitative spots on the upper surface of their bodies. Other environmental factors such as higher water temperature, oxygen deficiency, or behavioral factors like fear, aggression, and similar stimuli can cause color changes. An aggressive male stickleback can change its appearance at short intervals. A color change appears in many fishes (such as salmon) when they are ready for spawning. The male develops a very colorful display pattern, and the epidermis develops wartlike spawning eruptions (as in salmon, chondrostomes *Chondrostoma*, and European roaches).

The epidermis gives off a silvery mass consisting of fragments of decomposing material of the protein guanine; this forms an effective protective coloration, for a predator fish on the bottom sees a prey fish swimming above only as part of the glittering water surface. Bottom dwellers and deep-sea fishes generally are dark colored. Coloration is lacking entirely in blind cavefishes, and the most colorful species are those of tropical inland waters and lakes.

Some species, such as the electric eel, electric sheatfish, electric rays, mormyrids, and others, can produce electricity and deliver electric shocks. Electrical organs are located in diagonal muscles, and are modified skin glands only in the electric eel. They consist of thin plates of gelatinous consistency which lay beside and on top of each other and have nervous connections. When the fish rests, there is no activation of the organs. When excited, the side of the plates joined with the nerves becomes negatively charged and the free side positively charged. The electric current runs from the rear to the front in electric eels, in sheatfish from front to rear, and in electric rays from the belly to the back.

Some sharks, e.g., the spiny dogfishes (Squalidae), emit a greenish phosphorescent light created by small light organs which lie throughout the skin. The deep-sea scaly dragonfishes (Stomiatoidei) have light organs consisting of a convex or concave lens which rests in one of the dermal vesicles of the epidermis. The walls of this vesicle are formed by glandular cells which form the luminous material. If the walls contain black pigment, they act like the concave reflector of a dark lantern; in other cases the outer skin layer lies over the upper surface of the lens and has about the same effect as the shutter of a camera. The luminous organs are arranged in two rows on both sides of the body. In addition there are luminous organs on the head, jaws, and beneath or just behind the eyes. Fishes of genus *Anomalops* (see Vol. V) from the Indian Ocean have a large luminous organ beneath the eye which sits on a movable flap which can be turned inward and fits in a cavity under the eyes. Lanternfishes (Myctophidae) have luminous organs on their stomachs and fins which have a jewel-like glitter. The *Galatheathauma* caught in 1962 by the *Galathea* Expedition has a light organ inside its mouth on the foregum; any prey attracted by the light had merely to swim inside the mouth, which then closed. The toadfish *Porichthys* from the Californian coast has about 340 light organs on the head and body. These bright white spots consist of lens, gland, reflector and dye layer and they are so bright that a newspaper can be read twenty-five centimeters from them with no other light source present. The light comes from luminous bacteria or from the chemical reaction in mucus secreted by the gland cells. These function in lighting the water and providing species recognition.

The fish skeleton supports the musculature and provides good sup-

Electrical shocks

Luminous fishes

Skeleton

porting points for the muscles. Furthermore, it protects the brain and some of the sensory organs. In cartilaginous fishes (sharks and rays) it is composed entirely of cartilage, while in other orders it is mostly bony. The "substitute bones" arise as cartilaginous pre-formed bones, and the "covering bones" as dermal bones of tendon, which pass into the interior as ossified skin and displace the cartilaginous parts more and more. In bony (teleost) fishes the skeletal formation provides clues to the life of the fishes. Skilled, swift swimmers have a thin skeleton compared to slow bottom-dwellers; in deep-sea fishes the skeleton can be a very delicate structure. The spinal column, to which the trunk and tail muscles are attached, supports the entire body and develops from the chorda dorsalis, or notocord (see Vol. III). The notocord is a cartilaginous cord which originates in the skull between the auditory capsules. In sharks, rays, chimaeras, sturgeons and lungfishes it is present (as in lampreys) for the entire life of the individual and is protected by a sheath and skeleton-like layer; the cartilaginous skeleton develops from this layer and is the basis for the bony skeleton of later forms. In sharks and rays cartilaginous vertebral structures appear which confine the notocord more and more. In teleost fishes these vertebrae are fully ossified and contain only small remains of the notocord. The number of vertebrae varies from species to species, and thus puffer fishes have only fourteen, and the viviparous blenny (*Zoarces*) up to 117.

Skull

The fish skull consists of the neurocranium, containing the brain and sensory parts, and the visceral cranium, which consists of narrow clasping cartilage or bony pieces and which surrounds the mouth cavity and gills. With the exception of sharks, rays and chimaeras the skull is firmly associated with the vertebral column. The neurocranium has nose, eye, ear and occipital regions. Beneath it lies the skull of the face, consisting of numerous visceral arches; these are, from front to rear, the primary mandibular arch, the hyoid arch, and generally five branchial arches, between which the gill slits are located. Sharks, rays and chimaeras have a single skull element (primordial cranium); in sturgeons a second skull roof is located above the cartilaginous skull. This cartilaginous skull is increasingly ossified in teleost fishes, and only in some deep-sea fishes (e.g., *Argyropelecus*) are the skull and vertebral column completely cartilaginous.

Appendicular skeleton

The appendicular skeleton has the shoulder girdle and pelvic girdle as connections with the paired fins to the trunk, along with the fins themselves. In cartilaginous fishes and sturgeons the shoulder girdle has a paired cartilaginous arch on which the clavicle and a shoulder dermal bone, the cleithrum, lie. The cleithrum exists only in fishes. Teleost fishes have a corresponding system with three skeletal elements; the scapula lies on the rear side; and the belly side has the coracoid, procoracoid and clavicle bones.

The pelvic girdle is not connected with the vertebral column; in sharks it has a cartilaginous arch, which later disappears (exceptions: lobefins and lungfishes). In its place two stafflike or triangular pelvic elements appear, which develop from the fins. The fins consist of three cartilaginous pieces (propterygium, mesopterygium and metapterygium) in sharks, on which the thinly jointed fins rest. In sharks and lungfishes the fin seam is supported by thin horny fibers. In lobefins the propterygium and metapterygium are already ossified. Lungfishes have a cartilage rod, which rests on a cartilaginous base and bears the lateral fins. Teleost fishes do not have this base; their fins sit in the immediate proximity of the shoulder blade and normally have five fin rays (in anglerfishes, *Lophius,* however, only two, and in the eel eight). Their continuation forms thin dermal bone rays which support the free fin. Horny fibers are at the outermost fin seam.

The dorsal, caudal and anal fins are designated as "unpaired"; in sharks they consist of three-part cartilaginous fins and horny fibers. In lobefins and teleost fishes dermal bones appear in place of the horny fibers, and are united with daggerlike fin bearers in the musculature. These fin-bearing elements in part lie on the thorny processes of the vertebrae; in the caudal fin they are replaced by platelike vertebral arched processes. The dermal bones or fin rays are either undivided and thus hard and pointed or diagonally segmented, divided and bendable. In higher bony fishes (such as perch) the front segment of the caudal fin is hard, and in soft-rayed fishes (e.g., herring, carp, salmon) only at most the first three rays of the caudal fin are hard.

Fins

The dorsal and anal fins function in balance, the caudal fin for propulsion, and the pectoral fins for direction. The ventral fins can have varied locations: on the belly in tench, at the breast in perch, or at the neck in the burbot. Salmon and sheatfish have an additional "adipose fin" between the dorsal and caudal fins, without fin rays. Fins are often important distinguishing characteristics of species, and therefore the fin formulas are repeatedly included in systematic descriptions.

The muscle comprises a large part of the fish body. Its arrangement on the skeleton permits the very flexible fish movements. On the head is the facial musculature of the jaw and gill arches. The chewing muscles, and muscle groups of the gill cover, mandibular and gill arch muscles all belong to this. Together they form an independent muscle group proceeding from the sides. The strong lateral trunk muscles for propulsion of the fish are developed from the vertebrae and extend from the back of the head to the root of the caudal fin and are symmetrical on both sides of the vertebral column. They consist of numerous muscle segments one behind the other, which are separated by fine connective tissue partitions; they also consist of a dorsal part and a ventral portion, between which lies a wall of connective tissue. Corresponding to the arrangement of the axial skeleton, each part falls

Musculature

into small muscle packets (myomeres). The musculature is held to the vertebral column by connective tissue partitions (myocommata). The fin musculature developed from trunk musculature which had split off. Here the depressors of the fin rays, which fold them together, are differentiated from the lifters, which spread the fins out.

Nervous system

The fish nervous system consists of brain, spinal cord and nerves. The elongated brain has five segments: forebrain, diencephalon, midbrain, cerebellum and brain stem. The spinal cord proceeds from the brain stem down to the tail. It is a round cord which is enclosed by the vertebral column. Twelve nerve pairs (the "cranial nerves") are sent from the brain to the sensory organs of the head and to the head musculature. The spinal cord transmits nerve impulses from the brain and sends, on the basis of the number of vertebrae, an equal number of spinal nerves with a ventral motor root and a dorsal sensory root, which are associated with the sympathetic nervous system. This nerve system runs in the form of two fine nerve strands underneath and along the vertebral column. In cartilaginous fishes the forebrain is highly developed and both halves are not completely divided. All other fishes have clearly divided brain halves, which in lungfishes are large, and in teleost fishes are small and supplied with olfactory lobes. The brain weight amounts to about 1:1300 of the body weight (e.g., in gar). The brain cortex is, in contrast to higher vertebrates convoluted to just a small degree.

Eyes

Fish eyes are large, with an iris with a metallic sheen, a hard spherical lens and a flat cornea which is just slightly curved. Eyelids and tear glands do not exist in fishes. Sharks and short sunfishes (*Orthagoriscus*; see Vol. V) have nictitating membranes for protection. The eye of the fish is structured for near vision and when at rest is actually nearsighted. Accommodation of the eye for larger distances occurs by moving the lens by the lens muscle, which is returned to the rest position by a suspensory ligament. Fishes can differentiate colors, pictures and geometrical figures, and minnows have been trained to distinguish twenty color shades. Even ultraviolet light can be detected. Indeed a fish sees a body on the shore, such as a fisherman, in a somewhat distorted way. Small objects which are close by on the water cannot be seen because the light is completely refracted. The telescopic eyes of deep-sea fishes are elongated fish eyes with a giant lens which is extremely light sensitive; these are all adaptations to the prevailing conditions there.

Little is known about the scent and taste abilities of fishes. The olfactory organ consists of two cavities with blind ends at the back. These are not connected with the mouth. In cartilaginous fishes and lungfishes these nasal cavities lie on the lower side of the body, and in all other fishes on the upper side. They have a mucous membrane arranged in folds. With this, salmon can determine the quality of water against which they swim. A fisherman must realize what a great deal of time must be allowed for dispersion of odorous substances in water; bait with such substances only becomes effective when a fish in the area can detect it after it has spread sufficiently. Fishes have taste buds all over their bodies, concentrated mainly in the mouth, on the lips and barbs, on the ventral fin of gurnards and on the feelers of cod. Good-tasting objects stimulate swallowing, and unfavorable-tasting substances produce a reflex similar to our coughing. The terminal buds of the skin and fins detect chemical changes in the surrounding water. The skin sensory organs on the head are probably temperature sensitive, and they may also receive those stimuli which, together with temperature, stimulate migration. The low number of pain centers indicates that fishes have only a poor level of sensitivity in this regard.

The lateral line (Linea lateralis) detects water pressure which acts on the swimming or motionless fish. This extends from the head to the tail on both sides of the body and maintains contact with the water via its numerous pores. If the fish approaches an obstacle or a larger fish, the lateral line senses the resulting pressure change, and furthermore determines the direction and strength of currents so that those fishes which when spawning must swim upcurrent can find even the smallest streams. At night or in muddy water the lateral line protects the fish from bumping into objects, since these objects reflect the waves generated by the fish. Involuntary reflex equilibrium movements stimulated by the inner ear and lateral line are carried out by the paired fins and lateral musculature of the trunk and enable the fish to maintain its normal position in the water. **Lateral line**

With the exceptions of the cartilaginous fishes and flatfishes almost all fishes have a swim bladder. This skinlike sack always exists as a protrusion of the upper throat wall over the gut. It can be divided by constrictions into different chambers, is filled with gas and thus acts to equalize pressure in the water. The fish is thus able to adjust its specific gravity to that of the surrounding water and control its distance from the water surface. In Physostomi (e.g., sheatfishes, carp and herring) the swim bladder is connected to the gut throughout the life of the individual by an air passage, the ductus pneumaticus; in Physoclisti (e.g., codfish and perch) this passage is closed off in adults. Polypterid fishes have paired air passages. In some herring species there is a second passage from the swim bladder which passes behind the anus. The swim bladder gas is rich in oxygen produced by the gas glands on the inner wall (the so-called red bodies). In carp the swim bladder is connected with the inner ear by the Weberian apparatus. This structure has tiny bones which transmit pressure changes of the swim bladder to the inner ear. It has been shown that some carp can perceive sounds, e.g., minnows and dwarf sheatfish. The drum fish *Pogonias* (see **Swim bladder**

Vol. V), various gurnards (see Vol. V) and other species can create sounds with the swim bladder by vibrating its walls.

Respiration

Respiration occurs by internal gills which lie between the pharynx and body wall in pockets and take up oxygen dissolved in the water. The gills are thin layers of skin with blood vessels, which are free in sharks and protected by a gill cover in teleost fishes. Water enters the mouth and is expelled through the gills. It thereby runs over or through the gills, the surface of which is permeated by blood vessels. The blood takes up oxygen and releases carbon dioxide. The surface area of the gills of a carp measures about one-half square meter. The embryos of sharks and lungfishes have external gills like those of amphibians. When the oxygen supply is low, fishes snap at the surface for air to enrich their supply. Very sensitive trout die when the oxygen content declines to one and one-half cubic centimeters per liter, and the less sensitive carp do not perish until oxygen has dropped to one-half cubic centimeter per liter.

Fishes inhabiting dirty or shallow tropical waters or those that spend some time out of water (such as the climbing perch, Vol. V) can also breathe atmospheric oxygen. The labyrinth fishes (see Vol. V) have labyrinthine cavities in the upper bone of the first gill arch for this function. The South American silure (family Calriidae) has branched processes on one or both gill arches. Eels and pond loaches respire through their skin as well as with the gills. Gut respiration is found in pond loaches and a few silures (e.g., *Loricaria*). The lungfishes (see Vol. V) have so-called lungs which are sacklike appendages of the foregut which open into the lower side of the pharynx.

Blood circulation

The heart in fishes lies behind the gill arches and in front of the pectoral girdle in a pericardium with a partition between it and the visceral cavity. The heart has one atrium and one ventricle separated by valves. In sharks, rays and sturgeons a muscular element (the conus arteriosus) lies between the chambers and the aorta, the inside of which has from two to eight halfmoon shaped valves. Only a remnant of this structure is present in teleost fishes. In its place there is an enlargement of the aortic branch with a bulbus arteriosus which prevents a backflow of blood pumped through the heart. Lungfishes have an atrium which is divided into two parts by a partition, thus separating oxygen-poor (venous) blood (from the liver) from oxygen-rich (arterial) blood out of the lung. The heart pumps the blood through the aorta ascendens to the gills, and blood replenished with oxygen passes from the gills to the aorta descendens, which supplies all the organs with blood. Venous blood circulates back to the heart from the sinus venosus.

Fishes also have a lymphatic system which empties into the veins. A few glands which secrete internally empty their contents into the blood circulating through them. Among these the thyroid is responsible for metabolism, the thymus for growth, the pancreas for producing digestive juices and control of blood sugar levels, the pituitary gland for fat and carbohydrate metabolism, propagation of coloration material, blood pressure, stimulation of the smooth muscles, and other functions. Other significant organs include the gonads, which influence sexual activity; the adrenals, which influence breathing rate as well as heartbeat; the sympathetic nervous system and vascular muscles.

In contrast to the lampreys with their round mouths and rasping tongues, fishes have a jawed mouth adapted for swallowing food. It leads to the pharynx and from there into the muscular esophagus. The esophagus leads into the stomach without any striking transition, and from there the small intestine extends all the way to the anus. A short large intestine completes the system and has an external opening. The teeth are highly variable in shape and number; they correspond to the placoid scales of cartilaginous fishes. They lie on the jaw edges, on the palate and pharynx bones, and sometimes on the hyoid bone and gill arches. In predatory fishes they are pointed or conical, in other fishes wide and smooth or chisel-shaped. Species which feed on minute crustaceans (for example pipefishes) are toothless. Plankton feeders (e.g., salmon) have elongated thorny processes on the gill arches, which serve as sieves and retain the plankton. Carp have teeth only on the lower pharyngeal bone.

Digestive tract

In herring and perch the esophagus leads to an enlarged muscular stomach with a blind end which acts as a gizzard. The gut is winding and has a mucous membrane inside arranged longitudinally. In sharks, sturgeons and lungfishes the gut convolution system merges into a single screw-shaped spiral fold which increases the surface. There are no salivary glands. The liver is large and rich in fat. Sturgeons and many teleost fishes have several hundred short glandular blind tubes in the beginning of the mid-intestine. It is assumed that they increase the absorptive area. The anus is always located ventrally, but in genera *Gymnotus* and *Fierasfer* in the throat region. Sharks and lungfishes have cloacae; thus the intestine culminates with the urinary and genital tracts in a hollow cavity which leads to the outside through the anus.

The mesonephrons serve as excretory organs. These are dark, narrow, lobed structures, which lie just below the vertebral column and often extend on both sides from the head to the end of the body cavity. A pronephron is present during embryonic development, but later degenerates. In shark and ray males the front end of the mesonephrons, which become genital tracts, is connected with the gonad; such a connection is not present in shark and ray females or in teleost fishes. In the latter both ureters of the mesonephrons merge with each other at their ends and form a bladder which leads to the outside. A common urogenital opening occurs on male teleost fishes (exception: salmon), and the ureters of females end in a papilla behind the genital

Excretory organs

opening. Fishes urinate relatively small amounts. One excretory product, guanine, is utilized as the metallic shiny coat in the skin, eyes, and peritoneum.

Regulation of salt and water content

Kidneys do not only filter blood but also regulate the salt and water content of the body. The tolerance of most fishes to a greater or lesser salinity is highly limited (they are "stenohaline," from the Greek $\sigma\tau\varepsilon\nu o\zeta$ = narrow and $\alpha\lambda\zeta$ = salt). Thus most can live only in fresh or in salt water. The blood and connective tissues of fresh water fishes have a higher salt content than that of the surrounding water. The opposite is true of marine fishes: their blood and tissues have less salt concentration than the sea. Thus these fishes cannot lose too much water and still retain an excessive amount of salt. Marine fishes take in a great deal of sea water with many salts, and their kidneys and salt-separating cells in the gills retain the water and excrete the extra salts. The urine of marine fishes therefore is rich in salts and is excreted in small volumes. In contrast to the pure marine and aquatic fishes there are "euryhaline" forms (from the Greek $\varepsilon\upsilon\rho\upsilon\zeta$ = wide) that can make the transition from salt water to fresh water and have a great tolerance for changes in water salinity. These fishes (e.g., flounder and the three-spined stickleback) alter their excretory processes according to the nature of the water. The nervous system and hormones play an important role in this ability.

Genital organs

With the exception of a few hermaphroditic forms (e.g., from the genera *Serranus* or *Chrysophris;* see Vol. V), most species are bisexual and have paired genital organs which lie on both sides of the vertebral column in the visceral cavity. In a few cases they are united to form a single sack. In sharks, lungfishes and ganoid fishes the Müllerian duct, a part of the original ureter, takes up eggs released from the ovary and passes them to the outside. A part of the mesonephron serves as the spermatic duct. In salmon and eels the eggs pass into the visceral cavity through a break in the ovarian wall and are passed out through paired genital pores behind the anus. In most teleost fishes the testes and ovaries connect immediately with the spermatic ducts and ovidcuts, respectively, which together with the urinal tract terminate at the genital papilla behind the anus.

Male sharks and rays as well as a few teleost fishes have copulatory organs in the form of long cartilaginous appendanges on the ventral fins. Other fishes giving birth to live young (e.g., toothed carps) have copulatory organs formed from the anal fin. In the European bitterling female the genital pore is extended to form a five centimeter long egg-laying tube during spawning.

Fish hobbyists do not generally use the sex organs—or primary sexual characteristics—to recognize and differentiate the sexes, but instead rely on other derived differences between males and females, the secondary sexual characteristics. These include color differences or spawning colors of males, their spawning eruptions, and the increased size of particular fins during spawning. Occasionally the sexes are distinctively different, for example in the deep-sea fish genus *Ceratias*; the dwarf-sized males hang firmly to the females with their mouths, parasitize the females for nutrition and in essence are only sperm-creating creatures for the females. In black bass *(Centropristis)* a sexual transformation exists: females which have already undergone a breeding period become transformed to males. Parthenogenesis (development of young without fertilization) has been confirmed in the carp *Carassius auratus gibelio.*

The entirety of the eggs of a fish are designated as spawn. The spawning period is usually only once a year, primarily in spring, but in salmon in winter. Herring have several varieties according to the spawning period; each race spawns in spring, summer or fall at one specific time. At spawning time the male and female copulate and undertake a migration, which can be quite extensive, (as in herring, salmon and eel), to find a suitable nesting site. The fishes often move from salt water into fresh water or vice versa during this period. Salmon and sturgeons migrate from the sea upriver into tributaries (they are known as anadromous fishes, from the Greek $\alpha\vartheta\alpha$ = up and $\delta\rho o\mu\epsilon\tau\vartheta$ = migrate or wander). The eel goes from fresh water into the sea, and into the Sargasso Sea in the western Atlantic; such fishes are known as catadromous fishes (from the Greek $\kappa\alpha\tau\alpha$ = downward). Brackish-water species such as European smelt migrate into fresh water for spawning, and river inhabitants move into the upper courses; the European roach and the barbel migrate from inland lakes to the waters entering them or up streams, while grayling, common carp, and tench spawn right where they otherwise live.

Spawning

Egg development can take on gigantic proportions in fishes. Certain shark species lay only two eggs per year, and in the comb-toothed shark 108 embryos have been found in the mother's body; sturgeon females, as a contrasting example, can lay up to six million eggs, while the common cod and turbot lay six and one-half and nine million eggs, respectively. A ling (genus *Molva*) one and one-half meters long and weighing twenty-five kilograms has no less than 28,360,000 eggs in its ovaries. Among well-known fishes of Germany the carp lays up to 70,000 and the trout a maximum of 3000 eggs.

Number of eggs in various species

Those fishes giving birth to live young such as spiny dogfishes, hammerhead sharks, the great white shark, and many toothed carps, have internal fertilization and embryos develop within the womb. Spotted dogfishes, rays and chimaeras lay large eggs with horny shells after internal fertilization has taken place, so that development takes place externally. However most teleost fishes utilize external fertilization in which the eggs and sperm are released by the female and male after various species-specific foreplay behaviors take place. The male fertilizes the eggs immediately after they are released from the female. After being laid, the eggs either sink to the floor, as in most fresh water forms, or they are suspended like plankton (as is the case in most

Fertilization, development

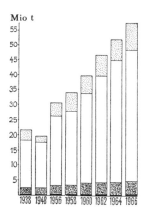

Fig. 3-8. Development of world fish catches. Fresh-water fishes (lightly shaded), marine fishes (white), crabs and mollusks (darkly shaded), and other (black). Figures (5 to 55) are in millions of tons.

Larvae

Commercial fishery and fish breeding, by K. Lillelund

Fig. 3-9. Fishery centers. Catches are given in millions of tons.

marine fishes). Care of the young takes place in many different species. The male stickleback (see Vol. V) builds a nest of plants and stones to protect the eggs. The female pipefishes (see Vol. V) attach their eggs to the ventral brood pocket of the male. Some cichlids (see Vol. V) or sheatfishes take the eggs into the mouth and the young hatch in the mouth cavity. The purpose of this is to protect them from predators and to insure a supply of oxygen-rich water. After hatching the young often stay together for up to two weeks and feed on their yolk sacs until they can feed independently.

The interval in which the embryo develops from a fertilized egg to a hatched young varies according to species and season. In sturgeon development takes 64-120 hours, in carp and perch up to two weeks and salmon of genus *Salmo* require as much as five months. The spotted dogfish *(Scyliorhinus)* lays up to twenty eggs in firm, hornlike capsules on the seaweed of the ocean floor and the young hatch after eight months. Generally larvae undergo a development which lasts as long as their yolk supply is available, which in herring is up to three weeks. In contrast the young of viviparous fishes and the pipefishes are fully developed when they hatch.

The flatfish (see Vol. V) and eel larvae are particularly different from the adult form. Flatfish larvae are symmetrical and only during the course of development as a bottom dweller does the larva lie on its side and the lower eye migrate over the back of the head onto the other side of the body. Some larvae have adhering fibers (such as in African characins) of the outer gill fibers (e.g., *Gymnarchus* and *Clupisudis*). Sexual maturity usually occurs in the third to fifth year of life, but in the female eel not until the twelfth year.

Artificial fertilization, heat, and nutrition yields many more viable eggs than can be fertilized naturally. Under these conditions destruction of the eggs by rotting or predations can also be avoided. During the last decades artificial breeding of fish eggs and raising fishes have increased more and more. Due to these efforts our fresh waters still have fish life in spite of the harm which has been wreaked upon them.

The steady increase of the world's population increasingly influences all aspects of our life. From 1850 to today alone, the human population of the earth has climbed from about 1,000,000,000 to 3,000,000,000. One factor among many which will be critically important for humanity is whether the production of food can match the explosive human birth rate. Since land occupies only twenty-nine percent of the surface of the earth, the seas with their nutrients take on increasing significance.

Fishery activity has increased continuously since the end of the last world war. Marine fish catches doubled between 1948 and 1966. Fresh water fishes, the catch of which has increased correspondingly, make up about fifteen percent of the world fish catch. Crustaceans and mollusks, at seven percent of the total fish catch, are also a significant portion. More than fifty percent of the world catch consists of herring and codfish. Marine fishes as such constitute over sixty percent of the total catch. If the quantities caught in the greatest fishery centers of the world are compared with the total world catch, these small centers make up about ninety percent of the total catch. Thus when compared with the total oceanic area, most fish caught are from a comparably small area. It is necessary to have a closer look at the production of commercial fishes.

At the beginning of the sea's food chain are microscopic, usually unicellular, plants. These primary producers metabolize the sunlight, carbon dioxide and salts. The smallest animals of the ocean feed on these producers and serve as the nutritional basis for other, primarily invertebrate, organisms. The small organisms suspended in the ocean are known as plankton. If plankton die they are decomposed by bacteria and transformed back into inorganic salts. If the water is not too deep, plankton sink to the floor and there they are fed upon by mussels and other filter feeders. Various fishes feed on members of various parts of the food chain. Thus, sardines, which have especially fine filtering systems, feed directly on the primary producers. Other species such as herring feed on copepods, shrimp or pteropods, all creatures from higher parts of the food chain. Those species which feed on invertebrate bottom dwelling organisms include especially the fishes which live continuously on the bottom, e.g., many flatfishes (see Vol. V). Codfishes are primarily predators which feed on other fishes. Some common cod change their diet daily, and may feed on the bottom during the day and move up to the surface at night where they prey on herring and other fishes.

The nutrient salts of the water, above all the dissolved phosphorus compounds, are rapidly exhausted by the activity of the unicellular plants; thus primary producers have a natural limit. This results in an overall limitation on the entire food chain of the oceans. If fishery industries are active in some particular area, one part of the life chain will be removed, which will reduce the total biomass production. Some particular area will therefore only be a valuable fishery if a steady influx of nutrient salts insures the renewal of primary producers. This renewal can proceed in three ways:

1. The water of the large rivers is generally rich in nutrient salts and forms a basis for fisheries on bordering seas.

2. Where cold and warm sea water currents meet a great mixing of water masses takes place with rapid changes of temperature and usually also of salt content. Many plankton organisms cannot adapt to these changes and they die and sink to the floor and thereby form the basis for a flourishing bottom dweller population. One such upwelling is the Newfoundland Bank, where the cold, low-salinity Labrador currents from the north meet the warm, salt-rich Gulf Stream

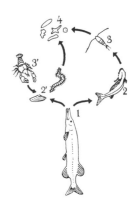

Fig. 3-10. Food chain: 1. Predatory fish; 2. Prey fish; 2'. Invertebrate bottom dwellers; 3. Plankton; 3'. Higher crustaceans; 4. Unicellular organisms.

Fig. 3-11. Ocean currents. Warm currents: 1. Gulf Stream system; 2. North Equatorial current; 3. South Equatorial current; 4. Brazilian current. Cold Currents: 5. Californian current; 6. East Greenland current; 7. Labrador current; 8. North Pacific current; 9. Alaska current; 10. Falkland current; 11. Humboldt current; 12. Benguela current; 13. Western wind drift.

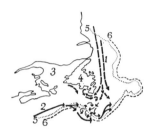

Fig. 3–12. Currents near the Newfoundland Bank. 1. Labrador current; 2. Gulf Stream; 3. St. Lawrence Gulf; 4. Newfoundland; 5. Depth line of 300 m; 6. 2000 m depth line.

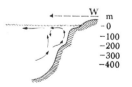

Fig. 3–13. Upwelling regions.

Fig. 3–14. Schematic cross-section of an up-welling region.

Marine commercial fishing; the trawl

from the south. The cod and rosefish (see Vol. V) populations of the Newfoundland Bank are fished by North American and European concerns.

3. Since plant growth is dependent on sunlight, plant production declines with increasing depth, thus also reducing the intake of salts at lower depths. Here decomposing processes play a major role and enrich the salt supply. In the presence of a steady strong wind blowing from the land into the sea a current from the shore outwards can be produced, which can equalize the seaward loss of surface water by upwelling deep water masses, rich in nutrients. Due to the direction of the great prevailing winds, the great majority of these regions are on the west sides of the continents. The largest such region extends along the northwest coast of South America. Not surprisingly, it supports a tremendous fish population.

A predisposition for food plankton development is thus primarily in regions with a fresh water inflow, in mixed regions of cold and warm sea currents, and in regions where waters from the depths are being driven up. These regions lie for the most part at the edge of the continental shelf or on it. The continental shelf is a lower layer of the earth which has been pushed forward and which lies just before the site where the continental slope inclines sharply into the ocean floor. The shelf seas extend to a depth of about 200 meters.

In contrast to the conditions in the shelf seas the nutrients in surface water of the open seas are quickly exhausted. The plankton content— and thus fish capacity—of the deep seas is thus much lower, for a given area, than in the shelf seas. With increasing plankton content the water color changes. While on open oceans, with a smaller production of plants and animals, the color of the water is a deep blue, a coloration from green to yellow appears as plankton development increases. This color difference is clear when one approaches land while aboard a ship.

Commercial use of fishes is not only dependent on quantities of available fishes, but also on body length, body weight and uniformity of the fishes utilized. In northern regions with their relatively low temperatures, fishes do not reach sexual maturity until a rather old age. Since most growth occurs before sexual maturity, it is evident that in the warm southern waters, where fishes rapidly mature, the body lengths are generally not as long. In its stead a greater variety of species is encountered. In northern seas the number of different species is low in comparison, but the fewer species that are present often grow greatly and produce a supply which can be used commercially. Similar favorable conditions for successful commercial fishing are in the warmer regions where small fishes like sardine and anchovies produce large, dense populations.

A further condition for commercial use of fishes is the potential use of effective catching equipment. The most important device for catching bottom-dwelling fishes is the trawl. Trawls are also used for catching fishes which are only temporarily on the floor. These include most cod-like fishes, but also some surface forms like herring, which come to the bottom when spawning. Use of trawls is commercially feasible at depths of down to 600 m. Beyond this, increasing technical difficulties appear. Thus trawls are used only in shelf regions. The floor must be free of obstacles, which is not the case in the shallow tropical waters with their coral growth.

A swim trawl has been developed for fishing surface ocean waters. The difficulty here lies in determining the depth of water in which the trawl must operate, and this is solved by echolocation. Ultrasonic transmissions will be reflected not only by the ocean floor but also by fish groups. Thus both schools of fish and the depth of the water can be determined ultrasonically. The catching equipment itself also has an echolocator, which reports steadily to the ship about the depth within which the trawl is working, and the trawl depth is altered according to the presence of fishes and water depth. A swim trawl pays when the fishes are in sufficiently dense concentrations in the water.

The ring net is especially suited for fishing sardines and tuna. This consists of a long net which is pulled along the surface by swimmers. A fish school is enveloped by it and pressed together by tightening the net, until the fishes can be caught. This device is independent of water depth.

Line fishing is for large predatory fishes, e.g., tuna, sword fish, barracuda, salmon or sharks. A so-called long line is used to catch floor-dwelling predatory fishes such as halibut (see Vol. V); it has numerous baited hooks. The line lies on the floor. Long nets, on which fishes catch their gill covers or get tangled in the webbing, are used almost exclusively in shallow coastal regions. Fish schools which move along coasts are often caught with nets in great baskets, as is done in the Mediterranean for tuna. All the conditions necessary for the success of the above methods are found in the shallow shelf seas. But shelf seas make up only ten percent of the total ocean surface. Thus it is evident that the greatest part of the world fish catch is from but a few regions with a relatively small area.

The northern hemisphere accounts for over seventy percent of the world catch. This is due to the great extent of shelf seas in the north, the lack of commercial species such as larger codfishes in the southern hemisphere, and the lesser technological development of the fishery industry in the south. Only two areas of significance are in the southern hemisphere. One is on the coast of the Pacific Ocean, primarily around Peru. The catch here consists almost entirely of one species of anchovy, the anchoveta *(Engraulis ringens).* In 1966, 9.6 million tons of anchovetas were caught off the Peruvian coast; over twenty percent of all marine fishes caught in the world are of this species. Anchovetas

Echolocation

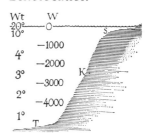

Fig. 3–15. Schematic cross-section through the continental base. K, continental slope; S, shelf; T. ocean floor; W, water depth; Wt, water temperature.

Line fishing

Catch sizes

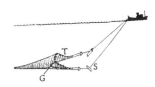

Fig. 3–16. Trawl. G, pulling cable; S, tow boards; T, supporting elements.

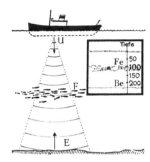

Fig. 3-17. Echolocation. Be, floor echo; E, echo; F, school of fishes; Fe, echo from fishes; U, ultrasonic frequencies. Black border area: reading on echo apparatus.

Fig. 3-18. With the ring net a school is surrounded.

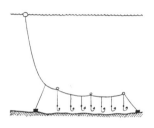

Fig. 3-19. Long line with baited hooks.

are processed into fish meal. In the coastal regions of South Africa and southwestern Africa the catch consists primarily of the South African sardine *(Sardinops ocellata)*. There are two other species of importance in this region, the anchovy variety *Engraulis japonicus* and the cape garfish, *Merluccius capensis*.

The greatest catches in the northern hemisphere are from the waters between the Philippines and the Japanese islands. Japan has been favored by its situation in shelf seas and the rich fish supplies in the adjacent waters, and is the greatest fishery country in the world. The total Japanese catch in 1966 was 7 million tons. The variety of fishes in Japanese catches is especially large. This is due not only to the varied species found in those waters but also to the highly developed state of fishery in Japan. Japanese fishing vessels today are found in practically all important catch regions in the world. The Japanese industry is especially oriented toward tuna (see Vol. V), and about half of the world tuna catch (which in 1966 was 1.3 million tons) was caught by Japan. The proportion of mollusks in Japanese catches is also very high. These consist primarily of copepods and oysters (see Vol. III). Another important fishery region is the north Pacific Ocean. There a cod-like fish, the walleye pollack *(Theragra chalcogramma)*, is caught by Japanese and Russian fleets.

The fishes in the north Atlantic and its bordering waters form the basis for European fishing. As in all fishing regions, only a few species determine the size of the catch, and in particular these are herring (in 1966 over 4,000,000 tons) and common cod (2,500,000 tons). Other important varieties are haddock, salmon, and the cod-like *Merlangius merlangus*, which together make only about half of the catch of cod varieties. Comparable to the Japanese in the east, the Norwegians in Europe have especially favorable conditions for a well-developed fishing industry. This is due to the fact that the Norwegian coast is kept ice-free throughout the year by the Gulf Stream. The Norwegian catch in 1966 amounted to almost 3,000,000 tons, and 1,000,000 tons of herring alone were caught. Norway and Denmark depend on fishing in the vicinity of their coasts, quite unlike Germany and England.

The chief fishery regions for Germany and England are the northern North Sea, and the waters around Iceland, Greenland, and Newfoundland with their cod, haddock and rosefishes. While the Norwegians and Danes could steadily increase their catches, Germany's fishing industry has leveled off to about 600,000 tons. Many fish varieties in European seas can only be caught at certain seasons. This is because fishing is only commercially practical when enough fishes are sufficiently concentrated, which may occur only when sexual maturity takes place. In temperature zones where a great seasonal temperature change in the water takes place, there are only a few days or weeks of spawning in each year. With the onset of sexual maturity the scattered fishes collect in dense schools at a common breeding ground. The Norwegian herring migrates in the late fall from feeding grounds around Iceland and in February and March of the following year is the focus of Norwegian late winter fishery. In the North Sea, fishing begins on the spawning concentrations of the North Sea herring in summer and reaches its high-point in September and October, when the fishes have collected at their main spawning area, the Doggerbank. The take of the Arctic-Norwegian cod is particularly impressive. The cod arrives in the winter months from the Barents Sea, moves along the Norwegian coast and then in February is the subject of intensive fishing in the Lofoten islands, which is carried out with hundreds of small vessels.

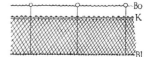

Fig. 3-20. The anchored net is supported by floaters (Bo) and a cork row (K). A lead line (Bl) opens the net.

Fresh water catches are obtained from lakes and rivers and from raising fishes. Therefore inland fishery is highly variable in various countries, due to differing quality of water and the proportion inland waters form of the total surface area of a country. As in the sea, there is a close relationship in fresh water bodies between the nutrient supply in the water, temperature relationships, and the amount of the catch. High nutrient content and temperatures (above 20° C.) are the impetus for rich development of plankton. Plankton that die sink slowly into deeper water and if the temperature is sufficiently high, decomposition sets in during this sinking. Oxygen is withdrawn from the water in this process. However in well-lit upper waters plant activity equalizes the oxygen loss. But in deeper waters which do not receive much light that is not the case; because of that, nutrient-rich lakes are practically devoid of oxygen at a depth of from eight to ten meters. The bottom is then not useful for feeding material. Clearly, production of commercial fishes is limited to shores and on the plankton-and oxygen-rich upper levels in open water. In those lakes with a low nutrient content and in which the sun penetrates fairly deeply (because few decomposing plankton develop) the oxygen content of the bottom permits fishes to flourish at those depths. However this supports only a relatively small number of fishes, and actually the best catches are from nutrient-rich, warm and shallow lakes only a few meters deep. These can yield 250 kg and more per hectare annually. The fewer nutrients and colder the lake, the smaller the yield.

Fresh-water fishing

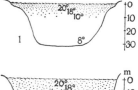

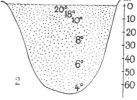

Fig. 3-21. Schematic diagram of the distribution of oxygen (shaded) and temperature in 1. nutrient-rich and 2. nutrient-poor seas during the summer in moderate climates.

The fish world has adapted to the particular environmental demands of the lakes by certain combinations of various species. The European lakes can be divided in terms of the most important prevalent commercial fish. Lake trout and charr are the most significant commercial fishes in nutrient-poor, cold high mountain lakes. Whitefish lakes are warmer and generally have more nutrients than the trout lakes. The whitefish lakes include the larger Alpine lakes, such as Mond Lake and Wörther Lake, but also the large lakes in northern Poland and a few larger north German lakes such as Selenter Lake. Whitefish are caught

Adaptation

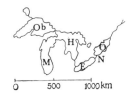

Fig. 3-22. The Great Lakes of the USA. Lake Erie (E), Lake Huron (H), Lake Michigan (M) Niagara Falls (N), Lake Ontario (O) and Lake Superior (Ob).

"Prey fish,"
"Predatory fish"

with anchored nets in these lakes. Lakes that are still richer in food supply and warmer often support breams as the most important commercial species, if the floor has a sufficient oxygen content even in summer. The bream lakes are distinguished on the basis of depth from the shallow European roach lakes and the still shallower gar-tench lakes, which are only one to two meters deep and support aquatic plants. One distinctive type is the pike lake (see Vol. V). The total catch is essentially determined by the amount of water surface. It comes as no surprise, then, that countries with the largest inland lakes also have the greatest catches. In North America almost one-third of the total catch (excluding salmon) is from the Great Lakes region. That is the most extensive interconnected lake region in the world, covering 250,000 km². The deeper ones are whitefish types. Lake herring and chubs are also among the most important commercial species found there. Shallow, warm water as found in Lake Erie supports perch and pike.

South America has few inland lakes. Lake Titicaca, with a mean summer temperature of 13° C., supports two species of genus *Orestias* as its most important commercial fishes, the prey fish *Orestias pentlandii* and the predator *Orestias cuvieri*. Since salmon species were not found in South America, the idea was developed to introduce rainbow trout there from North America. This trout killed off *Orestias cuvieri* in a few years, and the most important commercial fish was therefore lost, a loss not replaced by the rainbow trout.

The largest African lakes (and thus biggest catch areas) are in Zaïre, Tanzania, Uganda and Egypt. Conditions in Lake Tanganyika are especially well understood. This 650 km long, 30–40 km wide lake is in a deep trench. The water depth varies between 700 and 1400 m. Although the total nutrient supply, as measured from the visibility level of twenty-two meters, is relatively low, the upper levels (which throughout the year measure 23–24° C.) are subject to rapid metabolic activity and a high rate of oxygen loss. The lack of oxygen and presence of hydrogen sulfide exclude the presence of fishes below 100 m. The most important fishes are the many cichlid species of genus *Tilapia* (see Vol. V). Although the commercially fishable shore bank is only fifty meters wide, the fact that many *Tilapia* species feed on plankton in the open water has resulted in a maximum catch of 10 kg per hectare.

The Egyptian coastal lakes are highly productive. These are shallow lakes one to two meters deep connected with the sea. The water surface is labyrinthine due to many shelf islets. Eels, gray mullets, bass and *Tilapia*, which have adapted to brackish water, as well as *Tilapia zillii* and *Tilapia galilea*, thrive in these nutrient-rich lakes. They feed on freshwater and marine organisms. Even though these lakes annually produce 300 kg per square hectare and more, it is planned to drain them to create more farmland.

With the exception of the large Asiatic waters of the Soviet Union, Asia has few large lakes. The best fresh water catches are from large rivers and ponds. The Caspian Sea and the Aral Lake produce great catches in the Soviet Union. Both lakes are brackish with a salinity that is about one-third that of the oceans. Sturgeons are a valuable catch, for their eggs are made into caviar. Four-fifths of all the world's caviar is from Caspian Sea sturgeon. Yet in terms of quantity—almost fifty percent of the entire catch—another species in the Caspian Sea is the fish caught most often. That is a European roach, *Rutilus rutilus caspicus.* The relatively high salinity of these inland seas (which in terms of their lime and sulfate content are unlike the oceans, however) has made possible the introduction of marine fishes which have adapted to brackish water. Thus, from 1954 to 1956 eggs of an early spawning herring of the eastern Baltic Sea, *Clupea harengus membras,* were introduced and artifically fertilized. Fertilized eggs were then flown to Aral Lake and incubated there with Aral Lake water until the young hatched. In these three years a total of 19,400,000 eggs were introduced into the lake. The fishes developed extremely well and grew faster than their conspecifics in the Baltic Sea and after three years they were mature for spawning. Today these herring in the Aral Lake play an important role. The growing herring population also increased the pike and perch growth, since the latter prey on herring.

Since the end of the last world war the lake surface area of the Soviet Union, which without the large brackish lakes comprise 210,000 km², has been increased considerably through the construction of extensive dams. Damming has accounted for almost 50,000 km² of area. These artifical lakes were stocked according to fishery principles for optimum return. Preferred species for such stocking are whitefishes, sturgeon, carp, bream and pike. Stocking experiments have also taken place in natural waters so that the natural distribution of these species has been exceeded. Quite often these disturbances have had unexpected consequences so that the commercial result did not live up to expectations.

In European lake fishery the eel, gar, pike, perch and whitefish are the most important fishes. The nature of catching them depends on the particular type of water with which one deals.

River fishery forms another significant part of the industry. As the river is modified from the source to its mouth, the species associated with it change as well. In the sources of European waters salmon are predominant, for the fast-flowing cool water is favorable for them. This so-called "trout region", the river segment occupied by river trout, rainbow trout and other species, is followed by the "grayling region". Here the water is still relatively cool and rarely reaches 20° C. in summer. Although the many streams have formed a river, the current is still quite high at 60–120 cm per sec. Further down comes

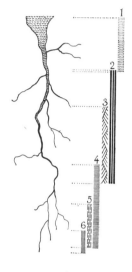

Fig. 3-23. The commercially fishable regions of a river. 1. Pope-flounder region; 2. Bream region; 3. Barbel region; 4. Greyling region; 5. Lower river trout region; 6. Upper river trout region.

Commercially important river regions

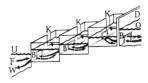

Fig. 3-24. Schematic diagram of a three-chambered fish stairway: B, bottom opening; D, dam top; F, fishes; K, Cutout section from damp top; O, water level at upper end; U, water level at lower end; W, direction of water flow.

Increasing catch in China

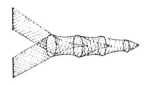

Fig. 3-25. Fish basket.

the "carp region", in the upper part of which are barbels, and in the lower part of which are breams, gars and eels. In this area the river bed becomes wider and wider, and the current decreases down to a few decimeters per second. The river shore, especially in still-water zones and in old arms with a rich underwater plant growth, is occupied by reeds, providing the best reproductive conditions for carp. The high salt content and summer temperatures exceeding 20° C. provide for the development of river plankton and a rich food supply on the river bottom. The last stretch of the river before the mouth into the sea is the "pope-flounder region" (see Vol. V). The influence of tides is already evident here. Eels are also of commercial importance here, particularly in the summer and fall months for the migrating white eel. In some lower river courses the smelt migrating in for spawning are fished in spring.

The annual salmon spawning runs in the rivers throughout the year are especially valuable for fishery. These are the European salmon on the Atlantic coast, and the Pacific salmon genus *Oncorhynchus* on the west coast of North America and northern Asia. The total world European salmon catch in 1966 was 12,000 tons, and the Pacific salmon catch amounted to 449,000 tons. The catches have sharply decreased in many rivers due to increasing pollution from industry and housing or from construction on water.

Attempts have been made to increase the salmon catch in the rivers of China since the last world war. The Yangtze-Kiang is so wide in some areas that larger and smaller lakes have developed. The river lakes, about 1,760 of them, comprise a surface of almost 33,000 km². They are seldom more than six meters deep. The requirements for an ample nutrient base and fish life are all present, for these lakes have high temperatures (in summer about 30° C.) and have a great food supply.

Generally 75–130 kg of fishes are caught per hectare in these waters. The idea to increase this catch stems from the consideration of stocking the lakes with fishes that use all of the available food present in them. Such species include four Chinese carp species, which complement each other in their food requirements: *Ctenopharyngodon idellus*, which feeds primarily on rooted aquatic plants; *Hypophthalmichthys molitrix*, utilizing floating plants; chubs (*Hypophthalmichthys nobilis*), which feeds on plankton; and *Cirrhina molitorella*, which feeds on bottom organisms. By removing competitive species it was possible to use rooted aquatic plants to feed large groups of *Ctenopharyngodon idellus*, for the plants are fed upon by very few species of fishes. Fertilization of the lakes by the manure of the carp resulted in increased plankton growth and provided conditions for thriving populations of the open-water species which had been introduced there. In certain river lakes the annual catch reached a record high of 600–1800 kg per hectare.

Variations in local conditions result in the use of quite different fishing devices. Their great diversity is actually derived from just a few basic plans. The most important devices are fishing tackle, anchored nets, movable nets and basket nets. Recently electrical fishing has increased on lake shores and in flowing water. One pole is set in the water and an electrical field is generated. Fishes swimming into the field are stunned for a short time and can easily be taken out. Electrical devices are particularly suited for testing population density in small bodies of flowing water, since the fishes captured can be put back in without suffering damage. In some countries this procedure is carried out by the use of poisons.

What possibilities does the fishery industry have to increase the annual catch even more? The first possibility is to find fish populations which have not yet been exploited. Efforts would have to be directed toward the shelf regions, the largest of which is the Patagonian Shelf which lies off the south Atlantic coast of South America. This shelf is very similar ecologically to the Newfoundland Bank, which is highly productive of fishes. The Patagonian Shelf is also an area of a mixing of cold and warm waters, the Falkland current from the south (which is cold) and the warm Brazil current from the north. In recent years German research vessels have investigated this area. It appears that a hake *(Merluccius hubbsi)* and a herring *(Clupea fuegensis)* are the prevalent forms. The anchoveta *(Engraulis anchoveta)* and Spanish mackerels *(Scomber japonicus;* see Vol. V) are also present and are fishable, but the large codlike fishes of the northern hemisphere are absent. Instead there exists a number of small fishes.

Other little-used (and narrow) shelf regions are on the east and still more on the west coast of Africa. A large upwelling region in the Arabian Sea is also still fishable. According to present studies this sea should support substantial fishery activity, primarily for herring species. But further extension of commercial fishing in the shelf seas and upwelling regions is not feasible. The possibility of finding commercial fish populations in the open ocean is small, since such fishes are usually spread over large areas of water and cannot be profitably caught with present-day apparatus. The lantern fishes (Myctophidae) present some possibilities. These fishes, which are generally only a few centimeters long, engage daily in vertical migration. During the day they dwell at depths of 300–600 m together with luminous shrimp, cyclopids (see Vol. I) and other invertebrates. These groups can be seen on echolocators as a dark horizontal strip. With the onset of darkness the organisms rise from this echo dispersal region into higher levels or even up to the surface, and shortly after midnight they return to the depths. Estimates on the degree by which the world fish catch could potentially be increased differ widely. It is generally agreed, however, that by exploiting new regions the total catch could at most be doubled, so that the maximum would amount to 100–120 million tons.

Inland fishing devices

Possibilities of increasing catches

Fig. 3-26. Patagonian Shelf.

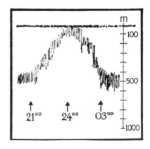

Fig. 3-27. Diagram of an echo dispersal region from a depth of 500 m reaching the water surface.

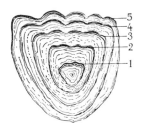

Fig. 3-28. Fish scale with five annual rings.

Fish markings

Fig. 3-29. Otolith of common cod with six annual rings.

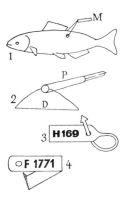

Fig. 3-30. Marking of fishes: 1. Marker (M); 2. Plastic tube (P) with instructions inside; Wire (D) or perlon for fastening the marker on the ventral musculature; 3. Plastic strip; 4. Plastic or metal marker for fitting to the gill covers.

It is evident that in addition to searching for new commercial fishing grounds, more efficient use of those already being fished is sought. The basis for this is the relationships between biological processes in fish populations and fishery, which has been gained from ichthyological studies. In these investigations the growth of the various commercial fish species and the age composition of the groups are determined. This is done with the help of growth rings on the scales or the otoliths (auditory bones).

The question also arises whether the fish of one species in some regions form a closed community or are in more or less independent groups with different spawning and feeding grounds. Catching such fish has, in the latter case, only minimal effects on the neighboring stand. To answer this question, marking experiments are carried out. Fish from the one region where the species is found are caught as quickly as possible, marked, and then released. Fish which will be sold fresh are usually marked externally. Such marking consists of an artificial substance placed on the ventral muscles, fins or gill cover. The markings carry a number and give instructions for those catching the fish. Schooling fishes of the open waters, such as herring, sardines or anchovies, which are processed to a great extent into fish meal, are better handled with internal marking. These are little metal plates which are shoved into the visceral cavity. The ground fishes move on a conveyor belt past strong magnets which pull the markers out. With the help of markers one can gain knowledge on whether the fishes of some region remain there or migrate and thus mix with conspecifics in other regions.

The great migrations of the Norwegian herring between the feeding grounds near Iceland and the spawning regions on the Norwegian coast were explained primarily through marking experiments. Such experiments also shed light on the various herring groups of the North Sea and the English Channel. This technique becomes difficult when various groups of one species mix in their distribution regions. In order to make a scientific division of the catches according to which groups the fishes are from it is necessary to seek characteristics by which groups can be differentiated. In herring this is often done by noting the number of vertebrae.

The effect of commercial fishing on fish populations can only be assessed when it is known how the mortality rate of the various ages of fishes changes from the larva to the commercially mature form. Such investigations have shown that the greatest part of the freshly hatched larvae die within the first weeks of life. Survival is essentially dependent on whether the larvae have a sufficient plankton supply on which to feed. Thus the determination of survival rates is dependent on food needs, and for this laboratory studies are necessary.

After the first weeks and months of high mortality, a period follows in which the young fishes are not greatly endangered and mortality is relatively low. The chief danger is now from predators. With further growth the fishes become big enough to be caught in nets, and the mortality rate of that respective age increases sharply at that point. The total mortality rate for commercially mature fishes can be determined in various ways. Marking has yielded useful data on the effectiveness of commercial fishing; here the number of re-caught marked fishes is compared with the total number of marked fishes under consideration. Another method consists of analyzing catches in terms of age composition. The proportion of some particular age group within the total population can thus be determined. The more one age group is subject to fishing and natural causes of death, the smaller will be its part in the catches of following years.

It has been found that in many groups the fishes have only scant survival chances upon reaching the commercially usable size. Out of every hundred fishes of one age in the heavily caught sole, cod and haddock, thirty to forty of them are caught annually by commercial fishermen, and another ten or twenty die from natural causes. The annual mortality rate is thus over fifty percent. In schooling fishes of open water (e.g., herring, sardines, anchovies, smelt and sand-lances of the North Sea and the northeastern Atlantic) commercial fishing has a generally smaller effect on mortality, since predators account for a comparatively high death rate.

Knowledge of the number and total weight of fishes in a population is important for fishery. Such a census can be taken by pulling a narrow-meshed net through a spawning area and determining the number of eggs per unit area of water surface. It is necessary that several catches be made in the immediate vicinity of each other and cover the whole spawning region. From the total number of eggs in the region and knowledge of the age composition of the population, sex ratio and average number of eggs laid per female at various ages, it is possible to calculate how many fishes belong to the spawning population and how they are divided according to age group. The total population weight can be assayed similarly. Such studies have shown that the number of different age groups is highly variable. Whether the growth of one season is strong or weak is above all dependent on the magnitude of mortality during the first weeks and months of life. The annually changing environmental conditions and ecological requirements play a decisive role here. One strong season's growth can be the impetus for a succession of good catches, and this accounts for the great catches which are being made.

One important task for fishery is to determine from what fish length and at what expense a population can be fished for optimal results over an extended period of time. The consequence of this is that a group cannot be continuously decreased in size. That is only possible when the annual catch does not exceed the annual growth. Annual growth is dependent on the middle age groups of the population; during the

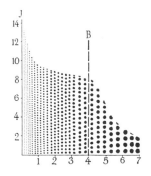

Fig. 3-31. Diagram of the decreasing numbers of commercial fish age groups (the size of the dots indicates the individual weight). B, beginning of commercial catching; J, numbers of each age group in billions; 1-7, age in years.

Census

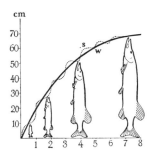

Fig. 3-32. Growth of a pike. Average growth (thick line); true growth independent of seasons (dotted line); S, summer; W, winter; 1-8, age in years.

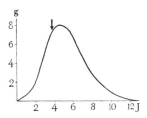

Fig. 3-33. Gain and loss of weight in various ages (2-12 = age in years) of commercial fishes.

first years of life growth increases rapidly, since the growth increase of the developing fishes is greater than the weight of the dying fishes. As growth rate decreases and mortality increases the losses begin to predominate. Thus at a certain age the maximum weight is attained. The optimal catches result by fishing that age group at the end of the year in which the highest annual growth rate has been reached.

The age at which sexual maturity is attained also plays a role here. To insure that enough eggs are laid the following rule is generally used: the minimum catch length should be such that the fish has been able to spawn at least once. The size of the smallest fishes caught is determined by the net mesh size. If the smallest mesh size is regulated, one can influence the smallest size which will be caught. Minimum length laws have been established for many species in the ocean and in fresh water. Marine fishes require international agreements. For this purpose many oceanic regions are subject to international treaties regulating commercial fishing activity. Agreements can be altered on the basis of the latest research finds so that an optimal long-term catch will be insured. The commercial interest of countries involved and the expense of changing mesh sizes often hinder effectiveness of scientific finds. Indeed, the cod near Greenland and in the Barents Sea, and haddock of the northern North Sea, as well as other populations of the European fishing industry, are being caught in the middle age groups, which in part is far beneath the maximum possible catch. Although increasing mesh size initially produces smaller catches during the first few years, since the population must first grow somewhat before being caught, the catch in the following years will increase because of greater annual growth rate.

Decline of a fish population is not always due to commercial involvement. Marine fishes are generally very sensitive to certain environmental conditions, above all to the salinity and mean temperature of their distribution region. If these conditions change, drastic results can follow, particularly in succeeding growths, for the food base is altered. The migrating routes and migrating strength, and the location of feeding grounds in spawning grounds are closely tied with ecological demands. Environmental changes are dependent on weather and can have long-term effects with corresponding long-term fish population losses. Such population variations have been confirmed in Norwegian herring. Such changes can be particularly prominent in upwelling regions. As a consequence thereof the inflow of nutrients from deep waters can be decreased for a short or long period of time. Such upwelling phenomena have been observed along the Norwegian coast. The fishes either died in great numbers due to a lack of food, moved into deeper waters, or migrated to more favorable areas. With the disappearance of the fishes came a mass dying out of the guano birds, which feed on these fishes.

Long-term ecological changes can also result in an alteration of the competition among fishes for food. One example of this is the decline of the South American sardine *(Sardinops sagax)* of California. At the same time the adjacent population of the North Pacific anchovy *(Engraulis mordax)* increased its numbers.

In fresh water, civilization has also been a part of the changing environment. Construction changes such as dams, gradients along streams, paving shores, locks, draining or damming lakes, can alter the original character of many bodies of water. Species which formerly existed there now find no means of further existence, and other fishes which are not as sensitive (but also of no commercial value) can increase.

Results of civilization

The slow enrichment of nutrients, especially addition of phosphorus compounds, as from fertilizers or detergents, can modify fish populations. After the war a rapid transformation was noted in many central European and North American lakes in which nutrient-poor whitefish lakes became nutrient-rich carp lakes of the types supporting bream or European roaches. The commercially valuable whitefishes disappeared and the less important carp species became the predominant species.

Commercial fishing in natural waters makes proper use of a wild fish population which can develop from the complex interplay of factors in the natural community. Any attempt which seeks to alter such populations for economic advantages can only proceed with great uncertainty. Such control can, however, be exercised when raising fishes. In fact the production is greater when fishes are raised than when working in nature. In correctly breeding fishes they are raised on the basis of selecting for those which can be developed into consumable condition. In a wider sense fish breeding also consists of raising young fishes and fattening them into a form suitable for consumption. The developing fishes are kept in ponds under controlled conditions. In order to better get the fishes out of the pond one can drain the water over a so-called "monk."

Considering all fishes which are bred for consumption it is surprising to note that most species are only important in certain regions and only three species are distributed world wide. These are the carp *(Cyprinus carpio)*, the rainbow trout *(Salmo gairdneri)*, and the Mozambique cichlid *(Tilapia mossambica;* see Vol. V).

Carp breeding

The major carp-breeding centers are in central, southern and eastern Europe, in the Soviet Union, and in southern and eastern Asia. Since carp inhabit warmer regions, ponds not over 130 cm deep with standing water are well-suited for carp breeding. The most favorable growing conditions are realized at temperatures of 22–24° C. In central European fish-breeding, four-to-six-year-old fishes spawn in ponds 80–120 meters square and 20–60 cm deep. These spawning ponds, the bottoms of which are covered with grasses, are filled just shortly before

Fig. 3–34. Schematic cross-section of a carp-breeding pond. D, dam; L, grass; Water depth: a, 20 cm; b. 50 cm.

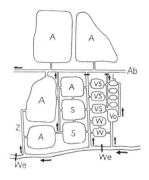

Fig. 3-35. Layout of a carp breeding facility; A, ponds for mature fishes; Ab, drainage ditch; L, breeding ponds; S, growth ponds; Vo, warmed ponds; Vs, growth ponds; W, winter ponds; We, locks; Z, course of current.

Trout breeding

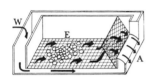

Fig. 3-36. Longitudinal section through a trout-egg breeding apparatus. A, drain; E. eggs; W, water current.

spawning, resulting in essentially parasite-free bodies of water. Two males are placed in for every female. Immediately after spawning, from the end of May to early June, the parent fishes are removed from the ponds.

Since the new fishes deplete the pond food rapidly, they are taken from the pond after at most a week and are placed into a larger pond. These larger ones are one-fourth to three hectares in size and can be over fifty centimeters deep. They are also filled shortly before occupation. After three to eight weeks the young fishes have a L of 3–8 cm. Due to increasing nutritional demands it is necessary to separate these fishes and place them in a series of ponds, each several hectares in size. In these ponds, which are 80–100 cm deep and hold about 10,000 fishes per hectare, the carp become nine to fifteen centimeters long and weigh twenty to fifty grams. In the following year they attain a weight of 250–500 gm and finally, in the third year (in other ponds), the carp weigh 1200–1800 gm. Recently the natural carp diet has been supplemented with plants and fish meal or meat meal. Enriched feed has been developed, so that the natural food of the ponds is becoming increasingly less important. The northern boundary for carp breeding is in the Soviet Union at about 60° N. In central Europe and in the Soviet Union carp are produced at 250–1000 kg per hectare.

In the tropics and subtropics carp grow faster, due to the higher water temperatures. After just one year they weigh 500 gm each. In east Asian ponds, which are generally only 500 square meters in size, 1000 kg per hectare and more can be produced; these ponds have stone walls and a strong water current. However the per capita weight is less there than in central Europe. Wherever warm water is accessible in sufficient quantity, such as from industrial plants or nuclear reactors, warm-water carp breeding can be carried on throughout the year.

For trout breeding, sufficient quantities of cool, clear and oxygen-rich water are necessary. This is restricted to corresponding climatic conditions. In contrast to carp, the rainbow trout, when ready to spawn, are divested of their eggs by hand. The eggs are pressed out of the genital opening by squeezing lightly on the stomach; they are then mixed with sperm in a dry dish to which water is added. Only then do the sperm cells begin moving. After at most three minutes the eggs are fertilized, at which time the sperm cell motion ceases. After rinsing the eggs thoroughly they are placed in covered containers, which are constantly supplied with fresh water from the bottom. Here they are kept until hatching. The young are put into small tanks where they are fed with ground spleen or artificial feed. When the fishes become five to eight centimeters long, they are removed to growth ponds, which are twenty to twenty-five meters long, five to ten meters wide, and have a continuous flow of water. The water temperature may not exceed 20° C., even in summer. These ponds can be filled very densely, up to 40,000 fishes per hectare, using exclusively artificial feed with meat and fish parts, or more recently with dried artificial food. After two years each trout weighs 250-350 gm and is ready for consumption. The annual production that can be realized in trout breeding is primarily dependent on the quantity and quality of the water used. Under favorable conditions the production can amount to 10,000-15,000 kg per hectare.

In tropical and subtropical conditions the Mozambique cichlid (*Tilapia mossambica*; see Vol. V) is particularly suited for breeding. This is carried on in many African countries, especially in Zaïre, but also in southeastern and eastern Asia. Two males are placed in spawning ponds for every three females. Immediately after spawning the females take up the eggs in their mouths and incubate them until they hatch. To separate the young from their mother the pond water level is reduced; the mothers spit out their young, which can then be fished out of the draining water. The Mozambique cichlid is sexually mature after two to three months, when it attains a L of 8-10 cm. The difficulty in breeding these cichlids is to prevent excessive breeding, since the ponds would become overpopulated. To implement this the sexes are separated as early as possible, which can be done when the fishes are three to five centimeters long. In the growth ponds they are fed with agricultural by-products. Each year the annual production in cichlid ponds can reach 4000 kg and more per hectare. The close relative of the Mozambique cichlid, *Tilapia melanopleura*, which in nature feeds only on filamentous algae and rooting aquatic plants, can also be bred. Studies have shown that five to seven kilograms of plants are needed to produce one kilogram of fish meat. Plankton combined with additional nutrients can be used successfully with *Tilapia macrochir*. Breeding all three species together permits efficient use of the natural nutrition in the ponds. Production figures in such mixed groups are thus exceptionally large and can be as much as 9000 kg per hectare annually.

In many southeastern and southern Asian countries fish breeding is combined with rice propagation. The rice fields are surrounded with low, thirty-centimeter high dams. Carp is the most important species for this type of breeding. If the rice is harvested just once a year, 18,000-25,000 young fishes (i.e., fishes five to six centimeters long) per hectare are put in immediately after planting the rice. When rice is harvested twice, half that many fishes are placed in. It is significant that damming the rice fields creates conditions in marshy areas which hinder the development of the snail *Oncomelania quadrasi* (see Vol. III). This three- to five-centimeter long snail is the host species for the schistosomid *Schistosoma* (see Vol. I), a parasitic trematode which in man can lead to bilharziasis, a tropical disease which about 150 million people have, according to the World Health Organization. Through

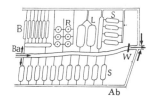

Fig. 3-37. Layout of a trout breeding facility. Ab, drainage ditch; B, breeding trenches; Ba, stream; L, spawning ponds; R, pools for feeding the young fishes; S, growth ponds; W, lock.

Rice growing and fish breeding

fish breeding, the *Anopheles* mosquito (the carrier of malaria) can also be combatted. In the tropics, then, fish breeding results not only in fish production but also contributes to improving health conditions.

The Chinese carp, which have already been mentioned in connection with the optimal use of river lakes, are along with the European carp the most important species for fish breeding in China. The breeding stock is captured in the large rivers. Other species of local breeding importance, particularly in Thailand and Indonesia, are the barbels *Puntius goniotus* and *Puntius orphiodes.* In marshy regions where the water has a low oxygen content the labyrinth fishes such as the gouramis *Osphronemus goramy* and *Melostoma temmincki* (see Vol. V) can be bred successfully, as well as a few threadfins (*Trichogaster;* see Vol. V), which with their respiratory apparatus are well-adapted to those conditions. In many coastal regions of the world, breeding in brackish ponds is carried on (e.g., for milkfishes).

Water pollution, by B. Grzimek

Where human populations are concentrated the resulting pollution of the streams, rivers and lakes has its primary effect on the fishes. Since the mass deaths in the Rhine River in 1949, a Rhine salmon is now caught only rarely. Since this time this great current has become no longer a river, but a drainage cloaca for hundreds of thousands of toilets and filthy industrial poisons. Twenty thousand tons of dissolved industrial salts flow past the Lorelei cliffs every day. It would require 950 railroad cars (equivalent to twenty-four freight trains) to transport that amount if it were undissolved. German rivers and streams receive 21,000,000 cubic meters of sewage daily, only a third of which is filtered. One liter of Rhine water in the vicinity of the city of Mainz contains twenty grams of dry matter. Even rain falling over industrial regions must penetrate air with 85,000 dust particles per liter, while above forests there are only 500–600 particles per liter. Two million tons of dust hang over the entire Federal Republic of Germany. In addition to that there are five million tons of sulfurous acid. In June 1969 the remaining fishes in the Rhine died by the tons after an insect poison was dumped into the river near Bingen. In Germany and Holland the tapping of rivers for drinking water had to be discontinued.

So today we can be happy that the "Rhine salmon" occasionally found on restaurant menus along the Rhine did not come from this filthy river, but generally made a long trip from North America or Scandinavia in the can or frozen.

The worst part is that not only do fishes live in river water, but we must drink it also. In West Germany, and probably in all other west European countries, underground water falls far short of supporting the needs of all the people. The larger part of what comes out of our taps is filtered water from rivers and lakes. Many samples of that water have been tested in the Hygienic Institute in Mainz. Even in water which by tasting seemed perfectly fine, traces of tars and related substances could be found, which in stronger concentrations are known to be carcinogenic. Furthermore traces of detergent materials were frequently found; the presence of these poisonous materials in water is steadily increasing. This is chiefly due to the fact that water filtration in West Germany is "regulated" by laws of the eleven various states. A unified federal law, which was presented to the German parliament in 1965, was rejected. Bonn has now had to allot 800,000,000 Marks just to eliminate the very worst poisons in the Rhine.

A blessing has become a curse! In 1880 salmon were caught on the lower Rhine, some of which weighed up to forty-five kilograms with a L of one and one-half meters. At the beginning of the last century domestic servants in the villages along the Rhine were assured they would not have to eat salmon more than twice a week. Today the same trend is being reported for villages along the Elbe River in Germany for sturgeon, which is also quite rare. At the end of the last century the Dutch pulled nets across the mouth of the Rhine and attempted to keep the entire catch for themselves. A conference was held by the two concerned countries and the Dutch soon reformed their activities. It had long been clear to them that the salmon had to swim the 1000 km to the sources of the Upper Rhine in Switzerland (i.e., to about 1000 m above sea level) in order to spawn. If no young salmon swam down the Rhine, no mature ones would swim up it from the sea. The Dutch were at once prepared to assume the costs for artificially raising a few million salmon to set into the upper courses of the Rhine.

In other rivers people also had difficulty mastering the situation. In 1827 more than 1000 salmon daily were caught at the mouth of the Memel River, of which the average weight was fifteen kilograms. Even at the price of one Mark per fish they could not all be sold and untold numbers had to be buried. In North America devices were once built which pulled salmon out with nets, and one such device could haul up to 14,000 salmon per day out of the water. Large parts of Alaska and Kamchatka would be totally uninhabitable for man if it were not for the salmon migrations. In some places the local inhabitants could work for just two weeks and what they caught lasted an entire year. In 1905, 59,000 kg of salmon were caught in British Columbia and the state of Washington. In the 1940s in Washington state only, 2,350,000 crates per year were caught, an insignificant amount (only one-sixth the salmon harvest of the 1910-1917 period). At that time it was believed that the salmon were hindered solely by dams and power stations, and as a counter-measure the dams and their associated structures were replaced by steam electricity plants, because at that time the salmon harvest was more important than using water for power.

The pollution of our waters is not just directly lethal for the fishes by poisoning; indirect damage is also caused by the high content of organic matter in sewage, which has a decomposing action. At high

Human folly

temperatures the decomposition of these materials proceeds at such a rate that the oxygen content of the water gets completely used up. The result in this case is also the death of the fishes; thousands of them often come to the surface of the water with opened gills, and dead.

Are emergency measures helpful?

Many harmful effects of sewage water can be avoided by the construction of purifying plants. Many progressive industries are now in increasing numbers recycling the water they use in their plants, purifying it and placing it back in circulation. But what help are all these measures when—as happened in June 1969—because of ignorance or a lackadaisical attitude, even those fishes which had adapted to their filthy conditions in the once-romantic Rhine were killed off?

Salmon migration

The various salmon species with their complex life as migrants and their resulting sensitivity have become a kind of test subject for the streams and rivers which have been modified for the worse by man's numbers and his ignorance. Thus a few words should be said here about their particular characteristics independent of the description of them which comes later in this volume. Earlier it was assumed that salmon could never be caught further than 100 km from the coast in the sea and that they had to stay in the mouth of their home river in order that they could find their way back. But A. Hartt marked 36,383 salmon in the north Pacific and then released them. In some Pacific salmon species up to seventeen percent were recaught within twelve months. Salmon which had been marked at the Aleutians were in part recaught a great distance away in the Amur River in Siberia. Baltic Sea salmon were found 1000 km from the mouth of their home river, and Pacific salmon were recaught at still greater distances. The longest migration route was that of a salmon which was marked in 1956 at Adak, halfway between Alaska and Kamchatka, and was recaught in 1957 in Idaho, 3800 km away.

Ichthyologists have long wondered how the salmon could find their way back to their home rivers, which they had left years before, and then even found the correct tributary and presumably also the stream and exact location where they had hatched. Of 100,000 Swedish salmon, which were marked before leaving their home rivers, only about 1000 were later caught (i.e., about one percent) in other rivers, and in North America the ratio was about the same for a similar test. The salmon are guided by smell. From other studies it has been found that salmon "recall" the composition of the water which they smelled just before migrating out of the river. If they are put in some other river just before migrating, they reappear in this "foster" river, not the true home river. The scent of a stream and of a tributary gets thinner as it mixes with other waters in the main current, and still thinner in the sea. The salmon must be able to detect incredibly small odor traces. Just how dilute the water can be has been demonstrated in other species.

The exertion involved in constantly swimming upstream over such long distances has almost unbelievable dimensions when one realizes that the salmon that once swam up the Rhine and those that still make the journey up other European rivers do not feed once they have gotten beyond the river mouth.

Because they fast, Rhine salmon could not be caught in earlier times. This was true also for those in other rivers. However in Scotland and in the rivers of Scandinavia salmon fishing is a popular sport, especially with artificial flies. In England good salmon grounds are leased. Most salmon are caught with nets at river mouths. In Aberdeen, Scotland, in 1952 10,000 salmon with a value of $160,000 were caught in this way. Salmon are now being poisoned in Scotland because the meadows along the streams and rivers are being treated with chemicals to protect the plants. Salmon stay in various age-class groups, which can be distinguished on the basis of size. Earlier in the Rhine definite salmon runs could be differentiated according to the season of the year.

Since the rivers have become so built up and plugged up the trend throughout the world has been to breed salmon by the hundred thousands and millions artificially, in great facilities with containers filled with flowing water. In Dorset, artificial breeding produced 20,000 two-headed small salmon which all died before they were one month old. According to French studies the migration of salmon into the sea is guided by the thyroid gland. This gland secretes a hormone with a high iodine content into the blood, which greatly stimulates the young salmon to pass into the river and swim against the current. In the sea, equilibrium is attained by the heightened intake of sodium.

The return route of the salmon is not only being cut off by detergents, human excrement and chemical wastes of the factories but also by the numerous dams, dikes and locks. Because the salmon catch is so valuable, steps have actually been built on many such structures so that the salmon can still swim upwards and jump over them. But these measures cost a tremendous amount of money, and water businessmen complain that water is thereby being lost, for it does not drive the turbines.

Fish elevators have even been built, in which the fishes are attracted by a water current and then lifted by an elevator up to the dammed water. In some places even that did not help, because the dammed water was too warm. Because of that they were put into tank cars and driven along the dam wall several kilometers upcurrent to places where both current and oxygen were in sufficient supply for the salmon. But all of these measures are very expensive and not always successful.

Even if the salmon have made it up to the water behind the dam they often encounter completely unnatural conditions. The lakes are very deep and have steep shores that are almost vertical, while salmon normally seek spawning grounds in water one to three meters deep. Even the streams often flow by steep waterfalls into the rivers. In Sweden, for example, power plants now have to breed that amount of salmon which would have reached the dam under normal conditions. Special salmon breeding facilities are constructed for this purpose. In the early 1960s, fifteen such Swedish breeding plants produced about 1,000,000 young salmon each year, of which every tenth is marked.

It is not so important any more that these salmon migrate up the rivers again, because many could not make it anyway. Today salmon are caught in great schools on the open sea with modern nets and sophisticated deep-sea fishing equipment. As marking has proven, fifteen percent of all salmon caught in the Baltic Sea by German, Danish, Russian and Norwegian fishers are from Swedish breeding facilities. That does not meet with great enthusiasm from Sweden, and the Swedes would like a bigger portion of the catch. The rivers which have their mouths at the Baltic Sea produce seven to eight million salmon annually.

So today we can utilize the expensive measure of artificially breeding millions of salmon because our rivers have become so built up and polluted the salmon can no longer tolerate them. But we ourselves must continue drinking the water of these rivers.

4   The Cartilaginous Fishes

The CARTILAGINOUS FISHES (Chondrichthyes) are just as much vertebrates as teleost (bony) fishes. They have internal skeletons, and the
notochord is largely replaced by a chain composed of individual vertebral elements. While cyclostomes have only a simple fin seam, the
cartilaginous (and teleost) fishes have two pairs of fins (in addition to
the unpaired fins: one or two dorsal fins, a caudal fin and anal fin),
which are borne by the girdles of the internal skeleton, and the ventral
fins, connected with the complex pectoral girdle. The cartilaginous
fishes have true jaws with teeth. However the entire internal skeleton
is composed entirely of cartilage; no traces of ossified bone can be
found.

Bony parts are found, however, in the scales. When stroking the skin
of a shark one generally has the impression of rubbing sandpaper, due
to many tiny, pointed teeth protruding from the skin in many species.
These dermal teeth (placoid scales) are distinguishing characteristics
for sharks and vary from species to species. The tooth tip is dentine
overlaid by dental enamel. The lower tooth part, which anchors each
tooth in the skin, is made of bone. The dermal teeth are more highly
modified in the rays. The skin is partially naked, but in some places
(particularly on the back or upper tail surface) the dermal teeth have
developed large and strong spines. The powerful teeth on the elongated snout of the sawfish shark and sawfishes are modified dermal
teeth.

The jaw teeth of cartilaginous fishes are also modified dermal teeth.
All sharks and rays lose their teeth when they become worn and continually replace them with new ones. Behind each tooth there is a row
of immature teeth, and whenever one tooth falls out the next one in
the row immediately fills the empty space.

Many sharks and rays have a large or small opening (the spiracle)
on both sides of the head behind the eye. The fishes usually take in
respiratory water through the mouth to extract oxygen from the water

Class: cartilaginous
fishes

in the gills. But in sharks which rest extensively on the floor and especially in rays (which even dig into the sand), taking up this water presents some difficulties. Each swallow of water is accompanied by substantial amounts of sand and mud, and the gills could eventually become stopped up. The spiracle has developed in these very species; it is on the upper side of the head. When such a species is on the floor, it sucks water through the spiracle, pumps it into the gill cavity and releases it through the gill slits.

On the basis of gill slits and other characteristics the cartilaginous fishes are divided into two subclasses, which are distantly related: A. Elasmobranchs (Elasmobranchii) with at least 5 gill slits and gills on each side, generally one spiracle behind each eye, dermal teeth on the upper body surface, a tooth jaw, and an upper jaw which is not firmly attached to the skull. There are two orders: 1. Sharks (Selachii) and 2. Rays and skates (Rajiformes). B. Chimaeras (Holocephali), with only one gill opening on each side, dentition in the form of tooth plates, and the upper jaw firmly attached to the skull. There is only one order.

There are sharks whose anatomy is very similar to that of rays, and some rays which resemble sharks to a high degree. They can be distinguished easily, however, by the location of the gill slits. If they are on the sides of the front of the body, the fish is a shark, and if they are on the ventral side and well behind the pectoral fins, we are dealing with a ray.

Some sharks appear to utilize their jaws as supplementary propulsion organs. Normally sharks swim with undulating motions of their body, whereby the caudal fin delivers the main propulsion force. Since a swim bladder is lacking in all cartilaginous fishes, sharks and rays of the deep sea and coastal waters alike must maintain swimming movements continuously in order not to sink to the floor, even when sleeping. Because of that, some sharks force the water of respiration out of their gills with sufficient force to produce a rocketlike rearward thrust, the force of which alone moves them forward. The pectoral fins have an important function in free-swimming sharks. While the dorsal fins primarily protect the body from toppling over, both pectoral fins are winglike surfaces which produce an upward force and function as depth rudders.

The swimming motion of rays is completely different. In association with the flattening of the body the rear spine became so hardened that a lateral undulating motion as sharks have is impossible to generate. Instead of that, the flattened sides which have become fused with the large pectoral fins, generate vertical undulations from front to rear. In the free-swimming rays (i.e., eagle rays and especially the manta rays, whose sides are elongated like wings) a beating motion has developed something like that of bird wings. The transition from the lateral undulation of sharks to the vertical rhythmic beating of the rays can be seen in some ray species, the front portion of which is flattened, as in other rays, but the rear of which is cylindrical, as in sharks.

The chimaeras also deviate from the other cartilaginous fishes in their swimming technique. In chimaeras, the rear end of the body is tapered and the tail is pointed. Both large pectoral fins are beat simultaneously like two paddles (or alternately to change direction). This pure pectoral fin swimming is a highly effective technique in such a "primitive" group; yet in teleost fishes the same method appears in various groups.

In all male cartilaginous fishes the pectoral fins have an important function: their rear part has been modified into a copulatory organ. All cartilaginous fishes have internal fertilization. However, the form of birth is highly variable and is related to the kind of environment which the different species inhabit. Many sharks which live near the floor lay eggs and are known as oviparous (from the Latin ovum = egg and parere = to give birth). In rays only the true rays (Rajidae) lay eggs. Chimaeras also lay eggs. To protect the embryos the eggs have a horny external covering with prominent twisted filaments. These filaments are still soft when the eggs are laid, but they wrap around the first firm object they meet as an anchor, much like the anchoring threads of grape vines. The period of development for the embryo varies and is in some cases unknown. In spotted dogfishes, this lasts about six to nine months, and in frilled sharks—two years.

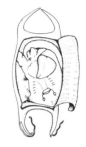

Fig. 4-1. *Raja:* egg capsule, opened, with embryo.

In other cartilaginous fishes which live near the floor, the eggs remain in the body of the mother until the embryos have become fully developed (these fishes are ovoviviparous). The young hatch in the mother's oviduct and then develop somewhat further until they are "again" born. Generally, the embryos live from the contents of their yolk sac, but sometimes a nutrient is generated in the oviduct, which is eaten by the developing young.

Deep sea sharks have internal development of the young and give birth to live young (this is designated viviparous, from the Latin vivus = living). The oviducts have a womblike cavity, and the yolk sacs have been modified to a form of placenta which furnishes the embryos with nutrients. Later the form of nourishment can be greatly altered, and the yet unborn young become predators within the mother's body. Shortly before birth, the larger young seize their smaller siblings and gradually eat them. In mackerel sharks *(Lamna)* and sand sharks *(Carcharias taurus),* the oldest young (still unborn) eat the unfertilized eggs or even the younger embryos. Thus, only two young of the litter remain. This represents the unusual case of pre-birth cannibalism. The birth of just two young in deep-sea sharks is in sharp contrast to the egg-laying sharks, which can have up to eighty young. In general it is very difficult to determine the life habits of the sharks, rays and chimaeras, because these species largely occur in regions where direct observation has not been possible. Only a few forms are suited for maintaining in captivity in aquariums or pools, and even there only a small portion of the total life activities can be followed: prey acquisition, copulatory foreplay, birth and growth. Thus, to a great extent, the only possibility has been to draw conclusions from freshly caught specimens. Analysis of stomach contents sheds light on seasonal variations in feeding preferences; and the development of sexual organs, eggs and embryos indicate the onset of sexual maturity, age at mating, and the nature and duration of embryonic development.

All cartilaginous fishes are predators in the broadest sense of the term; they feed exclusively on other animals. Their food varies from whale to plankton, depending on the species; furthermore a number of cartilaginous fishes live on hard-shelled floor-dwelling organisms. Sharks and rays are also eaten by man; many "primitive" peoples dislike shark, but enjoy rays. The French, who are famed gourmets, value the tender meat of mackerel sharks, and on the Côte D'Azur spotted dog-fishes are sold under the name of saumonette. Hammerhead shark meat is often sold as swordfish or tuna. In England, spotted dog-fishes are sold as rock salmon. In Germany, spiny dog-fishes are smoked, put in aspic and served as marine eel. The ventral flaps are also smoked in Germany. On menus, mackerel shark is listed as "carbonadexfish," "calf fish," or "sea sturgeon," for when boiled, it tastes like calf. The Japanese catch and eat blue sharks in great numbers. While rays are not often seen in German markets, they are considered delicacies in England, Holland and France. Occasionally the meat of thornback rays and flapper skates is smoked and marinated, and then sold as sea trout.

Cartilaginous fishes have commercial value in other parts of the world as well. Cod-liver oil is made from the livers of various shark species (i.e., Greenland shark, basking shark, and mackerel shark). Blue shark skin is used as a polishing agent, and the same is done with hammerhead shark and Greenland shark, the skin of which is used as leather for book binding.

Subclass: Elasmobranchii

Order: sharks

The first order within subclass Elasmobranchii is the SHARKS (Selachii). The body is elongated and torpedo shaped. In species adapted to dwelling on the floor, at least the underside is flattened. There are 5 to 7 gill slits on the side of the head in front of or above the pectoral fins, which are not continuous with the head. The eye is protected by a lower eyelid. There are 7 suborders with 19 families and about 250 species.

Suborder: comb-tooth sharks

COMB-TOOTHED SHARKS (Notidanoidei) have 6 to 7 gill slits, and one dorsal fin set well back. The vertebral column is incompletely developed; the lower jaw teeth are like little saw teeth. Prey which is too large to be swallowed whole is hacked up with the pointed upper jaw teeth and then sawed by the lower teeth. There is only one family, COW SHARKS (Hexanchidae).

The most frequent notidanoid, the six-gilled shark *(Hexanchus griseus)* with six gill slits, is usually coffee-colored to chocolate brown and only rarely dark gray. It feeds primarily on fishes, but also on crustaceans. The six-gilled shark attains a length of 5 m, but occasional specimens 8 m long are caught. These large examples could be dangerous for man, but no attack of one on man has been confirmed. Six-gilled sharks have been caught at depths of 1500 m (the deep sea) but also on the upper surface. During the day these sharks rest on the floor; they feed only at night. They are distributed on both sides of the Atlantic, around South Africa to the Indian Ocean and are also found in the Mediterranean Sea. In the North Sea, there are six-gilled sharks up to 4.5 m long. The SIX-GILLED SHARK *(Hexanchus corinus)* found in the Pacific Ocean appears to be another species. Although six-gilled sharks are occasionally caught, little is known of their life. They are ovoviviparous; 108 unborn young were found in the body of a four and one-half meter long female. Knowledge of other notidanoids, all of which have 7 gill slits, is also sparce. The NARROW-HEADED SEVEN-GILLED SHARK *(Heptranchias perlo;* L about 2 m) is found in the Atlantic and Mediterranean.

The FRILLED SHARKS (suborder Chlamydoselachoidei, family Chlamydoselachidae) look more like fat eels than sharks. The body is normally long and thin, with a dorsal and anal fin far rearward up to the caudal fin. The caudal fin has only an upper lobe, which runs straight back. The mouth opening is not on the lower part of the body, but nearer to the snout. The gill partition walls protrude like frills. The first gill slit extends right across the throat and has a collar-like covering (thus the name frilled). There is only one species, the FRILLED SHARK *(Chlamydoselachus anguineus).*

The frilled shark primarily inhabits the deep sea, where it feeds on cephalopods. The young hatch from eggs within the body of the mother, with a period of two years from fertilization until birth. One litter can have up to 15 young. Frilled sharks apparently are found in all oceans; they have been caught in the Atlantic, in the northern North Sea, off South Africa, near the Californian coast, and primarily off Japan.

A plump body with 5 gill slits, a thick head with rounded snout and a flattened underside characterize the HORN SHARKS (suborder Heterodontoidei, family Heterodontidae). They have two large dorsal fins, each with a powerful blunt spine on the fore edge. Several rows of teeth are used simultaneously; the middle teeth are small and pointed, and the lateral teeth have been developed as large, broadened chewing teeth. The mouth opening is well forward and only somewhat inferior.

Distinguishing characteristics

Fig. 4-2. Teeth from *Heptranchias perlo.* Upper and lower jaw, left side.

Suborder: frilled sharks

Distinguishing characteristics

Suborder: horn sharks

Distinguishing characteristics

Fig. 4–3. *Heterodontus japonicus.*

Suborder: sharks

Distinguishing characteristics

The upper lip has 7 skin folds. The spiracle is small and underneath the eye. All horn sharks are egg-layers, the eggs being in chitin-like capsules, on the corners of which are double-spiraled anchoring projections. The development period is at least 5 months. There is only one genus: HORN SHARKS *(Heterodontus).*

All horn sharks inhabit the shallow waters of the warm Pacific and Indian Oceans. They are not dangerous to man. Their meat is eaten in Japan. Oyster fishermen hate these bottom-dwelling tough sharks, because they can devour entire oyster beds. Among these species the best-known is the one meter long JAPANESE HORN SHARK *(Heterodontus japonicus).* It can be recognized by the bands saddling its body. The pectoral fins are particularly well developed. There are at least two species of horn sharks along the North American west coast.

The SHARKS (suborder Galeoidei), with their torpedo shape, are unmistakable. They generally have two dorsal fins, lacking spines, and always have anal fins. They have 5 gill slits. The spiracle, when present, is on each side of the head. The vertebral column is complete in its entire length. Development is ovoviviparous. There are small typical sharks hardly 1 m long and giant forms which are actually the largest fishes in the world. There are twelve families of sharks (sand sharks, goblin sharks [mackerel sharks], basking sharks, thresher sharks, whale shark, carpet sharks, dog-fishes, cat sharks, smoothhounds, false cat sharks, gray sharks, hammerhead sharks) with about 195 species, most found in the tropics. Distribution is such that some species may be found in all oceans. Some are even in fresh water.

The SAND SHARKS (Carchariidae) are rather elongated, with a pointed snout. They have two dorsal fins and an anal fin. The 5 gill slits are in front of the pectoral fin; the upper lobe of the caudal fin is much larger than the lower. The spiracle is small. The teeth are noticeably long, thin, smooth-edged and sharp, and there are one or two jagged teeth at the base on each side. Sand sharks primarily inhabit tropical and subtropical regions, in the Atlantic, Indian, and Pacific Oceans. They are usually found in coastal waters, but are not floor dwellers.

In 1810 the Italian ichthyologist A. Risso described a shark species which is seldom caught, the FIERCE SHARK *(Carcharias ferox,* L up to 4 m). The Italians call it "Cagnaccio". The fierce shark has a powerful body and a snout which is somewhat extended. Its back is gray, but sometimes also reddish-brown with large black spots. Practically nothing is known about the life and reproduction of this species. As a predatory long distance swimmer it appears to prefer somewhat greater depths, and is only found in the Atlantic near Madeira, from the Bay of Biscay to northwest Africa, and also in the Mediterranean.

The description of the closely related SAND TIGER SHARK *(Carcharias taurus,* L almost 3 m), which is also in the Mediterranean, is likewise from 1810. This darkly spotted species is widespread; it is found along the west African coast from the Canary Islands and Cape Verde Islands, off South Africa, in the west Atlantic, Caribbean Sea, and in the neighboring coastal regions north and south thereof. Other closely related species inhabit the Indian and Pacific Oceans.

The sand tiger shark has a yellowish-grayish coloring which affords effective adaptation to the sandy floor. It feeds primarily on various bottom dwelling fishes, but also attacks cephalopods (see Vol. III), lobsters, larger crabs (see Vol. I), and other suitable prey. Sand tiger sharks swim constantly but feed only at night. To lighten body weight and prevent sinking to the floor, air is swallowed and kept in the stomach, so that the stomach functions as a swim bladder. Although they have been found in shallow coastal water, no case of an attack on people in the Mediterranean Sea or Atlantic Ocean has been authenticated. However, they are considered to be highly dangerous along the South African coast, where numerous cases of attacks on swimmers are known. The close relatives of this shark in the Indian Ocean and Australian coastal waters are also said to be dangerous to man.

The female sand sharks are sexually mature when they reach a length of 2 m. Only 2 young are born per litter. The 2 oldest feed on the other eggs within the mother's body and also on the younger embryos (pre-birth cannibalism). At birth, they are already 1 m long. A well-known shark researcher recently found that they can use their teeth well, even in the prenatal stage: as he cut through the womb of a freshly caught female and reached inside in order to get at the young, he was sharply bitten by an unborn sand shark.

The GOBLIN SHARK (Scapanorhynchidae) can be recognized by the long snout, which has the appearance of a shovel-shaped nose. The mouth is protrusible and can be extended, a development which permits a motion unlike that of all other sharks. The teeth are needle shaped and very long; they protrude (partially above) out of the mouth. A spiracle is present. The fins are relatively small; two dorsal fins are located well rearward on the tail region.

The longest goblin shark caught thus far measures 4.30 m. The most well-known of the few species is the JAPANESE GOBLIN SHARK *(Scapanorhynchus owstoni)*, which has been captured near Japan's east coast, south of Australia and off Portugal. These sharks are definitely deep sea inhabitants and are probably distributed throughout the world.

The MACKEREL SHARKS (Isuridae, but also named Lamnidae), which have also been called "man-eaters," have a long streamlined body and a pointed snout. The first dorsal fin is large, and the second is noticeably small, as is the anal fin. The upper lobe of the caudal fin protrudes almost vertically into the air, while the lower lobe is almost as large and extends sideways. The entire caudal fin is in many species more or less symetrically half-moon shaped (a clear characteristic of a fast long-distance swimmer of the open sea). The tail shaft is very slender,

▷
1. Great white shark *(Carcharodon carcharias)*; 2. Whale shark *(Rhincodon typus)*; 3. Basking shark *(Cetorhinus maximus)*.

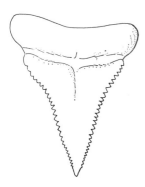

Fig. 4-4. Single upper jaw tooth of the great white shark (*Carcharodon carcharias*).

flattened from top to bottom, and the sides form a keel. Whenever spiracles are present they are always small. The teeth are large, triangular and sharp. The gill slits are conspicuously long. Embryonic development is ovoviviparous.

Mackerel sharks are primarily open sea inhabitants which do penetrate into coastal areas. They are found in all oceans, including in colder regions. Because of their frequent presence along coasts, their size and readiness to attack, they can become highly dangerous to humans. Man traps them often, not only to lower their numbers, but also for their good eating. Some species have considerable commercial value. Three genera exist: 1. GREAT WHITE SHARKS (*Carcharodon*) with only 1 species, 2. PORBEAGLE SHARKS (*Lamna*), and 3. MAKO SHARKS (*Isurus*), each with at least 2 species.

The mackerel shark most dangerous to man is the GREAT WHITE SHARK (*Carcharodon carcharias*), with a L of 5-6 m, but rarely 9 to 12 m long, and a weight of up to 3 tons. The pectoral fins are sickle shaped and prominent. The teeth are large, wide, triangular, and with saw-tooth edges.

In spite of its name, this shark is not completely white. The young have a brownish back, which in old specimens becomes grayish or bluish, but can also be brighter. Nothing further is known about the propagation of this species. Embryos 20-60 cm long have been found in some females.

The great white shark feeds not only on fishes of all species (including sharks up two meters long), but also on marine tortoises, dolphins, sea lions and other seals, and even any refuse which is thrown in the water. It is reputedly the most ravenous of all sharks; the great white shark hunts fish schools and follows ships, whereby they can enter coastal regions, where they are a great danger to swimmers. Throughout the world many people have fallen victim to the great white shark. They are especially feared in Australian waters. One great white four and one-half meters long bit a man in two, and a six-meter long specimen can swallow a man whole. They are known to attack small boats and thus bring the occupants into a dangerous situation.

The great white shark is found in all tropical, subtropical and moderately warm seas, in the Mediterranean and in the east Atlantic, south of the Bay of Biscay. It is often active on the surface of the water, but has also been caught at a 1000 m depth. On the basis of teeth found in the open sea it appears that there are still larger specimens there. Since the meat is inedible, the great white shark has no commercial value.

The PORBEAGLE SHARK (*Lamna nasus*) is less dangerous (the L is 3-4 m). Its long, slim tail shaft has a keel on each side and on the lower edge. The back is dark blue, and the belly white. The young develop within the womb. Since the yolk sac content is rapidly consumed, the growing

◁

Tope (*Galeorhinus galeus*).

embryos feed on the unfertilized eggs which lie in masses near the womb; the young thereby develop a fat, swollen "yolk belly." At the same time the chin and throat regions are greatly swollen and the caudal fin not yet symmetrical; such an embryo more nearly resembles a pug with a potbelly than a shark. Only 2 young develop in each oviduct, so that a litter contains 4 young. When the young first see the light of the underwater world, they are already about 60 cm long.

The porbeagle shark is a fast-swimming open sea inhabitant which occurs on both sides of the north Atlantic as well as in the Mediterranean, in the North Sea and in the Baltic Sea. Porbeagles follow great schools of fishes and feed primarily on mackerel, herring and sardines, but also on bottom dwelling fishes and even spiny dog-fishes and cephalopods (see Vol. III). Porbeagles wreak great damage commercially since they often rip nets and eat the catch.

The MAKO SHARK (*Isurus oxyrinchus*, L over 3.5 m; weight up to 500 kg), prefers warmer regions. The second dorsal fin and the anal fin are very small, and only one keel is found on the tail shaft. The back of the mako is dark gray to gray blue, and the belly is white. Little is known about embryonic development. The makos are apparently ovoviviparous, as are all mackerel sharks; the oldest young feed on the unfertilized eggs within the mother.

Fig. 4-5. Mako shark (*Isurus oxyrinchus*) upper and lower jaw teeth, front portion.

This shark inhabits the entire tropical and subtropical Atlantic and is also found in the west Mediterranean. Individual specimens have been found in the English Channel and along the English coast. As an exclusively open sea inhabitant, the mako swims just under the surface of the water. If the dorsal fin of a shark is seen from a ship, it is most likely that of a mako shark. It is a fast swimmer which hunts herring and mackerel schools and seizes its prey with powerful movements.

Since the mako avoids shallow water, it is generally not dangerous to swimmers. In spite of that, it has the reputation of being a man-eater, possibly rightly so, for it is said to attack people that fall overboard from ships. However, unequivocal proof has not been found for this. Since its white meat tastes very good, the mako is eaten in some regions. Sport fishermen along the coast of the USA catch makos; they fight for hours when hooked and jump completely out of the water.

The BASKING SHARKS (Cetorhinidae) are very similar anatomically to the mackerel sharks, but have much larger gill slits which extend from the upper side of the body almost to the middle of the chest. The teeth are unusually small and numerous. The cartilaginous gill arches have long, horny, closely spaced deciduous gill rakers extending into the throat. The dentition, gill apparatus, and gill slits are structured in a peculiar way which is related to the method of food intake (see below). There is only one genus (*Cetorhinus*), probably with two species.

The BASKING SHARK (*Cetorhinus maximus*) is one of the largest sharks of all, reaching a length of 14 m and a weight of about 4 tons, making it the longest shark of the European seas. The second dorsal fin and the anal fin are small. Only one keel is located on each side of the tail shaft. The sickle-shaped caudal fin has a somewhat longer upper part.

This open-sea inhabitant swims with its mouth open so that plankton pass into the filtering apparatus of the gills. In spite of its size, the basking shark feeds primarily on small free-swimming crustaceans, but also ingests everything else floating in the water such as eggs and larvae of fishes, squid and other similar organisms. With a swimming velocity of almost four kilometers per hour, an adult basking shark filters 1000–1500 tons of water per hour through its gills. Due to this feeding pattern, these sharks are not very aggressive and are not dangerous to man.

The basking sharks bear live young (viviparous). Only 1 or 2 are born, which at birth have a length of about 1.5 m. Basking sharks often swim in "schools" of 50–250 individuals. It is unknown whether this grouping is related to breeding activity. Basking sharks are found singly just as often, however. Distribution of these sharks encompasses the entire temperate and colder north Atlantic, in the east from North Africa to Iceland and northern Norway, and in the western Mediterranean as well. In the northern regions, however, basking sharks are found only in summer. In winter they withdraw to greater depths. Since the gill filter system degenerates during this period and does not become regenerated until the following spring, it is assumed that basking sharks do not feed during this period.

THRESHER SHARKS (Alopiidae) can be distinguished from all other sharks by the highly elongated tail. The upper tail element is almost as long as the rest of the body; the tail shaft is pressed together along the sides and thus is higher than it is wide. There is no tail keel. The body is slender and elliptical, the snout short and blunt. The two dorsal fins and anal fin are quite small. The teeth are very small with only one point. There is only one genus *(Alopias)* with four or five species found in the tropical and temperate waters of all seas.

In 1544 the thresher *Alopias vulpinus* was described scientifically for the first time by J. Salviani and due to its long tail was named Vulpecula (fox-tail). It attains a length of six meters. The embryonic development is oviviparous. Generally the female bears 2 to 4 young, which at birth have a L of 1.20 to 1.60 m. The characteristically long tail can already be seen in the embryonic stage within the horny egg capsule; however the embryos still utilize external gills and a yolk sac. The thresher inhabits subtropical and warm temperate seas. In the north, threshers are found as far as southern Norway and inhabit the central North Sea as well. They are found frequently in the Mediterranean and in the south along the Cape of Good Hope.

The thresher is an open-sea inhabitant, and approaches the coast only when preying on schools of fishes. In European waters it preys chiefly on herring *(Sprattus sprattus)*, other herring species, sardines, mackerels, but also tuna and cephalopods. When a thresher has caught up to such a school, it swims around it in ever constricting circles. At the same time, it strikes the fishes with powerful whip-like blows, from which its name is derived. The fishes become panic stricken and the school tightens even more. Perhaps the fishes are even stunned from the blows. In any case, the thresher attacks after this initial phase and according to fishermen makes a "true bloodbath"—a statement which is certainly unjustified in this exagerrated form.

The WHALE SHARKS (Rhincodontidae) have a gigantic, powerfully built body with a L from 15 to 18 m. The head is wide and blunt with a large mouth opening. The teeth stand in close rows and are small with the point facing backward. There is a small spiracle. The gill slits are very large, the last one lying over the pectoral fin. A high keel runs from the rear of the head over the back. Two or three other keels run along the upper half of the flank, the lower of which passes into the lateral keel of the tail shaft. The caudal fin is powerful, and its upper (vertical) lobe particularly big. The gill apparatus is structured in an unusual way; the cartilaginous gill arches (which bear the gills) are connected with each other by numerous cartilaginous diagonal shafts on top of which is meshed sponge-like webbing. Thus, the entire gill apparatus forms a sort of net or sieve. Respiratory water is pressed through this filter before passing into the gills, an adaptation to plankton feeding shared with basking sharks. There is only one genus with a single species.

The WHALE SHARK *(Rhincodon typus)* is covered with numerous round white or yellow dots on its dark back and flank. This species is found in most seas of the world, primarily in tropical regions, but not in the Mediterranean. Embryonic development is viviparous; one egg case measured 30 x 36 x 9 cm. Eighteen egg capsules with developed embryos have been found in the body of one female. The unborn young have the characteristic shape and coloration of the adult animals.

Whale shark swim in great herds and suck in the floating food while swimming. They also feed on smaller fishes and cephalopods which pass into the sucking apparatus of their giant mouth. Since the prey animals drift just underneath the water surface, whale shark have adapted the technique of "standing" in the water, that is, maintaining a vertical position with the head upward. When waves pass by, they float up and down like huge oil drums. The whale shark also "lie" motionless directly on the water surface. Since they are not shy, divers can come right up to them or propel a boat around them. They apparently sleep in this position. At night their bodies are occasionally rammed by boats, which usually leads to severe injuries or the death of the animal involved. These sharks are of no danger at all to man.

All CARPET SHARKS (Orectolobidae) have a rather plump body. The head is quite flattened and is wider than it is high. A skinlike barb is located in front of each nasal opening. The teeth are small and several rows are used at once. The eyes are small and the spiracle very large. The dorsal fin is elongated and extends all the way to the tail. Carpet sharks are oviparous or ovoviviparous. There are twelve genera with about twenty-five species, which differ considerably from each other anatomically; among them is the WOBBEGONG of Australia *(Orectolobus maculatus).*

Most carpet sharks dwell on the sea floor of shallow waters, but a few also appear in greater depths. They inhabit only tropical and subtropical seas and thus are not found along European coasts. Many carpet sharks have a prominent spotted or striped marking, which is interpreted to be a camouflage marking in rocky areas, seaweed growths or in eel grass. The skin flaps on the head also function as camouflage, for they have the appearance of algae; in some species these are also found along the body. Some carpet sharks can be kept in aquariums.

In the Red Sea, a scuba or snorkel-equipped diver can occasionally observe the ZEBRA SHARK *(Stegostoma fasciatum;* L up to 3 m). When young, this shark has a saddle stripe marking which gradually breaks up into various spots. The young can be kept in aquariums very well. During the day, zebra sharks can be found near the coast lying on the floor where they probably sleep. Even when touched they do not become excited. But at night they are active and search the floor for prey which consists primarily of crustaceans and mollusks. The zebra shark is among those species which lay horny egg capsules. The eggs adhere to their substrate by means of barb-like long filaments. The zebra shark is found in the Indian Ocean, the Red Sea and in the western part of the Pacific Ocean.

While the zebra shark, in spite of its size, is not dangerous to man, other carpet shark species attack swimmers and divers, but usually only when they are being irritated. Such a shark cannot be disregarded because it bites when a diver rudely attempts to finish it off with a harpoon. Such attempts do not succeed because of the thickness of the skin and the dermal teeth. The carpet sharks are definitely not "man-eaters".

The smallest sharks are the DOGFISHES (Scyliorhinidae). The body is long and quite slender. There are generally two dorsal fins, but only in *Pentanchus.* The caudal fin is always small and only pointed upward to a slight degree. Spiracles are present. The teeth are small and numerous, and several rows are always used at once. There are twelve genera with about fifty species, inhabitants of the coastal waters of tropical and temperate regions in all oceans. A few species inhabit greater depths. There are two genera with a total of three species found along European coasts.

1.5 m, but is usually smaller than that. It is found throughout the tropical Indo-Pacific region and is harmless to man.

The FALSE CAT SHARKS (Pseudotriakidae) have a prominent characteristic in their unusually long first dorsal fin. The body is long, slender and not greatly compact. The snout is of the typical shape; the second dorsal fin is much smaller than the first, and the caudal fin is small. The spiracles are quite large. The teeth are small, with a large middle point bent to the back and on each side a small lateral jagged edge. The six to thirteen rows of teeth, all used at the same time, form thereby a broad chewing area. False cat sharks are ovoviviparous. There is one genus (Pseudotriakis) with two species, both being open sea inhabitants.

The ATLANTIC FALSE CAT SHARK (Pseudotriakis microdon; L almost 3 m) is apparently quite rare; only nine individuals have been identified with certainty. They were caught in the eastern Atlantic along the Portuguese coast, off the Cape Verde Islands, near Madeira, Iceland, and off the north American coast, and at depths of 300-1500 m. The body form indicates that they are not long-distance swimmers, and the teeth suggest that they feed on floor-dwelling prey such as crustaceans and fishes.

The shark family with the most species is the gray shark family (Carcharhinidae), and its members (along with the great white shark) are known as man-eaters. The body is spindle-shaped; there are two dorsal fins, the first one considerably larger than the second. The caudal fin is not crescent shaped, but the upper part is pointed clearly upward and the lower part is quite big. The tail shaft is somewhat flattened and in some species is laterally widened. Spray holes can be present. The teeth are razor-sharp with large points; one or at most two rows are used at a time. They are ovoviviparous or viviparous with a well-developed yolk sac placenta. There are seventeen genera with about sixty species, found generally in tropical, subtropical or temperate seas, some in all regions. Most of these are harmless, but a few are indeed dangerous.

The GREAT BLUE SHARK (Prionace glauca) is among those "man-eaters" that have resulted in the gray sharks as a whole being considered so. The body is elongated and powerfully built, the tail flattened and quite long. The sickle-shaped pectoral fins are particularly long. The teeth are large, triangular, with sharp, jagged edges. The coloration of the upper body side is dark blue, and the flanks somewhat lighter.

The great blue shark is found in tropical, subtropical and temperate waters of all seas. It is found frequently in the Mediterranean, less frequently in the North Sea. In its great migrations, it follows schools of sardines, herring, mackerel, and tuna, its chief prey. But it also feeds on other sharks, cephalopods, and attacks anything it can perceive. In some regions, great blue sharks cause considerable damage to fisher-

▷
1. Great blue shark (Prionace glauca), 2. Common hammerhead (Sphyrna zygaena); 3. Thresher (Alopias vulpinus); 4. Wobbegong (Orectolobus maculatus); 5. Remora (Echeneis naucrates), (a teleost fish which attaches itself to sharks and other large fishes; see Vol. V).

men, since several sharks attack the filled nets simultaneously and rip them up to get at the catch. Whalers also have great problems with this shark, for large groups of them attack the dead bodies of whalebone whales being tugged through the water and bite chunks out of them. Whether great blue sharks also attack living basking sharks and kill them or severely injure them is not known. Of course a great blue shark can pose great danger for a human who has fallen into the water. But since it is not normally found in coastal water, attacks on swimmers are great exceptions. The meat of this shark does not have a good taste and consequently it has little commercial value. However, great quantities are caught in Japan and processed.

The TIGER SHARK *(Galeocardo cuvieri)* is greatly feared along many coasts of warmer waters. L averages 4.5 m, not infrequently is 6 m, and allegedly even 9 m. It is relatively easy to recognize from its markings; younger specimens have a prominent and regular spotted marking along the flanks, which pales with increasing age. A keel is on each side of the tail shaft, and a small crest is located in front of and in back of the first dorsal fin. The snout is relatively short; the spiracles are very small. Tiger sharks are ovoviviparous (i.e., the young have no placental connection with the mother). There are between 10 and 84 young within the mother (but usually 30–50). The young obviously are very small when born, as a result of this large number; they measure just under .5 m at birth.

The tiger shark is found in all tropical and subtropical seas. It is well known around the Canary Islands, but does not occur in the Mediterranean. It inhabits the deep sea and coastal areas as well, and even in the mouths of rivers. This voracious omnivore feeds on crabs and various other crustaceans, fishes, and other sharks, poisonous stingrays, marine tortoises, and even sea lions; it grabs gulls from the water surface and attacks crocodiles in river mouths. In the vicinity of tropical coastal cities it feeds on all sorts of garbage thrown into the water, even empty cardboard boxes and sacks. Since the tiger shark seeks very shallow coastal water, it can also threaten people. Fatal accidents have been recorded throughout the world, which can probably be traced back to the tiger shark. On the other hand the tiger shark has commercial value. Leather can be made from its skin and its liver oil plays an important role.

The TOPE *(Galeorhinus galeus)*, in contrast, is of insignificant commercial value. The L is about 2 m; the body is slender and development is ovoviviparous. The tope is found along the floor at depths of 40–400 m in the eastern Atlantic from Norway and the rest of the North Sea as well as from Iceland along the western European and African coast to South Africa; it is encountered quite frequently in the Mediterranean.

Topes feed chiefly on floor dwelling fishes, but also take anything else they encounter which is the appropriate size for them. They

◁
1. Porbeagle shark *(Lamna nasus);* 2. Larger spotted dogfish *(Scyliorhinus stellaris);* 3. Smaller spotted dogfish *(Scyliorhinus caniculus);* 4. Piked dogfish *(Squalus acanthias);* 5. Frilled shark *(Chalamydoselachus anguineus);* 6. Rabbit fish *(Chimaera monstrosa).*

are not dangerous to man. A closely related Pacific species, found along the west American coast, has played an important commercial role because of the vitamin A in its oil.

While most GRAY SHARKS (or REQUIEM SHARKS; *Carcharhinus*) inhabit only the tropics, the ATLANTIC GRAY SHARK (*Carcharhinus plumbeus*) is among those few which penetrate into temperate regions. The L is barely 2.5 m with a very robust body. The first dorsal fin is noticeably long, and is located well forward. The coloration of the back is dark gray, gray-brown or brown. This species is distributed throughout the warmer Atlantic, not infrequently in the Mediterranean, and along the European coast only south of the Iberian peninsula.

The Atlantic grey shark is ovoviviparous. After a development period of eight to twelve months 1–14 young are born which measure 45 to 65 cm. As a coastal inhabitant, the Atlantic gray shark does not enter depths greater than 50 m. It is found in shallow bays and even in river mouths, and is said to be found occasionally in the lagoons and large canals of Venice. This species, completely harmless to man, feeds on small bottom dwelling organisms, especially fishes, crustaceans and mollusks.

The same genus contains a large number of tropical reef sharks, the best-known of which is the BLACK-TIP REEF SHARK (*Carcharhinus melanopterus*; L up to 1.5 m, occasionally 1.8 m), which swimmers and divers quite frequently find in the Red Sea. This species is readily recognized from the black tips on all its fins. It inhabits almost the entire tropical Indian and Pacific Oceans. This shark is found along reefs alone or in groups, enters lagoons and occasionally is encountered in very shallow coastal water. Generally the black-tip reef shark does not bother man, and while it curiously follows each swimmer, it swims away if someone swims back to it. Yet even this species can be stimulated to attack man due to injuries, presence of blood in the water, or a fearful flight by man. In any case, a diver should keep a nearby reef shark in view to prevent unpleasant surprises.

The HAMMERHEAD SHARKS (Sphyrnidae) have the same characteristics as gray sharks, with the single exception of the hammerlike or T-shaped widening of the snout. The eyes and nostrils are located on the sides of this "hammer." The mouth is on the lower side and the gill slits behind the hammer protrusion. Spiracles are absent. Many species give birth to living young and have a yolk sac placenta. There are two genera with a total of 12 species. Most inhabit open seas of tropical and temperate waters, but are also found in the Caribbean Sea, off California, and even in shallower coastal waters. Three species have been observed in the Mediterranean.

The longest species is the GREAT HAMMERHEAD (*Sphyrna mokkaran*; L up to 5.5 m). It inhabits warm Atlantic regions, the Indian Ocean and the Pacific Ocean. It is found only rarely in the Mediterranean, and

▷
A reef shark (*Carcharhinus menisorrah*).

▷▷ and ▷▷▷
Large sharks often attack small cetaceans (which are mammals), as shown here, in which a great blue shark (*Prionace glauca*) attacks a bottlenose dolphin (*Tursiops truncatus*; see Vol. XI).

▷▷▷▷ and ▷▷▷▷▷
Unlike most large sharks, the whale shark (*Rhincodon typus*) is a docile species which feeds on small organisms. Divers can approach whale sharks without endangering themselves.

▷▷▷▷▷▷
Above: Eggs of the dogfish (family Scyliorhinidae) are rectangular; long winding fibers enable the egg case to hang onto seaweed. The embryo can be seen through the translucent case. Below: Dentition of the sand tiger shark (*Carcharias taurus*) shows the distinctive teeth which are characteristic of sharks.

▷▷▷▷▷▷▷
Smooth dogfishes (*Triaenodon obesus*) are shown in an inlet of the Red Sea. A remora (*Echenenis naucrates*; see Vol. V) has attached itself to one of them.

there, individuals are invariably much smaller. The forward edge of the hammer is straight in the adult shark. This rare species feeds mainly on fishes and cephalopods, but will swallow anything it can overcome, including empty tin cans. When in coastal areas, though seldom found there, this species can be dangerous to swimmers and divers.

The COMMON HAMMERHEAD (Sphyrna zygaena) exists in somewhat greater numbers. The L is about 4 m. The forward edge of its head is distinctly arched. It also inhabits all tropical and temperate regions, and can be found far out at sea or directly on the coast. The common hammerhead is apparently ovoviviparous. Pregnant females contain 29–37 young.

The SPINY DOGFISHES (Squaloidei) are very similar in structure to the species mentioned previously. There is no anal fin, and often a spine is located in the forward part of both dorsal fins. Spiracles are always present and generally are quite large. Embryonic development is probably ovoviviparous in most species. There are four families (angular rough sharks, spiny dogfishes, spineless dogfishes or sleeper sharks, and bramble sharks) with a total of 60 species.

The ANGULAR ROUGH SHARKS (Oxynotidae) have short, stout bodies with high backs. In cross section, the body is triangular with a flattened belly. Both dorsal fins have a strong thornlike spine. A long and quite tall dermal ridge is located on each side of the body from the chest to the pectoral fin. The caudal fin is small. There is only one genus with three species.

The HUMANTIN (Oxynotus centrina; L at most 1 m), is brown, dark gray or occasionally even reddish brown. As a pure bottom dweller, this plump, stocky shark is found at depths of 30–300 m, sometimes even at 500 m. With the wide rows of teeth (there are six of them) in each jaw, the humantin seizes and readily chews up any hard-shelled prey it encounters along the sea bottom. Embryonic development is ovoviviparous. Pregnant females have been found with from 3 to 23 large eggs or embryos. The humantin occurs in the east Atlantic from Great Britain to South Africa and is also found in the Mediterranean.

The SPINY DOGFISHES (Squalidae) have a spine on the forward edge of both dorsal fins. The body has the unmistakable shark-like elongation and slenderness. There are eight genera with about 50 species, among them the SPUR DOG or PIKED DOGFISH (Squalus acanthias). The L is at most somewhat more than 1 m. The upper side of the body is grayish, and the lower side white. When very young they have two rows of light or white spots on the upper body side, which disappear with adulthood. The spines are powerful, and each has a small poison gland on the back.

A spiny dogfish can reach an age of 20–24 years. It feeds on cod and herring, but also on many kinds of crustaceans and other invertebrates. Spiny dogfishes often gather into enormous groups when hunting or

Above: a wobbegong (Orectolobus maculatus); Below: the smaller spotted dogfish (Scyliorhinus caniculus) is one of the smaller shark species; it can be found along European coasts.

migrating, groups of up to 1000 individuals. In the northern Atlantic, the spiny dogfish is the most frequent shark, and its distribution region extends in the north to northern Norway, Iceland and southern Greenland, and in the south to northwestern Africa. It is also found frequently in the North Sea and the Mediterranean.

Spiny dogfishes are ovoviviparous; the embryos first develop in a common elongated egg capsule in the oviduct, until they hatch in the mother. The newborn young are 20 to 24 cm long. Each litter normally contains 4 to 8 young. The meat of the spiny dogfish is sold in Germany under the name of marine eel and has considerable commercial importance. In European waters the annual spiny dogfish catch is approximately 40,000 to 45,000 tons.

The LANTERN SHARK (*Etmopterus spinax;* L up to 45 cm) inhabits the Atlantic exclusively and is the smallest shark in this ocean. Numerous small light organs are embedded in the skin of its belly. This species dwells on the floor, feeding on crustaceans and small squid. The lantern shark is found primarily only on the edge of the continental shelf at a depth of about 200 m; individuals have been caught down to 2000 m depths. In the summer 8 to 20 young are born.

The spineless dogfishes (Dalatiidae) or sleeper sharks can be easily distinguished from the spiny dogfishes; as the name implies, there is no spine in front of the second dorsal fin, and often not even in front of the first. With a few exceptions, these are generally small sharks. There are six to eight genera with a total of just eight species.

One of the few giants in this family is the LARGE SLEEPER or GREENLAND SHARK (*Somniosus microcephalus;* L 3-4 m, or at most 8 m). It inhabits arctic waters, preferring depths of 200-600 m, living only in colder regions. Its distribution region ranges from the Arctic Sea to about the latitude of England. It feeds on cod, flat fishes, rays and other bottom dwelling fishes, and also attacks smaller dolphins and seals; it even catches birds from the water surface. I once investigated a freshly caught Greenland shark off of western Greenland and in the stomach of this specimen found two common porpoises (see Vol. IX) and two seals. The Greenland shark apparently reaches a ripe old age.

Among the smaller species is the SLEEPER SHARK (*Somniosus rostratus;* L 1 m), found in the Mediterranean. The DARKIE CHARLIE (*Dalatius licha;* L 1.5 m) is distributed worldwide. It is chocolate-brown and inhabits depths of 300-600 m. Occasionally it is also found at 100 m depths or as far down as 1000 m. Its chief distribution region is in the warmer Atlantic from Ireland to Morocco and in the western Mediterranean.

The BRAMBLE SHARKS (Echinorhinidae) are characterized by scales with large nail-like barbs on the entire body. The dorsal fins lack spines and are located to the rear of the shark with the rear edge tapering off toward the tail. The spiracles are small. There is one genus with perhaps two species, including the BRAMBLE SHARK (*Echinorhinus brucus;*

▷
1. Thornback ray (*Raja clavata*); 2. Common stingray (*Dasyatis pastinaca*); 3. Guitar fish (*Rhinobatos productus*); 4. Marbled electric ray (*Torpedo marmorata*).

L 1.5 to 2.5 m), which inhabits the Atlantic and parts of the Indian and Pacific Oceans.

The bramble shark is a bottom dweller inhabiting depths of 400 to 900 m. However, it is also found in the much shallower North Sea. In the eastern Atlantic, its distribution extends from the North Sea and Ireland to Senegal in western Africa, and it is not rare in the Mediterranean. The bramble shark feeds on crustaceans, various floor dwelling fishes and spiny dogfishes.

Suborder: saw sharks

Distinguishing characteristics

Fig. 4-6. *Pristiophorus japonicus* from above.

The SAW SHARKS (Pristiophoroidei, family Pristiophoridae) have the typical elongated body of sharks but with a sword-shaped elongated snout with a row of long, pointed, and a row of angular, teeth on the outer edges. A long fleshlike barb is located on each edge. There are always two dorsal fins lacking spines. An anal fin is also missing, but spiracles are present. There are five or six gill slits in front of the pectoral fin on the sides of the head. There are two genera with a total of four similar species off South Africa and in the area extending from Australia to Japan and Korea.

The saw sharks are not to be confused with the sawfish rays. They are rarely longer than 1.5 m and generally are found in deeper water. They dwell on the floor, where they plough the "sword" along the muddy bottom between algae and seaweed and use the long barbs for chemoreception of prey. Some species are ovoviviparous, and others allegedly bear live young. The long sword is developed before birth, but the teeth are enclosed in a skin fold to prevent injury of the organs of the mother; they erupt only after birth.

Suborder: squatinoids

Distinguishing characteristics

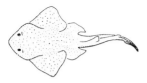

Fig. 4-7. The angel shark *(Squatina squatina)* from above.

◁

1. Manta ray *(Manta birostris)* with pilor fish *(Naucrates ductor;* see Vol. V); 2. Spotted eagle ray *(Myliobatis aquila)*; 3. Sawfish *(Pristis pectinatus)*.

The last suborder of sharks, the SQUATINOIDS (Squatinoidei) resembles the following order of rays and skates. The front of the body has the flattening seen in rays; the pectoral fins are broadened like wings, but are not connected with the head. There are always 5 gill slits in front of the pectoral fins and primarily on the head sides (thus distinguishing them from rays!). The rear of the body (actually the tail shaft) is cylindrical as in typical sharks. Both dorsal fins are located on this element. The caudal fin is a simple rudder, and the anal fin is absent. The spiracles are quite large. There is a terminal mouth opening. Embryonic development is ovoviviparous. There is only one family, MONK FISHES (Squatinidae), with 12 species occurring worldwide.

The squatinoids can also be distinguished from rays by their means of propulsion. Although they generally lie flat on the sea floor, the long tail shaft makes lateral undulating motions with both dorsal fins and the caudal fin to initiate swimming. Thus the tail shaft is the main propulsion organ although the pectoral fins carry out some winglike motions to raise and force the body up.

The ANGEL SHARK *(Squatina squatina)* inhabits the northeastern Atlantic from southern Norway to the Canary Islands, and the Mediterranean as well. This species attains a length of 2.5 m, but is usually much smaller. It lies in wait on the ocean floor and feeds on smaller floor-dwelling fishes, and various mollusks and crustaceans. In the winter angel sharks stay in deeper water, but in summer can be found in coastal regions, where their sexual activity also takes place. The females bear 7 to 25 young at lower depths between December and February.

In RAYS and SKATES (Rajiformes) the body is for the most part plate-shaped, although some species have a shark-like shape or even a transitional form between sharks and rays. There are always just 5 gill slits, always on the lower side of the body (that being the simplest and clearest distinguishing characteristic from sharks; compare *Raja undulata*). The mouth opening is always below the snout. The forward edge of the pectoral fins is fused with the body and generally with the head as well. The eyes are on top of the head, and large prominent spiracles are behind the eyes. There is no anal fin, but there are usually two dorsal fins on the tail shaft; a caudal fin is sometimes absent. The teeth are always used several rows at a time. Only true rays (Rajidae) lay eggs, the rest being ovoviviparous (i.e., the young hatch from eggs while still inside the mother). The embryos are initially elongated like sharks, with pectoral fins not yet fused to the head. Rays inhabit all seas, from the shallowest coastal regions to 3000 m depths, and some species enter brackish water and even fresh water. There are 5 suborders (sawfishes, guitar fishes, eyed electric rays, two sting ray suborders) with about 17 families, 47 genera and approximately 350 species.

Order: rays

Distinguishing characteristics

The body of the SAWFISH (Pristiodei; with the single family Pristidae) is elongated and similar to that of a shark. The head is somewhat flattened, and the snout has developed a sword form and is armed with long, sharp dagger teeth along the sides, like a saw. In contrast to the saw sharks, there is no paired barb on this "sword"; the gill slits are on the lower side of the body under the pectoral fins. The jaw teeth are small, placed in close rows, and each row forms a sort of paved surface. The fronts of the pectoral fins have become fused with the sides of the head. Sawfishes inhabit primarily tropical and subtropical regions.

Suborder: sawfishes

Distinguishing characteristics

These rays dig through the floor with their "saws," but can also kill smaller fishes with them and use the saws for defense. In the Mediterranean, but especially in the warmer waters of the east Atlantic, there are with certainty two species of sawfishes: *Pristis pectinatus* (L 6 m maximum) and *Pristis pristis* (L up to 2.5 m). They are both bottom dwellers as are all other sawfishes, and feed on small fishes and other organisms of the sea floor. They swim in the same way as sharks.

The GUITAR FISHES (Rhinobatoidei) have a ray-like flattened forebody. The pectoral fins are in their entire length fused to the sides of the head and to a part of the flanks. The hindbody including tail shaft is

Suborder: guitar fishes

Distinguishing
characteristics

cylindrical as in sharks and elongated. There are two dorsal fins of equal size on the tail shaft and the caudal fin is well developed. The large spiracles are located just behind the eyes. Five gill slits are on the underside of the body. Embryonic development is ovoviviparous. There are two families (Rhynchobatidae and Rhinobatidae) with 9 genera and 45 species.

All guitar fishes inhabit coastal waters in tropical and subtropical zones. They live in sandy and muddy floors, where they feed primarily on crustaceans, mussels, and snails. Three species inhabit the Mediterranean, of which only *Rhinobatos rhinobatos* (L about 1 m) is found frequently. This species is distributed in the eastern Atlantic from the Gulf of Gascogne to tropical western Africa. In the Mediterranean, it is found chiefly along the African coast, and off Sicily. Since it lives in shallow water and on the floor and feeds a great deal on mussels, it can become a plague for oyster fishermen.

While the above guitar fish is colored dark grey-brown to olive on the upperside, the GUITAR FISH *Rhinobatos cemiculus* (L 2 m maximum) has a light sandy color and is difficult to recognize on the sea floor. While preferring 10 to 80 m depths, it also enters brackish water. As in other guitar fish, this species has the unusual swimming motion with two "driving surfaces". If it swims slowly or hardly changes direction, then the pectoral fins beat just as in rays. When suddenly leaving the floor, the tail contracts and rebounds with a mighty push. When swimming rapidly away, the pectoral fins beat with the typical ray motion while the hindbody carries out undulating movements such as sharks use.

Suborder: electric rays

Distinguishing characteristics

The ELECTRIC RAYS (Torpedinoidei) have a low-set raylike flattened body with a powerfully developed tail. The body is plate-shaped, but in some species is elliptical or oval. There are one or two (or no) dorsal fins on the tail shaft, depending on the family. The skin is generally completely naked. The eyes are always small, and in deep sea forms regressive or absent. The teeth are small and arranged in wide bands. There are three families: 1. Electric rays (Torpedinidae), distinguished by two dorsal fins; the 7 genera are in all warm seas; 2. Narkidae, with one dorsal fin and pectoral fins modified into structures adapted for walking; found only in the Indian and Pacific Oceans; 3. Temeridae, lacking dorsal fins, and found exclusively in the western Pacific.

Touching an electric ray can mean receiving a strong electrical shock. Two large electrical organs are in the body and at the base of each pectoral fin. The musculature in these regions has been modified to form electrical plates, with a negative underside and positive upperside. Electrical discharges of more than 200 volts and 2000 watts have been measured in these rays. These shocks are a very effective weapon. A predator attacking an electric ray is driven off by a powerful discharge. If an electric ray discovers prey on the ocean floor, perhaps

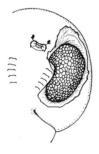

Fig. 4-8. The electric ray *(Torpedo)* with lower side opened, exposing the electrical organ.

a flatfish, it bends over the prey and issues an electrical shock to the creature. Thus electrofishery is a new invention for man, but an old innovation of nature.

Since the electrical organ takes up a great part of the body surface in the platelike region of the ray, the pectoral fins cannot function for propulsion, as is usually the case in rays. Thus, in spite of body flattening, propulsion is for the most part limited to the hindbody and tail. When swimming the caudal fin delivers powerful, if clumsy, rudder beats to the sides, and propels the body forward. The swimming technique of deep-sea rays is not yet known, we do know that their pectoral fins are used for walking across the bottom.

Only rays from family Torpedinidae (genus *Torpedo*) occur along European coasts, and then only in the Mediterranean and the western European region. The most frequently encountered species is the EYED ELECTRIC RAY (*Torpedo torpedo;* L up to 60 cm). It has five large prominent eye spots on its brownish back. The chief prey consists of fishes and crustaceans. This ray is found down to 50 m depths and inhabits the eastern Atlantic from the Gulf of Gascogne to Angola, as well as the Mediterranean. Like all electric rays it is viviparous. Between 3 and 21 young have been found in the bodies of mothers. At birth, the young rays are only about eight millimeters long. The MARBLED ELECTRIC RAY (*Torpedo marmorata;* L up to 60 cm) primarily inhabits sandy floors between five and twenty-meter depths. Its yellowish basic coloration is covered by numerous dark brown spots, which blend to form a marble pattern. It inhabits the eastern Atlantic from France to South Africa and is also found in the Mediterranean. The BLACK ELECTRIC RAY (*Torpedo nobiliana;* L up to 1.8 m) is distributed on both sides of the Atlantic from Scotland to the Azores and from Nova Scotia to Florida, and in the Mediterranean also. This species is chocolate brown to black on the back and is found at down to 250 m depths. Sixty embryos have been found in the body of one mother.

The RAYS (Rajoidei) are largely flattened and from above appear to have a shape varying from quadratic to rhombic longitudinally. The tail is clearly thickened, but nevertheless rather slender, and never bears a sawing spine. Both dorsal fins are located well rearward and taper off toward the tail; they are very small or absent altogether. The caudal fin has been regressively developed to form a skin fold. The back and upper side of the tail have numerous spines and thornlike protuberances. The jaw teeth are blunt or sharp, depending on the feeding style. Embryonic development is oviparous. The eggs are surrounded by a rectangular egg capsule, which has a long (initially soft) fiber on each corner. There are three families: 1. Rajidae; 2. Anachantobatidae (lacking dorsal fins and thorns the back); 3. Arynchobatidae (with one dorsal fin and soft skin on the back).

By far, most species belong to the SKATE (Rajidae) family, of which

Suborder: rays

there are numerous forms in the northern Atlantic, North Sea, Baltic Sea, and Mediterranean. The THORNBACK RAY (*Raja clavata;* L about 1 m) is found from northern Scandinavia and Iceland to the Moroccan coast, throughout the North Sea, western Baltic Sea, in the Mediterranean and in parts of the Black Sea. The upper side of its body has a great number of large thorns. It inhabits depths from 20 to 200 m. In the course of a summer, the female lays about twenty eggs. Embryonic development lasts from four to five months.

The FLAPPER SKATE (*Raja batis;* L up to 1.5 m or in exceptional cases 2.5 m) has almost no thorns on its back at all. A few can be seen close up, and there exists at least one row of thorns on the tail. The flapper skate inhabits the coasts off northern Norway and Iceland to Gibraltar, and the west and central Mediterranean including the Adriatic Sea. It is found from shallow coastal zones down to 500 m depths. The flapper skate feeds chiefly on flatfishes, shell fishes and crustaceans. The eggs, which are laid in late fall and winter, are astonishingly large, measuring up to fourteen by twenty-five centimeters.

Flapper skates and thornback rays are of tremendous commercial importance. Not only the liver is used, but also the meat, which in France is known as French turbot and is considered a delicacy. In Germany, the "wings" and tail segment are smoked and marinated, and sold under the name "sea trout".

Suborder: eagle rays

Distinguishing characteristics

The EAGLE RAYS (suborder Myliobatoidei) also have a greatly flattened body, which from above looks longitudinally oval or broadly rhombic. The tail is of medium length or whiplike, and a caudal fin is present in but one family. If a dorsal fin is present, it is located on the upper side of the tail. This structure is sharp, quite large, and has jagged edges on both sides or a hook; sometimes several spines are present. A poison gland is at the spine base. The highly distinctive groups are organized into four families by many researchers (sting rays, butterfly rays, eagle rays, and devil-fishes) with 17 genera and about 128 species. Some are found in warmer European waters.

The spine of this ray functions solely as a defensive weapon. It penetrates the body of an opponent, where it not only creates a heavily bleeding wound, but even breaks off and remains stuck inside. In man, injuries from such stinging spines cause very severe pains, serious poisoning symptoms, and illness lasting months.

Four species are cited here from the STINGRAY family (Dasyatidae): 1. COMMON STINGRAY (*Dasyatis pastinaca;* L up to 2.5 m, width to 1.4 m, spine up to 35 cm long). This coastal inhabitant also penetrates lagoons and brackish waters and is not found below a depth of sixty meters. The common stingray is ovoviviparous; six to nine embryos develop. Newly born young already possess the stinging spine. Distribution is in much of the Atlantic, in the eastern part of which it is found from the French west coast to Zaïre, and in the western Mediterranean

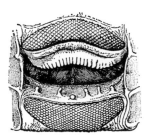

Fig. 4-9. *Dasyatis centroura* mouth opening with tooth plate.

more than in the eastern part (only rarely in the Adriatic); 2. The STIN-
GRAY (*Dasyatis centroura*; L about 3 m, width 2 m). Scattered large spines
are on the back, and the tail is considerably longer than usual in the
common stingray. The upper side of the tail is densely set with teeth
and spines. The back is olive-green. This species inhabits subtropical
regions on both sides of the Atlantic, and is frequently seen in the
Mediterranean. 2. BLUE STINGRAY (*Dasyatis violeacea*; L about 1.5 m). This
species is noticeably brown-violet and smooth; only the middle of the
upper body side has small teeth. The tail is smooth and almost whip-
shaped. This ray is found in the Atlantic, Indian, and Pacific Oceans,
but rarely in the Mediterranean. 4. *Taeniuya lymna* (L over 2 m); the
back is a medium brown color to sandy with large light blue spots,
and with a light blue longitudinal band along each side of the tail. This
species inhabits much of the Indian and Pacific Oceans, and at night
can be found in the shallowest water where it often encounters bathers
or divers. It is quite often kept in aquariums.

Fig. 4–10. Upper surface
of tail of *Dasyatis sabina*
with barbs.

The BUTTERFLY RAYS (Gymnuridae) are closely related to stingrays.
The body is considerably wider than long, and the flanks and pectoral
fins have wing-like appearance. The snout is short, and eyes and
spiracles are on the upper side of the head. The tail is noticeably short
and may bear a dorsal fin. Some species have a tail spine with jagged
edges. There are about ten species in the coastal regions of the subtropics and tropics.

The single butterfly ray species in the Mediterranean, *Gymnura altavela*, belongs to those species with jagged tail spines. While on the
west coast of Africa, specimens with a width of up to 4 m were found,
those of the Mediterranean are barely 1 m in length. They feed mostly
on fishes and squid, but are so rare that nothing further is known about
their life.

In all eagle rays (Myliobatidae) the flanks and pectoral fins are
greatly elongated as wing-like structures. The head with its beak-like
snout is clearly a distinct part of the body. The eyes and spiracles
are on the side of the head; the tail is whip-shaped, long, and thin,
and in some species has a jagged spine. There is always a dorsal fin,
while a caudal fin is consistently absent. There are four genera with
about twenty-five species; most of them are inhabitants of the tropics.

Only two eagle ray species can occasionally be observed along the
western European coast and in the Mediterranean: 1. SPOTTED EAGLE
RAY (*Myliobatis aquila*; L just under 1 m, width about 1.5 m, and tail
length about 2 m). The jagged spine is near the tail base just behind
the dorsal fin. The back is dark brown with a greenish or bronze sheen.
This species inhabits the eastern Atlantic and Mediterranean. Individuals have been seen off Scotland and Norway on occasion. 2. BULL
RAY (*Pteromylaeus bovinus*); this is only rarely encountered in the Mediterranean, since it prevails in southern regions.

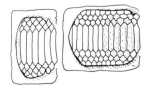

Fig. 4-11. Upper and lower teeth rows of an eagle ray *(Myliobatis)*.

Eagle rays resemble a swarm of pelicans or other large birds, the wings of which sweep out large areas as they slowly beat. For when these rays swim, waves no longer run over the pectoral fins and flanks; these parts of the body, which taper off to the sides, are beaten much more like the wings of birds. While eagle rays swim around in the open water, often just under the surface, they swim down to the floor for food. There they dig with their beak-shaped snouts for mussels and snails, the shells of which are broken with the broad ray dentition. Larger crustaceans, hermit crabs, and related armored crustaceans are overcome in the same way. As are perhaps all the members of the family, spotted eagle rays are ovoviviparous. Each litter contains three to seven young. The young resemble the adult forms to a great extent while still in the embryonic stage.

The MANTAS (Mobulidae) have a snout divided into two thin sponge-like elongations. These "horns" or "head fins" protrude horizontally forward and are highly movable. The dentition is not in the plate shape of eagle rays, but consists of very small teeth which stand in rows very close to each other. The mouth is usually terminal and very wide. The tail is considerably shorter than the body, and a dorsal fin is located at its base. A jagged spine is often present also, but a caudal fin is absent. Mantas are ovoviviparous or viviparous. There are two (or four, according to other ichthyologists) genera with an uncertain number of species.

These giants of the rays, which can have a span of up to 6 m, feed chiefly on plankton; thus their teeth play a very subordinate role. Searching for prey, the mantas "fly" in pairs or small groups through the water and hold their mouths open like great barn doors; thus, they shovel their food, which consists most of free-swimming crustaceans and small schooling fishes. The gill system has developed a fine sieve apparatus to hold onto the prey. Mantas can be observed wherever great plankton swarms are found in warmer to tropical waters, and thus on the open sea as well as along coasts.

One species is found along Europe's western coast: the DEVIL-FISH *(Mobula mobular;* L up to 2 m, tail length to 1 m, span up to 5 m and weight about one ton), which is found in the eastern Atlantic from the English coast to Senegal and in the western part of the Mediterranean. Devil-fishes generally swim in small groups. In spite of their great weight, they can make mighty leaps out of the water, such leaps being followed by a loud noise like a pistol shot when they slap back into the water. These great jumps are caused by copepods (see Vol. I) which parasitize these giant fishes and cause them great discomfort.

The largest of all the mantas, the MANTA RAY *(Manta birostris),* can be found only in tropical and subtropical waters. This species can exceed 7 m in width and reach a weight of 2 tons. The female apparently bears just one young at a time.

The CHIMAERAS (Holocephali) have a cartilaginous skeleton, but are distinguished from sharks and rays in so many other characteristics that they cannot otherwise be grouped together. Chimaeras have but one gill slit; there are two pairs of teeth plates in the upper jaw and one pair in the lower jaw. The upper jaw is firmly fused with the brain capsule. The vertebral column has no vertebrae, and thus the notochord has been retained without any constrictions in its entirety. The body is cylindrically shaped and is slightly flattened to the side. The head is relatively large, the snout blunt, but often beak-shaped at its terminus or with a hooked projection. The first dorsal fin is short and quite high, with a long spine at its forward anchoring point, the spine having poison glands. The second dorsal fin exists as a low, elongated seam, and the caudal fin is so small that it can create no propulsion force for swimming. The pectoral fins, however, function as powerful and large rudders. There are three families (Chimaeridae, Rhinochimaeridae, and Callorhynchidae).

Sometimes chimaeras are known in German as "sea rats" because their tail is long and thin like a rat tail, or because it ends as a whip. The sexes are clearly distinguishable. In the male, which is always the smaller, the last rays of the ventral fins have been modified to tubes which transmit sperm; the male furthermore has a long club-shaped projection on the upper side of the snout; this is filled with dermal teeth. With this curious structure, the male maintains a firm position during copulation. Chimaeras are egg-laying organisms, but the female lays only 2 eggs. As in the other egg-laying cartilaginous fishes, the eggs are enclosed by a horny capsule which can be up to 30 cm long. The capsule has up to 400 little openings through which the developing embryo can receive fresh water for respiration.

The chimaera family (Chimaeridae) has seventeen species, of which only one inhabits European waters; the RABBIT FISH (*Chimaera monstrosa*; L barely 1.5 m). The coloration is silver-gray with a violet sheen on the upper side and a dark marble pattern. The unpaired fins have black seams. This species is found in the northeastern Atlantic from Iceland and Norway to northern Africa, and in the western and central Mediterranean.

The rabbit fish is a bottom dweller which inhabits chiefly the continental shelf to depths of 200 m, but occasionally is found at 1000 m. There it feeds on various crustaceans, mollusks, echinoderms and also on small floor-dwelling fishes. The beak-like teeth plates of the jaw function to mash the hard-shelled prey. The rabbit fish has also been found off South Africa; it migrated down the tropical girdle into the deep waters and in this way passed down the continental base from northern Europe to South Africa. Man uses only the liver of the rabbit fish, which amounts to about one-third of the total body weight and from which valuable oil is produced. Fishermen dealing with rabbit

Subclass: chimaeras

Distinguishing characteristics

▷
The largest of all the rays is the manta *(Manta birostris)*, which attains a span of seven meters. The devil-fishes, to which this species belongs, inhabit the deep seas, unlike other rays.

Fig. 4-12. Rhinochimaerid male, head with pointed snout.

fish must proceed cautiously for the stinging spine of the dorsal spine is poisonous and lethal injuries can allegedly result from accidents with them.

The Rhinochimaeridae are characterized by a dagger-like elongated snouth and a long, whip-like extended tail. The dorsal fin has a large erectile poison spine. Males have processes on the snout and ventral fins, both of which serve to maintain a hold during copulation. As deep-sea inhabitants, these chimaeras are found at 600 to 2600 m depths. Only three genera with a total of four species are differentiated; these are almost worldwide. *Harriotta raleighana*, (L 1.2 m maximum) inhabits the northern regions of the Atlantic to western Africa and occasionally appears in European waters.

The Callorhynchid chimaeras have a short trunk-like extension of the snout which has a larger, highly movable skin fold. These species apparently dig through the floor with this structure. The single genus *Callorhynchus* has four species, which are all found in the seas of the southern hemisphere. Some of them occur in shallow coastal regions. None penetrates beyond 180 m, which means that these chimaeras inhabit the upper part of the continental base. Because of its good flavor, the "doodskop," "monkeyfish" or "elefantfish" (sic) of the South Africans (*Callorhynchus capensis*; L under 1 m) is frequently sold at market.

◁
Above: *Hydrolagus colliei* is a short-nosed chimaera (Chimaeridae) from the west coast of North America. Middle left: the underside of a skate *(Raja undulata)* is featured by the nasal opening located in the front of the mouth. Lower left and right: the stingray species *Taeniura lymna*, like many rays, generally burrows into the sand. In the photo on the right a wrass (*Halichoeris centriquadrus*; see Vol. V) is shown along with the stingray.

5 The Bony Fishes

In contrast to the cartilaginous fishes dealt with in the previous chapter, substantial parts of the skeletons in teleost (bony) fishes (Osteichthyes) are ossified. In the primitive groups within this class ossification is restricted to the skull, which initially is only covered with dermal bones. Vertebrae are not yet present in these species. Not all skeletal parts are ossified until we come to the true bony fishes (Teleostei). In those, the vertebral column with its vertebrae forms the actual axis of the entire skeleton and the skull has become a firm capsule composed of many various bones. The presence or absence of specific skull bones plays a major role in classifying these fishes. Even the jaws, which in cartilaginous fishes consist of two toothed cartilage strips, are composed of numerous bony elements in the teleosts and are also joined with the skull by a bone. The gill arches and the common mouth of the gills, the gill cavity, are protected by a series of firm bones, the gill cover.

The body of teleosts is covered by scales which overlap each other like roof tiles. In sturgeon the skin is partially naked, and partially covered with bony scales or ganoid scales. Polypterus and gar have an armored covering of ganoid scales.

Most bony fishes have a swim bladder, with the help of which they equalize their weight with the depth at which they swim and thus can adapt to the various water pressures they experience. They can increase or decrease their weight and thus rise or sink without requiring action by the fins. The presence or absence of a swim bladder and the configuration it takes are important elements in systematic classification of this fish group. Thus, they are divided into physostomes and physoclists, depending on whether the swim bladder has a connection to the throat or not; characteristics used in classification are the development of the so-called Weberian apparatus (a row of bony pieces which join the wall of the swim bladder to the inner ear), and the division of the swim bladder into various chambers or its use as a respiratory organ like a simple lung.

Class: bony fishes, by W. Ladiges

All teleost fishes are divided into two major development lines, which diverged in an early evolutionary period. Most of today's fishes belong to the subclass of SPINY-RAYED FISHES (Acanthopterygii) which are now in their flourishing period. Only a few species from the second line of development are still living. These are the lobefins and lung fishes, which are known as subclass soft-rayed fishes (Sarcopterygii) due to the development of the paired fins. Although they have maintained the primeval condition to a far greater extent than the "modern" true teleosts, we shall deal with the soft-rayed fishes as the last group of teleosts, due to their evolutionary relationships to land vertebrates (see Vol. V).

Subclass: spiny-rayed fishes

Four superorders are classed in Acanthopterygii: 1. Bichirs (Polypteri) with one order; 2. Sturgeons and paddle-fishes (Chondrostei) with one order; 3. Gars and bowfins (Holostei) with two orders; 4. Teleosts (Teleostei) with thirty orders.

6 Polypterids, Sturgeons and Related Forms

Due to their fin structure the polypterids form a single superorder with the single order Polypteriformes. The body varies from more or less elongated (in *Polypterus*) to eel-shaped (the reedfish *Calamoichthys*); it is covered with diamond-shaped ganoid scales which are very hard, consisting of three layers: isopedine, cosmine, and ganoine. Coloration is generally gray-green to yellow-brown, and depending on the species, interrupted by stripes or spotted markings which can be dark or black. There is but one family—Polypteridae—in Africa's inland waters, and it contains two genera: 1. POLYPTERUS (*Polypterus*), L as much as 120 cm, and nine species, including the BICHIRS *Polypterus bichir*, *Polypterus senegalus*, and *Polypterus ornatipinnis*; 2. REEDFISHES (*Calamoichthys*), L up to 90 cm, with a single species *Calamoichthys calabaricus*, the REEDFISH.

The polypterids can be distinguished from all other spiny-rayed fishes by their distinctive dorsal fin; it consists of five to eight flaglike "fins", of which each individual finlet is supported by a strong bony ray and a few soft fin rays. When carefully creeping up on prey, its tiny fins are raised and the pectoral fins are used as a rudder; however when fleeing from enemies the dorsal and pectoral fins are pressed close to the body and the fish propels itself forward with an undulating motion.

Numerous primitive characteristics indicate that the polypterids are very old forms, equivalent to living fossils. The fan-shaped pectoral fins, which contain a fleshy lobe with scales, resemble those of the lobefins (see Vol. V); the skeletal structure bears the same resemblance. These fins function in propelling the body when swimming slowly and in supporting the forebody. Although the caudal fin appears to be symmetrical, it is actually heterocercal, since the vertebral column passes into the upper fin lobe. Pelvic fins are present only in *Polypterus*.

Polypterids are predators which at least as adults feed primarily on

Superorder:
polypterids,
by F. Terofal

Order: polypterids

Distinguishing
characteristics

Fig. 6-1. Polypterids (Polypteriformes).

other fishes. But they also eat worms, insect larvae, crustaceans and other small organisms. During the day they are hidden within their resting recesses, and only at dusk do they become quite lively and begin to search for prey. One structural characteristic of their ancestors is the spiral fold of their intestine (compare with cartilaginous fishes).

Polypterids can be observed coming to the surface to breathe air, particularly in muddy, oxygen-poor waters. They possess gills with four gill arches, but have an additional respiratory organ, the swim bladder which resembles a lung. It is equivalent to the swim bladder of lungfishes (see Vol. V) and is structurally nearly as well developed as the lungfish swim bladder, with a primitive blood supply arrangement. If polypterids are prevented by meshed wire from reaching the water surface, they soon perish by "drowning". With their "lungs" the polypterids can survive drought periods in Africa by digging into the mud; they can even crawl over land for short distances to reach another water source.

As recent investigations have shown, polypterids have a highly developed olfactory sense. Their olfactory organ is basically different from that of the other spiny-rayed fishes and resembles that of the lobefins. Taste buds are confined to the mouth cavity in polypterids and are especially numerous on the upper side of the pointed end of the tongue. The "Fahrenholz" organs are presumed to function in a tactile sense; in species which have thus far been examined, they are confined to the head region and are particularly numerous on the top of the snout. Perhaps they communicate to the fish when it is approaching the surface of the water. The eyes are poorly developed. In 1968 Pfeiffer wrote: "While the eye of *Polypterus* can play a role in prey catching and avoiding enemies, it has but a subordinate function in the life of *Calamoichthys*."

Few observations have been made regarding the propagation of polypterids in nature. Spawning begins in the savanna with the onset of the rainy season in June or July and lasts until September. The sexually mature fishes depart the river bed and swim into the marshy flooded regions. The anal fin of the male, which is considerably larger than that of the female, swells during this period and the skin between the fin rays forms deep folds; the tail shaft, like that of pond loaches, becomes large and spongy. W. Armbrust has described courtship behavior in the bichir *Polypterus ornatipinnis:* "I accidentally discovered some peculiar behavior in one of them around noon. It was the male, which had become somewhat restless and swam to and fro. When it approached the female, it thrust its fin up and rocked back and forth with the head...Courtship behavior intensity of the male increased. Both fishes laid near each other. The male pressed its head against the female's. It appeared as if the male wanted to whisper something into the female's ear and convince it to come with it into the Cryptocoryne

Fig. 6-2. Polypterid larva.

forest. When the male then swam away, the female followed for a short distance, and this was repeated until the moment of spawning. Spawning did not only take place in the Cryptocoryne, but also on the open surface under the Najas plants. Here the spawning process could be observed to some degree. The female was completely passive. It lay supported on its pectoral fins with its head slightly raised. Occasionally it went to the surface to get some air. The male then swam with continuous, excited lateral head shaking, beside the female. The anal fin formed a sort of hollow hand and this was shoved under the female. It was not possible in the twilight to determine how many eggs were released...After exactly four days the first young hatched."

Young polypterids have external gills on both sides of the rear of the head during their first weeks of life, during which time metamorphosis takes place (as in salamander larvae). These are feather-shaped growths of the gill cover with a rich blood supply and they make it possible for the young to survive in the muddy, oxygen-poor water which often characterizes spawning areas.

The adults do not require good water conditions in their African habitat either. They are very resistant and fairly insensitive to water pollution, oxygen deficiency and temperature changes. Polypterids primarily inhabit the shores and flooded regions of the rivers (with their rich supply of plant growth), in Senegal, Gambia, Volta, Nigeria, Zaïre and along the Nile, as well as suitable portions of lakes, for example in Lake Chad and Lake Rudolf. The eel-shaped polypterids frequent muddy waters and the mouths of streams—as in the Nigerian delta—as well as brackish water. They are usually found near the coast from Dahomey to the mouth of the Congo.

STURGEONS and PADDLEFISHES (Acipenseriformes) are the single order in superorder Chondrostei. The body is elongated and spindle-shaped and the snout is more or less extended. The mouth is on the lower side of the head and is toothless or in the young has very small teeth. An upper jaw bone is present. The skin is nearly naked (in paddlefishes) or has five rows of large bony plates (true sturgeon); scales are only on the upper edge of the heterocercal caudal fin. The skeleton is primarily cartilaginous, with an intact notochord (Chorda dorsalis); vertebrae are not developed. Even the skull is cartilaginous, and in adults it bears a strong bony covering. Spiracles may be present or absent. Respiration occurs through the gills with five gill slits. Of the gill cover bones only the operculum is present. The foregut is short and is connected with the simple swim bladder by an air passage. The stomach has a strong musculature and the connecting intestine has a well-developed spiral valve. There are two families: 1. Sturgeons (Acipenseridae); 2. Paddlefishes (Polyodontidae). A third family is now extinct.

Superorder: Chondrostei

Order: sturgeons

Distinguishing characteristics

The STURGEON family Acipenseridae contains the world's largest fresh water fishes. The largest sturgeon known, *Huso huso*, was no less than 8.5 m long and weighed 1300 kg. The fish was claimed to be over 100 years old. This family has existed for about 200 million years and consists of fresh water and migrant species prevalent in the northern hemisphere, which can be recognized by the following characteristics: Five lateral rows of bony plates (which differ among various species in form, size and number), between which in some species small bony platelets are embedded in the skin at irregular intervals; the plates of the young are closer to each other, more strongly developed, and with a sharp keel terminating in a thorny projection. The mouth has thick wart-covered lips and can have a trunklike projection. Small soft teeth are found in the young of genus *Acipenser* and in no other sturgeon; this group possesses a yolk sac. Four barbs are in front of the mouth; their form and placement differs among the various species. The dorsal, anal and pectoral fins are located well to the rear. Coloration is generally olive green, brownish or gray on the back, and the underside of the lateral scales is white. Two subfamilies exist: 1. Sturgeon (Acipenserinae), with spiracles and a pointed snout; 2. Shovel-nosed sturgeon (Scaphirhynchinae), lacking spiracles and with a wide, shovel-shaped snout.

The sturgeon subfamily Acipenserinae contains the genera *Huso* with two species and *Acipenser* with seventeen species. One means of distinguishing the two genera is by the shape of the mouth and the mouth barbs; in *Huso* the large, half-moon shaped mouth opening extends to the end of the snout and the barbs are flattened, while in *Acipenser* the mouth does not extend to the end of the snout and the barbs are rounded.

The EUROPEAN STURGEON *(Huso huso)* is distributed in the Adriatic, the Black Sea and the Caspian Sea, and in Russia this species is called "beluga", the same name borne by a whale found on the northern coast of the Soviet Union (see Vol. XI). A migrating species, this sturgeon winters and spawns in the large tributaries of its home waters; summer and winter races may even develop, depending on the season and extent of migration. This giant fish was once very frequent in the Danube, and Schrank reported that in 1692 the species had even penetrated as far as Bavaria. Olah noted that in 1763 sturgeon were caught in such great numbers in Hungary that fishermen could realize no profit from their catch. Today this species is found only in the lower Danube and around the mouth of the Danube; as soon as the ice melts in the Danube the species swims upcurrent for spawning, usually in the vicinity of the Iron Gate in Romania. In the Soviet Union attempts are being made to preserve sturgeon populations by protective measures and artificial breeding; this is because they are the sources of the famous beluga caviar which is of great economic importance. The second sturgeon species of the *Huso* genus is the KALUGA ($\lozenge$ *Huso dauricus*) which inhabits the Amur River region from the Haff to the source rivers Shilka and Argun.

Sturgeons of genus *Acipenser* are distributed in Europe, Asia and North America. Of the seventeen species five are found in the Black Sea, Sea of Asov and the Caspian Sea, and these are all well-known caviar-producing species; they include the COMMON ATLANTIC STURGEON, STERLET, and others. Two other species inhabit the far eastern regions of the Soviet Union, the SIBERIAN STURGEON and the AMUR STURGEON. Five species are found on the Asiatic Pacific coast, four in North America and one in the Adriatic Sea.

One of the largest members of the genus is the COMMON ATLANTIC STURGEON *(Acipenser sturio)*, in which males attain a L of 2 m and females reach 6 m with a weight exceeding 200 kg. This species was once found in great numbers on the European coasts from the arctic through the Mediterranean into the Black Sea, as well as in the Baltic Sea, and in Lakes Onega and Ladoga. Predation, water pollution and construction of water works have almost led to the disappearance of this grand species in western and central Europe. Now only a few larger rivers (the Elbe in Germany, the Gironde in western France, and the Guadalquivir in southern Spain) are sites of sturgeon migrations in the spring, as they seek spawning grounds which must be protected. In the last fifty years few sturgeon have been sighted even off the mouth of the Danube, and large populations can be found only in the Russian tributaries to the Black Sea. With protection these groups could be maintained.

The STERLET *(Acipenser ruthenus)* was once found in southern Germany. The species can easily be recognized by the pointed, narrow snout which arches slightly upward, by the fringed barbs and the small lateral plates. It is distributed in rivers which empty into the Black Sea and the Caspian Sea and in the delta areas of those rivers. A subspecies inhabits Siberia. The sterlet is almost exclusively a fresh water species; only in the northern part of the Caspian Sea does it enter brackish water. In earlier times the sterlet migrated up the Danube past Ulm, Germany, but construction of dams has ended that, and today sterlets are a great rarity in the Bavarian Danube. In each of the years 1932, 1953, 1957 and 1962 one sterlet was caught near Passau, Germany; the last one was 45 cm long and weighed 1.25 kg. Sterlets in the lower Danube and in the southern Soviet Union, where larger populations exist, attain lengths of one meter and weigh 6–7 kg. Sterlets can be raised in ponds, and this is indeed done in Hungary, but such fishes are infertile. For breeding purposes wild specimens have to be caught from the Danube and its tributaries. Along the Volga attempts are being made to maintain the species through artificial breeding and by raising the young in incubators.

Fig. 6-3. Skeleton of the heterocercal caudal fin of lake sturgeon (*Acipenser fulvescens*).

The ADRIATIC STURGEON (*Acipenser naccari*, L 1.5 m) is less known. The species can be recognized by long barbs located near the tip of the snout. In the spring the species ascends the Po, Etsch and other rivers of the northern Italian lowlands for spawning.

Occasionally another sturgeon species, *Acipenser stellatus*, is observed in the Adriatic near Zadar in Yugoslavia. It is chiefly distributed in the Black Sea and Sea of Asov as well as the northern Caspian Sea. For spawning it ascends the tributary rivers, and in earlier times penetrated the Danube as far as Tokay and Komarom in Hungary. Today the species is seen only rarely in these regions and principally spawns at the mouth of the Danube. The species can be readily recognized by its long, daggerlike snout and the lateral plates which are separate from each other. Females attain a L of 2.20 m and a weight of 68 kg, but most of the sturgeon caught of this species do not exceed 12 kg. In southern Russia *A. stellatus* is second only to *A. gueldenstaedti* (see below) in terms of the amount caught.

The sturgeon *Acipenser gueldenstaedti* is also called the "RUSSIAN STURGEON" owing to its tremendous economic importance. It has a wide, short snout and lateral plates separated from each other, between which the lateral line is visible. The species ascends from the Black Sea, Sea of Asov and Caspian Sea during the spawning period into the rivers emptying into these bodies; in the Danube migration extends to Bratislava. Migration takes place up the larger tributaries of the above rivers. Prime specimens have a L exceeding 4 m and a weight of 160 kg. But today most are only 1.30–2.50 m long and weigh 20–30 kg. Three subspecies have been identified: *Acipenser gueldenstaedti gueldenstaedti* of the northern Caspian Sea, *Acipenser gueldenstaedti persicus* in the southern Caspian, and *Acipenser gueldenstaedti colchicus* of the Black Sea. The last species is still found frequently in the lower Danube.

Acipenser nudiventris is distinguished from other sturgeon species through the fact that its lower lip is not divided. This species is especially prevalent in the Aral Sea, in which it is the only sturgeon species. In the southern Caspian Sea and in the Black Sea, it is found along with the sterlet, *A. stellatus*, and *A. gueldenstaedti*. Although the species usually migrates from the sea into the rivers, those in the Danube remain there at all times. In spring these populations migrate up the Danube, as does the sterlet, and spawn in the tributary rivers Waag, Theiss, Maros, Drau and others, several days after the sterlet. The largest specimens yet caught were 2 m long and weighed 40–50 kg. They are less important economically than other sturgeon species in the southern Soviet Union.

The two species found in the far eastern Soviet Union, the SIBERIAN STURGEON (*Acipenser baeri*; L exceeds 2 m, weight up to 200 kg) which inhabits Siberian rivers from the Ob to the Kolyma, and the AMUR

STURGEON (⚥ *Acipenser schrencki*; L up to 2.9 m, weight 80–160 kg, rarely 200 kg), are both migrating species. They resemble each other structurally; the Amur sturgeon is somewhat more slender. In some large lakes and rivers varieties have developed which do not migrate into the sea any longer. An example of this is the Siberian sturgeon population in Lake Baykal.

Five other species are found on the Asiatic Pacific coast besides the Amur sturgeon: *Acipenser sinensis* and *A. dabryanus* from the lower Hwang-ho and Yangtze-Kiang in China, *A. kikuchii* and *A. multiskutatus* from the provinces of Sagami and Iwaki in Japan, respectively, and above all the sturgeon *Acipenser medirostris* (L 2.1 m, weight 159 kg) which also appears as a migrant along the North American west coast and which is closely related to the common Atlantic sturgeon. In North America the species ascends rivers in late fall from Monterey to the Columbia River, where it winters and spawns in the following summer, generally in July. This species has also become very rare.

The WHITE STURGEON *(Acipenser transmontanus)* of the North American west coast is somewhat more prevalent than the other two above species. The largest fresh water fish in North America, this species can be recognized by a blunter snout and brighter coloration. According to a report from 1897 which can no longer be verified, a female weighing 820 kg was caught in British Columbia. The second-heaviest female was caught by A. B. Chapman near Vancouver and weighed 583 kg with a L of 3.75 m. Today captured white sturgeon weigh only around 130 kg.

Two species are found along the North American Atlantic coast: the ATLANTIC STURGEON (⚥ *Acipenser oxyrhynchus*; L exceeds 4 m), which was earlier classed together with common Atlantic sturgeon as a single species; and ⚥ *Acipenser brevirostris* (L reaches 2.1 m), which is distinguished by a short, blunt snout which comprises only one-fourth of the length of the head. Neither species has great commercial value, since present day populations have been decreased greatly.

The fifth North American sturgeon species, the LAKE STURGEON (⚥ *Acipenser fulvescens*; L 2.4 m, weigh 135 kg), inhabits inland waters exclusively, those being the drainage area of the upper Mississippi, the Great Lakes (with maximum prevalence in Lake Erie and minimum in Lake Superior) and the drainage regions of Saskatchewan and Hudson's Bay. The species feeds chiefly on fresh water snails found in shallows, but it is not highly particular and in the vicinity of grain mills feeds on corn and rye grains. From May to the beginning of June the species ascends rivers for spawning, in areas of fast current and rocky areas near the shore. This species is another which has suffered a great decline and has almost disappeared in the St. Lawrence River. However newly introduced protective measures have already met with some

▷
Above: the bichir *(Polypterus ornatipinnis)* has the typical polypterid feature of division of the dorsal fin into a number of distinct finlike elements. Below: a sturgeon *(Acipenser stellatus)*

success. The rapid increase in lampreys presents a new threat to the lake sturgeon, for the lampreys inflict large wounds in the sturgeon.

The SHOVEL-NOSED STURGEON (subfamily Scaphirhynchinae) comprise two genera: 1. AMERICAN SHOVEL-NOSED STURGEON *(Scaphirhynchus),* in which the tail shaft is greatly elongated and has bony plates; there are three species; 2. ASIATIC SHOVEL-NOSED STURGEON *(Pseudoscaphirhynchus),* with a short tail shaft that is not fully covered with plates; three species are in this genus.

The SHOVEL-NOSED STURGEON *(Scaphirhynchus platorhynchus)* is distributed in the Mississippi delta and its tributaries. The species attains a L of 1.5 m but today specimens are rarely caught more than 90 cm long and heavier than 2.7 kg. Spawning migrations take place upcurrent from April into June. Food consists chiefly of invertebrates, the typical sturgeon diet. The shovel-nosed sturgeon *Scaphirhynchus albus* generally lacks the small bony plates on the belly. The species is found in the upper Mississippi and its tributaries, but prefers fast flowing current and thus predominates in the lower Mississippi. Only one *S. albus* is encountered for every 300 *S. platorhynchus.* Still rarer is *S. mexicanus,* which inhabits rivers emptying into the Gulf of Mexico.

The ASIATIC SHOVEL-NOSED STURGEON *(Pseudoscaphirhynchus)* are all found in the Aral Sea delta. Two species inhabit the Amu-Darya River: *Pseudoscaphirhynchus kaufmanni* (L 75 cm, weight 2 kg) and the rarer *P. hermanni* (L 27 cm). The upper lobe of the caudal fin in the larger species terminates in a fibrous structure. The smaller species may be distinguished from the larger by the lack of that structure as well as by a longer snout. *P. kaufmanni* reaches sexual maturity after six or seven years, and then from April to May spawning migration takes place toward an upcurrent location with a water temperature of about 16° C. The diet consists chiefly of fish, particularly loaches and young barbels. As the species is commercially valuable, its numbers have dwindled greatly. The natural history of *P. hermanni* has been studied very little. Due to its extreme scarcity it has no commercial value.

The same is true for the third Asiatic shovel-nosed sturgeon, *P. fedtschenkoi,* which inhabits the bed of the Syr-Darya River in the plains. Including the tail fiber the species is 36 cm long, and without it 27 cm. The large number of plates on its back distinguishes this species from the above two. Long-snouted and short-snouted specimens are known. A report by C. Grevé in 1896 gives the following account: Captain Borshchevsky traveled to Turkestan in 1876 on the Syr-Darya in the vicinity of the city Ak-metchet (white mosque) and there he saw a group of fishermen striking something with sticks to drive it into the river. When asked regarding their peculiar behavior the fishermen replied that they had gotten a net full of "shai-tan-dum-balyk" (devil's tail fishes) and it was fortunate that they were able to chase the horrible creatures back into the water. From their description Borshchevsky

◁

Polypterids: 1. Bichir *(Polypterus senegalus);* 2. Reedfish *(Calamoichthys calabaricus).*

concluded that these were probably *Scaphirhynchus,* for the fishermen distinguished two varieties, "usun" (long-snouted) and a blunt-snouted variety. All attempts to overcome the superstition of the fishermen, who considered even the smallest specimens dangerous, were unsuccessful in getting them to preserve just a few of the fishes.

Members of the second sturgeon family, the PADDLEFISHES (Polyodontidae), have existed since the Upper Cretaceous period about 80 million years ago. The skin is naked or has small bony platelets; a few plates are on the upper lobe of the caudal fin. The mouth is wide, and young have numerous small teeth. A single dermal gill ray is on either side of the head. The snout has developed into a thin, pliable, spoon- or sword-shaped flattened structure which can comprise one-third of the total L. Two very small barbs are located on the lower side of this structure. The eyes are relatively small, and spiracles are present. The tail is heterocercal. There are only two genera, each with one species.

The AMERICAN PADDLEFISH *(Polyodon spathula)* has a gill cover, the rear edge of which has a long lobe; when laid back, this lobe extends to the pectoral fins. Gill spines are long, very thin and numerous. The mouth cannot be protruded forward. The snout projection is flattened, spoon-shaped, and is two and one-half to four times as long as it is wide. In earlier times the L reached 1.83 m and the weight 76 kg. Originally distribution was throughout almost the entire Mississippi drainage area, from eastern Montana to Pennsylvania and New York, and south to North Carolina, Mississippi, Louisiana and Texas. The species is seen only rarely in the Great Lakes region and today has probably disappeared from there.

The body shape of this highly distinctive species is somewhat reminiscent of a shark, particularly the canted caudal fin. It is thus no surprise that in 1792 Walbaum described the paddlefish as a new shark species and that in 1820 Rafinesque wrote an extensive description of it as an "entirely new shark genus". Paddlefish predominantly inhabit large rivers, lakes and stagnant water, when these waters have a muddy sandy base and rich zoo- and phytoplankton supply. Paddlefishes generally are far from shore. The species travels great distances, but migrates upcurrent only during low water level and therefore an unobtrusive current. Even small dams are therefore an insurmountable obstacle for this fish. Dams, water pollution, predation and other causes have threatened populations north of the Mississippi with total extinction.

The species feeds on small crustaceans and other small floating organisms, which can be filtered from the water with the help of the long gill spines. The paddlefish swims with its mouth open close to the water surface as soon as plankton (which are light-seeking organ-

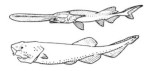

Fig. 6-4. Paddlefish young (above) and larva (below).

isms) have collected there. Kofoed called the paddlefish a "living plankton net" since with its flattened snout it acts much like a trawl.

Sexual maturity generally is reached in the seventh or eighth year, at which time the fish weighs nine to thirteen kilograms. The fish then gather in small schools and search for spawning sites. In the southern part of their distribution (e.g., in Louisiana), they spawn in February and March, but in the north not until the beginning of May. Spawning takes place both in the river bed and in shallow spots in lakes on sandy or rocky ground. Nothing is known of the actual spawning process. Pre-spawning behavior is occasionally seen by fishermen and is occasioned by jumping and splashing about on the river shore. The larval forms were not even known for a long time and were not discovered until 14 May 1932 when Thompson found young measuring 17–20 mm on a sandbank on the Mississippi in Grand Towers, Illinois. Originally he did not consider them paddlefishes. But a closer examination showed that they were indeed the pale, almost transparent larvae of paddlefishes, from which only the long snout portion was lacking. A single small swelling is on the nose, under which two strong barbs are located.

Into the 1920s in the United States paddlefish were caught at up to 1000 tons annually. The meat was eaten; the eggs alone or with sturgeon roe made excellent caviar. At present the catch has little commercial value; predation has also contributed to this great decline. In light of the dwindling numbers of this species it is necessary to take steps to prevent the complete disappearance of this notable primitive fish.

The CHINESE STURGEON *(Psephurus gladius)* may be distinguished from the American paddlefish by the relatively narrow and sword-like snout elongation. The gill cover lobes, laid back, extend to only half of the pectoral fin apparatus. Gill spines are short, strong, and less numerous. The mouth can be protruded. This giant colored species was reported by Ping in 1931 to reach a L of 7 m, the same as for the American species. The Chinese sturgeon inhabits the Yangtze-Kiang in the Chinese lowlands and feeds on other fishes. In China its meat is considered to be a delicacy. Even though the species is rare, it does have measurable (if modest) commercial value. Its natural history is practically uninvestigated.

7 Gars and Bowfins

The superorder LOWER BONY FISHES (Holostei) comprises the order of gars and that of bowfins. They are the latest descendents of fish groups from early geologic periods which in those times were widely distributed, well speciated, and quite prevalent. Today each of the two orders has but a single family, the gars with about ten species and the bowfins with a single species.

Superorder: lower bony fishes, by F. Terofal

The GARS (order Lepisosteiformes, the single family Lepisosteidae) have an elongated, pike-shaped body with an armored covering with close-set diamond-shaped scales. These scales do not overlap, but are merely closely set and are linked with each other. A layer covering them is as hard as porcelain and has the sheen of polished ivory. The middle ganoid layer is lacking. The head is protected by bony plates. The jaw varies in the different species and may protrude like a beak to a greater or lesser degree. With its long, sharp teeth, the jaw looks like the snout of an alligator. The dorsal and anal fins are supported only by fin rays and are located well to the rear; they are near the heterocercal caudal fin as in the pike. Pelvic fins are in the middle of the body and the pectoral fins are just behind the gill cover. The skeleton is bony but still contains many cartilaginous elements. Vertebrae are connected with each other by joints and are very similar to reptilian vertebrae. The front joint surface of each vertebrae faces outward and the rear surface faces inward (this is unique among all fishes). A well-developed joint structure enables nodding movements of the head. Spiracles are absent. The intestinal tract has a curved spiral valve. The swim bladder is lunglike and cellular with a special air passage to the foregut; the bladder serves in assisting respiration. There are about ten species in North and Central America as well as Cuba.

Order: gars

Distinguishing characteristics

Gars are primarily fresh-water inhabitants, but some migrate into brackish waters of river mouths and lagoons (as in some Guatemalan varieties). The species are distinguished by the length and form of the jaw and teeth as well as the shape and coloration of the body. Basic

▷
Sturgeon: 1. Common Atlantic sturgeon (*Acipenser sturio*); 2. Sterlet (*Acipenser ruthenus*).

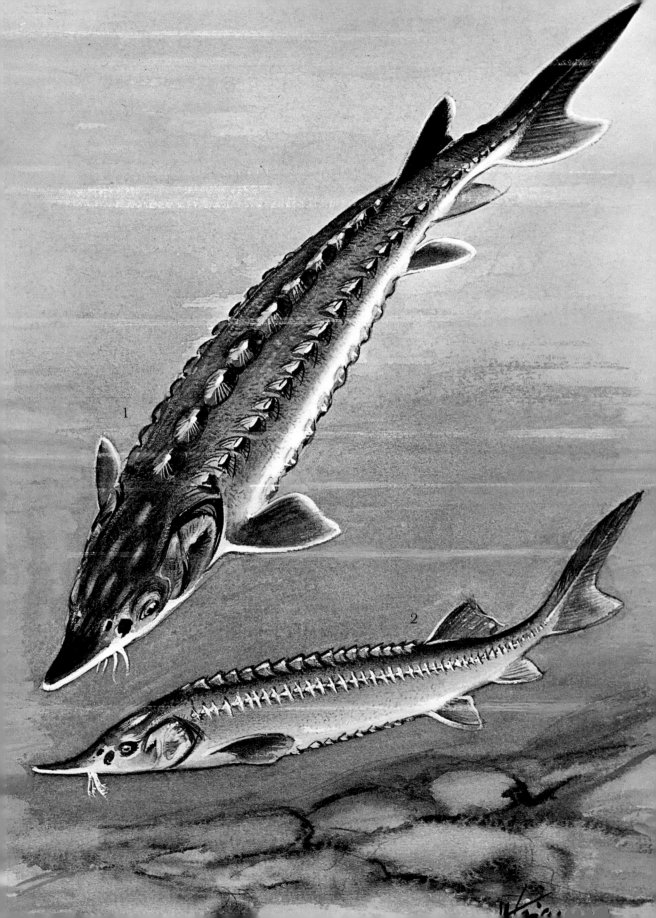

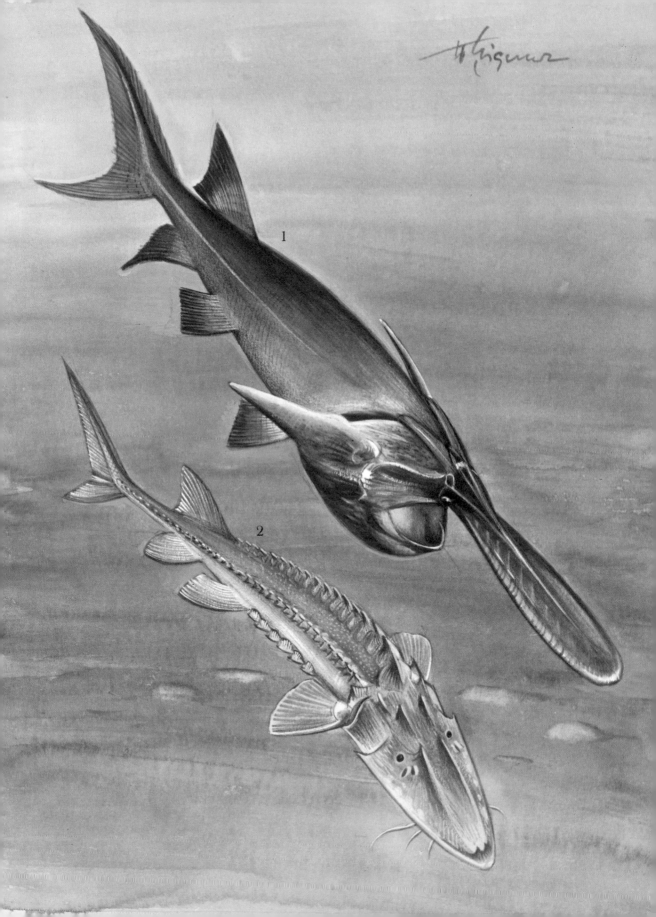

Fig. 7-1. Spatial relationship between swim bladder and gut in bowfins (left) and gars (right).

Fig. 7-2. Gar (Lepisosteiformes).

◁
Sturgeon-like species:
1. American paddlefish (*Polyodon spathula*);
2. Shovel-nosed sturgeon (*Scaphirhynchus platorhynchus*).

coloration is uniform: the back is more or less olive-green, and the bellyside has a bright silver sheen. Furthermore the body and fins have various spotted markings among the different species.

The ALLIGATOR GAR *(Lepisosteus spatula)* has been called the shark of fresh water. It is distributed in the rivers drained by the Gulf of Mexico, i.e., in the southern U.S.A. and in Mexico, but also in Cuba. In the Mississippi the species penetrates as far north as St. Louis. It can be recognized by its short, wide, blunt snout and by the two large rows of teeth in the upper jaw. As one of the largest fresh-water fishes of North America, the gar has an average L of 3 m, rarely 4 m and more. The largest specimen ever caught in the United States weighed 137 kg. Since this species often tears up fish nets and preys on other fishes, while they themselves are unsuitable for consumption, they are fished intensively by professional and sports fishermen. In spite of that, they are still quite prevalent in the rivers around the Gulf of Mexico.

Lepisosteus tropicus, a close relative, is the southernmost gar. It has been found in the Usumacinta basin on the east coast of southern Mexico and in the rivers flowing into the Pacific, from Chiapas, Mexico, to Costa Rica. In Costa Rica, the species inhabits rivers which empty into Lake Nicaragua. The species is distinguished from the above relative by the number of rows of plates.

The LONGNOSE GAR *(Lepisoteus osseus)* has the widest distribution, extending from Minnesota to Vermont and south to the Gulf of Mexico and the Rio Grande. The species has a L of 1.8 m and weighs as much as 7.5 kg. It can be recognized by the long, narrow jaw; the snout is three times as long as the head behind the eyes, while in other gars the snout is no more than twice as long as the remaining head element. The large teeth are in one row in the upper jaw. The body is more elongated than in other species. In contrast to the other species, the long-nose gar is also found in clear rivers, and thus does not apparently avoid fast-flowing water. In aquariums specimens reach an age of twenty years, but they have not been bred in captivity.

The gar *Lepisosteus oculatus* (L up to 1.2 m) has a striking spotted pattern on the upper side of the head, on the fins, and on the rear part of the body. The snout is short and wide. Distribution is in shallow, weedy waters from Iowa and Nebraska to the Gulf of Mexico. In Florida and southern Georgia, another species appears, *Lepisosteus platyrhincus* (L 75 cm) which has a still wider and shorter snout.

The gar *Lepisosteus platostomus* (L 90–120 cm) which is distributed in the Great Lakes and the Mississippi basin to the Rio Grande, is very similar to *Lepisosteus oculatus,* with the difference that the spotted markings on the upper side of the head are missing. Furthermore, this species has a greater number of rows of scales.

The best-known of all these species in the United States is the long-nosed gar, and its life is the best understood. Typical of all gars, this species is solitary and prefers quiet, shallow water offering good cover. It lies in wait in the underwater growths or under sunken tree trunks just as pike does; only the slow movements of the pectoral fins divulge the fact that this is a living creature. Its large eyes scan the surroundings attentively. If the prey is detected, the gar begins to creep up on it by slow undulating movements of the pectoral fins and the tip of the tail until the prey is at the side of the middle of its beak. Then the gar snaps in a way identical to that of an alligator with a fast sideward movement of the head and holds the prey with its needlelike teeth. The prey is turned about until it lies longitudinally in the jaw and with a short forward jerk of the head is swallowed. The floor of the mouth is flexible and the jaw arches can be bent to such a degree that even a large prey fish with a high back can be swallowed. The lower part of the head then resembles the sack of a pelican that has just swallowed a fish.

During the summer months, as soon as the oxygen content of the water has dropped considerably, the long-nosed gars can often be observed snapping for air at the water surface. In oxygen-poor water the gar can surface as often as six times in ten minutes, during which time it turns somewhat on its side and inflates an air bladder through its gill slits, with a gurgling sound. Then it swallows a large quantity of air with its mouth, whereby the jaws stretch far out of the water, and presses the air into the swim bladder, the wall of which has a rich blood supply and a honeycomb appearance. During the cold season, i.e., from October to April, gars seek greater depths in order to rest for the winter. At this time, they lie almost motionless on the floor, do not surface for respiration and cease feeding.

In late spring, from mid-May to early June (or in warmer waters as early as late April), the long-nosed gars move out of their wintering quarters and seek shallow shore locations with rich growths for spawning. Several males generally accompany a single female. After a vigorous display, the female releases the greenish eggs (about three millimeters in size) in groups, which are fertilized by the males. After fertilization, the eggs stick to the floor and on aquatic plants. The young hatch after ten to fourteen days, depending on the ambient water temperature. They bear no resemblance to the parents and thus are considered larvae. They are about 7 mm long, have a large yolk sac which makes movement away impossible, and an adhering disc in front of the mouth. The disc has numerous small warts and is used to adhere to plants until the greater part of the yolk supply has been expended. External gills are lacking in the larval stage. Parents apparently do not care for the young. After a week, the larvae are about 9 mm long and have already eaten part of the yolk supply. The body has grown longer; the upper jaw appears elongated in a trunklike manner, and the vertebral column extends like a fiber over the caudal fin. After about fourteen days the young gar can swim readily and has the appearance of an adult.

In some regions, e.g., around the Great Lakes or in the Florida swamps, gars are often so numerous that they become overpopulated and must be cropped in order to carry on any commercial fishing.

Order: bowfins

Of the six bowfin families (Amiiformes) which were most prevalent in the Jurassic and Cretaceous periods, only a single species remains today; the BOWFIN *(Amia calva)*. It has a bony skeleton and externally has the appearance of a "true" bony fish. However, several primitive features are still present in the species which determine its classification. Distribution is in standing or slowly flowing water east of the Rocky Mountains to the western edge of the Appalachians—from Minnesota, the Great Lakes, and Vermont in the north to Texas and Florida in the south. On the eastern edge of the Appalachians it penetrated to the Susquehanna River in Pennsylvania. Shallow lake and river shores with rich growths are preferred.

Distinguishing characteristics

The gar-like elongated body, which is laterally only slightly flattened, has relatively small, thin round scales with a thin ganoin layer. The heavy bony plates of the head also have a covering. The dorsal fin is long, spineless, and extends from the front of the back to just before the caudal fin attachment to the body. Each of its forty-two to fifty-three rays can be moved by special muscles. The caudal fin is rounded and appears to be symmetrical but is actually heterocercal. The anal fin is small and located well to the rear. Pelvic fins are approximately in the middle of the body and the pectoral fins are rounded and relatively small. The head is pointed and narrows somewhat from the upper to lower side, with a rounded short snout. The mouth opening is broad and almost perfectly horizontal; it extends to beneath the eyes. A large bony plate beneath both lower jaw bones protects the throat. Intermediate jaw bones are lacking. The intestinal spiral valve is not fully developed. An air passage leads from the esophagus to the chambered, lunglike swim bladder, which functions (primarily in summer) as a respiratory organ. Males have a round, black spot on the upper part of the caudal fin root, the spot being surrounded by an orange or yellow border. This spot, when present at all in females, is relatively small and subdued, without a border. Males are generally smaller than females and L reaches at most 55 cm; female L (based on catches near Montreal) reach 87 cm and weight up to 7.8 kg. Life expectancy in nature is about twelve years, while in captivity twenty-four to thirty years of age have been attained.

Since bowfins lead a secretive life outside the spawning season and hunt for food chiefly at evening and night, little is known of their natural history. Invertebrates are a chief part of the diet for young

Fig. 7-3. Bowfin (Amiiformes).

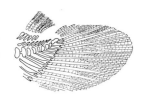

Fig. 7-4. Short, heterocercal caudal fin of a bowfin.

bowfins. Feeding on other fish increases with advancing age of the bowfin. In winter bowfins generally seek deeper water where, packed between aquatic plants in large groups, they rest for the winter.

As soon as water in shallow spots along the shore has reached a temperature of sixteen to nineteen degrees Centigrade (usually beginning in mid-April), the bowfins again appear in small groups in the quiet, weedy homeland waters. Males at this time have the most brilliant display coloration; all colors are more saturated, the belly being yellow to orange and the body sides bronze and green with a netlike marble marking, over which an orange sheen lies. The back is olivegreen. The fins are as bright green as young aquatic plants. Female coloration at this time is more subdued and darker, the bronze coloration tends to reddish; their fins are never green, but have a brownishred color.

Approximately three males for every female seek spawning grounds, and the males often arrive before the females. Females can only be seen during spawning. They wait in plant stands in deeper water until the males have completed nest-building. Females again disappear after completion of spawning. Males seek out nesting places and prefer locations thirty to sixty centimeters deep which are not too overgrown. After removing what plant life is present at the site, the males construct a nest out of roots, from which they form a thick mat. When possible the nest is protected by overhanging twigs and on one side by tree roots or a sunken tree trunk, so that the male has free access on the opposite side to the open water. This affords the male quick access to the outside in case of approaching danger. Since sites such as these are not so frequently available, the case is usually that a great number of nests are built next to each other, and that a kind of breeding colony as in birds develops. Reighard found seven nests close to each other in a six-by-nine-meter site in an inlet of the Huron River in Michigan. The nest itself is a ten-to-twenty-centimeterdeep saucer-shaped structure with a diameter of thirty to ninety centimeters, the floor of which has a fine covering or roots.

After a short time one of the females appears and cautiously approaches the nest which is guarded by the male and protected from neighboring males. As soon as the female is close enough, a wild circling motion begins. The male approaches its partner from the front, opens its mouth and protectively seizes the female's snout as if the male wanted to kiss the female. The male nudges and pokes the female lightly in the side and attempts to nestle the female by head to head contact. If the female is not yet prepared to spawn, it turns about rapidly and the chasing begins all over again. At intervals the fish swim away from the nest site or snap air from the water surface. At the conclusion of this pre-spawning behavior, which may last for hours, the female lies on the floor of the nest waving its fins lightly. The male,

Fig. 7–5. Bowfin male at the nest.

twitching its entire body, nestles the female's side with its head just behind the pectoral fins of the female. After about fifteen to twenty seconds the fishes separate, and the first group of eggs is laid on the floor and edges of the nest. Two females may spawn in the same nest, one after the other, and alternate between different males. More detailed descriptions of spawning behavior are not available, since this generally takes place at night and has only been observed by day on a few occasions.

After the females have once again disappeared, the males guard the nests. The male supplies the strikingly bright eggs with fresh water, keeps them free of mud, and drives off approaching sunfishes, which feed on bowfin eggs. The colorless larvae, seven millimeters long, hatch after about eight to ten days. The tip of their snouts has an adhering disc with numerous small warts, with which they can attach to and suck from aquatic plants. The larvae can swim when they have attained a length of nine millimeters. At this time, the body coloration is deep black, the adhering structure has degenerated, and the yolk supply has been two-thirds consumed. They form a dense school, which looks like a black cloud hovering in the nest. They feed on planktonic crustaceans when they have reached a length of eleven millimeters. The male remains constantly in their vicinity. The male drives off conspecifics which seek to feed on these young; the male strikes at these intruders with a powerful blow of the tail, ramming them in the side and shoving them away from the nest.

After approximately nine days the young have reached a length of about twelve millimeters and have become sufficiently developed to leave the nest, but they do not move far from it. They usually stay together in a thick school and remain under the father or in its shadow, and are led by the father through the water. If the male is driven off, the school of young does not disintegrate, but circles about collectively until the father is found again. If the father does not return, the young will be set upon by other fishes, particularly by sunfishes, and they will be destroyed unless they can become incorporated with the young of another male and thus gain a new source of protection.

Fig. 7-6. Bowfin male "walking" its young.

counts in a two-meter-long female weighing 63.9 kg indicate that this is a very fertile species. Nichols estimated the number of eggs to be no less than 12,201,984. On the basis of catches of specimens that have not yet spawned or that have just spawned, it is believed that the spawning season on the west coast of Florida extends from May to September. Preferred spawning sites are apparently along the coast and around the islands, generally in shallow water. Very young fishes have been caught along the coast of Texas, Alabama and Florida, as well as Puerto Rico, Haiti, Cuba and Trinidad. This indicates that the breeding area is rather extensive. Early tarpon larval stages have not yet been observed, either. Older larvae resemble those of eels as in the lady fish, bonefish and ox-eye herring.

Metamorphosis apparently takes place close to land and from there the young migrate and fall into the rivers. They make their way into brackish mangrove marshes along the lower courses of the rivers via flooded areas. Here they have a rich supply of crustaceans and young fishes and growth is correspondingly rapid during this period. Small tarpon thirty to forty-five centimeters long have been found in great numbers in the salt-and-fresh-water arms of the rivers. Later, when they have grown more, they move into the main rivers.

Older tarpon apparently do not undertake extensive migrations. They remain near the coast and frequently ascend rivers, and from there perhaps into Lake Nicaragua. Tarpon are found almost throughout the year in tropical American waters, but in the northern and southern reaches of their distribution they are found only during the warmer months. The species is very sensitive to cool weather. Cold waves killed great numbers of tarpon along Florida in 1885, 1894/95, 1905 and 1935.

Tarpon do not form schools. The fact that many are encountered together in the canals off the dams in the Panama Canal is probably due to the rich nutrient sources in this area. Occasionally great numbers of tarpon gather in pursuit of fishes, which they follow with great vigor. Their streamlined form is useful in this regard. Tarpon often carry out rolling motions on the water surface, a phenomenon well known to fishermen. Babcock's investigations on the lung-like structure of the swim bladder have clarified this behavior; presumably the rolling motion is accompanied by air intake.

The giant leaps out of the water are another tarpon peculiarity. The species can jump vertically two and one-half to three meters out of the water and can execute a horizontal leap of six meters. This is perhaps an attempt to escape predators and other irritants; but the jumping may also be just for fun. The leaps are made when the tarpon is hooked, possibly to free itself. Initial velocity is built up by striking with the tail against the water shortly before leaping out. The gill covers are opened so wide that the red gills are visible. The body bends in the air, and the fish falls back into the water on the side that is

Fig. 8-5. Tarpon leaping.

bent inward. Occasionally accidents have been caused by these leaps, in which people have been injured. Babcock relates a case in which a tarpon "jumped on a man sitting on a chair on the deck of a steamer; another knocked a ship's pilot overboard...the man being dazed, he drowned. In Galveston Bay, a leaping tarpon broke the neck of someone sitting in a boat."

The tarpon's diet consists of fishes, but also cephalopods. Sheat-fish and silversides have also been found in tarpon stomachs, and they have been seen following graylings, anchovies and other fishes. Young tarpon five to twenty centimeters long from a lagoon near Port-au-Prince, Haiti, were found with water ticks in their stomachs. The natural enemies of tarpon are sharks, but they are also apparently attacked by dolphins. Smaller tarpon are most certainly preyed upon by other fishes. The best defense which the tarpon has against its enemies is its rapid speed.

The tarpon is not valued for eating in North America, with the exception of the inhabitants of Panama and the West Indies settlers. Large specimens are said to be tough and full of bones. In contrast there is great demand for tarpon in South America. It can be found fresh, smoked and salted in the markets. Smaller tarpon are also eaten fresh in Africa. Fishermen dislike tarpon because of injuries which they occasionally inflict on nets. The only tarpon feature universally in demand is the large silvery scales which can have a diameter of five to eight centimeters. Jewelry can be fashioned from them and offered for sale in tourist souvenir shops for a price of from five to twenty-five cents.

But as a sport fish the tarpon is bettered by no other species. Its fame is known throughout the world. A great deal has been written about catching this fish, a task requiring strength, skill and endurance. Sport fishermen pursue small and large tarpon alike, wherever they occur in the western Atlantic from Nova Scotia to Argentina or in the east Atlantic. The most famous fishing grounds are in the Florida Keys, on the west coast of Florida, in the Rio Panuca in Mexico, and the Rio Encatado in Cuba. In 1938, the world's largest hook-caught tarpon was landed in Rio Panuco; it weighed 112 kg. *Megalops cyprinoides* is also sought by sport fishermen, especially on the east coast of Africa.

Living tarpon have been displayed in the New York Aquarium at various times. S. F. Hildebrand mentions one specimen which in five years grew 122 cm from 50 cm. This slow growth rate was due to the conditions in captivity. On the basis of scale studies, it is presumed that tarpon reach a length of 30 cm in their first year and by their third year they are 127–152 cm long. Sexual maturity is apparently reached at a L of about 120 cm.

The bonefish suborder (Albuloidei) comprises the two families of bonefishes (see below). Earlier the lack of a throat plate was considered to be a characteristic which distinguished bonefish from the tarpon

Suborder: bonefishes

and the lady fish, but recent studies by Nybelin and Whitehead have shown that the throat plate is present in the bonefish. It is simply poorly developed and is not externally visible. A lateral line is present. The mouth is on the lower body side and is very short. The dorsal fin is in the middle of the body. A swim bladder is present. The body coloration is silvery.

The bonefish family Albulidae consists of the two genera *Albula* and *Dixonina*. In *Albula* the last ray of the dorsal and anal fins are not elongated, in contrast to *Dixonina*. The pointed snout in both genera protrudes beyond the lower jaw, which in *Albula* does not extend to the eye and in *Dixonina* extends to the middle of the eye.

The single *Albula* species is the BONEFISH *(Albula vulpes)*. L is about 90 cm; the body is tapered and little compressed. The eye is almost completely covered by a thick, transparent skin. Coloration is shiny silvery, with an olive-green back and sides. Distribution is in all tropical and temperate seas with the exception of the Mediterranean; the species is particularly prevalent in the tropics.

Spawning period and location of spawning grounds of the bonefish are unknown. As in lady fish and tarpon, the young resemble eel larvae when freshly hatched. The larvae begin shrinking when they have reached a length of about 7 mm and when they are some 2.8 mm long they have attained the adult form. Adult bonefishes generally are found in shallow water while the larvae and freshly metamorphosed young are in shallow bays and mouths of rivers.

With its sharp snout the bonefish digs its food out of the floor, and the diet consists chiefly of worms, mussels, small crustaceans and copepods. Occasionally smaller fishes are also taken, evidenced by stomach findings. Due to the number of bones, the species is eaten in just a few areas, among them the Bermudas and Africa. Like the tarpon, this fish is of great importance to the sport fishermen, especially in Florida, the Bahamas and along the African coast. The average weight of captured fishes is between 900 gm and 2.25 kg and the record catch was made in the Hawaiian Islands, the fish weighing 8.1 kg.

Two species of bonefishes have been briefly described, one from the tropical Atlantic and the other from the Pacific Ocean off the North and Central American coast. Hildebrand considers them one species, *Dixonina nemoptera*.

The DEEPSEA BONEFISHES (Pterothrissidae, genus *Pterothrissus*) are two species found in cooler water along the Japanese and West African coasts. On the basis of their skull and body structure, they are considered to be close relatives of bonefishes. Externally they may be distinguished from the bonefishes by their very long dorsal fin (with 55–65 rays), while the bonefish dorsal fin has but 15 rays. The Japanese species, *Pterothrissus gissu*, reaches a L of 40 cm. Its spawning period is in spring and the species has commercial value.

Distinguishing characteristics

Fig. 8–6. A bonefish (Albulidae).

▷
Gars: 1. Mississippi alligator gar *(Lepisosteus spatula)*; 2. Longnose gar *(Lepisosteus osseus)*; Bowfins: 3. Bowfin *(Amia calva)*.

9 Eels

Order: eels,
by C. H. Brandes

The eels (order Anguilliformes) can be immediately distinguished from almost all other fishes by their snakelike, elongated body, which usually is more or less cylindrical, but can be laterally compressed. Ventral fins are absent in all present-day species, and it is from this that the earlier order designation Apodes (footless) was derived. Greenwood and his co-workers differentiate two suborders: 1. Anguilloidei, with twenty-three families and about 360 species; 2. Saccopharyngoidei, with three families and nine species.

Distinguishing
characteristics

Eel fins do not have hardened rays. The dorsal and anal fins have become united to a large part with the caudal fin to form a continuous fin seam. A caudal fin is not always present. The skin is very slimy and generally naked. Small cycloid scales are found in only four eel families. The swim bladder is joined with the foregut by an open canal. The stomach has a blind sack and lacks pyloric appendages. The head varies in proportion to the body. Unusual head forms occur through extension or shortening of the jaws. The intermaxillary bones are not free, but joined with the middle and both lateral ethmoid bones, often also with the vomerine bones to form a single unit. The teeth are in rows or bands and are formed in various ways: they may be brushy or velvety, in some cases bottle-shaped, and are generally slightly arched. There are large fangs, flat chewing teeth, etc., and the gill cover bones are small and the gill openings very narrow. The shoulder joint does not rest on the skull. There are a large number of vertebrae, as many as 260. Vertebrae in deep-sea eel families Cyemidae and Serrivomeridae have regressed to form thin bony cylindrical structures.

◁
Herring: 1. Wolf herring (*Chirocentrus dorab*); Order Gonorynchiformes (Chapter 13): 2. Milkfish (*Chanos chanos*); 3. *Gonorynchus gonorhynchus*;
Tarpons: 4. Ox-eye herring (*Megalops cyprinoides*); 5. Lady fish (*Elops saurus*).

All eels are marine organisms, with the exception of the freshwater eels of Anguillidae, which spend most of their lives in fresh water; but even their propagation occurs in the sea. Many species undergo metamorphosis, in which the larval stage can vary in length from one to three years. The larvae are transparent and have a leptocephalus form. Almost all adult eels are predators and feed on fishes, crustaceans, snails, mussels and worms.

Of the twenty-three families of the suborder Anguilloidei only the most important can be treated here, in order to present the diversity of forms and natural histories of this group. Those that will be treated are the freshwater eels (Anguillidae), the eels (Moringuidae), Myrocongridae, morays (Muraenidae), pike eels (Muraenesocidae), conger eels (Congridae), garden eels (Heterocongridae), snake eels (Ophichthyidae), and a few deep-sea families.

Suborder: eels

The freshwater eels (Anguillidae; the sole genus is *Anguilla*) include sixteen species, all distributed along sea coasts and in the rivers which empty into those seas, with the exception of the west coast of America and the coasts of the south Atlantic. The best-known and most commercially important is the EUROPEAN EEL *(Anguilla anguilla)*. It is distributed in the north through Iceland and the White Sea, in the south as far as the Canary Islands. The species is caught in Europe, Eurasia and North Africa on the coasts and in most of the rivers, streams and lakes.

The European eel

In classical antiquity the freshwater eel was already considered to be a delicacy and was a necessary part of every banquet. However the species remained a mystery to fishermen and naturalists because nowhere nor at any season were eels mature enough for spawning caught. A sort of eel research began at this time, which became fruitful only 2300 years later, and the case is by no means yet closed. The Russian researcher Nikolski wrote:

Fig. 9-1. Spawning grounds and distribution of the American eel (*Anguilla rostrata,* 1) and the European eel (*Anguilla anguilla,* 2). Numbers refer to larval size in mm.

"Even Aristotle was interested in reproduction of eels. He assumed that eels were born from earthworms which were created in mud. Later researchers were of the opinion that eels came from the viviparous small fish *Zoarces,* which in German is still called *Aalmutter* (eelmother). In 1777 the Italian Mondini found eel eggs and this showed that eels reproduced by eggs just like other fishes. But because the male eel remained undiscovered for a long time the belief spread that eels were parthenogenetic. It was not until 1837 that Syrski, working in Trieste, found eels in river mouths with distinctive lipped organs which proved to be male organs. Thus it was established that eels reproduce in the typical manner."

In 1788 Gmelin described an unusual transparent fish species, which W. Morris had captured fifteen years previously. In honor of Morris and on the basis of the thin head this species was named "*Leptocephalus morrisii*" (from the Greek τεπγοs = thin and χεοατη = head). In 1864 Gill realized that this "new species" from the English coast was nothing more than the larval form of the eel genus *Conger*. His opinion was confirmed in 1886 by the Frenchman Yves Delage in Roscoff. He succeeded in keeping a *Leptocephalus* in an aquarium for seven months. During this time he was able to follow the metamorphosis of the larva

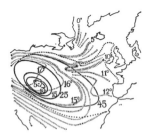

Fig. 9-2. Eel larvae (Anguilloidei, after Schmidt) indicating size in mm and annual temperature isotherm.

into the adult eel. In the years 1893 to 1896 two Italians, Grassi and Calandruccio, proved on the basis of aquarium observations, that the species from the Straits of Messina, which Kaup in 1856 named *Leptocephalus brevirostris,* actually metamorphosed onto a transparent young European eel. Furthermore people had occasionally found eels in the Straits of Messina with degenerated intestinal canals, more greatly developed gonads, and enlarged eyes. Grassi and Calandruccio concluded from these facts that the eels which migrated out of the rivers into the sea in fall moved somewhere off the coast into deeper water in order to spawn there. The leptocephalus larvae, which hatched from the fertilized eggs, thus completed their development and metamorphosis in deep water. In spring, as the researchers believed, the eels would then return to the rivers.

"But eel spawning grounds still remained undiscovered," continued Nikolski. "In 1904 the Danish ichthyologist Johannes Schmidt began searching for eel spawning grounds. First Schmidt searched for leptocephalus larvae in the Mediterranean, but here he could not find specimens under sixty millimeters long. He continued his search in 1910 in the Atlantic. The large-scale leptocephalus larvae catches which were conducted under his guidance revealed that the further southwest one came the smaller the larvae were. Finally, in 1922 in the Sargasso Sea, very small freshly hatched larvae were found. In this way the spawning grounds of the European eel were determined."

Summarizing what is presently known about the natural history of the European eel yields the following picture: its spawning sites are in the Sargasso Sea between the Bermudas and the West Indies above 6000 m depths and probably at a depth of about 400 m. An annual isotherm (an imaginary line connecting places with the same average yearly temperatures) of seventeen degrees Centigrade borders this region and the area also has a high salinity. Spawning takes place from March to April, about one and one-half years after migration out of the European rivers. Thus these are enormous distances which must be traversed: the mouth of the Elbe River in Germany is about 3500 nautical miles from the Sargasso Sea. The eels die after spawning. The eggs probably float freely in the water. The youngest larvae which have been found were about 6 mm long and were in a depth of 100 to 300 m. The leptocephalus larvae have long, needlelike teeth with which they seize their food, which consists of plankton. Larvae are driven more or less passively by the Gulf Stream eastward. With increasing age their back becomes higher and they become longer, taking on the leptocephalus form. In October of the third year of life they have an average L of 7 cm and reach the Spanish and Irish coasts.

Now the metamorphosis into the transparent form takes place. The larvae lose both height and length. The original teeth are replaced by bottle-shaped teeth. These "glass eels" are observed in the further removed mouths of the Weser or Elbe rivers in Germany in April and May, and not in the Baltic Sea until July. Ascent up rivers usually occurs at night.

With the beginning of pigmentation formation the eels ascend the rivers further. Some of them remain at the river mouth in brackish water but the greater part often migrates close to the shore upcurrent into the smallest watercourses and overcomes all hindering obstacles. Growth during this period depends on the food supply. In the first years the diet is chiefly composed of insect larvae, small mussels, snails and worms. Larger eels already begin feeding on small fishes. During the stay in fresh water the back of the eel is olive-brown, the sides and belly yellow. [At this time they are called yellow eels.] After five years males cease growing. They remain in the lower courses of rivers or in coastal regions and with a weight of 140-170 gm reach a maximum L of 50 cm. The females migrate further upcurrent and become considerably larger (up to 1.5 m) and attain a weight of as much as 6 kg. After seven years in fresh water the small degenerate scales begin to grow.

During the day eels generally remain in cover or hide in the floor. If the water becomes colder, they move into deeper, frost-free water, dig into the bottom and undergo a kind of winter resting period. In German waters two eel forms can be distinguished which have developed because of different feeding patterns: "spike-headed eels" which feed on insect larvae, worms, mussels, and crustaceans; and "broad-headed eels" which prey on larger fishes.

After spending nine to fifteen years in fresh water eels alter their appearance. From yellow eels they become so-called blank eels or silver eels. The back becomes deep black, and the sides and belly a shimmering silver-white. The flesh is now very firm and fatty; the eels cease feeding at this point. The fish-preying "broad-headed eels" become transformed into "spike-headed eels" because their strong jaw muscles are no longer needed and undergo a certain amount of degeneration. The lips are thinner, the pectoral fins more pointed and the eyes get larger. The eels prepare themselves for the long migration to the spawning grounds. In late summer and fall at night the blank eels move into the sea from the rivers. They overcome all obstacles in this movement, so powerful is the migration urge. They even leave isolated inland waters and snake across fields wet with dew, in order to arrive at rivers which will lead them to the sea. The gonads are not yet fully developed. Presumably the fat stores are utilized metabolically for the long migration route.

European eels have a very great commercial importance. In 1966 in Europe about 17,000 tons of them were caught. To support the natural populations one-to-two-year-old yellow eels and glass eels are placed in closed inland waters. Glass eels are caught all along the English coast. The most famous fishing site is Epney at Severn in England.

American and Japanese eels,

A close relative is the AMERICAN EEL *(Anguilla rostrata)*. The European and American species are hardly different in terms of coloration, size relationship of the sexes to each other and in the size of the glass eel form. But the American eel has only 103–111 vertebrae while the European species has 110–119. The American species migrates into the sea for spawning in fall and returns in the spring as a glass eel. Since it does not live far from the spawning grounds, the larval period is one year shorter. Spawning grounds are somewhat different for both species. In the American eel the center of the spawning grounds is more to the southwest; these were discovered by Johannes Schmidt. The American eel has less commercial value than the European species. In 1966 the catch was about 2000 tons.

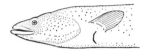

Fig. 9-3. Head of the yellow eel.

Of the other *Anguilla* only the JAPANESE EEL *(Anguilla japonica)* will be mentioned. Distributed along the western Pacific coast, it plays a significant role. In 1966 20,000 tons of Japanese eels were caught. Some authors consider all three species, the European, American and Japanese eels, to be three subspecies of the same species.

Other eel families

The eels (Moringuidae) have a scaleless body which is cylindrical and elongated like that of a worm. This family can be readily distinguished from all others by the anus which is located far to the rear. Dorsal and anal fins are generally composed of just a low seam on the tail. The middle of the dorsal and anal fins in *Moringua bicolor* is interrupted and the front portion seems like a stabilizing fin. In other species both fins are united with the caudal fin, which terminates in a projecting point. Moringuid eels have very small eyes and both upper and lower jaws protrude. The gill openings are narrow. The heart is located behind the gills. There are two genera (*Moringua* and *Stilbiscus*) with some twenty species distributed in tropical waters.

These eels inhabit loose sand, mud or fine gravely floors of the Atlantic, Indian and Pacific Oceans. They are found in coastal areas and often in brackish water. These eels can be seen only rarely, since during the day, they remain dug in the ground into which they can burrow very rapidly headfirst. Occasionally free-swimming Moringuid eels can be seen at night by having a source of light which attracts them. When sexually mature these eels become marine organisms; females alter their appearance greatly at this time. From yellow invisible "worms" with tiny eyes they are transformed into large-eyed, long-finned silver shimmering eels. W. A. Gosline has described such changes in *Moringua macrochir* from Hawaii; he notes that the male scarcely changes its appearance.

Fig. 9-4. Head of the blank eel.

The coloration change between the young and the adults is also a prominent one: for example, the young *Moringua javanica* widely distributed in the Indian and Pacific Oceans is bright orange when young, while older specimens are a hardly noticeable gray. Confusion has been caused systematically by this color change as well as by inconsistency in body proportions.

The eel *Stilbiscus edwardsi* from the western Atlantic often passes into fishermens' nets at night. Coloration above the lateral line is dark brown, and beneath it a bright blue-silver.

The Myrocongrid eels consist of one genus with a single species, *Myroconger compressus* (L about 56 cm) from the St. Helena area. The scaleless body of this whitish eel has traces of large, regularly arranged scale pockets on the throat and chest.

The MORAYS (Muraenidae) inhabit all tropical and subtropical seas. They are the most snakelike of all eels in terms of shape and behavior. L of most species reaches about 100 cm, with the largest species being *Thyrsoidea macrurus* (L over 3 m). The body is compressed somewhat laterally and is very muscular. The skin is thick, leathery, lacks scales, and is brilliantly colored with distinctive markings. There are no pectoral fins in morays. Gill openings are small and round. The mouth opening often extends far behind the eyes, so far that some species cannot completely close their mouths. There are usually powerful fangs arranged in one or more rows. Poison glands are located in the palate membrane and at the base of the strong teeth. The glands empty their contents into wounds which the morays inflict. There are twelve genera with some 120 species.

Morays generally inhabit rocky coasts or coral banks; they remain hidden in the various crevices and clefts of these structures during the day. They leave their hiding places at dusk to go feeding, and most species prey on fishes. Five species are said to be so poisonous that a human can die from their bite. Since many species are very aggressive, divers in warm seas should refrain from reaching into rock crevices for "sunken treasures" or marine animals. Such a diver could all too easily become a victim of his carelessness and make acquaintance with the sharp dentition of a moray. Only members of the genus *Echidna* are more passive. Their small teeth or blunt chewing teeth are used to crack the shells of the crustaceans, mussels and snails upon which they feed. The brown-yellow and white ringed zebra moray (*Echidna zebra*) is often kept in aquariums. The species inhabits the Indian and Pacific Oceans.

When various genera are compared, one characteristic is particularly prominent: the external part of the olfactory organs, the "noses." Generally the forward "nostrils" are tubular. The rear ones, which are in front of the eyes or somewhat above them, are simple round openings. In the two *Rhinomuraena* species, found in the Indian and Pacific Oceans, the front tubes have been developed into a leaflike element on the tip of the snout. In *Muraena pardalis* the front as well as rear

Family: morays

Fig. 9-5. Head shapes:

Rhinomuraena;

Muraena pardalis.

nasal openings are tubular structures. The rear ones, which are considerably longer than the front ones, stand up like horns or snorkels. The largest moray, which is also the biggest eel, is *Thyrsoidea macrurus,* which can be more than 3 m long. It is found on the coasts of the Indian and Pacific Oceans and has also been sighted in river mouths and migrating upcurrent in rivers. Although the species is not very aggressive, it becomes dangerous if it is driven into a corner.

Morays are in demand for eating in spite of their dangerous teeth. In Rome in 92 B.C. the first fish-breeding tanks (which were called vivaria) were built for producing the moray *Muraena helena.* Even today morays are frequently sold in Mediterranean fish markets. In order to avoid all danger, fishermen remove the head with the salivary glands as soon as the eels are caught. Moray fishing is more intense along the Asiatic coast than in the Mediterranean. In 1966, 46,000 tons of morays were caught in the latter area, while Chinese catches are not recorded; in Japan alone 32,700 tons of morays were caught.

The PIKE EELS (Muraenesocidae), which are classed with CONGER EELS (Congridae) by some authors (see below), are distinguished by sharp dentition on the vomerine bone. There is one genus *(Muraenesox)* with seventeen known species, among which three are grouped under the Malayan name "Putyekanipa": 1. BATAVIA PUTYEKANIPA (*Muraenesox cinereus;* L over 80 cm); 2. LARGER PUTYEKANIPA (*Muraenesox talabon;* L 150 cm); 3. INDIAN PUTYEKANIPA (*Muraenesox talabonoides;* L over 1 m). The first of the three species is sometimes found in fresh water.

The many genera and species of conger eels (Congridae) are distributed in almost all tropical and subtropical seas. The best-known of them all is the CONGER EEL (*Conger conger;* L up to 3 m, weight to 65 kg), which is almost worldwide, since it is caught in every ocean except the eastern Pacific. The species prefers rocky coasts, in the crevices of which it hides during the day. It is occasionally caught in brackish water of river mouths. The powerful teeth reflect that this is a predator which is dangerous; its diet consists of various fishes, crustaceans, and squid. In addition to size this species is distinguished from the European eel by the longer dorsal fin, which has its base shortly before that of the pectoral fins, and by the scaleless skin.

Pike eels and
conger eels

Not everything is understood about spawning in conger eels. Occasionally females have been captured which had not yet emptied all their contents. Females have also reached maturity in aquariums. After maturity it is estimated that the highest number of eggs which a female can lay is about eight million. Pathological changes take place concomitantly with maturation of the gonads, these changes being in the intestinal tract and other organs including skeleton and teeth. The adults cannot recover from these post-spawning changes and die from them. Eels probably spawn in the open sea at depths of about 2500 m, but specific spawning grounds have not yet been found for this group.

All other genera and species of this family are considerably smaller than the congers. Some are exclusively deep-sea inhabitants, as is for example the moray *Ariosoma balearica* (L 30 cm) at 2000 m depths west of the Azores, and *Bathycongrus mystax* (L 38 cm) at 800 m depths near St. Helena. *Promyllantor latedorsalis* (L about 32 cm) inhabits 950 m depths off the Azores.

The GARDEN EELS (Heterocongridae) are classed with congers by Greenwood, Gosline, Böhlke, Berg and others, while other zoologists treat them as an individual family. Pellegrin in 1923 was the first to describe the peculiar life of these unusual fishes.

Family: garden eels

Garden eels are some 30–50 cm long and live in tubes, which extend vertically about one-half meter deep in loose sand or fine-grained coral sand. It must be quite an experience for a diver to encounter such a garden eel "settlement." They often cover 100 square meters of sandy bottom and the garden eels inhabit the floor at intervals of twenty to sixty centimeters. With a slightly bent fore-end, which protrudes about ⅔ out of the tube, they sway to and fro with the head directed against the current, seeking the zooplankton upon which they feed. Such garden eel colonies have only been found at places constantly covered with water, and only where the current is uniform. The garden eels also avoid areas where breakers occur. Eibl-Eibesfeldt and Klausewitz further report:

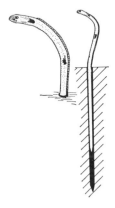

Fig. 9–6. Garden eel protruding from its tube.

"The Heterocongridae are strikingly timid. They flee from a diver by sliding into the soil when he reaches a distance of about three meters from them, and only their heads protrude; at a one-meter distance they withdraw completely into their tubes, and wait there for five minutes before shyly looking out again. All attempts to dig healthy garden eels out of the sand failed, since the animals can dig back into the soil very rapidly. With a poison solution they can be readily driven out of their tubes. As long as they were uninjured they swam headfirst with undulating motions over the sandy bottom, in a completely flat position. After swimming about one meter they turned about quickly and rapidly bored tail-first with powerful movements into the sand. The considerable mucus secretion of the eels was a striking phenomenon; the secretion was mixed with sand grains, with which the garden eel body came into contact."

SNAKE EELS *(Ophichthyidae)* are distributed in all tropical and almost all subtropical seas. Many inhabit coastal waters or coral reefs, and some stay temporarily in fresh water. The body is very slender, muscular and scaleless. A caudal fin is absent, and the tail end has thorny processes. The dorsal and anal fins—when present—do not form a continuous seam. The dorsal fins begin just behind the head. Pectoral fins are absent, but sometimes traces of these fins exist. There are some 200 species.

Snake eels are often brilliantly colored and may have bright bands

The other eel families

▷
Eels: 1. European eel *(Anguilla anguilla)*; 2. Conger eel *(Conger conger)*.

Fig. 9-7. Head of *Bathycongrus mystax.*

Fig. 9-8. Head forms:

Snipe eel;

Deepwater eel.

◁
Eels: 1. Moray eel
(Muraena helena);
2. *Echidna nebulosa).*

or various large spots. During the day they are hidden; they feed nocturnally. The diet consists chiefly of fishes and copepods. Some species, however, have a specialized diet, such as the snake eel *Ophichthys gomesii* (L about 75 cm) from the Gulf of Mexico and the parasitic *Pisoodonophis cruentifer* (L about 40 cm). The brown-to-yellow-hued eel chews an entryway into the body cavity of large fishes (e.g., halibut and cod) and feeds on the muscles of the host. The large species, *Ophichthys ophis,* from the West Indies, is greatly feared by fishermen due to its powerful dentition. L reaches 135 cm. Snake eels bore into the floor tail-first, like garden eels.

Synaphobranchidae members, about twelve species in the Atlantic and Pacific Oceans, are deep-sea inhabitants. The distinguishing characteristic of this family is the external gill openings. They lie in a common oval pit on the underside between the points of attachment of the long pectoral fins and are separated into two openings by a broad membrane. One species with very small scales is GRAY'S CUTTHROAT EEL (*Synaphobranchus pinnatus;* L about 54 cm). The species is found in the north Atlantic and western Pacific Oceans and rather frequently lands in fishermens' nets in the vicinity of the Newfoundland banks.

Ilyophid species inhabit the great depths of the eastern Pacific Ocean; there is a single species, *Ilyophis brummeri* (L 38 cm). The species has degenerated scales.

The SNUBNOSE EELS (Simenchelyidae) contain the single genus *Simenchelys* with two species. *Simenchelys parasiticus* (L reaches 61 cm) is widely distributed in deep waters of the north Atlantic from 700-1400 m, on the coast of South Africa, and in the Pacific Ocean. With their cutting teeth they gnaw through the skin of larger fishes (e.g., halibut), pull out the insides of the fish, and live off the firm muscles. The species has adapted to this type of life. The skin, with but degenerate traces of scales, has a rich supply of mucous glands throughout its surface; the short head is rounded and blunt and somewhat like a bulldog's. The gill slits are short longitudinal slits on each side of the "throat" in front of and below the pectoral fin.

SNIPE EELS (Nemichthyidae) are elongated deep-sea species. The body narrows from the tail and in one species, *Cercomitus flagellifer,* terminates in a whip-like structure. It is generally felt that these species cannot close their mouths, because the lower and upper jaws are bent away from each other. In 1916 Weber and de Beaufort referred to one snipe eel *(Nemichthys scolopaceus)* which Roule had caught in 1901 and in which the jaws were not bent. The two researchers conceded the possibility that such a "distinguishing characteristic" could have been caused by the effect of preserving chemicals.

The scaleless snipe eels inhabit all seas and have been caught at depths varying between 400 and 4300 m. For a long time it was assumed that they occurred only in tropical or subtropical waters, but in 1953 and 1954 *Nemichthys scolopaceus* was found at the fishing grounds called the "Rose Garden" off Iceland.

Serrivomerid eels are also scaleless deep-sea inhabitants. There are eleven species in the Atlantic, Indian and Pacific Oceans. Their vomerine bone has powerful teeth which are arranged in a sawtooth manner. Because of their often threadlike tail they have been called thread eels. The lower jaw is somewhat longer than the upper jaw and is only slightly arched. The mouth can be closed. The smallest species, *Spinivomer goodei* (L 13 cm) from the Gulf Stream north of the Bermudas, is particularly striking because of the beautiful silver sheen of the skin. *Serrivomer sector*, from the Indian and Pacific Oceans, is colored similarly but with small black spots as well. The species is four times as large as *Spinivomer goodei*.

The single species of the family Cyemidae is the DEEPWATER EEL (*Cyema atrum*; L 11–12 cm). The body is velvet-black, scaleless, and the dorsal and anal fins extend to the end of the body and look like a caudal fin with a cutout section. The actual caudal fin has regressed and has just five short rays. The species is widely distributed in the deep sea, but never about 2000 m. The leptocephalus larvae are about 56 mm long and are often found at higher water levels.

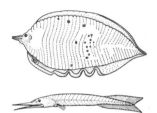

Fig. 9-9. Deepwater eel (natural size, 10.5 cm) and its leptocephalus larva.

The second suborder of eels, Saccopharyngoidei, consists solely of deep-sea species. The jaw is greatly elongated, and thus in two families the mouth is extraordinarily large. The eyes are very small, and a gill cover is absent. The gills are far behind the head. There are no ribs and no swim bladder. The dorsal and anal fins are very long; a caudal fin is absent, and the pectoral fin is small or missing also. There are three families.

Suborder: Saccopharyngoidei

Most species inhabit the Atlantic, Indian and Pacific Oceans at depths between 2000 and 5000 m. Nothing is known about their reproduction except that they undergo a leptocephalus stage. The SWALLOWERS (Saccopharyngidae) look like monstrous deep-sea creatures; there is one genus and five species. The largest species, the PELICAN-FISH (*Saccopharynx ampullaceus*; L 183 cm), has a long tail terminating in a thread. The tail is about four times as long as the body. The large mouth opening has many sharp, slightly arched teeth, indicating that the pelican-fish is a predator. Since the throat and stomach are flexible, the pelican-fish can actually swallow fishes which are larger than itself. The first pelican-fish was discovered in the north Atlantic on the water surface. Its body was distorted due to the fact that it had swallowed large fishes. A few still showed signs of life and tried to wriggle out of their predator.

Fig. 9-10. Head forms:

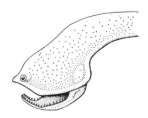

Swallower;

The GULPERS (Eupharyngidae) are considerably smaller; the group has one genus and two species. The slender, eel-shaped body is attached to the head, which is flattened, and actually consists of just the mouth opening with tiny teeth. The dorsal and anal fins are well devel-

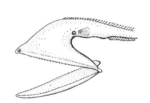

Gulper.

oped and extend almost to the tail end of the body. The pectoral fins are small. The stomach cannot be as distended as in the swallowers. Both species are colored velvet-black. L in *Eupharynx pelicanoides* is about 53 cm, and in *Eupharynx richardi* only 35 cm.

The gulpers feed on small invertebrates, especially crustaceans. According to Herald it is presumed that they swim through the water with their mouth wide open and open and close it like a great net, in order to catch plankton. This theory is supported by the fact that the fragile gulper body cannot hold onto convulsing, struggling prey.

The last eel family is Monognathidae (L 5-11 cm). They do not have pectoral fins and in contrast to members of the other families have only a small mouth opening. The skull structure is peculiar in that an upper jaw and many other skull bones are missing. Thus far only a few specimens 5-11 cm long have been caught in the Atlantic and Pacific. J. Böhlke, who has studied the Monognathidae in particular, considers it possible that they are nothing more than early larval stages of the swallowers.

Order:
Notocanthiformes,
by W. Ladiges

The order Notocanthiformes is closely related to Anguilliformes. Three families are known, which comprise twenty to thirty species. All are deep-sea inhabitants found at almost 3000 m. They differ from other eels in that the dorsal fin is not a unified structure. Thus the SPINY EELS (Notacanthidae) have a dorsal fin composed of individual spines (as many as forty) without a connecting skin. An unusually long anal fin, which is common to all Notacanthiform species, has 200 rays along the entire lower side to the sharply pointed tail. Some species have luminous organs.

▷
On their expedition on
the Xarifa, Hans Hass and
Irenaeus Eibl-Eibesfeldt
found the peculiar garden
eels (family Heterocon-
gridae), which live in
tubes they bore in the
sand. Shown here is a
colony of Gorgasia
maculata.

10   Herring

Herring (order Clupeiformes) are a characteristic fish group of the oceans, but also include many species inhabiting tropical fresh water. They form schools and are found near shores as well as in the open sea. Many of them are migratory.

Order: herring, by K. Schubert

Herring can be distinguished from other species by a number of characteristics; there are no rayed canals on the gill cover bones; lateral line pores are absent (except in denticipitoid herring); there are keel scales along the medial line of the belly. Noteworthy skull characteristics include: a suprabranchial organ with unknown function which joins the fourth and fifth gill arches; there is little dentition in the mouth, since most species feed on plankton (an exception is the wolf herring), and there are no teeth on the parasphenoid (a bone at the base of the skull). A mesocoracoid bone is present in the pectoral girdle. With the exception of denticipitoid herring, the caudal skeletal structure in all species is similar. A narrow connection joins the swim bladder with the inner ear and also with the intestinal tract. There are two suborders: Denticipitoidei and Clupeoidei.

Distinguishing characteristics

The denticipitoid suborder (Denticipitoidei) consists of a single family with one genus and one species, the DENTICIPITOID HERRING (*Denticeps clupeoides*). The species has the typical herring shape but may clearly be distinguished from other herring by thorns on the skull covering bone and on a few body scales, by the caudal skeletal structure (which resembles early herring larval stages), by the skull and jaw structure, and by the presence of a complete lateral line. L reaches 5 cm. The head is well compressed and the top of the skull is flat. Scales are large, the caudal scales being jagged and compressed. The caudal fin is short and is located well behind the anal fin, a very long structure. The pectoral and pelvic fins are small. Coloration is silvery tending to green on the sides; a network pattern is formed on the bases of the scales by blackish hues. A narrow dark green band runs laterally from the gill cover to the base of the caudal fin. A green-golden zone

Suborder: denticipitoid herring

Distinguishing characteristics

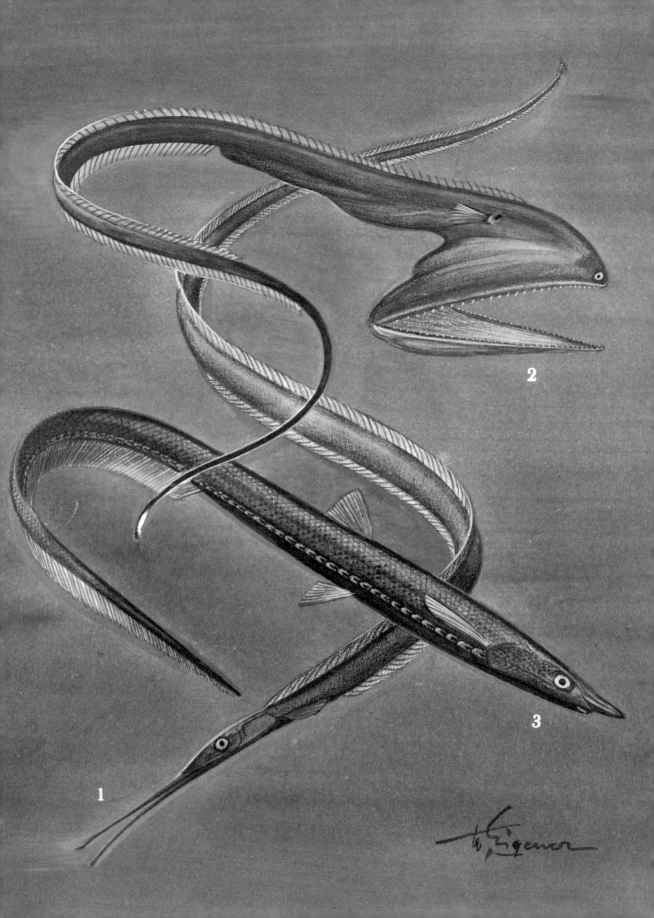

Fig. 10-1. Migrations of
the herring *(Clupea).*
D: Dogger Bank.

Suborder: Clupeoidei

Distinguishing
characteristics

Family: herring

◁◁◁
Morays are often found
in crevices but do not dig
them themselves. Here
zebra morays *(Echidna
zebra)* and a moray eel
(Muraena helena) have
withdrawn into an old
vat.

◁◁
Even the dangerous eels
(shown here is *Lycodontis
javanicus)* are groomed by
Labroides cleanerfish (see
Vol. V). Generally the
eels do not harm these
fish.

◁
Eels: 1. Snipe eels *(Nemi-
chythys scolopaceus);*
Gulpers: 2. *Eupharynx pelecanoides;* Halosaurid eels:
3. *Halosauropsis
macrochir.*

runs above it along the lateral line. Fins are colorless. The species is
only known in four rivers in southwest Nigeria.

This species is considered the earliest one extant. Its discoverer,
Clausen, notes that the species forms schools. The schools are generally found in the middle of rivers, where the current is strongest.
Because of its small size the species is often difficult to see. This skillful, tireless and speedy swimmer is not caught easily. Sexually mature
specimens have been caught in mid-September; young have been
found in mid-November, and subadults between February and July,
but occasionally in August as well. Females are plumper than males.
Due to its small size this species has no commercial importance. It has
probably long been confused with a similar-appearing carp species of
genus *Chelaethiops.*

All other herring are placed in the suborder Clupeoidei. The body
is "typically herring-like" with the exception of the head having scales.
Fin rays are united. The scales are large and thin; tail scales are along
the belly. The eyes have an adipose membrane. The uterus has the
typical structure. The stomach has numerous pyloric appendages.
There are three families: 1. Herring (Clupeidae); 2. Anchovies (Engraulidae); and 3. Wolf herring (Chirocentridae).

In most Clupeid species the body is slender, sometimes short and
wide, and generally compressed (but occasionally rounded). Toothed
or comb-like scales are present, often with tail scales, these frequently
having thorns. The mouth may be terminal or on the upper or lower
side of the body. The gill apparatus is extensive and is long and
slender. The dorsal fin is generally at the center of the body; the anal
fin is of moderate length or very long and the pectoral fins are well
developed. There are six subfamilies (Dussumieriinae, Clupeinae,
Pellonulinae, Alosinae, Dorosomatinae, and Pristigasterinae), with
fifty to seventy genera and 150–190 species. Of these some thirty-seven
genera containing 150 species are found in the tropics, six genera with
thirteen species are in subtropical regions, and seven genera with
about twenty species are in the northern latitudes.

Generally herring are sea inhabitants less than fifty centimeters
long. Typical for all free-swimming fishes of the open seas, coloration
is gray-green on the back and belly and silvery along the sides. They
feed chiefly on plankton; larger species also prey upon smaller fishes.
Propagation occurs in most cases through free-floating eggs; only
a few species spawn on the floor.

About twenty-five herring genera with some 100 species live in the
sea. A few migrate into fresh water, and others are found exclusively
in brackish water. Approximately fifteen genera with thirty species
spend their entire lives in fresh water. Thus eight genera with fourteen
species are distributed in fresh water in west Africa, four genera with
six species in the Malayan Archipelago, two genera with four species

in Indian fresh water, and just two species are in North America. Herring are commercially the most important fishes in the world, but they also form the essential nutritional base for marine animals and birds, tortoises, and predatory fishes. The portion of herring species from the total fish catch of the world in 1967 was 22,220,000 tons (33.4 percent); in 1938 it was just 5,000,000 tons (23.8 percent). This tremendous increase in catch size can be traced to a number of causes: new fishing grounds have been opened up (many of them sources thus far totally unexploited), for example on the Peruvian and South African coasts; these grounds were opened to create a source for producing fish meal and fish oil. Fishing activity on older fishing grounds became more intense and more sophisticated, such as through the use of deep-sea trawl nets and ring nets.

Fig. 10–2. 1. Atlantic herring *(Clupea harengus);* 2. Pacific herring *(Clupea pallasii).*

The best-known species belong to the genus *Clupea,* in which two species are distinguished: the ATLANTIC HERRING *(Clupea harengus)* and the PACIFIC HERRING *(Clupea pallasii).* Some researchers doubt the validity of placing these into two separate species. They are free-swimming slender marine species with a L reaching 45 cm. The belly of the Atlantic species has a keel, while the Pacific herring has no keel in front of the pelvic fin. There are also differences in the dentition of the vomerine bone. Both species occur only in the northern hemisphere in temperate and cold waters of the north Atlantic and the northern Pacific Ocean.

For centuries the ATLANTIC HERRING *(Clupea harengus)* has been the most important commercial fish in the northeastern Atlantic. Thus it comes as no surprise that since the end of the 19th Century researchers in many European countries have dealt intensively with the natural history and ecology of this species. Originally it was presumed that the entire herring population in the northeastern Atlantic was a unified group inhabiting the region from the Arctic Ocean to the English Channel. From here this group presumably migrated extensively to the north and south (and back) during the course of the year. But since Heincke found structural differences in herring caught in different regions, herring became the "classic" species for subspecies investigations. The studies clearly showed consistent differences in certain characteristics in well-isolated herring groups living in various parts of the northeast Atlantic. In Copenhagen in 1956 herring researchers came to the conclusion that it would be useful to distinguish a number of large "biological groups" on the basis of different spawning periods, spawning grounds and spawning conditions. These groups, between which no significant intermingling apparently occurs are: A. Herring inhabiting the open Atlantic Ocean and spawning on the Atlantic coasts of northern Europe in mid-winter, spring and possibly also in early summer (Atlanto-scandian herring). These fishes reach an appreciable size and are characterized by an intermediate number of

The Atlantic herring

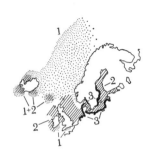

Fig. 10-3. Distribution and spawning grounds of the genus *Clupea*. 1. Atlantic herring *(Clupea harengus)*; 2. Shelf herring; 3. Coastal herring.

vertebrae (57 or more). B. Herring inhabiting the North Sea, on the shelf west of the British Isles, in the transition zone between the North and Baltic Seas; they spawn between August and January along the coasts (these are known as shelf herring). They reach a smaller size and have an intermediate number of vertebrae (between 56 and 57) in all regions outside the Baltic Sea, where they have less than 56 vertebrae. C. Herring distributed inside coastal waters of the North Sea, in the transitional area between the North and Baltic Seas, and in the Baltic Sea, and which spawn in shallow water during winter and spring. Body size and vertebral number are smaller than in previous herring (the vertebrae numbering generally 55). D. Herring in the northeastern part of the region, which until now were classified with the Pacific herring *(Clupea pallasii)*. They share various characteristics with group C in terms of coastal spawning during winter and spring.

The composition of herring populations is complex. Furthermore, many characteristics which until now have served to differentiate the various groups have taken on a great variability due to various ecological pressures. The result of this is that the genotypic (inherited) and phenotypic (external appearance) differences cannot be recognized immediately. On the basis of some particular inherited characteristics such as fertility, egg size, blood groups and others it now appears that there are indeed genetic differences between the three "biological groups"; thus, they should maintain the status of subspecies.

Herring populations in the western Atlantic, an area which has been of lesser commercial importance, are accordingly less well understood. One large population in this region is characterized by summer-fall spawning; it apparently consists of several groups which are comparable to the second group of the northeast Atlantic populations.

Commercially the most important herring for Germany are those in Group B (shelf herring). They inhabit the North Sea, eastern English Channel, the west coast of Scotland, the northern Irish Sea, the Kattegat, Sund, Belt, and the southern Baltic Sea. This is the group on which the most scientific investigations have been made, and the group is broken down into a number of components: 1. North Sea bank herring (summer-fall spawning; spawning grounds are along the coast of Scotland and England and the Dogger Bank); 2. Eastern channel herring (or downsherring) (fall-winter spawning; the spawning grounds extend from the Dogger Bank through the southern North Sea [to Sandettie] to the eastern English Channel) (Cape Antifer); 3. Northeastern Kattegat or Kobbengrund herring (summer-fall spawners with spawning grounds along the Swedish Kattegat coast); 4. Sund, Belt and southern Baltic Sea herring (summer-fall spawners; the spawning grounds are in the Sund and Belt in the southwest and southern Baltic Sea); 5. Scottish west coast herring or Minch herring (summer-fall spawners, the spawning grounds being in the northern and southern Minch, the channel between the Outer Hebrides and the west coast of Scotland); 6. North Irish herring or herring of the Isle of Man (summer-fall spawners; spawning grounds are off the southeast coast of the Isle of Man).

The eggs of the Atlantic herring have a diameter of .99–1.9 mm; this considerable variation depends on the group to which the eggs belong. The number of eggs laid also varies according to group. In general the winter-spring spawners have a relatively lower fertility and larger eggs, while summer-fall spawners have smaller eggs but greater fertility. Among winter-spring spawners between 22,000 and 40,000 eggs may be laid; summer-fall spawners lay 48,000–70,000 eggs.

Embryonic development time is chiefly dependent upon the ambient temperature. The larvae of the various spawning groups have a size and yolk supply according to the egg size. Incubation studies have shown that larger larvae hatch from the larger eggs of the winter-spring spawners than from the fall-spawner eggs of Baltic Sea herring. Atlanto-scandian herring larvae have a length of 8 mm, while those of the eastern Channel herring are 7.5 mm. Larvae of the smaller, yolk-poor eggs of the shelf and Baltic Sea herring are only 6.0–7.2 mm long. These differences in egg and larval development among the various spawning groups can be understood as adaptations to seasonal variations in food supply. Larvae which hatch in winter and spring have a smaller food supply and a smaller number of predators. This is advantageous to produce larvae which are fewer in number but vigorous. Larvae hatching in summer and fall have a greater food supply available but must also deal with more predators; the greater number of eggs laid thus increase their chances for survival.

Freshly hatched larvae are found in tremendous masses on the spawning grounds and in the vicinity thereof. They are transparent and very slender. As soon as they have attained a length of two centimeters, the chief herring characteristics can already be observed; the dorsal fin is far to the rear; the caudal fin has already formed; and the rays in both fins can already be distinguished. The number of vertebrae has also been established by this time and the last segment of the vertebral column arches upward. With a length of two to three centimeters the swim bladder migrates underneath the stomach. Metamorphosis from the larval to adult stage in spring spawners is complete with a length of 3.1–4.4 cm, and in fall spawners at 4.4–6 cm. At this period the young herring generally inhabit coastal regions, at all depths. This stage is attained in from four to nine months. Developmental time for the spring brood is longer, and for the fall brood shorter.

The larvae drift from the spawning grounds by the force of the current. Bückmann has followed the distribution of the larvae from

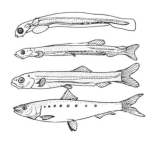

Fig. 10–4. Developmental stages of herring (top to bottom), all brought to equal size.

the spawning grounds at the Dogger Bank and from the southern Bay of Heligoland. In October larvae from the western Dogger Bank move southeastward over the southeast corner of the bank. In November and December they were found above the oyster grounds, where they spent three months in a circulating current system. In March these larvae began invading the Bay of Heligoland and in April they disappeared into the open sea, in which they migrated to coastal regions on their own strength. Recently the distribution of the Atlanto-scandian herring has been thoroughly researched by the Soviets. Generally these herring drift along the Norwegian coast to the north, and some pass into the fjords.

The availability of suitable plankton is the most important determinant for the development and ultimate survival of the larvae. The larger larvae undoubtedly have a larger mouth opening and thus can feed on larger organisms. The prey must be at a distance of one-half centimeter from the herring larvae in order to be perceived and eaten. The larvae feed only on moving organisms; this results in selection for the prey organisms. As in adult herring the larvae hunt only when it is light. Digestion time is dependent upon temperature and in young herring is somewhat longer than in adults. For the larvae the plankton must be fairly dense. Competition for food among the larvae would hardly be expected, for there are barely more than 100 larvae without yolk sacs in a cubic meter of water. Studies on intestinal contents of 300 larvae from the Clyde region (the larvae had a length of ten to twenty millimeters) showed that larval stages of lower crustaceans formed the greatest part of the total diet.

Precise determination of growth and age plays a major role in commercial fishing. Scales or auditory bones known as otoliths are used for this purpose. Nutritional changes in the fishes are indicated by alternating transparent and dark rings. The narrow, transparent zones indicate the poorer food supply in winter, while the broader, dark zones reflect the better feeding conditions in spring and summer.

Summer and fall spawners of the North Sea reach sexual maturity in their third or fourth year, at which time they have a L of 23–24 cm. Since approximately 1952 sexual maturity in species throughout the North Sea has set in during the third year and is apparently dependent upon the size of the fish. The first spawning may appear at three years of age, but in some cases not until the ninth year. In late-winter spawners of the Norwegian coast the beginning of the first spawning varies between the northern group in the Lofotens and the southern group. The average age of northern herring for spawning is six to seven years, and for the southern group four years. Life expectancy of the summer-fall spawners is from twelve to sixteen years, while late-winter spawners of the Norwegian coast live from twenty-three to twenty-five

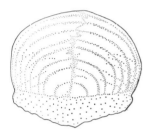

Fig. 10-5. Herring scale with eight narrow winter rings.

Fig. 10-6. Otoliths of an Atlanto-scandian late winter spawner with translucent (above) and glass-like (below) central field.

years. On the American east coast ages of as much as nineteen years have been recorded.

In the northwestern North Sea spawning begins from the end of July to the end of September in the coastal region off the Orkneys, the west coast of Scotland and on the banks off the eastern Scottish coast. Shortly thereafter, from September to October, the herring spawn on the west edge of the Dogger Bank. Since the beginning of the 1960's the herring have sought out their spawning grounds on the Dogger Bank less and less, but increased spawning activity has been observed in September off the east coast of England near Whitby. Another spawning period from November to January occurs at Sandettie and in the eastern English Channel at Ailly. In November still more small spawning grounds are located in the region from Halsborough and Downsing and in December near Galloper.

The spawning grounds of the Norwegian late-winter spawners have become known due to Runnström, who carried on the first quantitative estimation of a spawning ground by collecting eggs. Spawning on the Norwegian coast takes place from February to March, but occasionally in April as well. It takes place along the entire coast from Lindesnes to Trondheim; earlier the center was located in the region between Bergen and Lista. But during the last decade the arrival of the mature herring has been delayed by about one month. At the same time the spawning grounds also became displaced to the north of Bergen at around Alesund. On the American east coast spring and winter-fall spawners are both encountered.

Spawning grounds and periods can be designated on the basis of distribution of freshly hatched larvae, stomach analysis of haddock and other fishes which feed on herring eggs, and from the catches of commercial fishing vessels which caught mature herring. Attempts to collect herring with dredges had been unsuccessful for a long time. Later Polster and Bridger successfully collected herring with a trap on spawning grounds near Sandettie; they had laid the trap on a layer of flint in the coarse sandy bottom. Parrish and others have thoroughly investigated the subsurface of a spawning ground in the Firth of Clyde. They found that spawning was restricted to an area of small stones and coarse sand. Groups of eggs laid together formed carpetlike layers as much as eight eggs thick. Water depth varied between twelve and fifteen meters. Hemmings confirmed the findings of Parrish at these grounds by taking underwater photographs. He found that the spawning covered a surface area of 2350 m. The thick egg layer was five eggs thick, and Hemmings estimated a minimum of 1.69 billion eggs in the area. With the exception of Pecten mussels (see Vol. III), no organisms preying on the egss were found at the spawning grounds.

While on expedition with the fishery research vessel *Anton Dohrn* in September 1967 I found several spawning clumps three centimeters thick. They were on the northeast banks of the North Sea at a depth of approximately seventy meters on small stones and piles of mussel shells and snail shells. Coastal spawners lay their eggs at the mouth of the Elbe River, Germany, in spring, on dams in brackish water. Generally the spawning grounds of summer-fall spawners in the North Sea and in the English Channel are at twenty-five to forty meter depths, where the temperature is ten to thirteen degrees Centigrade and the salinity is twenty-four to thirty-five percent. Selection of a spawning ground is apparently dependent on the quality of the floor. The first larvae hatch after three weeks, and they have a L of 5-7 mm. Spawning apparently begins with the onset of daylight and pre-spawning displays have not yet been observed. On the basis of these investigations fertility can be considered to be quite high.

Herring were found in schools ranging from hundreds to thousands of individuals, in all sizes from young to sexually mature adults. Those individuals within a particular school are generally of equal size and age. It is not known how long they remain together. Herring which were marked together at the same place have been recovered later at various far distant locations. In aquarium studies Mohr found that large schools of young herring swim faster and encounter obstacles less than smaller schools. When feeding the school drifts with the current. When they are migrating the fish swim side by side in the same direction. Fridriksson and Aasen noted that living herring kept in nets without disturbances traveled seven to eight meters per minute. Marking studies have shown that they are capable of undertaking great migrations, such as from the east coast of Iceland to the Norwegian coast and back.

By means of echolocation experiments it was found that herring in the North Sea spend the day at the floor. With the beginning of dusk they ascend to 30-50 m depths into a warmer temperature zone. Light plays an important role in this movement. During darkness they seem to remain scattered at this level or lower. It is with the onset of dawn that they again collect and by the time it is daylight they are once again on the floor.

Recently herring behavior at darkness has been studied closely. Soviet researchers have observed daily activities from a submarine. They reported that the Atlanto-scandian herring spent the night motionless at the surface of the water in an oblique position, as if they were sleeping. They became active shortly before dawn and began moving into greater depths. Other researchers believed that schooling behavior ceased at night, but Craig and Priestly reported just the opposite and have provided photographic evidence.

In recent years the behavior of commercial fishes has been studied more intensively with a view toward developing better fishing methods. One finding of these studies has been that vision plays a significant role in herring. Herring did not avoid plastic sheets if they were transparent; herring in which the eyes were kept shut could not perceive a net, while those with sight did detect the obstruction. During the day a wall of air bubbles acted as an obstacle to herring, but they swam right through it at night. Herring can also detect noises and vibrations, to which they respond with fright behavior. This has been confirmed with echolocation tracking. When a ship moved over a herring school, the school immediately sank to a depth of fifty meters. Fright can also be induced in the aquarium by tapping on the wall.

During a few months of the year herring can be found in certain regions in tremendous quantities, while at other times these same areas are completely devoid of herring. On the other hand, herring can be caught in some places throughout the year, but the catch varies from year to year. Because of the great commercial importance of herring it is no surprise that migratory and stationary behavior in herring of the north Atlantic have been subjects of great research interest. Recently herring have been tracked with internal or external markers. Marking studies have shown that Atlanto-scandian herring migrate between feeding grounds off Iceland and spawning grounds on the Norwegian coast. Using echolocation Devold tracked migrating herring from their spawning grounds to the feeding grounds near Iceland and later to wintering areas off eastern Iceland; he thus clarified the migratory routes of these herring.

Today it is known that this major herring population has three major growth and development areas: in the Norwegian fjords, the Barents Sea and in the southern and eastern parts of the ocean off northern Europe. Young herring which have developed in the fjords migrate into the sea at an age of two to three years, where they meet those herring which have been developing there. Later, when they are older, they become distributed from the central part of the northern sea off Europe to the polar regions. With the onset of sexual maturity they are found in the southern part of their distribution, where they migrate for the first time for spawning. After that migration to the feeding grounds in the polar region and to the spawning grounds on the Norwegian coast regularly takes place. Between both regularly occurring migrations the herring winter off eastern Iceland.

Since World War II spawning herring have been marked in the North Sea. It now seems clear that the majority of North Sea bank herring winter in the northeast North Sea and in the Skagerrak. In early spring they move along the Norwegian coast to about Utsira. In the beginning of March they then move to the feeding grounds in the northern North Sea, for example to the Fladen grounds and the Bressay Shoal. The various groups on the feeding grounds separate in mid-July. One group moves in early August to spawn on the east coast of Scotland, while the other probably migrates from the Fladen grounds and from the Gat to spawning grounds on the Dogger Bank and off the English east coast. After spawning they do not migrate further, as was earlier supposed, east of the Dogger Bank to their wintering region in the northeast North Sea, but instead they apparently retrace the same route they took on the spawning migration, up to the northern North Sea (Gat/Fladen, Bressay); from here they migrate to wintering areas on the Norwegian channel and in the Skagerrak.

The Downs herring of the southern North Sea apparently does not undertake such a lengthy seasonal migration. Its wintering grounds are presumably between the eastern English Channel, the Dutch coast near Texel, and the western edge of the Dogger Bank. During feeding migrations the Downs herring apparently mixes with the North Sea bank herring in the Fladen grounds and in the Gat. At the onset of the spawning period one part of the population—consisting of the older members—migrates to the Dogger Bank. The rest seek spawning grounds in the southern North Sea and at the eastern end of the English Channel.

The migration routes of young herring in the North Sea, which have a growth and development region east of the Dogger Bank (another lies in Moray Firth), are not as well known. It appears that the older herring from the growth and development regions east of the Dogger Bank migrate around the northern part of the Dogger Bank, while the Downs herring move to the southern edge of the bank. Still less is known about migratory routes for the herring in the western Atlantic, since coastal fishing is the predominant activity. Since European fish steamers have also become active there, scientists from nations engaged in fishing are now studying these herring populations, so that in the near future our understanding of these groups will also be increased.

Herring feed on plankton, which is not simply filtered but selected, a phenomenon found by stomach content investigations and aquarium studies. Blaxter and Holliday reported that herring select food visually, and then again in the mouth, in which materials which are useless and of bad taste are rejected. Vision even plays a role in the latter case; this has been shown by the fact that feeding ceases at night, with the exception of moonlit nights. Johnson found that herring cease feeding at temperatures below 4°C.; other researchers recorded this threshold temperature to be 1°C. in bank herring, and according to Soviet scientists it is 0.4 degrees C.

In extensive investigations in Plymouth, Lebour found that larvae had ingested the following organisms shortly before the disappearance of the yolk sac: snail larvae, mussel larvae, crustacean larvae, and other crustacean juvenile stages, as well as green algae. At a length of 12 mm

the herring also eat a few small copepods; from that point until metamorphosis the diet consists solely of copepods, and after that of
copepods, large crustaceans (decapods, amphipods) and fishes.

The primary diet is the crustacean *Calanus*. Other important organisms include fish larvae, euphausids, *Oikopleura* and *Temora* (see Vols.
I and III). These organisms occur in different amounts seasonally.
In general heavy feeding activity begins in the spring, but declines
steadily until the end of July. Some researchers report a second high
point in fall. During development of the gonads feeding decreases
sharply; the body fat stores may function as nourishment at this time.
However spawning has also been accompanied by another feeding
period. Stomachs were completely stretched full with spawn. Soviet
researchers report that the Atlanto-scandian herring feeds the least
during the spawning period (from February to March). Only a few
studies have been made on the diet of herring on the American side
of the distribution. Since there are fewer crustaceans here, the herring
are thought to feed on other planktonic organisms. Stomach contents
have also included small fishes of various species, small cephalopods
and polychete worms.

The chief intestinal parasites of herring are nematode larvae (see
Vol. I). They are not regularly found in the body cavity, but can
be so numerous that the fish is unsuitable for human consumption
because of them. Recently investigations have been carried out on
nematode larvae in North Sea herring, since lightly salted herring eaten
in Holland caused diseases from the larvae. Two species of parasites
were identified. One is a larvae, probably of *Contracaecum aduncum*,
which is unrelated to disease, since it is sensitive to higher temperatures. All these larvae had died before the temperature reached 37°C.
and have not been found in marine mammals. Another larvae belongs
to the genus *Anisakis*, which in culture was very similar to *Anisakis
marina*. The life cycle of this organism has not yet been determined.
However all sexually mature *Anisakis* have been described in marine
mammals, and presumably they are related to the disease which has
occurred in humans.

Tapeworms and trematodes (flukes) (see Vol. I) have occasionally
been found as well, as for example the trematode *Octobothrium harengi*
on the gills. A parasitic copepod (see Vol. I) from the German coast
attacks the eye of the herring and the sprat. The head of this crustacean
(*Lernaeenicus sprattae*) generally presses so far into the eye as to blind
the object fish. Sporozoans (see Vol. I) have been found in the musculature of young herring. Skin diseases caused by *Ichthyosporidium* and
other protozoans and bacteria have also been found in herring on the
American coast. In some cases these have led to mass deaths.

The herring has a great number of enemies. Eggs are eaten by haddock and other gadoid fishes, flat-fishes (see Vol. V) and echinoderms (see Vol. III). When haddocks are stuffed with herring spawn they are known in England as "spawny haddocks," and are prized as such. The larvae are also eaten by jellyfishes, comb jellyfishes (see Vol. I), sagittoid worms (see Vol. III, *Sagittoidea*) and many fish species. Both the young and adult herring are prey to mackerel, tuna, cod, salmon, ling, spiny dogfish, mackerel shark, and the Greenland shark; numerous birds such as guillemots, grebes, gulls, northern gannet (see Vols. VII and VIII); and marine mammals such as the finback whale, the lesser rorqual whale, bottle-nosed whale, pilot whale, killer whale, common porpoises, dolphins (see Vol. XI), and seal species (see Vol. XII). In the Arctic Sea and occasionally in the North Sea, great collections of birds, whales and dolphins can often be seen when the weather is calm. They are pursuing great schools of herring swarming up to the surface. This is a characteristic sign for the fisherman searching for herring.

The herring catch has been subject to great annual variations. As age investigations have shown, the various age groups appear in different strengths. Thus one speaks of "rich," "normal" or "weak" vintages. Rich vintages for North Sea bank herring were the years 1921, 1924, 1927 and 1929; poorer vintages were 1923, 1925 and 1928. Richer vintages also appeared after the war, but since that time the number of poorer vintages has increased. These variations in the vintage strength are even more distinct in the Atlanto-scandian herring. The causes of these variations can be found in the increasing factors acting on the larval stages. The transition from yolk sac nourishment to self-feeding has been determined to be a dangerous developmental phase. English studies showed that larger amounts of herring broods perish when the larvae are driven into unfavorable growth and development regions by the winds.

For man the herring is probably the most important fish in the world. Herring fishing is one of the oldest occupations of coastal peoples. In England herring fishing was first mentioned in A.D. 709 in an old chronicle. It was already known in Norway in the 11th century and in Holland since the 12th Century. It was in Germany that a herring trade was developed. Toward the end of the 12th Century Hamburg was an important herring trade center, and its business increased as the Hanseatic League engaged in deep-sea fishing. In 1425 the first commercial herring fishing establishment opened in Heligoland. The city of Emden was carrying on commercial activity in 1552; it ceased at the beginning of the 17th Century and began anew in 1769 after East Frisia became part of Prussia. On other parts of the German coast on the Elbe and Weser Rivers commercial herring fishing activity also began. The catch was secured by drift-nets pulled by special ships and was originally seasonal. It was carried on from May to October. After the herring drag-net was introduced to steam fishers in 1913, the size of the annual catch increased greatly. Since the end of the second world war a trawl net has been introduced which permits herring fishing throughout the year. More recently a free-swimming trawl net has been utilized. German commercial fishery is now carried on by three major types of techniques: 1. The large commercial activities of big ships; 2. Steamer fishers; 3. Cutter commercial fishing.

The oldest apparatus used for fishing were probably hooks, fences and baskets, as they are used today in coastal fishing. Present-day deep-sea fishing also uses the ring net (see Chapter 3) very often; this device is not used by German fishermen, however.

Herring has such a prominent position among commercial species because it appears in such great quantities and thus can be sold at low prices. It has become the food of a great many people of the world. In addition there are the many ways of preparation which permit the herring to become a tasty, nutritious food rich in vitamins. It has a moderate protein content (18%). The fat content varies according to development of the gonads from 8% to over 20% and averages 18%. Herring has a high mineral content, especially in calcium and phosphorus. It also contains a number of valuable vitamins. Among edible fish the herring has the highest nutritional value after eels.

Herring may be eaten either fresh or as salt herring. On land they are processed with smoke, vinegar, salt, oil and spices in smokehouses, marinating plants and food plants into the various end products. In some countries herring is used primarily for fish meal and oil.

The PACIFIC HERRING (*Clupea pallasii*) inhabits the coasts of the northern Pacific Ocean from the Bering Strait to Korea and in the Arctic Sea to the mouth of the Lena River. On the North American coast its distribution extends from California to Nome, Alaska. Herring in the White Sea and from Cape Kanin to the Kara Sea are very similar to the Pacific herring. This species differs from the Atlantic herring by the smaller number of vertebrae, among other features. The keel scales are more or less developed under the pectoral fin process, and the vomerine teeth are weak. Biological differences also exist between the two forms. Herring with fewer vertebrae are spring spawners which spawn in a narrow region along the coast in shallow water ten to fifteen meters deep. Generally the eggs are laid in brackish water on plants. Pacific herring form spawning groups whose distribution is limited to very specific narrow zones. They apparently do not migrate to a great extent. On the Asiatic coast ten spawning groups are known in the region from Korea to the Sea of Okhotsk. Of these the Hokkaido-Sakhalin herring is commercially the most important. Two other populations are known in the Asiatic region of the Bering Sea off Kamchatka.

There is also a series of spawning groups on the American and

The Pacific herring

Canadian west coast. Since 1936 Canadian researchers have marked over 500,000 spawning or recently spawned herring. The recaught specimens showed that herring regularly spawn where they had spawned the year before. Newer investigations, as by Jones, indicate however that in every spawning group there are up to twenty percent strayers.

The herring in the White Sea and in the Petchora region also form various spawning populations. In the White Sea there are two major groups: the smaller "yegoro" herring (L 12–20 cm) and the large "Ivanov" herring (L 20–30 cm). The Petchora herring are similar to the large White Sea herring. Compared to the oceanic groups of Atlantic herring the Pacific herring reaches sexual maturity faster; it also grows faster. On the Asiatic side and in the northeast Atlantic herring spawn from the middle of March to the beginning of June; in the southern distribution on the Californian side they are said to spawn in December.

Spawning temperatures vary between 0° and 9°C. according to the geographical location of the spawning ground. Eggs are laid on algae (*Zostera, Fucus, Cladophora, Phyllophora* and others) between two- and fifteen-meter depths, generally from two to four meters. They are only occasionally found on stones. Fertility is greater on the Asiatic coast (averaging 50,600 to 72,200 eggs) than on the American coast (average is 20,000 eggs). Under natural conditions, at a water temperature of 6°C., development lasts 26 days. Freshly hatched larvae are 6.0–6.2 mm long. At the end of November they have reached the stage of a young herring, and they can fend for themselves and separate from the adult herring. The adults also separate after spawning; they do not undertake a great north-south migration like the Atlantic herring, but move from the open sea to the coast and vice-versa.

Feeding migration routes of this herring are still unknown. The adult herring feeds on small crustaceans (copepods and krill). From May to August it feeds primarily along the coast. In the Pacific Ocean the herring reaches a length of over 40 cm and can attain an age of 15 years. In the White Sea the larger variety measures up to 34 cm, the smaller to 20 cm, while the herring from the Petchora region is up to 32 cm long. In the Pacific Ocean and White Sea the Pacific herring plays an important role for the commercial fishing fleets of Japan, Canada and the Soviet Union. It is generally caught during the spawning period and during the chief feeding period along the coast. In the Pacific the total catch varied from 1961 to 1967 between 473,000 and 778,000 tons. This herring is also prepared in many different ways.

The sprats   Sprats (genus *Sprattus*) are close relatives of herring. Six well-defined and one equivocal species have been identified. The majority are found in the southern hemisphere. There are only minor differences between the various species. L does not exceed 20 cm; some differences

Oceans and on the west and east coasts of tropical America (and chiefly in rivers).

The sixteen species and subspecies of the genus *Caspialosa* from the Black Sea, Sea of Asov and the Caspian Sea play an important role commercially. These are large and medium-sized species which differ from their relatives by the large mouth opening with vomerine teeth. The four species in genus *Pomolobus* distributed in the Atlantic Ocean from Nova Scotia to northern Florida and the Gulf of Mexico migrate regularly and may then be found in the rivers of the southern United States and Central America.

The four genera *Spratelloides, Jenkinsia, Dussumieria* and *Etrumeus* are characterized by a rounded belly. The *Jenkinsia* herring are small fishes distributed on the tropical American coast of the Atlantic and off South Africa and the west coast of India and Ceylon. One member of genus *Etrumeus* (L reaches 30 cm) lives on the Atlantic and Gulf coast of the United States, while other relatives inhabit the eastern and western Pacific, the coast of South Africa and southeastern Australia. The annual Japanese catch of *Etrumeus* herring is some 10,000 tons.

The anchovy family (Engraulidae) differs from the herring family by the prominent protruding upper jaw. Fifteen genera have been identified with a total of some 100 species in the tropical and temperate regions of the northern and southern hemispheres. Distribution is chiefly in the Indian and Pacific Oceans. Anchovies school along the coast and some are also found in fresh water. Due to the great masses in which they occur the family has considerable commercial importance in several countries for the production of fish meal and oil.

Seven species in the main genus *Engraulis* have been identified from the Pacific and Atlantic Oceans. This includes the ANCHOVY *(Engraulis encrasicholus)*. L reaches 20 cm, but is usually 12–16 cm. In addition to the protruding upper jaw the species has a very large mouth opening. The pectoral fins lie in front of the dorsal fin. Coloration resembles that of herring, with silver lateral stripes. This is a schooling species of the open sea. Distribution is chiefly in the Mediterranean Sea and Black Sea as well as on the Atlantic coast of southwest Europe and north Africa. In the north the distribution extends to the English Channel, and to the south along the African west coast from Togo to Dahomey. Anchovies are also found in the Sea of Asov, in the southern North Sea and in smaller numbers as far as Bergen, Norway. Five subspecies are known, which are distinguished by a number of characteristics, particularly in terms of L.

In their chief distribution region, e.g., in the Mediterranean, anchovies migrate little. In spring and summer they appear in great schools at the surface of the water both in the open sea and along the coast. They spawn in the Mediterranean from April to September. During

Family: anchovies

▷
Herring: Atlantic herring
(Clupea harengus).

Fig. 10-16. Anchovies
(Engraulis).

this time they are also fished. After spawning, at the beginning of winter, the adults and the subadults from the spring and summer spawn move into depths of 100–150 m. The schools probably dissolve at this time, and the fish remain in some small region at the floor of the sea; at least this is what stomach content studies indicate. The fall brood, however, remains in the upper water levels off the coast.

Anchovies in the bordering regions of their distribution are migratory, however. They migrate from wintering grounds in the Black Sea in early spring to the Sea of Asov, where they spawn and return to the Black Sea in fall. In the northern regions anchovies also migrate in great schools to the north and northeast. They move through the Bristol Channel into the Irish Sea and to the west coast of Scotland, where they can be found from May to September. Spawning grounds are presumably located in this area, since specimens found here are mature for spawning.

In spring the anchovies migrate in great numbers through the English Channel to the North Sea. They move along the French-Belgian-Dutch coast to the East Frisian coast. The spawning grounds were located here in 1930, and considerable fishing activity developed here, particularly in the Zuyder Zee. Since this time and particularly after World War II it was observed in the southeastern North Sea that the numbers of anchovies decreased and that they also spawned there. Eggs and larvae were found as far as the North Frisian islands. The spawning period was in warmer brackish water from June to August. Climatic changes probably acted as a factor in the spread of anchovies to the north, as was the case with sardines. In the fall the anchovies migrate through the northern part of the English Channel along the English coast to wintering grounds off the west exit of the Channel.

The number of eggs varies between 13,000 and 20,000; they are laid in the open water in groups. After a year the fish are 9–10 cm large and spawn for the first time in the following summer at a length of 12–13 cm. The Sea of Asov anchovies, the smallest variety, grow at a slower rate. Anchovies feed on small plankton, chiefly on crustaceans of various families. Fish eggs have also been found in their stomachs. The anchovies are prey to many predatory fishes and marine birds.

The annual catch of anchovies in 1958 amounted to some 180,000 tons and by 1967 had climbed to about 340,000 tons. The chief countries fishing anchovies include the Soviet Union, Spain, Italy, Turkey, Yugoslavia, Greece, France and Portugal. Anchovies are caught with drift-nets, baskets, ring nets and to some degree with trawl nets. They are usually sold in a salted form, in which the head and insides are removed. After an aging period of four to eighteen months, during which time the flavor improves, the anchovies are ready to be sold.

◁
Herring: 1. Shad *(Alosa
alosa)*; 2. Twaite shad
(Alosa fallax).

11 Osteoglossids and Mormyrids

The OSTEOGLOSSIDS or BONYTONGUES (Osteoglossiformes) are pure tropical fresh water species (with the exception of North American mooneyes). The dentition rests on the parasphenoid, glossohyal and basihyal bones. There are generally bony paired shafts at the base of the second hypobranchial arch. Bonytongues are predators, explaining the well-developed dentition (exception: the African bonytongue). There are two suborders: 1. BONYTONGUES, containing osteoglossids and butterflyfishes, and 2. FEATHERBACKS, with mooneye and featherback families.

Order: bonytongues, by W. Ladiges

Distinguishing characteristics

The osteoglossid suborder Osteoglossoidei contains the BONYTONGUES (Osteoglossidae), characterized by an elongated shape, well-developed scales, large eyes, a bony covering on the head, long dorsal and anal fins, and deeply oriented pectoral fins. Distribution is peculiar: three of the six species inhabit South America; one is in Africa; and two are found in the Malayan-Australian region.

Family: osteoglossids

The most well-known member is the GIANT ARAPAIMA (*Arapaima gigas*), distributed in the Amazon and its tributaries. Its common name stems from the Indians of the Guianas. In Peru the species is known as "paiche"; in Brazil it is called "pirarucu". This is one of the world's largest fresh water species, being exceeded in size only by sturgeon and a few tropical catfishes. Actual size estimates of arapaima vary considerably. Thus in 1836 Schomburgk, in his description of the fishes of British Guiana, reported a L of 1.50–4.47 m and a weight of 27–186 kg. The latest work by Lüling on this species (from 1964) cites the largest observed specimen to be a male 2.32 m long with a weight of 133 kg. No doubt the intense fishing of the area has led to the disappearance of this species in wide regions, especially in the major Amazon currents, and thus the overall size has decreased considerably. In 1942 the American researcher Allen wrote:

"The species has therefore disappeared from inhabited regions and even in uninhabited areas is rarely encountered. I still found them in the Rio Pacaya, but not in great numbers even in this desolate area". Lüling also made some observations here. These consequences of human destruction do not surprise us, since in 1870 Marcoy authentically reported "that the catch by a single village amounted to 10,000 pirarucu and 4,000 sea-cows in a quarter-year".

The harpoon is generally used to catch these fishes. The hunters generally lie in wait in their canoe, often on an artificial obstacle which slows the water passing from the main current to the tributary waters. The fish is dried, and is marketed in rolled bundles. The species is important throughout the Amazon region as a commercial fish and has important nutritional value, even though according to European standards the flavor of the dried fish supposedly tends to be rancid and soapy. However, it is excellent when fresh. The scales and bones, especially of toothed species, are important to the indigenous Indian population, and the tongue is used as a file.

The behavior of this giant, which raises its own young, has only recently been investigated by Sanchez and Lüling, and only some aspects of its behavior are understood. Lüling found that the arapaima inhabits the flooded region of the "Varzea" in areas dense with aquatic and shore plants. As a predator the species avoids the acid waters of the so-called "black water zones" where other fishes are not prevalent. The chief spawning period occurs in shallow lakes of the flooded region in October and November. The female spawns on places free of plants, which are worked into nests. "Mouth incubation" has not yet been confirmed in this species, but cannot be refuted either. The father guarding the eggs is known by Fischer at least to take the eggs in its mouth and move them to another location. The young are led by the male in a group once they are able to swim. Lüling notes that they remain close to the head of the father.

Whether and to what extent the secretions of the glandular areas of the head are related to caring for the young has not been clarified. Lüling has described the secretion: "This substance is suspended in the water and functions as a means by which parents and young are kept together, whether swimming for longer stretches or if they become separated". On the contrary, Sanchez feels that the secretion is related to marking the territory. It is secreted when the arapaima makes its noisy leaps into the air, which so characteristically betray its presence.

The arapaima has the ability, along with many fishes found in oxygen-poor water, to breathe atmospheric oxygen with the swim bladder. It is delicate to keep in an aquarium, but recently individual arapaima have been kept successfully. The light reflecting from its large scales makes an interesting play of hues.

Another species found in the same region is the ARAWANA (*Osteoglossum bicirrhosum*); L reaches 60 cm. In 1966 another species was discovered: *Osteoglossum ferreirai*. This elongated species with its metallic shimmering hue is greatly compressed along the sides. It generally swims about just below the water surface and prefers the cover of shore plants. Both barbs on the lower jaw are extendable and can touch objects about the fish. This is a predatory fish, but as a juvenile shares the characteristic with arapaima of feeding on all kinds of small organisms, particularly shrimp. The species has often been kept in aquariums. It has recently been confirmed that mouth breeding takes place, using a pocket-like area between the lower jawbones. Mouth breeding also occurs in the closely related Malayan and Australian species described below.

Fig. 11–1. Distribution of bonytongues (Osteoglossiformes).

These species are the MALAYAN BONYTONGUE (♂ *Scleropages formosus*) and the AUSTRALIAN BONYTONGUE *(Scleropages leichhardti)*. In 1905 Fuhrmann in Paris first described mouth breeding in the Malayan species. The species is distributed in Sumatra, Borneo and the Malayan peninsula, and its shape resembles that of *Osteoglossum*, but is longer and less compressed. The barbs are also shorter. The colorful hues of its scales with their golden green sheen equal those of *Osteoglossum*. The Australian species has been studied very little; photographic evidence indicates that it is also distributed in New Guinea. The Australian and Malayan species are characterized like *Osteoglossum* by a deep mouth opening which is directed diagonally upward.

The AFRICAN BONYTONGUE (*Clupisudis niloticus*; L at most 90 cm) resembles the arapaima in shape. This species preys on small organisms and has less dentition as a result of this behavior; this is distinctive among osteoglossids. For breeding, the species builds a nest with a surface diameter of as much as 120 cm; the aquatic plants removed from the nest site form thick walls around the edge. The eggs are relatively large (approximately 2.5 mm). Initially the larvae have external gills. As in the other species of osteoglossids the swim bladder is modified to form a suprabranchial organ for respiration; the gill apparatus has regressed considerably.

The BUTTERFLY FISH family (Pantodontidae) consists of a single species, *Pantodon buchholzi*. L reaches 15 cm. The winglike pectoral fins are considerably enlarged. The thread-shaped elongated rays of the pelvic fins function as tactile receptors. The head and back are flattened, and the dorsal fin is located well to the rear of the body. In contrast to other members of the suborder, the scales lack ornamentation. Distribution is limited to tropical west Africa in a few river regions (primarily the Niger and Congo systems).

Family: butterfly fish

The butterfly fish is found at the surface of the water, where it remains as if hung there. The diet consists chiefly of flying insects; small fishes may also be eaten. In aquariums meal worms are readily eaten. The species has been successfully bred in captivity. Eggs float freely on the surface. Raising young butterfly fishes is difficult; they themselves resemble small insects. They can only be kept by feeding them small springtails, aphids, and later, fruit flies. The flight ability of the species has been a subject of great controversy. Accounts by Buchholz traveling through Africa that the fish flies about like a butterfly are pure legend. The large pectroal fins enable it simply to leap into the air, rarely as much as two meters out of the water. Thus any aquarium in which butterfly fish are kept must be well-covered.

Suborder: featherbacks

The MOONEYES (Hiodontidae) in the featherback suborder (Notopteroidea) look like salmon. On the basis of anatomical evidence they belong nevertheless in the featherback group, which they do not resemble externally. The two identified species inhabit North America. The MOONEYE (*Hiodon tergisus;* L 40–50 cm), with strikingly large eyes, is a popular sport fish. *Hiodon alosoides,* distributed in North America from Saskatchewan to the Gulf of St. Lawrence and Ohio, resembles *H. tergisus.* The third species, *H. selenops,* is now considered identical with *H. tergisus.*

The FEATHERBACK family (Notopteridae) is characterized by an elongated, laterally compressed body which tapers to the rear. The anal fin is very long, beginning under the head and joining the caudal fin. The dorsal fin is very small, and is absent in *Xenomystus nigri.* There are two genera with four species, distributed in the tropics of India, southeast Asia and Africa.

Fig. 11-2. Mormyrid (Mormyridiformes) distribution.

The featherback *Notopterus chitala* is a fairly well-known species from the Indian-Malayan region. This important commercial species attains a L of as much as 75 cm. The male guards the eggs, laid on wood or plants, and also watches over the young for a short time. *Xenomystus nigri,* from tropical west Africa, is a smaller species and is kept in aquariums. The large members of *Notopterus* are predatory in the adult stage; they have deeply slit mouth openings. The exceptional *X. nigri* feeds on small organisms, and is active at night and shuns light.

Order: mormyrids, by J. Géry

The MORMYRID order (Mormyridiformes) is most probably a highly specialized branch of the mooneyes and featherbacks. Dentition resembles the osteoglossids, but they feed chiefly on insects and bottom-dwelling organisms. The sides are generally well compressed and coloration is dark. There is great variety in size and shape among the various species. The mouth opening is quite small with conical, two-pointed teeth on the jawbones (in some species on the parasphenoid). The palate and pharynx lack teeth. The bones about the eyes are greatly flattened. The swim bladder is connected with the foregut by a canal, but in larval stages is connected to the inner ear. There are two families: mormyrids and gymnarchids. Distribution is limited to fresh waters of tropical Africa.

The striking quality of mormyrids is that they possess an electrical "system", which is characterized by a series of distinctive anatomical features: an organ oriented toward the rear which generates electricity, and the top and bottom of which are puzzling "Gemminger bones"; the highly modified skin with unusually many cell layers; the numerous small lateral line organs on the front of the body; and the striking cerebellum, which is overdeveloped and covers almost the entire remaining parts of the brain (not the case in gymnarchids). Mormyrids have four electrical organs: one rear and one bellyside pair in the form of long cylinders, which are as long as the tail shaft or one-fifth the entire length of the fish. Each individual organ is insulated by tissue and at the same time is divided diagonally by that tissue into a number of cells, in which the current-generating elements are located. Investigations have disclosed that these elements arise from modified, greatly flattened muscle cells. Each organ contains some 120–200 plates—thus a total of 480–800. These plates are extremely thin, measuring about .01 mm each. Each plate has nerve connections on the rear surface, the nerves coming from the upper and belly sides of the central nervous system; their connection with the plates is a highly complex one.

A single species, the GYMNARCHID, which is classed into an entire family, has eight narrow tube organs which are of different lengths. They run in four pairs beginning in the middle of the body between the tail musculature. One pair is located on the belly, one on the back, and two in the middle. Each tube is divided into various numbers of electroplates, never more than 140. Nerve supply is from motor roots of the central nervous system and through four strands running parallel to the axis of the body, from which small nerves branch off. All these nerves originate from approximately 400 nerve cells in the central nervous system.

In both families the epidermis is covered by thick mucus. This is composed of several cell layers, the uppermost of which being of flat, six-sided cells in three or four rows. Small openings can occasionally be seen between these cells; they lead to sensory organs in the skin known as mormyromasts. They are particularly prevalent in the head region. The function of these was long an open question. Today it has been established that these are reception sites for the electrical field. The gelatinous mass which comprises their hollow space is connected with sensory cells at the base of the hollow area, which has a rich nerve supply. This area conducts well relative to the outside of the skin. This system, which is commonly found in fishes and tadpoles as part of the lateral line system, and the particular development of the nerves and the nervous system (the brain weighs one-fiftieth as much as the entire body!), have resulted in great sensitivity on the part of this fish: variations as little as three-millionths of a billionth of an ampere can be detected! A gymnarchid can react to an electrical field of 0.15 mV/cm.

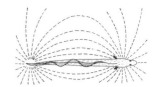

Fig. 11-3. Electrical field of a gymnarchid.

Bonytongues: 1. Arowana (*Osteoglossum bicirrhosum*); 2. Arapaima (*Arapaima gigas*).

to their hiding places, often in schools, and here the electrical impulses probably function in forming the schools. These impulses are perhaps of even more importance in those species (like *Petrocephalus*) which do not conceal themselves but swim in schools near the bottom. For them their electrical fields are more important in terms of living together than in locating objects.

Members of Mormyridae have various forms, but generally the sides are compressed and the body is relatively long. The rear part of the body and the caudal fin are less developed than in other fishes. All fins are present and lack spines. The dorsal and anal fins can have quite different lengths, and usually one is considerably longer than the other. This is a similar situation to that of the gymnarchids and eels, in which one of the two fins is completely absent and the other takes up most of the L. The mouth opening is small, and a tubelike projection is sometimes found at its terminus; four stages of development of this projection are found in *Mormyrops, Mormyrus, Campylomormyrus* and *Boulengeromyrus*. The hyoid bone arch and tongue often have teeth. The swim bladder is simple and rather long. Gill openings are very small, and breathing movements are quite fast. The Mormyridae mormyrids are nocturnal fishes dwelling near the floor. There are thirteen genera with more than 150 species.

Those living in groups include *Petrocephalus* and *Marcusenius*; they live in great schools. Some others are very territorial. Practically all species feed on worms, small larvae and insects dwelling on the floor; they also take refuse and plant matter. The method of reproduction is hardly known; some guard the eggs. The sexually mature males have a greatly twisted anal fin, which perhaps is important in mating.

The *Marcusenius* mormyrids are possibly the earliest group, but not on the basis of anatomical characteristics. There are more than 65 species in this genus, and all have few teeth in both jaws. The dorsal and anal fins are not large and both are about the same length. A round projection is often located behind the mouth. The arched line of the head led to the German name of this fish, meaning PARROT MORMYRID. The size of *Marcusenius* mormyrids varies between 6 and 45 cm; the largest species include *M. stanleyanus* and *M. montede*. There are elongated forms such as *M. sphecodes*, which are similar to the *Isichthys* mormyrids (which also have elongated bodies); and small, short species also exist (e.g., *M. isidori*, which resembles *Petrocephalus* members). Due to this great diversity this large genus may someday be divided up.

The *Petrocephalus* group, rather primitive forms with 16–18 species, and *Stomatorhinus* mormyrids, with 12 species, are small fishes less than 20 cm long; they are compressed and short and often have the mouth opening on the lower side of the body, as in "parrot" mormyrids. These groups differ in rather inconsequential details, such as the

Family: mormyrids

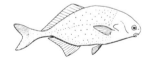

Fig. 11-4. *Marcusenius* ("parrot") mormyrid.

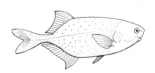

Fig. 11-5. *Petrocephalus* mormyrid.

12 Salmon

Order: salmon

The order Salmoniformes consists of eight suborders, with some quite unlike others. The last five consist almost exclusively of deepsea fishes: 1. Salmon (Salmonoidei) with three families; 2. Galaxiids (Galaxioidei) with four families; 3. Pike (Exocoidei) with two families; 4. Argentines (Argentinoidei) with three families; 5. Stomiatoids (Stomiatoidei) with eight families; 6. Deepsea slickheads (Alepocephaloidei) with one family; 7. Deepsea smelts (Bathylaconoidei) with one family; 8. Lanternfishes (Myctophoidei), with fifteen families.

Suborder: salmonids, by L. Karbe

Salmonoidei contains a number of the most well-known species of this fish order. A rayless dorsal fin is always present; parapophyses are not joined with the vertebral column; oviducts are absent or are incompletely developed. These are primarily migrating species and fresh water fishes of the northern hemisphere. Breeding takes place in cold, oxygen-rich water. There are three families present today: 1. Salmon (see below) with three subfamilies; 2. Ayu; 3. Smelt. There is also one extinct family, the thaumaturids (+ Thaumaturidae; Middle Eocene to Lower Pliocene, 40 to 25 million years ago).

Distinguishing characteristics

Many of the salmon species are good eating not only because of their flavorful, fatty meat but also because they lack those bones which in most fishes are embedded in the cartilaginous walls between the muscular segments.

Family: salmonids

Salmonidae contains such familiar fishes as salmon, trout, and charrs. Medium-sized fish, these species are rounded with numerous small scales (with generally more than 120 in the fully developed lateral line). The mouth opening is wide and has powerful teeth. The vertebral column arches upward at the base of the caudal fin. An additional bone (the supraoperculum) is located at the gill cover, and the dermosphenoticum (which belongs to the auditory capsule) is absent. There are three subfamilies: A. Salmon; B. Whitefishes; C. Graylings.

Distinguishing characteristics

Subfamily: Salmoninae

The salmon subfamily (Salmoninae) contains five genera: 1. Salmon (*Salmo,* which also contains trout); 2. Pacific salmon (*Oncorhynchus*); 3.

Fig. 11-6. *Stomatorhinus* mormyrid.

number of teeth and the position of the front nasal opening, which is near the mouth, but not as close to the mouth as in many eels (where it is in the upper lip). All these small species live in large schools, like "parrot" mormyrids, either near the floors of large rivers or in plant growths of smaller rivers. They stand there motionless, head downward. They feed on small prey; stomach contents have consisted of Tubificid worms (see Vol. I) and midge larvae (Chironimidae; see Vol. II), mixed with sand and refuse.

Fig. 11-7. *Myomyrus* mormyrid.

The *Myomyrus* genus, with just a few species, contains members reaching a L of 35 cm. They are very slender, but are also characterized by the enlarged lower teeth which point upward and the rather long dorsal fin, which also exists in *Mormyrus* species. *Hippopotamyrus* mormyrids have similar dentition which perhaps belong to another evolutionary line. In *Mormyrus* the dorsal fin is much longer than the anal fin; this development reaches its culmination in *Mormyrus.* The fifteen species in this genus contain relatively large fishes. Some have a L of 70 cm. They often have a long snout, albeit not as long as in *Campylomormyrus.* One recently described species, *Boulengeromyrus knoepffleri,* is particularly distinctive in that it looks like a "parrot" mormyrid on which a long *Mormyrus* nose has been placed. All these species have a somewhat large mouth opening. They feed (as does the "parrot" mormyrid) on large insects and shrimp which they find in the mud. *Hyperopisus bebe,* which has a L of just 46 cm, has a small dorsal fin but a long anal fin with up to 71 rays. On the basis of the cerebellum, *Hyperopisus* and *Mormyrus* can be considered to be highly developed forms.

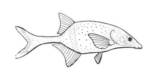

Fig. 11-8. *Boulengeromyrus knoepffleri.*

The genera *Isichthys* and *Mormyrops* contain one of the most striking adaptations in the order, for these species are elongated like eels. *Isichthys* contains one species and is distributed along the coast of western Africa. It is not found in the Congo basin as are the other genera. The dorsal and anal fins are well developed. *Mormyrops* is represented by 16–18 species, all of which have long snouts, unlike *Isichthys.* Sometimes the snout is elongated by as much as 80 cm, as in *Mormyrops boulengeri.* Furthermore they possess more teeth. These highly specialized, large species are predators and feed on shrimp, insect larvae and bottom-dwelling fishes. They guard a definite territory and are not social creatures. As in *Isichthys* they most resemble the pike eels (Muraenesocidae).

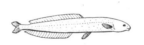

Fig. 11-9. *Isichthys* mormyrid.

The mormyrid genera *Genyomyrus, Gnathonemus,* and *Campylomormyrus* have chin appendages as adaptive structures, which are sometines very long and can be pointed. They apparently play the same role as the barbs of carp.

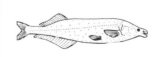

Fig. 11-10. *Mormyrops* mormyrid.

Genyomyrus donnyi (L up to 45 cm) is the single, very rare species in the genus. In contrast to all other mormyrids this species has a brushy dentition.

Gnathonemus (L 20–25 cm), with its few species, resembles the "parrot" mormyrids in dentition and body form, but has a long barb. In spite of its nocturnal habits, the most prevalent species (*G. petersi*) is a fascinating aquarium fish. Meder reports that this mormyrid plays with a small ball much as young mammals do. "Playing" is definitely unusual in fishes and this is an indication of a well-developed brain in this species, which it indeed possesses.

Fig. 11–11. *Genyomyrus* mormyrid.

The fourteen species in *Campylomormyrus* (L in some species is 60 cm), which were once classed with *Gnathonemus*, are the most unusual of all the mormyrids. Most species have a "trunk", the length of which can sometimes equal half the L and which is more or less curved. With few teeth and the touching or tasting appendage on the end of the snout they are capable of finding larvae in the smallest crevices.

The GYMNARCHID family (Gymnarchidae) is closely related to but clearly distinct from the mormyrids. There is but one genus and species: *Gymnarchus niloticus* (L up to 1.6 m). The large eel-shaped fish inhabits swampy districts of the Nile, the Chari, and the Niger Rivers. A nocturnal creature, the species preys upon fishes and shrimp. It is a solitary species which guards its territory and which builds a nest in the grass during the spawning period. There it also guards the eggs. The gymnarchids have evolved some distinctive structures as an adaptation to their bottom-dwelling life. It is the longest of all mormyrids and has a thread-shaped tail lacking a caudal fin. Only the dorsal fin is well developed and it runs along the entire length of the body with 200–215 rays. There are numerous small scales. The head has an elongated snout portion and the large jaws have many cylindrical teeth in each side. Teeth are absent on the tongue and palate. The gill opening is large and the gill skins are joined with the throat. The eyes are extremely small.

Family: gymnarchids

Fig. 11–12. Distribution of the gymnarchid (*Gymnarchus niloticus*).

The gymnarchid swim bladder, described by Hyrth more than 100 years ago, has many blind alveoli. This permits the fish to respire by breathing air when the marshes have dried out. The cerebellum is the most poorly developed of all mormyrids'. However it can perceive changes in electrical fields, an ability found in many members of the animal kingdom.

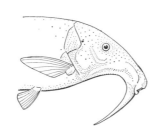

Fig. 11–13. *Campylomormyrus* mormyrids showing the elongated nose.

Hucho; 4. Charrs (*Salvelinus*); and 5. *Brachymystax*. Many of the species tend to form local populations, so that classification is often difficult.

Before our waters were polluted by waste materials and altered by various structures, the ATLANTIC SALMON (*Salmo salar*) was one of the most prevalent fishes in the Atlantic drainage areas. Its distribution extended from Kara in northeastern Russia along the coast of Europe to Douro in the northwestern part of the Iberian Peninsula; and on to Iceland, the southern tip of Greenland and across Newfoundland to Cape Cod in the northeastern United States. Salmon migrate. The early part of their life is spent in the upper courses of large rivers; then they migrate into the sea, where they grow relatively quickly, and then return to swim up rivers for spawning. During their stay in the oceans salmon also traverse great distances. Thus, salmon marked off the European coast were recovered in waters off western Greenland. Generally they stay near the shore. Feeding grounds are primarily in the southern Baltic Sea and off of northwestern Norway. When preying upon other fishes they are found in the upper water levels to a depth of 10 meters, but also penetrate deeper.

The Atlantic salmon

Some studies indicate that their distribution at various depths depends upon daily and seasonal changes. During their period in the sea salmon grow at a remarkable rate, often exceeding one kilogram per month. They spend one to three years in the ocean before returning to the rivers to spawn. During this period they have stored great quantities of fat, so much that their skin is orange-red. They had left fresh water when they were 10 to 20 centimeters long. After a year in the ocean they measure 50 to 65 cm in length and weigh 1.5–3.5 kg; after two years their L is 70–90 cm with a weight of 4–8 kg. After three years salmon are 90–105 cm long and weigh 8–13 kg. Salmon probably reach a maximum age of 10 years. Occasionally old males 150 cm long weighing as much as thirty-six kilograms are caught. Females are generally smaller, and rarely are they longer than 100 cm. Their greatest weight is twenty kilograms.

Fig. 12–1. Distribution of the Atlantic salmon (*Salmo salar*).

Salmon from various rivers meet at the feeding grounds. When the spawning season comes, they separate once again and each salmon seeks out the river in which he was born. The exact manner in which salmon find their way back to their home river is not understood. It is only known for certain that their olfactory sense plays a crucial role in the second phase of their ascent up the river. This ascent takes place throughout the year in some rivers, while in others it is found only at certain seasons. Often larger salmon are seen ascending a river at one period, while smaller ones are found at some other time. Four major types are distinguished: large and small summer salmon and large and small fall or winter salmon.

When they meet at the river mouths the salmon can be distinguished by the development of their gonads. The individual groups then seek

Fig. 12–2. Salmon migration routes, determined from marking experiments.

out their various spawning sites. If these sites are far from the mouths of the rivers, it is generally the large fall salmon which ascend the river. Their germ cells are still immature when they begin their ascent. The ascent is interrupted by the onset of frost; at this time the salmon winter somewhere upriver and will reach their spawning sites in the following fall. One large portion of the Rhine salmon, which once came in droves from the Rhine across the Aare to the upland sources of the Reuss, Limmat and Linth, were of this type. From the beginning of the ascent they spent fifteen to sixteen months until they spawned in the Rhine. The large summer salmon behave differently. They begin ascending in summer with mature germ cells; they spawn that fall. In the Rhine the salmon cover the distance from Holland to Switzerland in 45 to 60 days. This corresponds to a daily distance of twelve to fifteen kilometers. Similar figures have been derived for other rivers.

While the energetic large salmon are capable of swimming great distances upstream, the smaller species generally find spawning sites near the river mouths. Each day these salmon cover greater distances. In small Scottish rivers daily distances of up to fifty-four kilometers were recorded. One must also consider that the salmon only migrate for five to six hours per day, during which time they swim with great strength and duration. Their highest speed has been estimated at sixteen kilometers per hour for short stretches. Over longer distances they can maintain a speed of thirteen kilometers per hour. They can also swim through rapids. Against a current of six meters per second a salmon can still push ahead at a rate of over one meter per second. Small waterfalls are passed by jumping over them, and leaps of up to three meters high and five meters long have been observed.

To jump out of the water, the salmon swim up through the water surface on a slant, and with a particular strong beat of the tail they gain additional acceleration at the surface. While jumping salmon typically show a distinct lateral arching of the body. If the first attempt is unsuccessful, the leaps are repeated constantly. This often results in skin injuries in stony waters; these injuries can quickly become infected and result in death before the spawning site is reached.

During this entire migration the salmon virtually cease feeding from the time they enter the river. However the biting reflex is maintained for several weeks, particularly in the northern and eastern parts of their distribution. During their journey their fat reserves are converted into energy, and the orange-red hue of the skin disappears. As the germ cells mature the salmon also alter their appearance. In the ocean salmon have rather plain coloration, with a gray-green back, silvery sides and white belly. X-shaped spots are above the lateral line and round black spots mark the head. However when migrating upstream a brilliant coloration develops. The back becomes considerably darker, while the sides take on a bluish shimmer, and the stomach becomes

Fig. 12-3. Salmon with plain coloration.

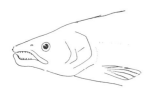

Fig. 12-4. Salmon with upward arched lower jaw.

reddish. Purple-red spots appear beside the black ones, and even the lower sides of the pectoral, anal and caudal fins take on reddish hues. This display coloration is most striking in the males. In fact the color change is accompanied by a major anatomical modification of the lower jaw. The jaw points upward and develops cartilaginous growths from which a hook-shaped appearance develops.

The salmon spawning period in central European waters occurs generally from mid-November to mid-December. In the north the onset of spawning is in mid-September, and in some groups may last until February. Spawning sites are located in regions with clear, cold, oxygen-rich, fast-flowing water and with a clean gravel bottom. Salmon typically seek gravel banks in the upper water levels at a depth of about half a meter. If such sites cannot be found, spawning may also take place in the upper courses of the major rivers at depths of two to three meters.

Once the female arrives at the spawning site she prepares the nest. The female digs up the floor with powerful rump and tail motions and constructs a depression ten to twenty centimeters deep and often well over one meter long. Bohemian fishermen once said that a horse would fit in one of these nests. Building this nest can last a long time and is independent of the spawning act itself. During the entire period there are usually several males in the vicinity of the female, but they never assist with the construction of this pit.

The construction of the pit and the subsequent spawning are carried out in a very similar manner in all salmon species. Spawning is always preceded by a courtship display. The male which is victorious in the vigorous rivalry contests approaches the female repeatedly from behind, shaking and spreading its fins, and it shoves the female's side with its mouth. The next sequence is side-by-side swimming. Finally both partners are pressed close together with mouths open just above the spawning pit, and the eggs and sperm are released. Eggs are released several times, interrupted by additional rivalry fights and more courtship display behavior. One female frequently spawns with several males, one after the other. At the conclusion, the female closes up the spawning pit, and then may proceed further upstream to spawn once again. The entire spawning process may last several days, or even more than a week. Spawning behavior is most prevalent in darkness.

A single female lays a total of from 10,000 to 30,000 eggs—it lays about 500-2000 eggs per kilogram body weight. This is actually a low number compared to other fishes; carp, for example, lay about 100 times the number of eggs per unit body weight. At the conclusion of spawning the salmon are so exhausted that many of them (primarily males) perish. They have lost thirty to forty percent of their body weight since leaving the ocean. Those surviving fishes winter in deep pools, while others drift along with the current into the sea. Those few which return alive to the ocean recover rapidly, and may gain one kilogram in a week. After one or two years they return to the rivers, but out of 100 salmon only four to six are able to spawn twice and at most one will spawn a third time.

The yolk-rich, sticky eggs are five to seven millimeters large and according to the water temperature will lie between the stones in the spawn pit for 70–200 days. The young hatch in April or May. As long as the larvae still have yolk upon which to feed they remain hidden in the pit. Once the supply is exhausted they begin feeding independently. They move into the water and initially feed on small crustaceans and insect larvae. They have a juvenile coloration which is common to every species in the family: a row of eight to ten dark diagonal stripes or spots are on the sides. As they get older, the young salmon increasingly feed on fishes, and at the end of the juvenile period they feed exclusively on small fishes such as minnows, loaches and skulpins. The young stay in fresh water for one or two years (in the north up to five years). Slowly the juvenile coloration disappears and the typical oceanic coloration appears when the salmon are ten to twenty centimeters long.

Fig. 12–5. Juvenile coloration in a salmon.

Some of the male salmon already reach sexual maturity in this initial period in fresh water, without migrating into the sea. They are ten to fifteen centimeters long at this point. Pure inland populations have even developed in lakes which have no access for salmon to the sea. These salmon remain in smaller, nutrition-poor lakes (in Norway, northeastern North America, on Labrador and in the state of Maine) and do not grow as much as migrating salmon do. In some large inland lakes, such as Vänersee in Sweden and the Onega and Ladoga lakes in northwestern Russia, the inland salmon grow as well as migrating species, with maximum lengths of almost one meter and a weight of twelve kilograms.

Wherever salmon still migrate up rivers, they present a most fascinating process. These fishes have almost completely disappeared from the Rhine, Weser and Elbe Rivers in Germany. A few isolated salmon still appear in the lower Rhine. They are inedible, for their fat has been contaminated with phenol wastes in the river. In Europe, the decline of the salmon began in the mid-1890's as the rivers became polluted. In England industrialization began earlier, and thus the Thames, once a salmon river, became contaminated still earlier. The last salmon caught in the Thames was caught in 1833. The only remains of the once rich growths of salmon in so many rivers are the old reports which we still have. One well-known case is a memorial which Ludwig of Hessia set up on the old town hall of Kassel in memory of a record catch in the Fulda River: "Anno Domini MCCCCXLIII auf Bonifacien Tag hand unser gnaediger Herr von Hessen 800 Laesse gezogen mit einem Zug un 2 Laesse un ein Hecht also guth als der Laesse einer."

Until recently the region off the East Prussian coast was the most productive one for German commercial fishing. At the turn of the century German fishermen repeatedly took more than 1,000 tons per year of salmon. Such catches cannot be made today, although recently the salmon population in some tributaries of the Baltic Sea (e.g., the Oder River) has increased somewhat. Today most European salmon are caught in Norway and Denmark. Canada has the greatest catch in the world, annually amounting to 1500–1800 tons. Salmon are generally smoked.

The BROWN TROUT (*Salmo trutta*) is much more prevalent in European waters than salmon. It differs from salmon in appearance as well as behavior. Many subspecies have been described, and in some cases it is still unclear as to whether these are subspecies or individual species in their own right. Basically three types of trout may be distinguished: 1. OCEANIC TROUT (*Salmo trutta trutta*), which migrate like salmon; 2. LAKE TROUT (*Salmo trutta lacustris*), which inhabit large inland lakes and ascend tributary rivers to spawn; 3. RIVER TROUT (*Salmo trutta fario*), small fishes which remain in one area. Salmon and oceanic trout are often confused. The differences between the two species are so slight in the Weichsel River that it is indeed difficult to separate the two. In other rivers there are distinct coloration differences; the trout can be distinguished by the marks on the dorsal fin. Generally trout are plumper than salmon; they have a higher tail shaft; their head is less pointed, and the eyes are more forward, so that the upper jaw extends to the rear edge of the eye. The vomerine dentition of the trout is a relatively good distinguishing mark as are the gill rakers on the first gill arch: in salmon all gill rakers are rod-shaped, whereas in trout only the middle ones are so shaped.

Migrating trout are found on the European coast from the Tscheskaya Bay in the White Sea to Douro, Portugal, around the British Isles and down to the South Islands. During the Ice Age trout were also prevalent in south European seas. Today the Mediterranean is too warm, but populations are maintained in the Black Sea, Caspian Sea and Aral Sea. Trout is apparently better adapted to fresh water than salmon. Thus they grow better in fresh water than do salmon, but do not do as well as salmon in salt water.

In the ocean trout generally stay near river mouths. Well before the onset of spawning they move into the lower courses of rivers, but the migration itself extends over a long period of time. The ascent extends much further than in the case of salmon, and reaches into the upper courses of rivers (some sites are chosen in the middle courses). Thus oceanic trout seldom ascended the Rhine further than the mouth of the Main River. Today oceanic trout have completely disappeared from the Rhine. During the spawning period the belly and sides of the males are yellow and orange. Oceanic trout are distinguished by a pronounced spotted marking, which in contrast to the salmon court-

The brown trout

Fig. 12–6. Distribution of the trout (*Salmo trutta*).

Fig. 12–7. The trout vomer bone, front view (left) and side view (right).

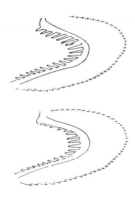

Fig. 12-8. Gill structures: salmon (above) and trout (below).

ship coloration extends beyond the lateral line. One portion of the spots is yellow or brownish-red to red.

The trout spawning period is somewhat later than in salmon, from December to March (in colder regions from October to November). Spawning behavior resembles that in salmon, with the exception that the eggs of one female are typically fertilized by sperm from but a single male. The number of eggs is up to 10,000 (e.g., about 1500 per kilogram of body weight), and the development period is shorter. It is noteworthy that oceanic trout do not become as weakened during their migration as salmon do. Thus most of the trout reach the sea again and spawn the following year.

Generally trout are smaller than salmon (L 80-100 cm; weight 10-15 kg at an average age of four to six years). The oceanic trout are considerably larger; at least those that ascend the Kura River from the Caspian Sea. In this region trout measuring 1.5 m with a weight of fifty-one kilograms are no rarity. Unlike other trout, these species spawn only once at an age of five to nine years. Other Caspian oceanic trout are smaller, become sexually mature in their third year, and spawn five to six times. In the Kura oceanic trout ascend at two seasons: in fall small trout with mature germ cells are found ascending; they go to sites not far from the mouth of the river; and in winter large trout with immature gonads begin migrating to sites far upriver. The sites are not even reached until the following fall or winter. These trout are behaviorally very similar to salmon. As in salmon there are dwarf males which become mature in the rivers without migrating in the ocean.

The lake trout which inhabit large, deep, cold inland lakes are rather similar to the oceanic trout. In central Europe they are solely distributed in the numerous lakes of the alps, where they can live at altitudes as high as 2000 meters. Even in the high mountain lakes they can reach a considerable size. The various populations have diverse growth conditions; generally lake trout are larger than oceanic trout. Also they are generally migratory and ascend the tributary rivers of their lakes to spawn. A few populations also spawn in rivers emptying the lakes. Those in Lake Geneva migrate first up the Rhone, and then in a tributary, the Arve. In many lakes, such as the Koenigssee, Schliersee and Chiemsee (all in Germany), and others, spawning takes place in the lake itself, usually at greater depths where underwater springs rinse the gravel bottom. Typically lake trout spawn earlier than their oceanic relatives: from September to December. In Lago di Garda in northern Italy a summer spawning group has developed; they are classified as an individual species, *Salmo carpio*.

The young lake trout stay in the spawning waters for one to three years. In the lakes they initially live in the upper water levels preying on small fishes; later they live at greater depths. Trout growth differs considerably in different lakes. In some lakes trout are never caught weighing more than five kilograms, while in others they commonly grow to a length of almost 1.5 m with a weight exceeding 30 kg. The time of transition from living at the upper depths to moving deeper also varies and is related to the age in which sexual maturity is reached (i.e., between three and seven years). Unfortunately the lake trout populations in most alpine lakes are constantly decreasing. The reasons for this include construction in and pollution of the spawning waters, heavy fishing, and noise pollution from motorboat activity.

The river trout are distinguished from oceanic and lake trout by their small size (L at most is about 50 cm) as well as their behavior. As inhabitants of cold, oxygen-rich streams and upper courses of some rivers they are distributed over all of Europe and Eurasia in many local forms. An entire series of river trout, the systematic position of which is still unclear, lives south of the Mediterranean in streams of the Atlas Mountains. Young river trout cannot be distinguished from oceanic and lake trout of the same age. The adult oceanic and lake trout have a relatively inconspicuous silvery coloration initially, interrupted by a few blackish spots. Stronger hues appear in the courtship display coloration. In contrast river trout assume the coloration of sexually mature adults when they are about ten centimeters long, which is at the end of the juvenile period. They have numerous blackish spots above the lateral line and eye spots surrounded by reddish edges along the lateral line and beneath it.

River trout coloration is very diverse—it may be characterized by blackish, dark forms in which the spots are barely noticeable, to pale species lacking spots entirely. These differences seem in part to be related to differences in diet, but primarily because of the floor of their home waters. Species with highly developed markings are generally residents of fast-flowing streams with clear water. The dark species are found in ponds with a muddy or marshy bottom. And a river with an ample gravel bottom produces river trout which have the silver-gray hue interrupted by blackish spots as in lake trout. In one case the trout in a stream altered their color when the stream was polluted by modeling clay. In fact trout have been called the chameleons of fishes, even though their change in color must take place over a long period of time. By setting out marked river trout it has been established that they not only develop the plain hues of the lake trout but also grow more rapidly. One can assume that such changes also take place without man's interference.

Like lake trout the river trout are found exclusively in waters which are oxygen-rich and cold, which in summer have a temperature of 10–18° C. They are not found where the water is colder because of the lack of sufficient food. In some cold mountain lakes with glacial formations trout have been found which grow extremely slowly and

▷
Trout breeding facility. The rainbow trout *(Salmo gairdneri)* has been artificially bred in Europe in great numbers.

▷▷
Rainbow trout shown squeezed together in the aquarium of a restaurant. This should be avoided!

▷▷▷
From top to bottom: Arawana *(Osteoglossum bicirrhosum;* see Chap. 11); pike *(Esox lucius)* in a mountain lake; the mormyrid *Gnathonemus tamandua;* the deepsea viperfish *(Chauliodus tamandua),* a deepsea predator with powerful teeth.

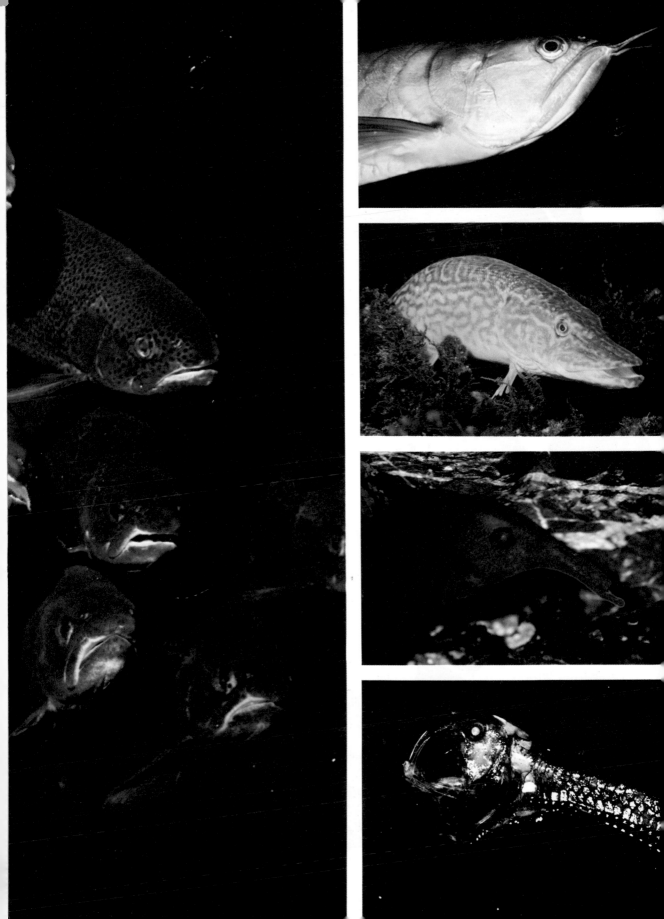

tip of Asia to southern California along the Pacific coast, the rainbow trout has also developed numerous local forms. The steelhead trout of America behaviorally resembles the oceanic trout. It generally attains a L of about 70 cm and weighs some seven kilograms. Steelhead trout are spring spawners, while other groups of rainbow trout spawn in the winter beginning in December. Their spawning behavior corresponds to that of other salmon species. It is interesting to note, however, that generally two males participate in spawning. One male is usually larger, and the other smaller, than the female. Like oceanic trout, the steelhead trout also survive the ascent into fresh water and spawning, in contrast to the Pacific salmon *(Oncorhynchus)* which has yet to be discussed, which ascend the same rivers yet which practically never survive spawning.

American investigators have shown that a calcification appears in salmon and trout while ascending rivers toward spawning sites that is similar to the human condition of arteriosclerosis (hardening of the arteries). This calcification also attacks the blood vessels of the heart and can kill salmon. In order to better understand this condition in steelhead trout, trout of all ages were caught in the ocean and in fresh water and the condition of their blood vessels was examined. Young trout and those which were in the ocean but still immature showed no abnormal conditions; it was at the end of the maturation period, partly just at the entry into the rivers, that the disease first set in. During the migration upstream the calcification increased steadily. Finally, all trout at spawning sites showed calcification. The astonishing fact is that apparently these pathological conditions disappear again when the trout re-enters the sea. At any rate, about half of all trout which were caught on their way to a second or third river ascent had completely normal cardiac blood vessels. That blood vessel diseases should disappear is otherwise unknown in nature. This regression of calcification has been shown to be due to the diet (fasting is accompanied by the most rapid appearance of arteriosclerosis) as well as a change in the hormonal state. Still, we cannot yet explain why trout and Pacific salmon behave differently in this regard.

In addition to migrating trout there are localized species in the mountain streams of the Sierra Nevada. It was from these that the rainbow trout which have been imported into Europe since 1880 have developed. The trout from the McCloud River in California's Shasta mountain range are particularly valuable. The great demand for rainbow trout has led to breeding steelhead trout and CUTTHROAT TROUT (an American trout which has been classified as an individual species *Salmo clarki*) and introducing them to Europe. Some failures in introducing rainbow trout to natural waters may have led to the fact that breeding migrating species ceased. Introduction of non-migrants has been very successful in many cases. Today many of the trout streams in Europe contain only rainbow trout, because they have forced other trout (the European stream trout) away. Both species prefer slightly different home waters so that they can be kept near each other.

Rainbow trout are not so much dependent on having a shelter as river trout are, and are less sensitive to polluted water. Therefore they can still survive in some water which has become intolerable to the river trout. Since rainbow trout can also withstand greater temperature variations than the river trout, they have successfully been introduced into water in which the summer temperature exceeds 20° C. for a long period of time. Sports fishermen also enjoy the rainbow trout because it is less shy and bites more readily. Rainbow trout eat good quantities and grow rapidly; in breeding stations they grow to lengths of 25 cm and a weight of 200 gm or more in two summers, at which time they are ready for consumption. The river trout, on the other hand, do not reach this size for another year. In natural waters old rainbow trout of the pure Shasta group are caught weighing several kilograms.

Pacific salmon

The PACIFIC SALMON (Oncorhynchus) form a single genus with the salmon group. They may be distinguished from salmon and trout of genus *Salmo* by their very small scales (150–240 in the lateral line) and a long anal fin (14–15 rays). Distribution is along the coasts of the Arctic Ocean, from the Lena River in Siberia to Colville River in northern Alaska, along the Pacific coasts of Asia and America from Taiwan to San Francisco. There are six species: 1. DOG SALMON (*Oncorhynchus keta*); 2. PINK SALMON (*Oncorhynchus gorbuscha*); 3. SOCKEYE SALMON (*Oncorhynchus nerka*); 4. CHINOOK SALMON (*Oncorhynchus tschawytscha*); 5. SILVER SALMON (*Oncorhynchus kisutch*); and 6. MASU (*Oncorhynchus masou*).

The masu is found only off the Asiatic coast of the Pacific Ocean from Korea and Japan to Kamchatka. All other species ascend rivers in America as well as Asia. Pink salmon has the greatest commercial value. Second place is held by dog salmon in Asia and sockeye salmon in America. Attempts have been made to introduce Pacific salmon at places all over the world. The largest of them, chinook salmon, was successfully introduced to New Zealand. Dog salmon and pink salmon have been introduced by Soviet fishery biologists in the White Sea and in the Baltic Sea. It remains to be seen how large the populations will be in these groups or whether they will spread to other European waters. Both species have already appeared off the British Isles.

The Pacific salmon are the most highly adapted salmon to marine life. They fast while making their spawning migration in fresh water. Many of them die before reaching the spawning grounds from exhaustion, or from external or internal injuries. The exhausted salmon are easy prey for any predators. Thus bears are found in great numbers in some Canadian rivers during this migration time. Chinook salmon seem to suffer the least from this spawning migration, although they undertake the greatest migration distance.

Frequently chinook salmon are caught which have reached the

Fig. 12-9. Distribution of the Pacific salmon *(Oncorhynchus).*

ocean after participating in the complete spawning process. However even in this species no case has been found in which one individual survived a second ascent for spawning. Most salmon die shortly after spawning, right on the spawning grounds. Shores can be lined with the corpses of salmon for long distances.

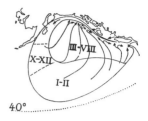

Fig. 12-10. Pink salmon *(Oncorhynchus gorbuscha)* migration route from British Columbia during the 1.5 years it spends in the seas (the numbers indicate calendar months).

The life of the Pacific salmon is much like that of its Atlantic relatives. Chinook salmon, sockeye salmon and silver salmon spend the first two to three years in their rivers. A few males reach sexual maturity while still in fresh water, as happens with the Atlantic salmon as well. Dog salmon and pink salmon have a slightly different life cycle. The young begin migrating toward the sea just a few weeks after leaving the nest; they are only three to four centimeters long. The young pink salmon stays in fresh water for a short time and does not develop the juvenile coloration with the dark spots on the sides.

In many rivers different kinds of salmon are seen ascending, one after the other. Even within one species there may be summer and fall forms. The further a river must be ascended, the earlier the migration begins. The larger species generally undertake the longest migrations. Chinook probably holds the record in terms of distance traveled; in the Yukon it migrates nearly 4000 km upstream as far as Caribou Crossing and Bennett Lake. Chinooks reach sexual maturity at an age of four to seven years. Adults weigh 25-45 kg. The pink salmon has the shortest life span. Sexual maturity is attained in two summers, at a L of 40-50 cm and a weight of 2-3 kg. Pink salmon also undertake the shortest migration of all the salmon, both in rivers and in the sea.

Fig. 12-11. Migration of sockeye salmon *(Oncorhynchus nerka)* during its three-year stay in the ocean.

Generally those salmon which go furthest up the rivers also migrate the furthest into the ocean. Marking experiments with salmon have shown that the summer forms of the dog salmon, which spawn in the small rivers on the Okhotsk coast and in west Kamchatka primarily have feeding grounds in the Sea of Okhotsk, while the large fall salmon of the Amur (which are also dog salmon) travel far into the ocean to the Bering Sea and the coast of Japan.

Of all salmon those in the Pacific undergo the greatest alteration of their appearance during the spawning migrations. In pink salmon one can almost speak of a metamorphosis. After the ascending salmon have spent about one week in fresh water the process begins, and is concluded in thirty-five to forty-five days. As in Atlantic salmon, males undergo an elongation of the teeth and a complete change in the jaws, which in some males take on such proportions that they cannot close their mouth. Male pink salmon also develop a prominent hump. The fish takes on a brilliant red hue. In other species coloration and structural changes are not as pronounced. Chinooks change the least and also have the highest survival rate.

Fig. 12-12. Pink salmon outside the spawning period (above) and during the spawning period [a male] (below).

Pacific salmon differ only slightly from other salmon species in their spawning behavior. Spawning pits are dug exclusively by the females.

Only one male participates in spawning, unlike rainbow trout. The female covers up the eggs after they are fertilized. She defends the eggs for several days against any intruders until she is exhausted and drifts off with the current. The number of eggs from which young are produced which eventually reach the ocean is extremely small, generally about two percent of all those laid. Some drift off when the female lays them and others are distrubed when other salmon build their spawning pits near previously established ones. Many charrs, trout and graylings feed upon the eggs.

Because there are so many natural pressures bearing upon the survival of salmon, the number of fish which survive in a particular year can vary drastically. Today, with heavy fishing of salmon, construction on spawning sites and the general pollution of streams and rivers, salmon are suffering a severe decline. Wherever it is feasible, steps are being taken to reverse this trend. When passageways cannot be built for salmon to get past large dams, mature salmon are caught in tremendous quantities and the eggs and sperm are pressed out by hand so that they can be artificially fertilized. Great costs are expended to insure maximum success with this method. In British Columbia alone, where the spawning grounds in the Columbia River have been largely destroyed by dams, 21 breeding stations built from 1909 to 1961 are breeding 70,000,000 young chinook salmon and 30,000,000 silver salmon per year. The chinooks are set out when they are 100 days old (known as fingerlings) and weigh 5 g; silver salmon must be kept for fourteen months until they weigh twenty to thirty grams before being released. The success of artificial breeding has been due to use of good water and supply of suitable food. The chinook fatality rate in breeding stations is about ten percent, and the rate is higher in silver salmon. Another method which has been employed is the construction of artificial spawning canals, in which the salmon are provided with a suitable, pure gravel bottom and the optimum supply of water. Two such canals were built on Seton Creek. The smaller of the two has been in operation for quite some time; the larger one was opened on the 6th of October, 1967, and by the 12th of October there were already 21,000 pink salmon spawning there. Each spawning pair is provided with about one and one-half square meters of area. Both canals together should produce 13 to 15 million young salmon a year.

How do salmon find their spawning waters?

The spawning migrations are probably the most interesting aspect of the life of salmon and trout; swimming thousands of kilometers, they consistently find the place where their parents spawned. For a long time it was argued whether these fish actually could sense their home waters or found them accidentally. Today, after many experiments with marked fish we have found that salmon indeed have an amazingly keen sensing ability for their home waters. In 1939 in British Columbia 499,326 young salmon were caught and marked by cutting certain fins; they were then released in a tributary of the Fraser River. Large box cages were placed in all neighboring and distant spawning grounds, and years later nearly 11,000 of those original fishes were recovered as they made their way up to their spawning sites. With insignificant exceptions, they ascended only that river from which they had swum into the sea. Similar studies were repeated in the years following this initial work, and the results were very consistent. Apparently a number of mechanisms are at work in this process. Fishes have an internal clock (a mechanism found in many animals) and also have the ability to perceive the position of the sun and to direct their swimming direction on the basis of the sun. With these two abilities, salmon can orient toward their goal with "clock and sextant" as is done on our ships. It has not been shown that these mechanisms also guide the salmon out to the sea, but this could be true.

More is understood about how salmon find their way once they have reached the river. Here the olfactory sense plays a crucial role. Laboratory studies have shown that trout and salmon have a very high olfactory ability and react to all kinds of scents, even when they are very faint. A field study was undertaken, whereby many silver salmon were caught in two arms of the Issquah River. The nasal pits were stopped up with cotton in one group of fishes; these serve olfaction entirely and have nothing to do with breathing. All the fish were then released at a point further downstream, and when they re-ascended it was found that only those fishes with unstopped nasal pits could detect the difference in the two arms of the river; they returned to the river where they were caught. The salmon with stopped-up noses ascended both river arms randomly. Thus we must assume that young salmon know the "scent" of their home waters so that years later they can return to the river mouths (where the scent is much fainter) and follow it back to their spawning grounds.

All salmon species treated thus far have a high oxygen requirement, and generally prefer cold water. This is particularly true for CHARRS (*Salvelinus*), which are distributed in various bodies of water in northern Scandinavia, Siberia and North America, and further south only in high altitude streams and lakes or lakes with cold water at substantial depths. Of all freshwater fishes, charrs have penetrated the furthest into pure Arctic regions; they are found at the New Siberian Islands, Novaya Zemlya and Spitzbergen. In these areas they have commercial importance since they are the only freshwater fish found there. In terms of diversity of localized forms, which sometimes are strikingly different, charrs exceed such diversity in salmon and trout by far. Apparently most characteristics of the individual charr species are only to a small degree genetic, so it is not surprising that the systematics of this group has not yet been fully clarified.

Charrs are characterized by a short vomer. The teeth of this bone

Charr

Fig. 12–13. Vomer of the charr: front view (left) and side view (right).

(vomerine teeth) are separated from those of the palate by a small intermediary space. The scales are very small and the mouth opening is a wide one. A white band is on the front edge of the pectoral, pelvic and anal fins as well as on the lower edge of the caudal fins; this is often accompanied by a blackish stripe. During the first years of life charrs are quite slender and graceful. Larger localized populations have a tendency to become quite massive and fat. At an age of ten to twelve years charrs attain a weight of ten kilograms and a L of over one meter. Other charrs remain very small during their entire life, with a maximum L of 10–15 cm. Many charrs are quite colorful, while others appear totally lacking in color.

All European charrs are classed as a single species, *Salvelinus alpinus*. In the northern part of its distribution this species is represented by migrating forms which ascend tributaries of the Arctic Ocean to spawn. Migrating typically begins in September or October, and spawning itself takes place in late fall or winter of the following year, on gravel bottom, usually in lakes or near the upper courses of rivers. The males in particular have a brilliant red color on their fins and bellyside during this time. In fact, they are among the most brilliantly colored freshwater fishes. At the conclusion of spawning, which proceeds similarly to that in the other salmon, the fishes winter in the lake and return to the sea the following spring. These migrating charrs attain a L of 80 cm and a weight of 8–10 kg in ten to twelve years. Generally charrs only reach a L of 70 cm and an average weight of 4 kg.

During the Ice Age migrating charrs probably inhabited the central European rivers and lakes. Today the warm lower courses of these same rivers are an insurmountable obstacle. Non-migrating charrs are found in the cold alpine lakes. It was in those that highly distinct varieties developed. Practically each lake has its own charr which is unlike any others. Spawning sites of lake charrs are in the large lakes which have some warm water at the surface; the spawning sites are at greater depths. In high altitude mountain lakes spawning sites may also be in shallow water near the shore. The eggs can only develop in the purest gravel bottom which is rinsed by clear water. In some lakes, in which increasing pollution has muddied the spawning sites, gravel is actually put in every few years to protect the charr population. Charr manage to live in mountains as well as at great depths in lakes under very unfavorable diet conditions. Thus it is no surprise that many charr populations consist exclusively of small stunted forms.

There are four major types of alpine charr among the many which occur. The common form lives in the Alps, primarily in lakes at intermediate altitudes. These feed chiefly on plankton and bottom dwelling organisms. They are usually much smaller than migrating charr; they are caught when they weigh 150–200 gm, sometimes less in certain lakes. Because they are popular for consumption they have been introduced in many lakes at more southern latitudes where they did not originally occur. Thus they have been in Lake Lugano since 1895, where they quickly became commercially important.

Charr which inhabit high mountain lakes and are somewhat stunted in their growth form the second major type. In the Koenigssee, at an altitude of 600 m, they still attain an average weight of about 100 gm. In the Gosausee, 300 m higher, charr were particularly valued until the group succumbed to a power plant built there. The heaviest specimens weighed 50 gm. The highest lake in which charr are found is the Doessnersee in Kärnten at an altitude of 2218 m. Here charr are barely larger than minnows. Even as adults they maintain the typical juvenile spotted marking seen in the young of all other salmon species. Like common charr these feed mainly on plankton and small bottom dwelling organisms, and sometimes take flying insects.

A third group of charr are solitary, unlike the common and mountain forms. As adults they prey on other charr, whitefishes, and other fishes. They are particularly brightly colored and are strikingly beautiful fishes. They can reach a considerable size with a weight of 10 kg and more. Often these solitary forms are found together with normal or mountain charr.

The fourth charr type has pale coloration and large, protruding eyes which reveal that it inhabits great depths. The German designation "Hungersaibling" ("starving charr") aptly describes the species. These are invariably extremely small fishes and those which are fifteen centimeters long are already an exception.

Other charr forms are distributed in northern Siberia and America. Charr play the same role as river trout on the spawning sites of Pacific salmon as spawn predators and enemies of young salmon. One species, the BROOK TROUT (Salvelinus fontinalis), which originally occurred in the northeastern part of Canada and the United States, has been introduced into other areas along with rainbow trout. Since 1884 it has been bred in Germany. Like rainbow trout, brook trout also grow faster than river trout. Since they are not very dependent on shelter and in fact prefer open water, they are suitable for occupying canalized streams. Generally they move upstream toward the coldest parts. Here they thrive in temperatures which are too cold for other trout. Since river trout and charr resemble each other in terms of spawning behavior, and since their spawning periods occur at the same time, hybrids occasionally appear. These are called tiger trout in German, and are infertile. This is not the case, incidentally, with "Alsatian charr", a hybrid between river and lake charr, which was once introduced into small alpine lakes such as the Soiensee in Upper Bavaria.

In one respect the large American LAKE TROUT (Salvelinus namaycushi) occupy a special position. They were once popular with fishermen in the Great Lakes and southern Canada until they suffered a decline when the sea lampreys spread. This species is characterized by an irregular yellowish spotted marking. They generally are found at greater depths beginning at 120 m, where they grow to a stately one meter long with a weight of seven to eight kilograms. But occasionally even lake charr have been found which are 1.5 m long and weigh 30–50 kg. Lake trout have also been raised artificially; these attempts have been without success in Europe. One group inhabits Lake Tahoe in California, but these introduced fishes are eliminating the original lake trout.

Hucho

One of the largest freshwater fishes is the salmon genus *Hucho,* closely related to charr. They have a short vomer. The vomerine teeth and palate form a closed row. These are particularly elongated fishes which are compressed only slightly along the sides. In cross-section they are also round. The head is somewhat flattened as in pike. There are three species, including the Danube salmon *(Hucho hucho)* and *Hucho taimen.*

DANUBE SALMON *(Hucho hucho)* is known as the redfish in southeastern Alpine countries because of its red display coloration. It is sometimes confused with the sturgeon genus *Huso.* Danube salmon is distributed in the Danube from Ulm to Romania and in some tributaries of the Danube. Today it is a rare species. The only evidence of its once great prevalence are older reports of the "Danube salmon". Once they were commonly caught with a L of 60–120 cm (15-year old specimens were up to 1.5 m long) and a weight of as much as 52 kg. *Hucho taimen* is still larger; it inhabits European-Asiatic waters from the Volga and Petschora eastward to the Amur. This species attains a weight of 30–60 kg, sometimes 80 kg.

Fig. 12-14. Vomer of *Hucho hucho:* front view (left) and side view (right).

Not much is known about the natural life of the Danube salmon. Generally it remains in one area, but apparently undertakes a short migration in spring after the snow melts. Like other salmon they lay their eggs in spawning pits, usually on gravel banks in the major rivers or at lower depths of tributaries. Because of the relatively high water temperature the eggs develop in about thirty-five days. Young Danube salmon initially feed on invertebrates. But in their second year, when they have grown to 15–20 cm, they begin feeding on fishes. As adults they have a diverse diet: primarily fishes, frogs, small mammals and birds. Earlier, nase were the main part of the Danube salmon diet; nase were found in great quantities in the Danube but construction of waterworks led to a great decrease in their numbers. In fact the present decrease in Danube salmon is due to the disappearance of its food supply. Perhaps the decrease is more immediately due to pollution of the water and construction work in the Danube and its tributaries. They prefer deep pools with cover; they dart out of their hiding places when pursuing prey.

Fig. 12-15. Distribution of *Hucho hucho.*

The closest relative of the Danube salmon is the LENOK *(Brachymystax lenok),* the only member of its genus. It occupies a distinct position within the salmon group; while all other species have wide mouth openings, the lenok is distinguished by a small mouth which does not extend beyond the rear edge of the eyes. Lenok are prevalent in the Siberian rivers from the Ob to the Kolyma and on the Asiatic Pacific coast southward to the Yalu River. Lenok is found especially in waters just before mountain ranges. The diet consists of May fly larvae, amphipods and other bottom dwellers. In the Amur it feeds on great quantities of dog salmon eggs when that species is spawning, and later it preys upon the young dog salmon. In some parts of Siberia the lenoks are the most important commercial fish. They attain a maximum weight of 3–4 kg, rarely 6 kg.

The WHITEFISHES (Coregoninae) are also slender, somewhat compressed fishes with rather large scales; there are less than 120 in the fully developed lateral line. The mouth has few or no teeth and the mouth opening is generally a narrow one. The vertebrae on the base of the caudal fin are arched upward. The dermossphenoticum in the auditory capsule is fully developed (unlike other salmon), and a suprapraeoperculare (gill-cover bone) is absent. There are two genera: Inconnu *(Stenodus)* with a wide mouth opening and the whitefishes *(Coregonus),* with a narrow mouth opening. Species of the latter genus have many localized forms, the systematics of which are unclear.

Stenodus contains a single species, INCONNU (⟩ *Stenodus leucichthys).* It is considered to be in a subfamily in the whitefishes because of its skull structure, large scales, coloration and other characteristics. Actually in shape it resembles other salmon more than the other species of this subfamily. One subspecies of inconnu is migratory and inhabits the Arctic Ocean and its tributary rivers from the White Sea to Mackenzie; another subspecies is found in the Caspian Sea and its northern tributaries (the Volga and Ural). As adults inconnu prey on other fishes in brackish waters of the Arctic Ocean. They reach sexual maturity at an age of seven to ten years (in the Kolyma River not until eleven to twelve years of age). They move far upriver to spawn; in the Yenissei the migration route is 1500–1900 km, and in the Ob 3500 kg and more. The ascent generally begins in summer.

Spawning occurs in fall. The number of eggs laid varies between 125,000 and 325,000. They are about 2 mm in size. The eggs are laid on the bottom between stones and are fed upon by young *Hucho* species, graylings, tadpoles, and other caviar lovers. After spawning inconnu remain in fresh water for a long period of time until returning to their brackish waters. An individual inconnu returns to spawn again after three or four more years. Inconnu can become quite old and may exceed an age of twenty years; such specimens are over one meter long and weigh up to forty kilograms.

The inconnu in the Caspian Sea are fast growers. Their feeding

Subfamily: whitefish

The inconnu

grounds lie primarily in the central part of the Caspian Sea, and sexual maturity is reached after five to seven years. The chief spawning sites are in the Ufa, a tributary of the Volga. The Caspian inconnu are greatly endangered at the present, due to waterworks construction in the spawning waters and other construction works which hinders the spawning migrations.

Whitefishes

WHITEFISHES (*Coregonus*) are far more important commercially than inconnu. Whitefishes are in fact the most important commercial fish in the lower courses of many Siberian rivers, and they support a considerable amount of commercial fishing in a few central European lakes. Their scientific name originates from the shape of the whitefish pupil, the front edge of which is somewhat pointed (*Coregonus* is derived from the Greek χορη = pupil and ϑογυ = knee or corner; e.g., "pointed pupil"). Almost all whitefish species have numerous local forms which can vary greatly from one another. Transport studies have shown that the environment also greatly affects growth rate, number of scales and the shape of the mouth.

Fig. 12-16. Whitefish (*Coregonus*) distribution.

Of all characteristics thus far used to distinguish the species, only the gill filtration apparatus is genetically determined and thus relatively independent of the environment. Transplanting the fishes into new water did not change the number or shape of the gill structures; sometimes a change could be made by artificial selective breeding. Thus from a whitefish population with thirty to forty-two gill rakers, two pair with forty-one and thirty-two gill rakers, respectively, were chosen and their offspring were separated in small lakes. The first generation already showed the effect of this artificial selection: in one lake an offspring population was found with thirty-five to forty-two rakers, and in another was one with twenty-seven to thirty-eight. In other cases introduction of whitefishes from various populations with divergent number of rakers led to a complete hybridization under the prevailing environmental conditions and hybrid schools with a new, intermediate number of gills developed.

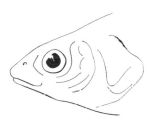

Fig. 12-17. Head of a whitefish.

Two chief kinds of whitefish are distinguished: bottom-dwelling and free-swimming whitefish. The former feed chiefly on small bottom-dwelling organisms and are found near the shore or in shallow water. The latter are open water species which feed upon plankton. Interestingly, these two main forms differ in the number and shape of their gill structures: bottom-dwelling whitefishes are characterized by a few powerful gills, while free-swimming whitefishes have long, fine structures.

An attempt had been made to explain this difference on the basis of the feeding patterns in the two whitefish forms. It was felt that the free-swimming whitefishes used their long, dense gill rakers as a filter apparatus when feeding. This viewpoint was sustained for a long time, but it fails to explain how plankton trapped in the gills could

The Grayling Region

The trout regions of European waters merge with grayling regions. Streams have merged to form a small river with cool, fast-flowing water, which is thus rich in oxygen. There are still trout species in such regions, but they are no longer the predominant fish. Fishes: □ Salmon: 1. River trout *(Salmo trutta fario)*; 2. Rainbow trout *(Salmo gairdneri)*, a species which has been introduced from America to Europe; 3. The European grayling *(Thymallus thymallus)*, the characteristic fish of this environment. □ Carp: 4. The gudgeon (see Chap. 15) *(Gobio gobio)*; 5. The gudgeon *Gobio uranoscopus*, found only in the Danube; 6. *Noemacheilus barbatulus* from the carp group; 7. The loach *Cobitis taenia*. □ Scorpionfishes (see Vol. V): 8. Bull-head *(Cottus gobio)*. Other animal groups: Cyclostomes: 9. Fresh water lamprey (see Chap. 2), which migrates through the grayling region only when spawning. □ Birds: 10. House martin *(Delichon urbica;* see Vol. XI). Insects: 11. Demoiselle *(Calopteryx;* see Vol. II).

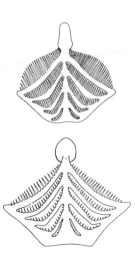

Fig. 12-18. Gills of a free-swimming whitefish (above) and a bottom-dwelling whitefish (below).

Tugun and tschirr

be transfered to the mouth. Furthermore the physiological capabilities of the species (for example its swimming ability) could not be supported solely by the amount of plankton which could be trapped by the gills. Today aquarium studies have shown that free-swimming whitefishes also feed on their prey piece by piece. Thus the great incongruity between the composition of plankton in the water and the stomachs of the fishes is explainable. Whitefishes consistently prefer particular organisms, generally large ones, which they can perhaps more easily recognize.

Whitefishes are very widely distributed. They are found in the outlying parts of the Arctic Ocean, the north Atlantic and the north Pacific as well as in many inland waters in Europe, Asia and North America. They are either migratory or localized inland forms. Those in Siberian waters show the greatest diversity; they have presumably propagated from this region and penetrated Europe and North America from Siberia. The majority of the Siberian species are migratory and their principal feeding grounds are in the lower courses of rivers and the brackish water off the Arctic Ocean. Only a few whitefishes penetrate pure sea water when searching for food. A few live exclusively on plankton, while others feed chiefly on bottom-dwelling organisms such as lophogastrids, amphipods, isopods, and mollusks. Actually they feed on anything they can grab with their small mouth, up to and including the young of their own species.

Some whitefishes remain relatively small. Thus the TUGUN (*Coregonus tugun*) in the rivers and some lakes between the Ob and Yana has a maximum L of only 20 cm and a weight of 80 gm. On the other hand, some species grow to considerable lengths, such as the TSCHIRR (*Coregonus nasus*), in which the Kolyma specimens grow to weigh 16 kg. All Siberian species grow relatively slowly. They do not attain sexual maturity until they are twelve years old, and they can reach an age of fifteen to twenty years. The migratory forms begin their ascent in the last half of summer. Their spawning sites do not usually lie as far from the river mouths as is the case with other salmon species. However among whitefishes there are also varieties which migrate 1000 km or more to their spawning sites. There are also pure inland forms. These, too, undertake more or less extended migrations during the spawning period. Whitefishes in Lake Baikal and other inland lakes move at least to the mouths of the tributary rivers or still further upstream.

Spawning generally occurs in October or November at a water temperature below 4° C. or at 7° C. over stony bottom. The number of eggs varies greatly and depends on the size of the fish. Thus the small tugun lays only 1500–6000 while the tschirr lays 13,000–135,000 depending on its size. Unlike the yolk-rich eggs of other salmon, those of whitefishes are small, about the size of shot. At the conclusion of spawning the whitefishes are heavily exhausted; although a few die, most return to their feeding grounds and return to spawn again. Larvae hatch in the spring. Generally they are carried downstream by the current before the yolk sac has been completely eaten. In a few white-fishes the young remain in the river one to two years.

Migrating whitefishes were once prevalent in the tributaries of the North Sea and Baltic Sea. Thus *Coregonus oxyrhynchus* formerly pene-trated the Rhine to Speyer and Strasbourg, the Weser River to the lower courses near Werra and Fulda and the Elbe River to Magdeburg and Torgau. These whitefishes have almost completely disappeared. Another variety in the Baltic Sea is also a rarity. These were once popular for consumption and were caught in great quantities during the spawning period. Migrating groups of the *Coregonus artedi* (see below) live in the Baltic Sea and ascend the Neva River from the Gulf of Finland. However these do not penetrate Lake Lagoda, in which many conspecifics live; they are also prevalent in many Baltic and Karelian inland waters. Whitefishes have all but disappeared from central European rivers. They have practically no commercial value today; this is in contrast to the inland whitefishes which still inhabit many Baltic lakes and the northern Alpine lakes.

In the Middle Ages these whitefishes were already a popular dish. A report from 1150 states the number of whitefishes which the prior of Saint-Jean near Geneva had to deliver to the Bishop of Aosta. Even long ago these valuable fishes were shipped to other lakes and intro-duced there; a runic inscription from 1000 A.D. reports of the artificial introduction of fishes in a Norwegian lake. These shipments were continued until recently and have contributed to the difficult position the systematist now has in classifying these numerous whitefish forms.

Practically all whitefish lakes harbor one or several forms which can be differentiated from those in neighboring lakes on the basis of the general shape. When several varieties are present in one lake they do not hybridize even though this can be carried out artificially and the young thus produced are viable and fertile. Natural hybridization is limited by the use of different spawning sites and somewhat divergent spawning periods. However if the resident whitefishes are trans-planted to another lake, or if the environment changes considerably in their home waters, the fishes are forced to alter their spawning habits; this results in hybridization and the development of a distinct population with characteristics of both parent groups.

One distinctive central European whitefish is *Coregonus albula* (L 20 cm, maximum 30 cm). It can be fairly easily distinguished from other whitefishes by its mouth opening which curves upward. Distribution extends around the northern hemisphere from the British Isles to North America. In central Europe the southern boundary of their dis-tribution corresponds to the border of the last great glacial movement.

The European whitefish *Coregonus albula*

They are not present in the lakes along the northern edge of the Alps. They have been successfully bred in dammed up lakes, such as the Naiental and Möhnetal lakes in Germany. The species resembles herring in appearance and behavior.

The shallow water cisco, an American species

The American species, the SHALLOW WATER CISCO (*Coregonus artedi*) probably belongs to the same species and thus should be called *Coregonus albula artedi*. It is also known as LAKE HERRING. During the day lake herring are found in large schools far from shore. At night these large conglomerations dissolve into small groups. Their chief diet is composed of plankton which is picked up piece by piece. Studies in Plöner Lake in Germany showed that a single 16 cm-long specimen ate 35,200 damsel flies and 25,000 cyclopids. Feeding is heaviest during the summer. During the spawning period they cease feeding. Their prevalence is affected by that of plankton, and thus they may be found at one time in one area and at another time somewhere else; and the day is spent at different depths than the night.

They appear in shallow water only during the spawning period. Here they lay 800–8000 eggs (depending on size) in sandy or gravel bottom. The larvae hatch in spring. Since the yolk sac is completely eaten in three to four days, they must begin feeding independently at a relatively early age. The food supply at the time of hatching probably influences the success of that generation and perhaps in large part causes the great variations seen from year to year. In individual lakes growth of these whitefish varies considerably. Generally this is a short-lived, slow-growing species. Sexually maturity is attained in the second summer of life. They seldom are more than five years old.

The Siberian whitefish

The Siberian species *Coregonus albula sardinella* is considerably larger (L up to 42 cm; weight about 500 gm). They live to be eight to eleven years old. Being larger, fertility is also higher than in their western relatives; up to 24,000 eggs are laid. Among all central European whitefishes, *Coregonus albula* varieties have an extraordinarily high number of gill structures (35–55).

The large free-swimming whitefishes

Systematics of the remaining central European whitefishes is still uncertain. Some zoologists tend to group them all together as the whitefish *Coregonus lavaretus;* others claim that several distinct species exist which can be distinguished by the number and shape of gill rakers. There are four general types: two free-swimming whitefishes and two bottom-dwelling varieties. In some lakes, such as the Bodensee, all four types are found together. Local fishermen can distinguish them on the basis of age. Even one Mangold, in the fish book published in 1557 by Conrad Gesner, the cleric, wrote "Of the nature and characteristics of fishes," about the four types found in Lake Constance. Gesner was the first individual to attempt to classify these particular whitefishes with those in other lakes; this undertaking has since been attempted unsuccessfully by many zoologists.

One of the four types is considered to be a member of the species of HOUTING (*Coregonus lavaretus*). This is a pure free-swimming fish. Generally their schools are far from shore in the upper twenty meters of the lake, rarely deeper, and at most forty meters deep. The claim by many fishermen that these fishes were at greater depths when the fishing was poor could not be substantiated. They feed almost exclusively on plankton, but at certain times also on perch three to four centimeters long. Their own young are even eaten, up to a length of five or six centimeters. They do not feed on any larger fishes. As plankton feeders these fishes stay in the same area. During the day, however, they are found at different depths than at night. Furthermore they carry on seasonal migrations across the water.

From the end of November to the end of December, when the water temperature has sunk to less than 7° C., these fishes collect above the deepest parts of the lake, before they have reached complete sexual maturity. They initially collect in deeper water above their actual spawning sites. The older fishes of both sexes already have had a row of small white dots on their sides for a few weeks. Spawning takes place in the upper water levels. Males generally appear before the females. Since they are capable of releasing sperm numerous times they also generally stay there longer than do individual females. Therefore there are always more males than females at the spawning sites. Spawning usually occurs late in the evening or at night, but it has also been observed by day. Spawning is preceded by an elaborate. display ritual. The fish swim in pairs, touching their sides together, just underneath the water surface. The splashing of water from the fishes swimming about can be heard repeatedly. Finally the eggs and sperm are released simultaneously, and the eggs are fertilized while they sink to the floor. If they are not eaten by burbot (*Lota lota*) or their near relative *Coregonus acronius* or die from a lack of oxygen, the eggs develop on the lake floor.

Houting is characterized by a great adaptability to quite diverse environmental conditions. In large food-rich lakes they grow to lengths of 60–70 cm and weigh up to 10 kg. However they are also found as dwarfs in lakes which are cold and lack sufficient food supply; this is particularly true when they are also competing with other whitefishes. The effect of environmental pressures on growth has been well studied recently in Lake Constance (there have also been transplant studies in Sweden). Due to the increasing supply of sewage water, which serves as food, there has been a steady increase since 1950 in the number of plankton in Lake Constance. This higher food supply led in turn to an unusual acceleration in the growth rate of the whitefishes. Today these fishes reach a length of thirty centimeters in two summers; this length earlier took four summers to reach. This fertilization of Lake Constance initially increased the fish catch tremendously, but in the early 1960s almost led to a catastrophe. The greatest part of the whitefishes were caught when they were one year old, which was before they could spawn. The legal size limit was raised to 35 cm from 30 cm and the mesh size of the nets was raised from 38 to 44 cm. This has preserved the spawning population of this variety of whitefish in Lake Constance. In addition to this variety there are also other free-swimming whitefishes in the Alpine and Baltic lakes as well as migrating forms in the Baltic Sea.

The small free-swimming whitefishes

One of the other whitefish varieties, *Coregonus oxyrhynchus,* is found in Lake Constance and is also a free-swimming variety. However this fish inhabits the shallow parts of the lake near the Halde River. They appear for spawning at the connection of the lower and upper lake in fall somewhat before houting but in some parts of the lake appear later than *C. lavaretus,* perhaps in January. Both free-swimming varieties are only found together in larger lakes. The smaller species, *C. oxyrhynchus,* is usually under pressure from the larger and does not usually grow as rapidly. However wherever the competition conditions permit it, this small free-swimming whitefish attains a L of 50 cm and a weight of 2 kg.

The large bottom-dwelling whitefishes

The fastest-growing variety is the large bottom-dwelling whitefish *Coregonus fera.* It is not rare to find individuals 50 cm long and weighing more than 1 kg. They have a gill raker number which varies from 20 to 29, and is usually 24. This variety is found in lakes in the West German state of Holstein and the East German state of Mecklenburg. The tschirr *(Coregonus nasus)* from the Ob and east of the Petshora to Alaska is probably not a close relative, even though it has the same number of gill structures. In contrast to the social free-swimming whitefishes these large bottom-dwelling varieties come to the surface singly searching for snails, mussels, insect larvae and worms in shallow waters of the shore. They form small groups only in spring and during their spawning period in fall (which may occur from October to December, depending on locality, and in the Alpine district at the beginning of December).

The small bottom-dwelling whitefishes

As with the bottom-dwelling whitefishes, the variety known as the small bottom-dwelling whitefishes cannot clearly be classed together with a number of Siberian forms which have the same number of gill rakers. This whitefish has been named *Coregonus acronius* or *Coregonus pidschian.* They have 19–22 gill rakers, or 15–28 when hybridized with other whitefishes. These whitefishes are always small when they are present together with another whitefish variety. The small bottom-dwelling whitefish in Lake Constance is just rarely over 30 cm long. The Alpine members of this group are much unlike all other whitefishes.

The Lake Constance small bottom-dwelling whitefishes are found together with another variety in the shallow water offshore. They feed largely on plankton. However in fall they migrate to deeper waters along the Halde River. In October they are found at a depth of 30–60 m and in November and December at 100 m depths and more. In January they are found at still greater depths. Here they feed on the eggs of the large free-swimming whitefishes. The small bottom-dwelling whitefish spawning sites are also at great depths. Because of the lower water temperature this fish spawns considerably earlier than the other varieties, often in September. In the Ammersee lake it is believed that spawning occurs in June and July.

As in other salmon species, whitefish populations are being maintained to a large extent by catching mature specimens and artificially extracting sperm and eggs from them. The eggs are then artificially fertilized and the young are raised in breeding stations. Generally breeding takes place in cold water so that development takes longer. This allows the young to be put into the lakes at a time when the food supply is large enough to support them properly.

The GRAYLINGS (Thymallinae; the single genus is *Thymallus*) are medium-sized fishes with a prominent large dorsal fin. The mouth is narrow and has small, well-developed teeth. The scales are relatively large. Graylings have an odor resembling thyme. There are five species and many subspecies.

Subfamily: graylings

In their distribution, which extends over wide parts of Eurasia and North America, graylings live in localized areas in clear, fast-flowing water. Since they are not as sensitive to water temperature as river trout and feed on streaming aquatic plants in deep pools, they primarily are found in water below the river trout region, which is therefore designated as the "grayling region". However, in the upper courses of rivers, they are found together with river trout. They avoid mountain rapids. In some Alpine lakes they are only found at the mouths of streams where the water still flows. Pure lake varieties are found in a few Scandinavian and Siberian lakes, in some cases as the only fish in those lakes. Graylings are found in brackish water along the Swedish and Finnish cliffed shores.

Fig. 12-19. Distribution of grayling *(Thymallus)*.

All European graylings belong to one species, the EUROPEAN GRAYLING *(Thymallus thymallus)*. Graylings grow fairly quickly. After two years they measure 15–20 cm, with a maximum L of 50 cm and a weight of 2–3 kg. Such specimens can live to be seven to fourteen years of age. Some Alpine rivers once contained graylings one meter long, weighing ten kilograms. Graylings, unlike river trout, are often found in small groups. They feed on the larvae of caddis flies and May flies, which they grab from the stones on the river bed. Other food includes the amphipod *Rivulogammarus pulex*, snails, mussels and other organisms. They seize flying insects in the air. They also feed on salmon and trout eggs which are either floating through the water or are in the open on the bottom, but graylings have never been observed digging up the ground to get at eggs, so it cannot be said that they ruin the spawn.

European graylings

Larger graylings also feed on smaller fishes and small mammals. But otherwise they are a peaceful species.

Graylings spawn in the spring. According to the prevailing water temperature they reach sexual maturity in March or as late as May as two- to three-year old fishes. They resemble salmon in their spawning behavior. They do not generally undertake spawning migrations, however. The northern lake inhabitants spawn either on shallow shore sites in the lakes themselves or they ascend short distances up the tributary rivers. During the spawning period members of both sexes are colored darker and more brilliantly. The entire body has a light red shimmer. Rows of distinct purple eye spots appear on the dorsal fins. The males are characterized by thickening of the skin on the sides of the back and tail. As in salmon and trout the female grayling digs the spawning pit herself; it is located on shallow sites with a gravel floor. Rivalry fights typically break out among the males, but they are not carried out with the vigor which characterizes those rituals in salmon. While releasing eggs and sperm both partners stir up the water and bottom soil with powerful beats of their tails. This mixes the eggs and sperm; furthermore the sticky eggs thus come into contact with the floor at once, to which they adhere and sink into the spawning pit. Depending on the size of the female, 3000-6000 eggs 3-4 mm in diameter are laid; they are covered in the spawning pit with gravel. The small yolk sac is eaten shortly after the young hatch, and the young grow relatively quickly; after one summer they are 7-12 cm long. Like salmon and trout they have a distinctive juvenile coloration with a row of dark spots along the lateral line.

Graylings are very popular with sport fishermen, for they put up a good fight when they are hooked. They are also popular for eating. However they do not have great commercial importance since they cannot be shipped and kept easily. Their meat loses its good flavor soon after their death. Unfortunately graylings have become uncommon and have completely disappeared from some former grayling waters, since the species is very sensitive to pollution.

Ayus

The AYU (*Plecoglossus altivelis*) is the only species in the family Plecoglossidae. L is up to 30 cm; the species is small, elongated, and is distinguished from the salmon by a number of structural features: the vertebrae at the base of the caudal fin are not arched upward; a row of short, wide jagged teeth line the upper jaw and gums of the lower dentary bone, and they are moveable; the premaxillary has conventional teeth. The skull shows many distinctive features. At the pylorus, the exit from the stomach, there are up to 400 fingershaped appendages; the mucous membrane of the mouth forms a sacklike paired fold on the front end of the lower jaw. Distribution of this migratory species is in the Pacific coastal waters from southeastern Hokkaido south across the north Chinese coast, Korea, and east to Taiwan.

The small fish with the shape of a trout has a short life expectancy

Fig. 12-20. Distribution of the ayu *(Plecoglossus altivelis).*

of one, rarely two or three years. In fall the ayu ascends rivers in great schools. On the basis of age some are caught in certain rivers, with trained cormorants; rings around their necks prevent them from swallowing the fishes. Ayu eggs adhere to all sorts of objects and develop in three weeks at a temperature of 10-19° C., in two weeks at a temperature of 15-20° C. The young migrate to the sea immediately after hatching and collect at the river mouths in spring. Here they feed on diatoms and blue algae until the spawning period in fall. This is an unusual form of food for salmon species. Generally the entire life cycle of the ayu is complete in one year. In some lakes pure inland populations have developed, and because of a lack of food these ayus are always very small, being less than ten centimeters long.

The SMELT family (Osmeridae) contains small, slender, silvery fishes. The scales are soft and rest loosely. The last vertebra does not arch upward. The stomach has a blind end. There are four genera: smelt *(Osmerus)* with five or six species, capelin *(Mallotus)* with one species, *Hypomesus* with two species, and *Taleichthys* with one species.

Most members of this family live in the northern Pacific Ocean and its bordering waters. Only the European smelt and the capelin have invaded Atlantic waters. Despite their small size smelt have particular commercial value, as a food base for other fishes and in their own right as edible fishes. They are processed industrially or are used as bait. Their odor is somewhat like that of fouling pickles, not to everyone's delight.

The EUROPEAN SMELT *(Osmerus eperlanus eperlanus)* is generally considered to be a subspecies of a larger group widely distributed in North American and Siberian waters. As migratory forms or inhabitants of inland lakes near the coast the smelt in Europe are found along the northern coasts from the Bay of Biscay to Petschora. Migrating smelt feed in shallow waters, usually not too far from river mouths. Here they feed on small crustaceans and fishes, including their own young. Most smelt are sexually mature at the end of their second year of life. In fall they collect before the river mouths and in spring ascend the rivers following the thaw. In the Elbe River (Germany) the ascent generally occurs in mid-March to mid-April. The spring storms apparently elicit these migrations, for the ascent is usually accompanied by rainy, stormy weather. The older smelt first appear at the spawning sites, which in the Elbe River lie around the Hamburg harbor; then the younger ones follow, and within individual age groups males come before the females. When the weather is favorable the females lay their eggs in a short time and then return, while the males spend a longer time at the spawning site. Because of that there are four times as many males as females at the spawning sites toward the end of the spawning period.

Smelt

Fig. 12-21. Distribution of smelt *(Osmerus)*.

European smelt

Smelt usually spawn in the vicinity of the shore where a fast current flows over solid bottom. Females 20–22 cm long lay 40,000–50,000 eggs 0.6–0.9 mm in size; smaller smelt produce still smaller eggs. The greater part of both fertilized and unfertilized eggs drift off at once with the current, while others sink to the floor and adhere to obstacles with their sticky coating. This outer, sticky envelope breaks under the force of the current, but the egg hangs on what remains of it and swings freely in the current. Eventually even these eggs are torn loose or are driven about under the influence of the tides and slowly they float downstream. Depending on water temperature, smelt eggs require two to five weeks to develop. Eggs and larvae can only develop normally when they are in fresh water or brackish water with a salinity less than 1%. Thus many of the smelt eggs at river mouths, where the salinity is higher, do not survive.

Young smelt first spend a few months in the lower river courses. Often those three to four months old are caught with fine mesh nets for bait. Many of those that spawn die in their first spawning period, while others spawn repeatedly and live to be nine years old if they are not fed upon by predatory fishes which often choose smelt at river mouths.

Inland populations of smelt have developed in coastal lakes. These are distinguished from the migrating forms by a smaller body size (L 6–12 cm), a still more elongated body, a shorter life cycle and lower fertility. In the Baltic Sea they owe their existence to the recession of the former Yoldia Sea about 12,000 years ago. In some lakes the smelt spawn in shallow parts of the lake, while in others they migrate up the tributary rivers. The inland and migratory smelt are subject to great fluctuations in their population size. "Smelt years" alternate with those in which hardly any smelt are caught at all. The seasonal production is dependent upon storm conditions, the condition of the water during the spawning period and the amount of plankton present when the young hatch. Furthermore, high production results in a self-regulatory mechanism whereby the following season will not again be excessive.

Like other commercial fishes, smelt have been introduced into new waters. Their introduction into the Great Lakes was particularly successful. In 1912 the eggs of a coastal population of an American smelt (*Osmerus eperlanus mordax*) were introduced to Crystal Lake, and in 1918 the first mature smelt were caught. By 1924 they had spread throughout Lake Michigan. In 1932 they had invaded Lake Huron and into Lake St. Clair; after three more years they also inhabited Lake Erie and finally, in 1940, the smelt were found in Lake Ontario and Lake Superior. Today smelt is the most prevalent fish species in the Great Lakes.

Capelin

The CAPELIN (*Mallotus villosus*) is a pure marine species. It is found in the Arctic Ocean and in the northernmost parts of the Pacific Ocean.

In summer schools of capelin are found far from shore at depths down to 150 m, while their nights are spent near the surface. They feed on the masses of plankton present in these upper levels. They themselves form the diet for many predatory fishes and whales. Sexual maturity is attained at an age of two to four years, by which time they have a L of 15-20 cm. In spring or summer they then move in great schools along the coast, darkening the water. They spawn at 50-100 m depths or in shallow water just offshore at a water temperature of 2-4° C., like the Pacific subspecies does. The males are larger than females and can be recognized by the longer anal fin and tufted growths on the scales around the lateral line (which are fully developed in contrast to smelt). A female often spawns with two males; 8000-12,000 sticky eggs measuring 0.6-1.2 mm are laid in a sand furrow, in which they firmly adhere to the bottom. Most males die after the spawning period; some females survive and spawn two or three times the following year.

Fig. 12-22. Distribution of capelin *(Mallotus villosus)*.

The icefish family (Salangidae; see below), New Zealand trout (Retropinnidae), galaxiids (Galaxiidae; see below) and the apolochitonids (Aplochitonidae) are all grouped together on the basis of their external and internal structure and particularly the jaw, into the galaxiid suborder Galaxioidei.

The ICEFISHES (Salangidae) are very small species; L is 8-10 cm at most. The body is elongate, very slender, with no or very few, loose scales. The head is greatly flattened from the top. The digestive tract is simply a straight tube (this indicates an early fish form). There are about six genera distributed throughout the fresh and salt water of the Pacific coast from the mouth of the Amur to Vietnam.

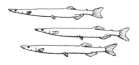

Fig. 12-23. Icefishes.

The Soviet scientist Berg considered icefishes to be in a continuous larval stage; that is, that they remain in this stage and even reach sexual maturity as larvae. These short-lived fishes spawn near the shore and die thereafter. In some places, as in Siberia, China and Korea, their very large spawning schools are commercially significant. The icefishes are closely related to the NEW ENGLAND TROUT (Retropinnidae) with dorsal and anal fins projecting to the rear.

The GALAXIIDS are trout-shaped, fully naked species lacking scales; the dorsal and anal fins are directed to the rear and are located just before the caudal fin. There is no adipose fin. Distribution is unusual: there are about eight species on New Zealand and neighboring islands, seven in New South Wales, about four in South Australia, one in Western Australia, two in Tasmania, about seven in South America from Chile southward and one in southern Africa; finally there is one individual species found both in New Zealand and Australia.

Family: galaxiids

Distinguishing characteristics

This peculiar distribution has been used as evidence by those who have maintained that there was once a continuous southern continent. However, the fact that these fishes are very salt tolerant and during their life constantly move between salt and fresh water (or at least during the

Fig. 12-24. Distribution of the galaxiids (Galaxioidei).

Fig. 12-25. A galaxiid.

spawning period do so), or live in regions where the salinity of the water is constantly fluctuating, permits other interpretations. The thirty to thirty-six species which have thus far been described are of various sizes. While most are only 10–15 cm long, *Galaxias alepidotus* in New Zealand grows to be one half meter in length. In a little-known study Fisher in 1962 made a number of interesting remarks about the two Chilean species *Galaxias maculatus* and *Brachygalaxias bullocki*. He reported that both species feed on small insect larvae; *Galaxias maculatus* supplemented this with land insects as well and *Brachygalaxias bullocki* fed additionally on small crustaceans, and primarily on bristle worms in salt or brackish water. Fisher writes the following about the prevalence of *Galaxias maculatus* in Chile:

"The distribution of this species according to all available information lies exclusively in southern and central Chile between Concepcion and Punta Arenas. Most sightings of the species are made in fresh water regions, including lakes far from the ocean. Often the species is found in the lower courses of rivers, and sometimes even at the mouth itself. Eigenmann has already observed the migration of transparent young from the ocean into river mouths. ...Through Eigenmann's observations it seems justified to suppose that *Galaxias maculatus* spawns in the ocean or at least can do so when the opportunity is there. It would be fruitful to compare the similar species of Australia and New Zealand, *Galaxias attenuatus*, in this regard. The possibility that these species are one and the same has already been made [Stockel 1943]. *Galaxias attenuatus* was also described as being in Chile, but thorough collecting by Eigenmann in southern and central Chile failed to find the species. Several papers have been written about the biology of *Galaxias attenuatus*: MacCulloch (1915) described the larvae migrating from the sea into the rivers. MacKenzie (1904) found that spawning activity takes place in the tidal zone during the spring tides and that the eggs are unprotected until the next spring tide, when the young hatch. Later it was found that such conditions are not essential ones for propagation of *Galaxias attenuatus*; evidence exists that *Galaxias attenuatus* spawns at the mouth of the Selwyn River, which is not subject to tides; this river has its mouth at a brackish water pool and the fish spawns in the shallow water at the edge of shore where the eggs are constantly covered by a layer of water."

A related species, *Neochanna apoda* of New Zealand, has no pectoral fins. It can live in mud for long periods of time when water has retreated. These few examples speak for the great adaptability of this fish family.

The aplochitonids (Aplochitonidae) are very similar to galaxiids and differ only in the jaw structure; the maxillary has been pressed forward toward the skull by the premaxillary; this development has been taken up in steps by the other families. Of the few species known, the scaleless genus *Aplochiton* has one species in South America and on the Falkland Islands; *Prototroctes,* with small scales, has one species in Queensland, South Australia and New Zealand, respectively.

The pike suborder Esocoidei has a mouth lacking the moveable premaxillaries and a toothless upper jaw (maxillary); the three skull bones known as the mesocoracoid, orbitosphenoid and mesethmoid are absent. There are two families: pikes (Esocidae) and mudminnows (Umbridae), the latter containing the blackfish, which were once considered to be an individual family Daliidae, and the Olympic mudminnow once classed as Novumbridae.

Suborder: pike

PIKE (Esocidae) have a long snout which resembles that of a duck bill. The joint of the lower jaw is behind the rear edge of the eye. The NORTHERN PIKE *(Esox lucius)* inhabits flowing and standing water throughout Europe with the exception of Spain, southern Italy and Dalmatia; it is also found in Siberia and North America. The northern pike is a powerful predator which can reach a considerable size.

Family: pike

The pike

Fishing magazines always carry stories on the catch of a "record pike". The giant specimens weighing 35–70 kg are probably restricted to the rivers in the east and there they are rare. In central Europe pike can be caught which weigh fifteen kilograms. One powerful pike, perhaps the mightiest in western Europe in recent times, was supposedly caught by accident shortly after the first world war. A farmer in Dachauer Moos wanted to get rid of some hand grenades which had been left on his property be retreating soldiers. He picked them up and threw them into a marsh. But after the explosion the water did not calm down, but was being stirred up. He found the cause to be a dying giant pike which was thrashing about. As the farmer pulled the pike into town by a rope strung over his shoulder, the fish was allegedly larger than the man and its tail dragged on the ground. The preserved head is said to have been kept for years in a restaurant in the vicinity. These giants are most often females, which grow to be considerably larger than the males.

Fig. 12–26. Distribution of pike (Esocidae).

When the solitary pike lies in wait for prey, it sits quietly in the upper water levels at the edge of plants in order to grab passing fishes with a rapid, darting movement which is usually headfirst. This jump is so powerful that a 40 cm long pike in the Hellabrunn Zoo (Munich) aquarium rushed at the wall of its tank and literally broke its neck. One Hamburg fisherman told that he had just hooked a 30 cm long pike which had swallowed his fish bait, and when the fisherman tried to land the fish a still larger pike appeared and swallowed the hooked specimen.

In addition to fishes pike also prey upon rats, mice, birds, frogs and other small vertebrates. Since they feed on commercially worthless minnows and carp (Cyprinidae), which then become converted into valuable pike flesh, the pike is a popular fish for eating and is even bred for this purpose. Four to five grams of minnow meat suffice to produce one gram of pike meat. The pike spawning period occurs in the first spring months, approximately from March to May. The fishes move into meadows which have been flooded by the high spring waters. During this time the otherwise very cautious fishes are easy to observe; there are generally one large female and two to three smaller males. The sticky spawn adheres to objects and plants, perhaps on the feet and wings of birds as well. By this mechanism pike get carried into small bodies of water where their presence would otherwise be completely unexplainable. The number of eggs laid varies depending on the size of the female between 100,000 and 1,000,000. Often pike are bred in artificial stations and the young are then placed into natural waters. The young pike initially feeds on plankton, but when three to four centimeters long changes to a fish diet. Of a group of young pike in an aquarium there will soon be just one animal surviving, this one having eaten all its brothers and sisters.

The pike in Europe is one of the oldest and most well-known fishes along with the eel and the carp. Fishermen relate all sorts of stories about them. It is probably the most popular sport fish and it is caught in a diverse variety of ways. The simplest, but illegal method is to catch pike with a snare during the spawning period. Unfortunately the method of using live fish bait, which is repulsive to many animal friends, is widely used. According to legend pike can live to be several hundred years old; however even authoritative scientists such as Regan consider an age of sixty to seventy years as possible.

Other pike species

In Siberia a second species, the AMUR PIKE (Esox reicherti) is distinguished; its distribution is limited to the Amur region and Sakhalin. The American northern pike, which is the same species as the European variety, is accompanied by four other pike species, of which the MUSKELLUNGE (Etox masquinongy) is the largest. Weights of 35–50 kg have been reported. The muskellunge has a very limited distribution which extends only slightly beyond the Great Lakes. The species is suffering a great decline. Three small species of pike include the CHAIN PICKEREL (Esox niger) from New England, the pickerel Esox americanus from the Atlantic coastal plains and the gulf states, and finally the GRASS PICKEREL (Esox vermiculatus) distributed from Iowa eastward to the Ohio valley and down to Texas.

Mudminnows

MUDMINNOWS (Umbridae) are characterized by a short snout and a lower jaw joint which lies in front of the rear eye edge. These are small fresh water species which prefer standing or slowly flowing water. They generally live quietly or hunt slowly near the floor. When the water level is very low mudminnows can dig into the mud and withstand those dry periods; they can even breathe atmospheric oxygen. Their distribution is unusual and has an Upper Tertiary pattern. The genus Umbra is represented by one species, the EUROPEAN MUDMINNOW

(Umbra krameri) in the lower Danube and Dniester and by two species, the CENTRAL MUDMINNOW *(Umbra limi)* and *Umbra pygmaea* in North America. The North American species have been introduced in Europe. They are now found here and there in central European waters and they are bred as aquarium fishes. All mudminnows are small predators which feed on fish eggs, insects and other invertebrates. The spawning period in the European and American species falls in spring from March to April. Females guard and care for the small number of eggs (about 150) laid in furrow nests.

The OLYMPIC MUDMINNOW (⊕ *Novumbra hubbsi)*, the only species in its genus, is found only in western Washington; in contrast to the dirty-brown Umbra mudminnows it is a beautifully colored fish. The ALASKA BLACKFISH *(Dallia pectoralis)* inhabits the inhospitable regions of the Chukchi Peninsula and Alaska. It inhabits the small bodies of water in this region which are almost never ice-free: pools, moors and small channels of thawed water. It is only active for a short period of time, and often winters frozen in the mud. In the Soviet Union interesting freezing studies have been carried on with this species, as by Borodin in 1936. The spawning period occurs in the summer from June to July. In some places the species is used as food for dogs and man.

Little is known of the ARGENTINES (Argentinoidei). It is believed that among deep-sea salmon species they are the closest relatives to the *Salmoninae* subfamily. During the course of their development the salmon invaded northern latitudes and deepsea regions, and those which penetrated the deep seas developed into the argentines. The Argentinidae members inhabit the north Atlantic and Soviet coast along the Pacific, where they are found along with DEEPSEA SMELTS (Bathylagidae). The third family, the BARRELEYES (Opisthoproctidae) contains deepsea bottom dwellers (bathybentonic fishes) with telescope eyes directed upward.

Deepsea salmon, by W. Villwock

Suborder: argentines

Deepsea smelt are in the light scattering zone of the deep sea; it is assumed that their eyes look through the upper water levels for food sinking toward them. Other researchers maintain that the telescope eyes are associated with the phenomenon in many deepsea fishes of luminous organs being pointed upward. The advantage of upward oriented eyes is that they can orient toward prey and predators without forcing the fish itself to move, allowing it to remain hidden. At this time it cannot be determined which of the two viewpoints is correct. However it is not true that barreleyes just have a small mouth opening and thus feed primarily on tiny organisms. Marshall's observations confirm the assumption that barreleyes feed on larger prey, for his stomach investigations of the rare barreleye *Opisthoproctus grimaldii* revealed thousands of nematocysts (stinging capsules) of siphonophores (*Siphonophora,* Volume I) which had formed a wadlike mass.

Fig. 12-27. An argentine.

Fig. 12-28. A barreleye.

The stomiatoids (suborder Stomiatoidei) are aptly called "Grossmünder" (big mouths) in German. These bathypelagic (deepsea) fishes are practically distributed worldwide. Within the suborder the various dragonfishes, consisting of five families, form a closer relationship. In addition to their large mouths they carry a single threadlike barb on the lower side of the chin. This characteristic is found in most dragonfishes, and in some it is longer than the head. However there are a few, such as the deepsea loosejaw *Ultimostomias mirabilis*, which have a barb much longer than entire body.

Three families in this group of five have an eel-shaped elongated trunk. The DEEPSEA SCALELESS DRAGONFISHES (Melanostomiatidae) consist of over 100 species. The most unusual members are in genus *Bathophilus* and a few related fishes; the pelvic fins have moved so high on the side of their bodies that they are about as high as the lateral line.

The second family, DEEPSEA SCALY DRAGONFISHES (Stomiatidae), has just eight to ten species. They are found in open waters of the deep sea throughout the world.

The DEEPSEA STALKEYE FISHES (Idiacanthidae) with just five species have the fewest of any family in this group of eel-shaped fishes. Some of the stalkeye fishes resemble the scaleless dragonfishes. However their larval form warrants classifying them as an individual family. Development proceeds by metamorphosis. The larvae have long, stalked eyes, and the rear part of their intestines hangs outside the body. Their form is so divergent from that of the adults that in the old Brehm's Animal Life Encyclopedia of 1914 they are depicted with their own scientific name, "*Stylophthalmus paradoxus*". They were known to be larvae, but at that time it was unknown to which adult they correspond. Today there is no doubt that they are stalkeyes. The more the larvae grow (initially 1.5 cm in size), the shorter the eye stalks become, and at the end of development they have assumed their lateral position. The intestines undergo a similar transformation and eventually are in the characteristic adult position.

Fig. 12-29. A deepsea snaggletooth.

The remaining two families are the DEEPSEA SNAGGLETOOTHS (Astronesthidae) with thirty-five species and the DEEPSEA LOOSEJAWS (Malacosteidae) with just fourteen species. The snaggletooths have dangerous teeth both in their mouth and on the bony gill arches. The loosejaws have a gill slit on the hyoid bone. Furthermore five species lack the otherwise typical barb. The barb in *Ultimostomias mirabilis* has the greatest relative length in all the dragonfish varieties, measuring nine times the length of the body.

The DEEPSEA BRISTLEMOUTHS (Gonostomatidae), a closely related group to the dragonfishes, are very small species. The giants among them are at most 7–8 cm long. The somewhat over thirty species all resemble a miniature edition of our herring, but equipped with luminous organs. In fact they were originally considered to be herring. But since some of them have an adipose fin they are considered to belong to the salmon group. Thorough studies have revealed that the dragonfishes are among the most prevalent fishes in the world, although the average man never sees a single one. Herald writes that "only marine biologists examining the stomach contents of larger, deepsea predatory fishes or fishing nets cast in the great depths see them".

Among the bristlemouths those in genus *Cyclothone* are particularly prevalent. In terms of distribution and prevalence the next two positions are occupied by the 3–5 cm long members of *Vinciguerra* and *Maurolicus* with their silvery eyes and protruding luminous organs. Despite their worldwide distribution and great numbers little is known about the life of the bristlemouths. This is perhaps because they are delicate fishes, and even when caught with a plankton net they are so fragile that few can be examined. Some notes on the life of *Vinciguerra lucetia* from the eastern Pacific Ocean have been made. According to the Americans Ahlstrom and Count in 1958 the eggs of this species, which float in the ocean, have a diameter of just 0.8 mm. The freshly hatched larvae resemble sardines; and the luminous organs do not appear until metamorphosis is complete.

The DEEPSEA HATCHET FISHES (Sternoptychidae) form another stomiatoid family. Genus *Argyropelecus* in this family has been known for a long time and has telescope eyes directed upward. The species actually look like tiny hatchets, an impression which is reinforced by the silver coloration. There are large numerous photophores (luminous organs) along the lower side of the body; even at great depths the hatchet fishes glow brightly. The family includes over a dozen species and is not nearly as prevalent and widely distributed as the bristlemouths. Distribution is primarily in the great depths of tropical and temperate seas. The small fish, which are at most 9–10 cm large are an important food source for tuna.

The last stomiatoid family is the DEEPSEA VIPERFISHES (Chauliodontidae). Their name is derived from the extremely long, poisonous teeth reminiscent of those of a snake. Their striking shape, highlighted by the prominent pattern of their photophores has led to their being one of the most frequently depicted deepsea fishes. Only a few species have been identified of this 25 cm long fish. They are found on both sides of the tropics in the Arctic and Antarctic Oceans and they carry out vertical migrations. At night they can be caught near the surface, but during the day they are found at depths between 450 and 2800 m. Like stalkeye fishes they are active predators. They occasionally swallow prey which is much larger than themselves, and the question has been asked how this is possible without rupturing the gills and their blood vessels. Tschernavin investigated the process of feeding and swallowing in the viperfish *Chauliodus sloani* and, quoted by Marshall, writes:

▷
Mudminnows: 1. Alaska blackfish *(Dallia pectoralis)*; Whitefish: 2. The European whitefish *Coregonus albula;* 3. The whitefish *Coregonus oxyrhynchus;* 4. Inconnu *(Stenodus leucichthys);* Graylings: 5. European grayling *(Thymallus thymallus).*

"When attacking prey the strong muscles between the vertebral column and the back of the skull contract, which pushes the skull and the front vertebrae upward and the joint between the upper and lower jaw forward. At the same time the jaws are opened and the pectoral girdle, with which the pericardium and the heart are joined, is pushed to the rear and below by special muscles. Other muscles cause a similar movement of the gill arches. During these motions the hyoid processes, from which the gill arches develop, are rotated through an angle of 180° so that their outer upper surface is now turned inward. The heart, aorta and gill arches are thus removed from the mouth cavity and access is paved for large prey. After being caught the prey is forced through the throat by a series of moveable throat teeth. Then the skull structures return to their original positions and normal breathing returns." Tschernavin writes further: "These creatures behave differently from snakes, which rest after a huge meal. Resting is impossible in the great ocean depths. When the stomachs of deepsea fishes are investigated, they are empty in over half of all cases. This has led to the conclusion that digestion proceeds extremely quickly". However one must also remember that there are many cases in which captured viperfishes have ruptured their stomachs while being brought up.

If deepsea fishes are brought to the surface too rapidly, the swim bladder swells due to the reduced pressure, finally bursts and drives the internal organs (primarily the stomach and intestines) out through the mouth. Even in these cases any remains within the stomachs cannot be assessed with certainty. However in connection with the catching and swallowing of prey another viperfish structure has caught our attention. The first vertebra of the vertebral column is rather elongated, about as long as the five following vertebrae. A number of researchers feel that this first vertebra is used for shoving; indeed, the viperfishes storm their prey with opened mouth. However it seems just as reasonable to assume that this elongation is related to swallowing of prey; an elongated first vertebra could be used as a lever when seizing prey and as support for the specialized muscles to open the jaws.

The "big mouths" are in general small to at most medium-sized fishes. However most of them have developed sexual dimorphism (a difference in appearance between the sexes). Males are typically much smaller than females. In stalkeye fishes of genus *Idiacanthus* males are 4 cm long, and females can measure as much as 27 cm. Male stomiatoids undergo a degeneration of their digestive organs and apparently do not feed as adults. Their only function at that time is to reproduce. The teeth are also correspondingly degenerated in many "dwarf males" and in some they are entirely absent. Even the photophores are not as well developed; as a rule the dwarf male possesses just one large photophore behind the eyes. Marshall adds, "The smaller males are not as strong swimmers as the females. It is entirely conceivable

◁
Salmon species: 1. Dog salmon *(Oncorhynchus keta)*; 2. Sockeye salmon *(Oncorhynchus nerka)*; Gadoid fishes (see Chap. 18); 3. Burbot *(Lota lota)*; Perch (see Vol. V); 4. Striped bass *(Roccus saxatilis)*; 5. Pirate perch *(Asphredoderus sayanus)*; Carp (see Chap. 15); 6. Longnose sucker *(Catostomus catostomus)*; Catfish (see Chap. 16); 7. Channel catfish *(Ictalurus punctatus)*.

that the females pursue the males during the spawning period, led by flashes of their bright cheek photophores". Presumably the males have developed into dwarf forms because of the lack of nutrition and difficulties of seizing prey in the deepsea conditions.

A number of other deepsea salmon are in the suborders Bathylaconoidei and the DEEPSEA SLICKHEADS Alepocephaloidei. These are small species which are only known in museum collections. Somewhat more is known about the latter group than the former.

Two families of deepsea slickheads are distinguished: Platyproctidae and Searsiidae. Their common distinguishing characteristic is their smallness, since they are slender, dark hued species; the swim bladder and typical adipose fin are absent. This absence of the adipose fin has recently led to putting both suborders into the herring order. They are in fact also commonly called DEEPSEA HERRING. The deepsea herring *Dolichopteryx* is of particular interest; it has upward oriented telescope eyes and the rays of its pectoral fin are almost as long as its body. Some deepsea herring have photophores.

The Searsiidae deepsea herring are species whose natural life is very poorly understood. They bear some features reminiscent of herring and also seem to have a form of luminescence. A glandular structure is located just behind the pectoral girdle, about which Marshall has written the following: "This organ consists of a sack filled with soft tissue lined with black pigment, which opens with a tube running above and toward the rear of the body on the lateral line. Parr noted that this is undoubtedly a secretion organ which is probably functioning in producing a luminescent substance". Unfortunately this hypothesis has not been demonstrated in living deepsea herring. If it does prove true, this would be a parallel structure comparable to similar structures in deepsea copepods and some deepsea shrimp, which can also secrete luminescent substances into the water. More precisely, this has to do with a secretion "which becomes activated as a luminescent substance upon contact with water", as Marshall writes. It would appear that the value of such a secretion would be to protect the deepsea herring in flight. Thus it would have the same functional significance as the "ink cloud" of cuttlefish.

The largest group of deepsea fishes in the salmon group are the LANTERNFISHES (Myctophoidei). There are fourteen families in this group. The LANTERNFISHES (in the narrower sense) (Myctophidae) comprise 150 species of primarily small to very small deepsea fishes; the smallest measure only 2.5 cm, while the largest are about 15 cm in size. Hjort has appropriately designated the lanternfishes and other bathypelagic species as "Lilliputian fauna". Their common name is derived from the pearllike strings of photophores which form particularly beautiful patterns. They differ from species to species and from genus to genus and are especially characteristic of the individual species on the head and the root of the caudal fin.

Deepsea suborders: Bathylaconoidei and Alepocephaloidei

Suborder: lanternfishes

Like other deepsea organisms lanternfishes also carry out daily migrations, which are vertical and dependent upon the time of day. During the night, especially if the moon is not out and the sea is calm, they ascend out of the depths and feed at the surface of the water. They assemble just below the surface and look like a lantern overboard or even like the lights of a ship. Their innumerable blinking organs and reddish, shimmering eyes betray their presence. At daybreak they return to the great depths, where they can be caught with fine-meshed nets during the day at 500–800 cm depths.

Lanternfishes are widely distributed. One of them, *Lampanyctus leucopsarus*, penetrates into the north Atlantic and is one of the most important prey of gadoid fishes. Only a few lanternfishes have been more closely studied. Among these better-known species is *Myctophum punctatum*, which is prevalent in the Atlantic and Mediterranean. Adult specimens measure almost 10 cm. The females are characterized by numerous lateral photophores and three to five luminous plates on the lower side of the tail base. Males have just one to three such structures on the upper side of the tail shaft. *Myctophum* begins reproducing toward the end of winter and into spring. During their first weeks and months of life the larvae are transparent and until they are 2 cm long stay near the water surface. As soon as metamorphosis sets in they migrate toward the depths and begin the characteristic lanternfish way of life.

Other lanternfishes also have these plate-shaped luminous bodies at the tail base; the number and arrangement of these plates are associated with sex differences. The fact that there is much more development of glandular tissue in these plates on the tail than in the button-shaped photophores on the rest of the body leads us to believe that the plates serve more functions than simply recognition of the sexes. In this connection Marshall has written, "One would expect the light emanating from the tail glands to be much brighter than that from the lateral organs; Beebe's observations have indeed shown this". With reference to the lanternfish *Gonichthys coccoi,* Beebe wrote, "In the darkness it would be quite possible to distinguish species of lanternfishes from the hieroglyphics on their sides and likewise the sexes on the basis of the caudal lights being directed upward or downward. I have never seen the tail organs to light up when a fish swam alone in an aquarium". Beebe feels that these tail flashes are related to more than distinguishing the sexes; they apparently have the important function of blinding enemies: "When the belly lights go out, they do so gradually, so that we still can perceive the fish for a certain time after its lights go out. But one's eyes are so blinded from the sudden flash of the caudal photophores that when they go out we are completely blinded for several seconds. One could hardly conceive of a more effective means of defense and flight."

Beebe accidentally confirmed this with another lanternfish, *Myc-*

tophum affine. The fish responded with light flashes from its tail gland when Beebe brought the luminous face of his watch to the aquarium wall. Repeated trials with the luminous watch face consistently produced the same response in the fish; however it did not flash back when Beebe shined a flashlight on it. Free choice tests between the flashlight and luminous dial showed that *Myctophum* would only respond when the light source resembled what it would encounter in natural conditions.

The next closest relatives are the lanternfishes Neoscopelidae. They have photophores on the external body surfaces and even on the tongue (as in *Neoscopelus macrolepidotus*). This is a highly practical situation, for the light serves to lure prey into the mouth of these lanternfishes.

Practically all lanternfishes have some very striking features. That is least apparent in species of family Aulopidae. This group contains less than ten species. They are distinguished by a relatively large, saillike dorsal fin which has a single thread-like elongated fin ray. The pectoral, pelvic and caudal fins are also pointed. These lanternfishes are found primarily in the upper parts of deep water, e.g., beginning at 100 m and higher, and they are colored bright red to purplish. This coloration is shared by many other organisms found at this depth.

One of the smallest aulopids is *Aulopus japonicus,* which is 20 cm long and is fished in 90 m depth off the southern parts of Japan. It is considered to be a delicacy. One of the largest species is the SERGEANT BAKER *(Aulopus purpurissatus),* which is caught off the coast of Australia. Stomach content studies have shown that this species prefers mollusks, particularly mussels and snails. *Aulopus filamentosus* inhabits the Atlantic and Mediterranean. All aulopids are distributed in warm and temperate seas.

LIZARD FISHES (Synodontidae) also occupy the upper water levels. L is 30–60 cm. However the bathysaurid family, in which this group is now included, also have pure deepsea fishes such as those of genus *Bathysaurus;* Herald still lists *Bathysaurus* with the BOMBAY DUCKS (Harpodontidae). The lizard fishes are called so because of their elongated, rounded bodies. They even resemble lizards in their predation technique; supported by their pectoral fins, they lie in wait, and attack their prey in a sudden leaping snap, the prey being swallowed whole. The many lizard-like teeth on the jaws and tongue prevent even the most agile prey from escaping. Lizard fishes also dig into sandy bottom with the help of their pectoral fins; they can dig in so deeply that only their eyes are visible.

The approximately thirty-five species in this family live in shallow water at depths of 30–40 m. They inhabit warm seas; although members of *Synodus* have also been found on both sides of the tropical Atlantic. *Synodus foetus,* for example, has been observed between Cape

Cod in the north and Brazil in the south. *Synodus synodus* is likewise found outside tropical regions. The *Synodus* species of the Pacific Ocean, such as *Synodus lucioceps*, may move as far north as the coast of California. Another Pacific species is *Trachinocephalus myops*.

Recently the genus *Saurida* has been made an independent family and is separated from the other lizard fishes. This group has lizard fishes from the Indian and Pacific oceans which externally are barely distinguishable from the *Synodus* species. They are somewhat larger and are found at somewhat greater depths. One typical species, a member of *Bathysaurus*, reaches a length of 60 cm and is one of the largest lizard fishes known. It differs from the *Synodus* and *Saurida* species in the absence of an adipose fin and the considerably longer snout. It is probably distributed worldwide at depths between 900 and 2800 m.

Fig. 12-30. The Bombay duck.

The BOMBAY DUCKS (Harpodontidae) have been classified as members of the suborder of lanternfishes even though they have no deepsea species. This unusual group consists solely of the genus of Bombay ducks *(Harpodon)* with five species. They look like blunt-nosed lizard fishes, the pectoral and pelvic fins of which have become greatly elongated. Bombay ducks are exclusively distributed around India and are already absent by the coast of Ceylon. They are found in the mouth of the Ganges and in other tidal zone rivers, which they occasionally penetrate.

The designation "Bombay ducks" comes from the fact that in some parts of India they are brought to market by fishermen, and are eaten fresh or dried and mixed with curry rice. This curry rice dish is known as "Bombay duck". With time this name has gone over to the fishes themselves.

The remaining ten families of lanternfishes are true deepsea species. One family, Chlorophthalmidae, resemble the aulopids with the exception that their fins have the typical shape of other fishes. They have unusually large eyes which indicate that they can perceive light at the 500 m depths they inhabit. A few species have been found in the Atlantic.

Fig. 12-31. The lantern-fish in the family Chlorophthalmidae.

The family Bathypteroidae is characterized by a completely divergent body shape. This contains about twelve species, most of which belong to the genus *Bathypterois*. These fishes, with an elongated body and strikingly small eyes, are caught with nets at 500 m depths. In most species the first or first several rays of the pectoral and pelvic fins are extremely elongated. They walk on these "stilts" along the muddy floor while searching for food. These rays may also function as tactile organs and perhaps also for tasting. According to another opinion they are a kind of landing apparatus; when the fish leaps, it comes thereafter to rest on these rays, using them like a tripod. At any rate it is certain that because of these structures *Bathypterois* forego creating a disturbing turbidity, which would normally come about by stirring up the floor mud. The length of some of these rays seems to differ with sex differences. Whether this is true sexual dimorphism has not yet been ascertained.

The closely related Ipnopidae lanternfishes have greatly modified eyes. In some of them the cornea is flattened while the retina is greatly developed. This provides these species with an extremely sensitive visual apparatus which seems particularly useful in detecting the weak emanations of other deepsea fishes. There are also ipnopids which have degenerate eyes or are completely blind. In place of eyes these latter have a pair of long translucent bones which are found on the front of the skull. Each bone has a bright, and probably luminous, tissue.

The paralepidid barracudinas (Paralepididae) inhabit 3000 m depths and are worldwide. They are the most prevalent deepsea fish wherever they appear, with the exception of the Arctic and southern Atlantic regions. The mouth of these small, slender and pale fishes is surprisingly well-supplied with tiny teeth; in German these fishes are called barrakudina. Like their namesakes, the barracudas of tropical seas (see Volume V), they must be extremely adept swimmers. Otherwise one could not explain why practically no adults of this species have been caught. The largest specimen ever caught, a *Paralepis barysoma* from the Antarctic, measured 60 cm.

Fig. 12–32. A "barracudina".

With the exception of genus *Lestidium* the "barracudinas" lack the typical photophores of deepsea fishes. *Lestidium* members carry on regular vertical migrations at night and day, whereby they follow copepods and crustaceans (and other floating organisms), which ascend to the surface at night and retire to greater depths during the day. During these migrations the barracudinas are subject to predation and to capture by man. Man exploits their tendency to follow light by attracting them at night with bright lamps. *Lestidium* species are often caught, primarily on Luzon in the Philippines; they are sold in Manila and neighboring towns. As far as individual observations indicate, the barracudinas swim in a peculiar manner: head-down. Their tails face the water surface.

Herald has expressed one member of another family, *Anotopterus pharao*, to be a divergent barracudina. Quite unlike the barracudina, these fishes lack a dorsal fin; it is replaced by an adipose fin. A distinct feature of these fishes is that they have developed certain bones within their muscles. The longest *A. pharao* which has been caught was 86 cm long. Only about fifty individuals of this species have been observed, mostly from the stomachs of tuna, halibut, toothed whales and other large fish predators of the oceans. As far as is known, the species is distributed at great depths in the southern hemisphere.

The PEARLEYES (Scopelarchidae) are also somewhat similar to the barracudinas. They are apparently also excellent swimmers, since adults are rarely caught and are only seldom found in the stomachs of the larger predatory fishes. The longest pearleye thus caught was 20 cm long; all specimens were captured between 180 and 3000 m depths. Pearleyes resemble the lancetfishes (see below) in numerous internal aspects and in two external features: like lancetfishes they have large hooked teeth on the tongue (in addition to the two teeth on their jaws) and telescope eyes which are arranged differently in different pearleye species. In some pearleyes they are pointed forward, while in others they point upward and in a third group they are oriented to the side.

Evermannellid lanternfishes are among the most predatory deepsea fishes even though they are only up to 20 cm long. They have paired large saber teeth and in German are called SABERTOOTH FISHES. Another distinctive feature they possess is grooves in the head which are believed to contain photophores. The eyes are telescoped and are directed upward. Sabertooth fishes are apparently distributed worldwide.

The lancetfishes (in the larger sense) contain the little-known Scopelosauridae and Omosudidae. These fishes are found at depths of 750–1800 m; the younger specimens live in the upper levels, the older ones more toward the lower border of their distribution. The latter family is distributed in all tropical and subtropical seas. The lower jaw has been modified specifically for the predatory life the species leads. Since the omosudids hunt prey which is larger than themselves, the hammerlike widened lower jaw is not only useful but necessary. The structure of the shoulder girdle serves the same purpose; a few bones in it are pointed to the outside so that they are not in the way when extremely large prey are swallowed.

The LANCETFISHES (Alepisauridae) are among the most exciting deepsea fishes. With a L of 1.8 m they are the largest known deepsea species in the salmon group. They have a giant dorsal fin which runs from the neck to just before the base of the tail and leaves little room for the small adipose fin behind it. The lancetfish teeth have an ominous appearance; they are long (like venomous teeth) and are bent inward. It would seem that they would lead a life much like that of shark or barracudas. According to stomach investigations, lancetfishes are "the wolves of the sea" as Herald terms them; they feed on whatever they encounter. However if there were no lancetfishes, a number of deepsea species would still be unknown to science. The stomachs of this species are an inexhaustible source for deepsea researchers. Lancetfishes often seem to gorge themselves. Their trunk may be extended like a balloon, whereby they float to the water surface and can be easily caught. A severe attack of parasitic worms can also cause them to leave the safety of the deep and come to the surface. Another means of catching them is by actual deepsea line fishing, as is done off Portugal and Japan.

Fig. 12-23. A lancetfish.

13   Cetomimiformes, Ctenothrissiformes, and Gonorynchiformes

Only recently was a new deepsea order of fishes, Cetomimiformes, introduced. In German the group is known as "walköpfige" ("whale-headed") fishes. As late as 1961 this group was put with the berycoids (now Beryciformes); together with some related families such as the barbourisiids and the rondeletiids. In 1966 the actual relationships were discovered, and the new order Cetomimiformes was developed, with four suborders: 1. Cetomimioidei; 2. Giganturids (Giganturoidei); 3. Deepsea ateleopids (Ateleopodoidei); 4. Mirapinnatoidei.

Cetomimiformes, by W. Villwock

The ateleopids (suborder Ateleopodoidei, with the single family Ateleopodidae) and the mirapinnatoids (Mirapinnatoidei, with the families Mirapinnidae, Kasidoridae and Eutaeniphoridae) are still not very well understood. The mirapinnatoids ("wonder finned fishes") have derived their name from their unusual fin shapes.

Suborder: ateleopids and mirapinnatoids

The "whale-headed" fishes (Cetomimoidei) are probably the most interesting ones in this group. The suborder contains three families (Cetomimidae, Barbourisiidae and Rondeletiidae). Of course the whale-headed fishes have nothing to do with whales, which are marine mammals; however their silhouette bears something in common with that of whales. Members of Cetomimidae and Barbourisiidae look like miniature baleen whales (see Volume XI). This impression is increased by the position of the dorsal and anal fins, which are well to the rear, and, as in whales, by the absence of a pectoral fin.

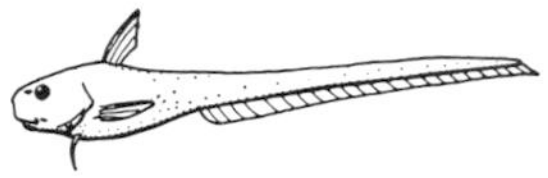

Suborder: cetomimids

Fig. 13-1. The deepsea ateleopid.

Cetomimid species are small to very small fishes (L 5-15 cm) which inhabit depths devoid of light; they are usually colored darkly. Red and orange red hues are found solely around the mouth and the fin bases. Some species lack eyes and are blind, while others have small degenerate eyes. The barbourisiids also have small eyes and thus have correspondingly poor vision.

The large, wide mouth openings of the Cetomimidae indicate that they are predators; their stomachs have been known to contain prey as large as they are. The stomach and abdominal wall are thin enough

▷

Pike: 1. Muskellunge *(Esox masquinongyi)*; 2. Grass pickerel *(Esox vermiculatus)*; 3. Chain pickerel *(Esox niger)*; 4. Northern pike *(Esox lucius)*.

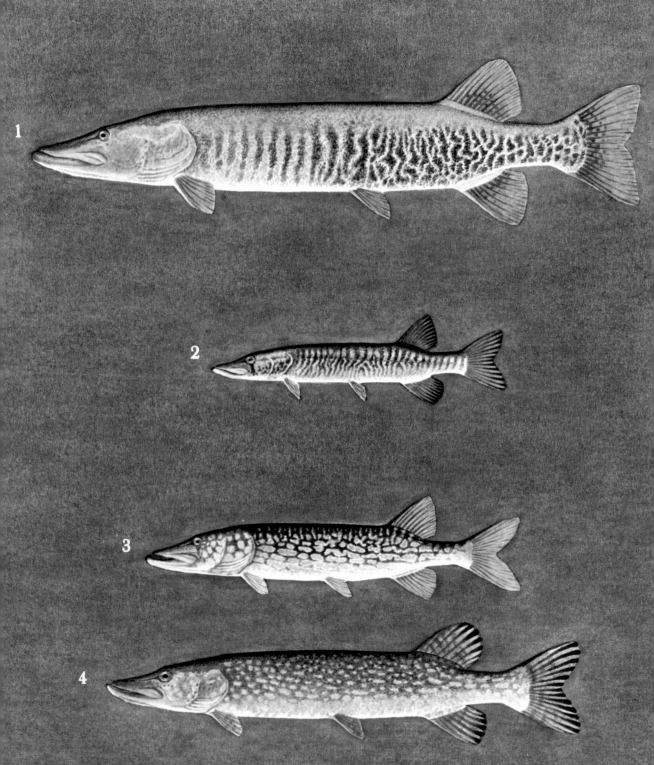

Fig. 13-2. A "whale-headed" fish.

Fig. 13-3. A rondeteliid.

◁

Stomiatoids: 1. Deepsea stalkeye fish (*Idiacanthus fasciola*); 2. The bristle-mouth *Cyclothone signata*; 3. The bristlemouth *Diaphus elucens*; 4. The deepsea hatchet fish *Argyropelecus affinis*; 5. The bristlemouth *Muurolicus muelleri*; Lanternfishes: 6. The bathypteroid *Bathypterois ventralis*; 7. *Omosudis lowei*; 8. *Myctophum phengodes*.

and sufficiently flexible so that large chunks of prey can be swallowed without hazardous consequences. Luminous spots have been found on the rear edge of the dorsal and anal fins. Presumably this substance is produced by glands at the base of the fins, in a way similar to the gulpers (Eupharyngidae).

Cetomimids have not developed a true swim bladder. Swim bladders function as a hydrostatic apparatus to keep the organism floating without having to make swimming motions. Recent studies have shown that the cetomimids have solved the hydrostatic problem in an entirely different way. The lateral line organ forms a continuous tube with a cavity which contains a series of chambers throughout the entire length of the tube. Capped processes have also formed along the tube, and the entire system functions like a swim bladder. The cetomimids are very small species, and the relationship between a small size and a missing swim bladder is not limited to these fishes, but also exists in numerous other deepsea species. This has led some to postulate a definite, predictable relationship between body size and absence of a swim bladder. Careful studies have revealed that these swim bladderless midgets also have comparatively lighter organs than larger related species with swim bladders. For example, the musculature is less developed in the cetomimids and their skeleton has less massiveness than in species with a swim bladder.

Lacking a swim bladder has another physical effect. The small swim bladderless species chiefly inhabit great oceanic depths in which the temperature is lower than in the upper levels. However as temperature decreases, the density of the water increases, along with its buoyancy (up to a limit). Thus the small deepsea fishes can carry on a free-swimming life without a swim bladder, while solving the hydrostatic problem. Fishes in the upper levels which have no swim bladder (e.g., tuna) must be in constant motion to prevent sinking. The deepsea fishes could not produce the necessary energy with its correspondingly high metabolic turnover needed for constant swimming, because they have a much smaller food base upon which to draw.

The barbourisiids (Barbourisiidae) from the Atlantic and the area around Malagasy have few species; they resemble the cetomimids externally. There are no blind species, however, and they always possess paired pectoral fins.

The rondeletiids (Rondeletiidae, genus *Rondeletia*) are characterized by a deviant lateral line. Instead of running the direction of the longitudinal axis of the body, the lateral line consists of about twenty diagonal papillary rows which presumably also have a hydrostatic function. Mucous tubes are found at the base of these rows. This arrangement of the lateral line organ is basically like that in gibberfishes (Gibberichthyidae; see Volume V), which is one of the berycoid species.

It was for this reason that the rondeletiids were formerly classed together with berycoids.

The giganturid suborder includes the giganturids (in the narrower sense) (Giganturidae) and the rosaurids (Rosauridae). Actually it is not known for sure whether the latter are simply larval forms of the giganturids. The systematic position of the giganturids is also unclear. For a long time they were put between lizardfishes and the gulpers (Chapter 9), and only recently have they been classed under the Cetomimiformes. The problem is that only a few specimens have ever been caught of this group.

The mighty telescope eyes of these fishes have led to their German name, "telescope fishes". The cylindrical lens, typical of all fishes, rests on a long stiff tube which is pointed forward. Light sensitive organs fulfill their purpose to the extent that enough light is present for them to be of practical value. At depths of 450–1800 m, at which the giganturids are caught, a little light can still penetrate (at least into the upper levels). In deeper areas the various deepsea organisms illuminate their environment by luminescent organs. These faint emanations are sufficient for giganturids to properly orient themselves; the tube-shaped optic shaft permits the giganturids to perceive light with enough precision to orient toward the source of light.

Fig. 13-4. A giganturid.

Giganturids thus far caught have been small organisms 6–11 cm long. Pectoral fins are absent, and the dorsal and anal fins are well rearward. The lower part of the caudal fin is elongated, forming a whip. The relatively large pectoral fins have their base immediately above the gill cover, an unusual position. Giganturids are predators, and it is believed that the pectoral fins fan fresh water to the gills while prey is being swallowed. When such a fish has a "mouthful", it is hardly possible to breathe through the mouth. An 8 cm long *Gigantura vorax* has been observed swallowing a 14 cm long viperfish of genus *Chauliodus*, which was held in a V-shape. This closed off access to respiratory water. Swallowing such large prey involves so much time that unless the giganturid had some other source of air it could not survive. Thus the peculiar location of the pectoral fins is actually an effective adaptive mechanism which is related to the method of feeding in this species.

Since the technique of feeding is often very unusual and "unconventional" compared to fishes in the upper levels, many such unusual mechanisms and adaptations exist in the predatory deepsea species. In addition to jaws which have been developed in all sorts of ways for specialized swallowing of food, structures are also present for keeping the captured prey from escaping. In the giganturids these structures are represented by numerous dagger-shaped teeth.

The Ctenothrissiformes contain a single species, *Macristium chavesi*.

Order:
Ctenothrissiformes,
by W. Ladiges

Distinguishing
characteristics

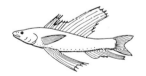

Fig. 13–5. *Macristium
chavesi.*

Gonorynchiformes,
by N. Peters

Milkfishes

Only one specimen has ever been found; it was recently caught off the Azores and had severe injuries. So nothing is known about the natural life or prevalence of the species, but it is undoubtedly a deep-sea form. To get a general picture of the species we must use Regan's description from 1903:

The body of the fish is elongated like a herring and is flattened along the sides, lacking scales. The length to the beginning of the caudal fin is 11 cm. The head is small and forms about one-fourth of the L. All fins except the caudal fin are greatly enlarged. The dorsal fin is about half as long as the entire back and of its twelve rays the eighth to twelfth are twice as long as the first seven and amount to about half the L. The pectoral fin begins about the same distance along the body as the dorsal fin, but it is still longer. The anal fin, with three rays, extends to the end of the caudal fin. There are small sharp teeth on the lower jaw, vomer and tongue. The upper jaw is wide and lacks an additional bone. The present systematic classification of this species is based on this description.

There are a number of quite unusual families grouped in the Gonorynchiformes, which have very different systematic histories. Externally the families resemble each other to a very limited degree. Thus the milkfishes (Chanidae) look like herring, the kneriids (Kneriidae) and cromeriids (Cromeriidae) something like minnows, the phractolaemids (Phractolaemidae) like osteoglossids and the gonorynchids (Gonorynchidae) like mormyrids or sturgeon. There are hardly any external characteristics which the above-mentioned families share in common.

Because of common skeletal elements these groups are placed into one order. They resemble each other in terms of the development and arrangement of bones in the tail and the upper jaw; furthermore, teeth are absent in practically all members of the order. A few features indicate that this group is closely related to carp from the evolutionary viewpoint.

With the exception of the kneriids all the families have just one species. Thus these are probably species which have changed little from the lower Tertiary period and are the survivors of what was once a much larger group. Thus the milkfish *Chanos* of today has been known since the Eocene (about 50 million years ago); and the closest relatives of the *Gonorynchus* genus are *Notogeneus* from the Eocene and Oligocene (40 million years ago). There are two suborders: 1. Milkfishes (Chanoidei; see below) with five families and 2. Gonorynchoids (Gonorynchoidei) with one family.

The MILKFISH (*Chanos chanos*) is the only member of the suborder Chanoidei. L reaches 100 cm; the shape is like a herring with pointed fins. The dorsal fin is small; the caudal fin has a great indentation, and the anal fin is poorly developed with a heavily scaled base. The pectoral fins are beneath the dorsal fin. The paired fins are above and below finlike, partially scaled folds of skin (which are one-third to two-thirds the length of the fins). The lateral line arches up slightly toward the head, while it is otherwise straight and extends to the very scaly base of the caudal fin. The mouth is terminal and lacks teeth. The medium-sized cycloid scales have a smooth rear edge. The eyes have "glasses", i.e., beneath the actual skin. Coloration is silvery to milky white. Distribution is along the tropical coasts of the Pacific and Indian Oceans, in salt, brackish, and fresh water.

Although milkfishes are inhabitants of the open water, they spawn near the shore in shallow water. The spawning period occurs during the monsoon season and may be from April to July, depending on the geographical location. One female can lay up to 9 million eggs. The young remain for some time immediately offshore, and during two months grow to a L of 5–10 cm; at this time they move away from the coast.

Young and adult milkfishes feed chiefly on plant material, especially plankton and algae. The young feed on green and blue-green algae near the shore. As plankton feeders, milkfishes have a very fine gill filtration apparatus. The heavy dependence on plant material for the diet necessitates a very long intestinal tract. The so-called suprabranchial organ is possibly related to the plant diet, for it is found in all members of *Gonorynchiformes.* This organ consists of a pair of bags off the upper rear gill cavity, a structure also found in some salmon which feed on plants. The function of this organ has not yet been clarified. It is perhaps a supplementary filtering apparatus for the fine food particles.

In southeast Asia the milkfish has considerable commercial importance. The fishes are not caught in the open sea; the 1–2 cm long young are caught by the millions with simple fine-meshed nets in shallow offshore waters and they are sent to brackish breeding tanks with a depth of 10–30 cm. The milkfish is ideal for such artificial breeding because it grows extremely rapidly, is insensitive to changes in salinity and temperature, and is very resistant to disease, even when the tanks are extremely crowded. Furthermore plant material, on which all animal life must ultimately depend for food, can be exploited in such tanks. Generally the plants, as primary producers, provide the nutrients for fishes which feed on plankton which in turn are preyed upon by still larger fishes; the end result is the so-called "food chain" passing through these various steps. This means a considerable loss of nutrients, since each animal is able to convert only a small part of its food into its own body tissues. However, the milkfish lives immediately off the plants, making the food chain very efficient and short.

The shallow ponds for breeding are flooded a short time before introducing the young, after the muddy ground has been fertilized with horse or cow manure, with another organic material, or (most recently) with inorganic fertilizers. This shallow water, which is seldom more than thirty centimeters deep, produces a rich growth of algae (primarily blue-green) under the influence of the high temperatures. Under these conditions thirty to fifty young fishes per square meter grow to be five to ten centimeters long in one month. When the algae have been exhausted, the fish are put into a new breeding pond. Small insect larvae, which are otherwise good fish food (such as in carp breeding ponds) are only competitors for the milkfishes, and attempts are made to eliminate these insects in milkfish ponds by the use of insecticides. Unfortunately the quantities of insecticide used are toxic to the fishes. The milkfish breeding system thus employs measures which contradict the fundamental rules of European fish breeding.

Under the uniformly favorable conditions of the equatorial region, the pond-bred fishes attain their market length of 30–40 cm in nine to twelve months. At this time they weigh 300–600 gm. The flesh is firm and tasty, although the larger fishes have many bones. Each hectare of water surface produces 500 kg of fish per year; carp breeding can be equally productive only when additional food is put in the tanks. In the course of a few years of artificial development the milkfishes can reach a length of almost one meter. However they do not reproduce under these conditions, and apparently do not even reach sexual maturity in the ponds.

The kneriids (Kneriidae) are minnow-shaped. L is 4–15 cm. The head is interrupted by a neck fold separating it from the rest of the body. The pectoral fins have a broad, horizontally placed base. Pelvic fins are beneath the dorsal fin. The lateral line is straight, but in genus *Kneria* runs above the pectoral fin in a slightly downward arch. The mouth is clearly on the underside of the body, and is toothless. However the edges of the jaws are armed with a horny covering. The gill opening has been reduced to a short upper slit. The species have very small cycloid scales. Coloration consists of various gray tones with dark spots or stripes along the lateral line; sometimes a dark marbled effect is on the dorsal and caudal fins. There are two genera, *Kneria* and *Parakneria,* each with several species in fast-flowing rivers and streams of central Africa.

Parakneria is distributed in the regions bordering the Congo basin; the species is found in the fast-flowing streams of the Congo tributaries. The distinctive flattened head and foretrunk and the large horizontal pectoral fins enable them to grab hold of cliffs or stony ground in the fast current. No specialized grabbing organs have been developed; the body is held against the floor by these pectoral fins alone.

Kneria is distributed in the regions south and east of *Parakneria*, e.g., in Angola, Rhodesia, and east Africa. This species also inhabits fast-flowing streams and the fishes prefer stopping at sites with stones and roots, for they avoid the strongest currents. Small groups of *Kneria* can be seen tumbling about in the agitated water.

The horny jaw edges indicate that both genera are primarily plant feeders. While *Parakneria* feeds on the algal covering on cliffs and rocks, *Kneria* chooses the grassy algae found on the leeside of stones and roots. The food is scraped off and thereby masticated. The fine gill filtration system retains the tiny food particles. *Kneria* does not feed exclusively on plants, a fact which has been established in aquarium observations. Under these conditions they may be kept for a long period of time feeding on marine crustacean larvae, small insects, tubificid worms, and even dry food.

Little is known about the natural life of *Parakneria. Kneria,* however, has been the subject of recent investigation, largely because of a striking feature found only in adult males. They have a bowl-shaped process on the gill cover, behind which an epaulette-shaped, cross-ribbed structure is located. Both parts of this occipital organ are raised. The bowl-shaped formation on the gill cover consists of a ring of firm, thready tissue which is supported by another ringlike edge of the gill cover bone. The "epaulette" is formed from very densely packed enlarged scales which mutually cover each other. Both parts have horny skin on their exteriors, the arrangement of which is somewhat irregular in front; the back portion, however, runs parallel to the back edges of the large scales. The general impression one has is that the fishes have giant ears and this weird structure has no equal in any other species.

Fig. 13-6. The occipital organ of a kneriid.

The occipital organ was the subject of great puzzlement for a long time. All sorts of theories were postulated as to its meaning; it was related to sensing the current, care of the young or adhering to stones or plants. Detailed studies have finally shown that it is a male sexual recognition sign; it first appears with the onset of sexual maturity (at an age of six to nine months) and has become significant by the time of copulation. Its firm construction and partially horny surface indicate that it has some sort of mechanical function; this is in agreement with courtship behavior observations. A stimulated male attempts to rub against a female which it pursues. Side by side the pair moves through the water, sometimes for as much as thirty centimeters, at which point they separate again. This has been observed in the aquarium, but the release of eggs and sperm has not yet been seen during this behavior. However it can be presumed that the occipital organ has the function of uniting the two fishes during the copulation process, which is of very short duration, The bowl on the gill cover may serve as a suction cup during this time. The horny processes may prevent the round animals from sliding away. *Parakneria* have no occipital organ, which leads us to feel that this bottom-dwelling species has a different method of reproduction.

The kneriids do not care for the young. In the aquarium the fertilized, 1.5 mm large eggs are strewn about irregularly on the bottom and are no longer handled by the parents. At a temperature of 23° C. the young hatch after four days. The yolk body of the freshly hatched young is club-shaped, something which is otherwise found only in carp. This has been used as evidence of the relationship of these two groups.

Another characteristic ties the kneriids with carp. Males have so-called pearl organs on their heads and backs. These are processes which can be seen in kneriids only with the aid of a magnifying glass, but which are visible to the naked eye in some carp. Since they are particularly prominent in European carp during the spawning period they have been called spawning processes. They are much more numerous and more developed in males than in females and are only found in sexually mature fishes. In *Kneria* the pearl organs and the occipital organ probably both assist in copulation. During pre-copulatory courtship display they probably serve a stimulating function. There is more than a functional relationship, however, between the pearl organ and the occipital organ in *Kneria*. The horny processes of the occipital organ develop from pearl organs which multiply and become merged; there is a direct relationship between them.

In 1914 Karl von Frisch discovered an astounding behavioral pattern in minnows; this was later found in most carp and recently in *Kneria*. All these organisms produce an alarm substance in specialized cells of the epidermis. If the skin is injured by an enemy, this substance is passed into the water; it alarms the other conspecifics (members of the same species) and sends them into flight. The fishes can also perceive the alarm substance of closely related species. Even salmon and minnows react to the alarm substance of *Kneria*. Indeed, the kneriids would be considered relatives of minnows if it were not for the fact that the kneriids do not have the Weberian apparatus which is characteristic of all carp.

The cromeriids (Cromeriidae) contain just a single species, *Cromeria nilotica*. L is 3-4 cm. The shape is like that of kneriids but is somewhat more slender and leaf-shaped. The body is transparent, so that individual muscles can be identified. All fins are clearly developed. A lateral line is absent. There are no scales at all on the species. The mouth is on the lower side of the body and has no teeth. The gill slit is very narrow. This species is known solely from the White Nile and the Niger in Africa.

Fig. 13-7. A cromeriid.

It is caught only rarely. The fish has the appearance of a larva, but it can hardly be the larval form of some other fish, since its skeleton is fully developed. Some authors place this fish with the kneriids.

It is still the custom to name systematic groups of animals after famed researchers, although nomenclature should actually serve to characterize the animal in some way. Thus one fish recently discovered in Gabon, Africa, was named *Grasseichthys gabonensis*, after Grassé. It is the GRASSÉ FISH from Gabon. The species closely resembles *Cromeria* in many ways: it is also transparent, lacking scales, lateral line, and teeth; the gill slit is also a very constricted one. However this unusually small fish, 2.5 cm long, differs from *Cromeria* and *Kneria*, most strikingly by the terminal, slightly protrusible mouth. One could put this newly discovered species into a single family, Grasseichthyidae; but *Kneria, Cromeria,* and *Grasseichthys* could also be made members of a single family.

The phractolaemiids (Phractolaemiidae) also have just one species, *Phractolaemus ansorgi*. L is 20 cm; the body is rounded and elongated. The head is small and flattened, with a small mouth a bit toward the upper side of the body. The mouth has few teeth and is trunk-shaped and protrusible. There is one nasal opening on each side, in front of which stands a small barb. The caudal fin has furrows in it. The pectoral fins are somewhat in front of the dorsal fin. The lateral line is straight and distinct. The fish has large prominent scales. Coloration is uniformly brownish gray, with dark fins. Distribution is in the lower Niger and central Congo basin in Africa.

The species inhabits weedy, muddy water. Since it cannot get all the oxygen it needs by gill respiration (a problem for most fish in such water), it can breathe atmospheric oxygen. However it has not been satisfactorily resolved whether the elongated swim bladder or the suprabranchial organ functions as an accessory respiratory structure. The diet consists of small organisms which the fish gets with its protrusible mouth from the mud and weeds. In the aquarium it feeds on a variety of material, including dry food and decayed plants. The sexes are distinguishable in that males have prominent whitish pearl organs on the head and the scales of the tail shaft. Nothing is known about reproduction in this species.

Gonorynchus gonorynchus, the only member of the family Gonorynchidae and the suborder Gonorynchoidei, has a spindle-shaped elongated body. L is up to 60 cm. The head is pointed and fully covered with scales. The mouth is on the lower side of the body and has few teeth. A single barb is located on the lower side of the head. The dorsal, anal and pelvic fins are situated well to the rear, with the pelvic fins beneath the dorsal fin. The species has small ctenoid scales. It has a complex coloration pattern: the throat is reddish white; the gill region is purple to black; the base of the caudal fin is deep black; and the

Gonorynchoidei

rest of the caudal fin is reddish and white. All scales are blue in the middle and are edged with brown. Distribution is off the coasts of Japan, Australia, New Zealand, South Africa, and the Hawaiian islands.

The species lives in sandy shallow water. It is found down to 150 m depths and can dig into the ground in a short time with its pointed head. Presumably it also digs for its food in this way; the diet consists of many kinds of small organisms. In New Zealand the species is caught by sport fishermen using marine isopods as bait. The New Zealanders consider the fish to be a delicacy. In South Africa the species is often caught accidentally in nets and is considered worthless.

14 Char
acins and Electric Eels

Carp (Cypriniformes) and their distant relatives, the catfish, belong to the so-called Ostariophysi group (from the Greek οοταριου = bone and φγσα = bubble skin, which here means swim bladder); that is, they possess a row of tiny bones which E. H. Weber discovered about 150 years ago. This structure unites the swim bladder with the auditory organ and is comparable to our auditory ossicles which join the eardrum and the inner ear; however in carp they have developed from the first vertebra and not from bones of the lower jaw as in humans. Recent findings have revealed that these Weberian bones show what is known as convergence in their structure and their function in comparison with other vertebrate ears. Organs which fulfill the same function yet are of different origin are considered to be convergent. It is very probable that members of Ostariophysi (particularly the characins but not the electric eels) are highly sensitive to high frequencies, which correspond to high tones. In this connection the swim bladder functions as an amplifier and sound transmitter to the ear. This sensitivity is increased by the fact that in most characins the side of the body under which the swim bladder is located is thinner and forms a triangular surface; this might be called a pseudo eardrum.

The ability to perceive higher tones may seem to be of no value to fishes. However quite the contrary is true; in the life of the carp group these high frequencies play a major role. It has recently been found that these fishes are not only able to detect high frequencies, but also emit them, probably continuously. This takes place by means of a valve in the throat which is activated by vibrations of the swim bladder, which is connected with the throat by an elastic canal. The swim bladder delivers air pressure and thus operates here like our lungs. This sender-receiver system has a certain resemblance to other well-known mechanisms, such as those in bats and dolphins. These high frequency emanations give information on the other animals near the individual fish and may also be a form of echolocation. This may be

Order:
carp, by J. Gery

vitally important for fishes like characins and other carp species which live in great schools. This is an entirely new subject which was completely unknown not many years ago, and which in the future will provide us with a great deal of knowledge.

Carp are divided into three suborders: 1. Characins (Characoidei) with fourteen families, 2. Electric eels (Gymotoidei) with eight families, and 3. Carp (in the narrower sense) (Cyprinoidei) with seven families.

Suborder: characins

The CHARACINS (Characoidei) are considered to be the most important bony fish group from the scientific point of view. The group offers the greatest possibilities for studying questions of very general biological importance, such as those dealing with the evolution and adaptation of organisms to their environment. The group might also help clarify the origin of the Ostariophysi (e.g., carp in the larger sense) fishes and, in connection with that, help clarify the famous Wegener theory of continental drift. Characins are distributed in tropical-subtropical America and in Africa, but nowhere else.

Fig. 14-1. Distribution of the characins (Characoidei).

Characins are best characterized by the complete development of the Weberian apparatus, which in some respects reaches its fullest development in the characins. The species have scales, with a few exceptions. Generally an adipose fin is present, and it corresponds to such a structure in salmon, a distantly related group. All species bear teeth during the juvenile stage, while members of the suborder Cyprinoidei have no teeth (this is excepting those with teeth in the gullet). The shape and number of teeth vary from group to group; thus characins are characterized as being heterodont, and in this respect they are the most variable group of fishes after the sea breams (see Volume V). The nature of teeth is an important systematic key. Distribution is primarily in the tropics, with a few species in the subtropics. In the Americas from Texas to Argentina they are the most prevalent fishes in terms of absolute numbers and number of species. In this part of the world they comprise nearly half of all freshwater fishes, and this part of the world has more fishes than any other region. They are not as prevalent in Africa, but still represent about one-fourth of the freshwater fishes in the so-called "Ethiopian Region" (see map Volume VII, Chapter 1), which is tropical Africa. This fish group has one of the largest numbers of species in the world; there are probably more than 1000 species in the tropical Americas and a little under 200 in Africa.

Family: characins

The large CHARACIN family is divided into fourteen subfamilies: Agoniatinae, Rhaphiodontinae, Characinae, Bryconinae, Clupeacharacinae, Glandulocaudinae, Tetragonopterinae, Stethaprioninae, Cheirodontinae, Rhoadsinae, and four less well-known subfamilies.

Agoniatinae contains the single genus *Agoniates* which is in the Amazon and the large rivers of Guiana. It lives together with freshwater herring. Despite their distant relationship these two groups actually resemble each other to a large degree. Both inhabit the open waters. The Agoniatinae species have strong, sharp, hooked teeth and are probably predatory.

Rhaphiodontid species externally resemble the Agoniatinae and freshwater herring; this characin family has two genera, *Hydrolycus* and *Rhaphiodon*. The actual relationship between them, however, is not as great as it appears. Both characins inhabit large rivers, principally the Amazon and the Rio de la Plata, and they have similarly large jaws armed with sharp, powerful teeth, as well as compressed bodies. Some specimens become more than half a meter long. One species, *Rhaphiodon gibbus*, has an unusual appearance reminiscent of a deepsea fish with a shortened body. It almost looks like it has a vertebral column malformation.

Fig. 14-2. Distribution of the characins (in the narrower sense) (Characidae).

The characins (Characinae), with twenty or more genera, are a very important group. The majority of genera are clearly related to the subfamilies thus far discussed. Characins in the genera *Charax*, *Cynopotamos* and *Acestrocephalus* are predators with sharp teeth and a long toothed upper jaw, a compressed body, often with a hump and a long anal fin. They rarely reach a L of 20 cm. The genus *Charax* has the oldest scientific name of the subfamily and in fact gave the subfamily its name. A few small species (e.g., *Gilbertolus* and *Gnathocharax*) are found at the water surface; their large pectoral fins and the keeled chest probably permit them to glide along the surface. One small genus with the single species *Hoplocharax goethei* was just recently discovered. This fish is equipped with a whole series of weapons: within its body are hidden a collection of gill cover-, pectoral fin-, anal fin- and caudal fin-spines; it can practically be called a museum of defensive weapons.

Fig. 14-3. An agoniatin characin.

At least three genera have conical, short, bony nodules on the outer side of the mouth (on both jaws and the upper side of the upper jaw). This is found in the genera *Exodon*, *Roeboides* and *Roeboexodon*. At first it was believed that they burrowed through the floor with these nodules, but stomach content studies showed that they had ingested fish scales; these bony nodules are therefore used to scrape off the scales from larger fishes. This form of mild parasitism could be considered as a deviant form of fish predation. It is also known in cichlids from the large African lakes.

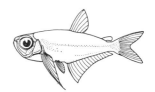

Fig. 14-4. A *Gnathocharax* characin.

Another characin group is shaped like pike and reaches a L of 20–30 cm; these are the genera *Oligosarcus* and *Acestrorhynchus*. They even have the same predation pattern as pike: they hunt alone and suddenly fall upon a school of prey from some concealed position. With the exception of piranha, they are probably the most important predatory fishes; they are equipped with powerful teeth and an impressive throat and are very prevalent in many parts of South America. Their African counterpart is the genus *Hepsetus*.

The subfamily Bryconinae, with six genera, resembles salmon. This is particularly true for the genera *Brycon, Chalceus* and *Salminus. Triportheus,* which looks somewhat more like herring, has well-developed high-set pectoral fins. These, together with a keeled belly enable these fishes to jump out of the water and leap several meters in the air (this is also true of the flying characins). The bryconins can be recognized by their dentition; the teeth are numerous, multiple-crowned, and often in two or three rows. *Brycon* contains forty or more species from Central America and northwestern South America. They are considered good eating, as are both species of *Salminus* (from northwestern and southeastern South America) which grow as large as salmon. They are popular with sport fishermen for the fight they put up when hooked.

Clupeacharacinae contains a single species, *Clupeacharax anchoveoides.* This characin also is salmon-shaped, but its dentition and skull structure are the keys to its characin relationship. A few smaller, insignificant subfamilies might be mentioned here: Iguanodectinae with two genera, Paragoniatinae with five or six genera and Aphyocharacinae with one genus.

Glandulocaudinae, with about twelve genera, contains the unusual ones *Tyttocharax, Xenurobrycon* and *Corynopoma.* They have a divergent reproductive method and have developed specialized organs for this purpose. Recently K. Nelson investigated the internal and external structure of these fishes. He found that they are adapted to feeding at the surface; some have a belly keel and enlarged pectoral fins, and their entire behavior (including raising the young) is adapted to this surface life. In most species fertilization of the eggs is internal; the male puts a small sperm packet (spermatophore) in the body cavity of the female. The female lays the eggs on a leaf at a much later time and without interaction on the part of the male. In some species the males possess a soft projection on the gill cover which is widened at its end; according to Wolfgang Wickler it is used as a lure for the female. This is most strikingly demonstrated in *Corynopoma riisei.*

Besides such structures or similar ones, which only appear in a few species, all male glandulocaudins possess organs variously developed at the base of the tail. They probably produce glands which secrete an alluring substance. Many characins have a large number of hooks on the lower fins which serve to anchor the partners together during copulation.

Stethaprioninae, with three genera, have a spine in front of the dorsal fin. A disc-shaped body is also present. Such a spine in connection with a disc-shaped body is also found in prochilodins (Prochilodinae) and especially in Serrasalmidae.

The largest characin subfamily now follows: the TETRAS, as aquarium keepers call Tetragonopterinae. It contains several popular aquarium genera (*Hemigrammus, Hyphessobrycon* and others), of which some [such as the RED TETRA FROM RIO *(Hyphessobrycon flammeus), Hyphessobrycon ornatus,* and the HEAD-AND-TAIL-LIGHT *(Hemigrammus ocellifer)*] can be easily kept by beginning aquarium enthusiasts. Most tetras have a very wide food base. They feed primarily on small insects and their larvae, which fall into the water or are driven in during floods; but they also can feed on plant material when other food is scarce.

Tetras are such a large group that it is impossible here to treat them thoroughly. They comprise 40–45 genera with very small species (L 1.5–15 cm), of which some (e.g., *Astyanax, Moenkhausia, Hemigrammus* and *Hyphessobrycon*) have over fifty species each. The few species cited here are not typical of the structure and behavior of the subfamily, since most tetras are fairly uniform in these respects. This fact, however, gives them the ability to conform to new conditions. They are in the flourishing period of their development in the truest sense of the word; their numerous species are not only a source of enjoyment for aquarium hobbyists but also a burden for every systematist who must describe new species, which are always being discovered. The proportion of unknown new species in every characin collection which one South African museum receives, amounts to 5–10% of the entire collection. Most are tetras and their close relatives. Some, as for example *Astyanax,* which can live almost anywhere, are more variable than others, as for example *Hemibrycon,* which need a lot of oxygen and thus are limited to rapids and fast-flowing waters.

Fig. 14–5. Courtship in swordtail characins: a. Male displays to the female; b. Male shaking; c. Male (lower animal) shivering.

One of the most unusual exceptions among the tetras is *Anoptichthys jordani* from caves in Mexico. It is completely blind. It has moreover been shown that this blindness is a recent condition, because all sorts of transitions between seeing and blind fish have been found. Kosswig and his students have made interesting courtship studies with the species. One of Kosswig's students, N. Peters, has written the following about them:

The MEXICAN TETRA *(Astyanax fasciatus mexicanus),* distributed throughout Central America, includes a few blind varieties from central Mexico. These cave fishes are so different from their overground ancestors that until their relation to the *Astyanax* species was ascertained they were classed as an individual genus, *Anoptichthys* (fish without eyes). They are not only completely blind; their skin is entirely lacking in pigmentation so that they look flesh-colored, pale red. They also behave completely differently from their above-ground relatives. While the river-living *Astyanax* is a schooling species, which engages in predatory behavior, the cave variety is a solitary form which interacts little with conspecifics and just unceasingly searches the underground waters for food. Its diet consists primarily of animal and plant remains which have been swept into the caves, or included in

Fig. 14-6. Glandular scale of *Mimagoniates microlepsis* male.

bat droppings which fall on the water surface. Thus it comes by quickly when one splashes his hand on the water.

The regression of the eyes and pigments is not a direct result of the darkness in which the species spends its entire life. If it is kept under daylight conditions in the aquarium and bred under these conditions for many generations, its vision still remains poor and the eyes degenerate. One must conclude that the degeneration of the eyes and pigments is an inherited trait. In spite of this great discrepancy between the river fishes with normal vision and the blind cave fishes, they can be crossed (which appears not only unusual but amazing). Thus the courtship behavior in both forms must correspond to a high degree. The hybrid from such a cross is a mixture between the river inhabitant and the cave fish. It has small eyes, is clearly colored and completely fertile; it can be crossed with one of the parents or with another hybrid. In the latter case a second generation is produced which varies from species with full vision to those which are completely blind. Coloration varies tremendously also in this third generation. Interestingly, there are pigmentless forms with well-developed vision and blind but fully colored fishes. Geneticists have concluded the following from this:

1. Development of pigmentation and the eyes proceed independently and are inherited independently. 2. The differences between river fishes and cave fishes arises from mutations. In the transition to cave life those characteristics which have become useless degenerate by changes in the gene structure. This process in the fishes is a model for the general degeneration of organs throughout the animal kingdom, if not solely for the degeneration of pigmentation and eyes in other cave-dwelling animals.

We can even estimate how long it must have taken for the colorless, blind fishes to develop from the normal fishes. *Astyanax fasciatus mexicanus,* which lives above-ground, is originally from South America. It could have penetrated Mexico toward the end of the Tertiary period (about 1,000,000 years ago) when the land bridge between South and North America was formed. It could not have invaded Mexico any earlier because it is a freshwater fish. However the caves into which it moved were formed during the Ice Age rainy period 500,000 years ago through an outgrowth of the calciferous stone deposits. Thus it could have taken at most 500,000 years for the blind varieties to develop from the normal fishes. The earlier cave rivers dried up during the drought of the Ice Age, so that these present-day cave fishes inhabit just a few scattered grottos.

Until recently this cave fish was considered the sole example of a blind characin, although there are several blind carp and catfish species. However in 1965, while digging a well in Brazil, another blind characin was discovered when drilling reached 30 m: *Stygichthys typhlops.* The first to describe it, Britton and Böhlke, have pointed out the unusual fact that in this species not only the eyes and normal pigmentation are absent. They also lack the lateral line organ, which is so important for blind forms in particular. The bones which normally cover the eye region have disappeared, along with most pores on the head and important sensory organs. One must ask what substitute organs the species has developed to make up for this loss. Britton and Böhlke note a remaining ringed canal on the head and a leaf-shaped adipose fin, which perhaps play a special role.

Fig. 14–7. The blind characin *Stygichthys typhlops.*

Another special case is found in the southernmost characin of all: *Gymnocharacinus.* It is completely naked, without the slightest trace of scales. This was also believed to be a special exception, and some authors referred to this condition as an adaptation to the cold Argentinian climate. However, another completely naked characin was discovered, this time in a tropical region in Ghana by Roberts.

Some characin species live an extremely difficult life in the infertile waters of South America with the lowest salinity of the continent; one such area is the Rio Negro. R. Geisler recently visited this region and believes that one of the most popular aquarium fishes, the RED NEON (*Cheirodon axelrodi*), which has been taken by the millions every year from the Rio Negro, never reaches the size of 4 cm there that it does in aquariums. It probably dies of starvation after the first spawning as soon as the rains come and the food supply becomes scarce. Since the spawning period occurs at the start of the rainy season, these characins, along with the toothed carp *Cynolebias* and *Aphyosemion*, can be characterized as "annual fishes." The toothed carp die when their home waters dry up. A few American tetras are also seasonal, and they, too, become much bigger in the aquarium than in nature; one example is the TETRA PEREZ (*Hyphessobrycon erythrostigma*).

The red neon belongs to Cheirodontinae, which is a closely related group to the tetras. This is a very artificial group; all tetra-like species with just one row of teeth in the upper jaw are put into this group, while tetras in the narrower sense have generally two rows—this is even done when each of these rows has just one or two teeth. *Cheirodontinae* contains thirty genera, some of which have barely been observed at all. *Hyphessobrycon* resembles the small genera *Megalamphodus* and *Pristella,* with just slight differences in dentition. They are often caught together with tetras, and one can assume that one group imitates the other. *Cheirodon* and *Odontostilbe* are characterized by the comb-shaped arrangement of the palmate teeth (*Cheirodon* is from the Greek χειρ = hand and οζους = tooth).

Cheirodon and *Odontostilbe* are closely related to the neon characins or belong with that group; these characins are a joy to aquarium hobbyists and a problem for scientists. Their name is derived from the blue-green sheen along the sides of the body which looks like neon lights and in brilliance has a counterpart only on the wings of some *Morpho* butterflies. In addition the belly of neons is a rather brilliant red—it is no wonder that neons are among the most popular aquarium fishes. However the scientific problem begins with trying to systematically classify the neons.

There are three well-known species of neons, all with practically the same coloration, but which by reason of their dentition must be placed into different genera: these are the NEON TETRA *(Paracheirodon innesi)*, the RED NEON *(Cheirodon axelrodi)* and the recently discovered, very rare FALSE TETRA *(Hyphessobrycon simulans)*. The last species belongs to another subfamily, the tetras. It appears that the false tetra is an example of protective mimicry, imitating the red neon. This cannot be the case with the other two species, since they occur in separate areas. The remarkable similarity in shape, systematic details and color indicates that they belong to the same line of development; as a result of differences in diet, however, their respective dentition differences have arisen. They have maintained their brilliant coloration, since this is an important means of recognition in the dark South American waters. The coloration has the same functional role as the luminescent organs of deepsea fishes.

Rhoadsinae, with two genera, stands between the tetras and Cheirodontinae. In the juvenile stage the subfamily has a row of teeth in the upper jaw, while two rows are present in the adults. The upper jaw thus differs at the two age levels, so that juveniles and adults have little resemblance.

The serrasalmids

Piranhas

The Serrasalmid group is known both in the scientific world and to the public in general for the almost legendary predatory species, the PIRANHA. Piranhas have been known since the discovery of South America by the conquistadors (e.g., by the first European to travel down the Amazon, Francisco de Orellana). Alexander von Humboldt observed them on his well-known trip to the New World. Since that time they have been the continual subject of amazing reports. Even today it is difficult to separate fact from legend. Some reports are not without humor. Thus the late Harald Schultz, known as "Indian" Schultz, and one of the individuals best acquainted with South America, relates the following:

"As my father was just fifteen years old, he fled before attacking Indians in a small fragile canoe. The boat tipped over, and he swam away, but as he climbed out of the water he was just a skeleton; that could never happen to him again!"

It has been found that piranhas live chiefly on fishes. But most species have such powerful teeth that they could tear the strongest mammal to shreds. They are one of the few examples of mass predation in the bony fishes. It must still be remembered that the danger they present for man has been greatly exaggerated. Harald Schultz

Fig. 14-8. Distribution of the serrasalmids (Serrasalmidae).

wrote, "For more than twenty years I have made trips to many parts of Brazil for the purpose of studying Indian life in association with a Brazilian institute. In all these years I have never had a bad experience with these greatly feared piranhas. Among the thousands of people I have met, only seven of them had been injured by piranhas. These were all insignificant bites. I did hear of one boy whose boat capsized in the middle of a river and who became the victim of the piranhas. His spirit allegedly is still at that site. Some people feel that the boy was eaten by the piranhas after he had already drowned."

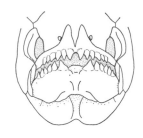

Fig. 14-9. The mouth and teeth of a piranha.

It is undoubtedly correct that some species (for example the "true" piranhas, known in Brazil as "piranhas verdadeiras") can truly be dangerous for injured or otherwise weakened people who have fallen in the water, if this occurs near densely populated areas at a time when the piranhas have a lack of normal food. Such species include the piranha *Serrasalmus piraya* in the Rio São Francisco, the piranha *Serrasalmus nattereri* and related species in and around the large island Marajo at the mouth of the Amazon or to the south in the Rio Guapore. To a certain degree their attack can be compared with the "group effect" such as in the behavior of a few alligators, who also attack when they are driven into a restricted area. However it is also possible that certain chemical substances play a role. Dieter Backhaus observed in the Frankfurt Aquarium that piranhas attack each other when they stay in the same water for more than one month; however their compatibility returns when the water is exchanged.

According to well-founded reports by the local population, piranhas also attack in the tributaries of the Rio Paraguay only when waste bits from the large slaughterhouses or cattle farms are thrown into the water. Piranha just a few kilometers away in water which is not polluted by blood or flesh are considered so harmless that people can swim among them. Some writers, e.g., Richard Gerlach, have described piranhas as "the wolves or hyenas of the water"; they play an important role in the South American ecology. They prey primarily upon wounded and diseased fishes, and they themselves are probably not invulnerable to diseases from parasites. Thus it seems that they serve as a dam for fish diseases which would otherwise become epidemic. As with most predators in the animal kingdom, they can in no way be considered as harmful. Thus recent attempts to eliminate them from certain areas, such as in the Rio São Francisco, are highly questionable. This poisoning proceeds in the deceptive hope of increasing certain valuable fishes. Perhaps the precise opposite will occur, due to an increase in epidemics which were previously controlled by the piranhas.

However even the commercial value of piranhas, particularly in remote parts of the Amazonian jungle where animal proteins are scarce, should not be underestimated. They are easy to catch and are

▷

Osteoglossids: 1. Butterfly-fish *(Pantodon buchholzi)*; Milkfishes: 2. *Kneria polli*, a kneriid; 3. *Phractolaemus ansorgei*, a phractolaemiid.

edible. They are put to an interesting use as well: the toothed jaws are often carried around the waist by Indians using blowpipes for hunting. The teeth are used to make a notch in the poisoned dart so that it breaks into the wound and lets the curare poison enter the body of the prey. In general these teeth are used as scissors or razor blades, and among some Indian tribes, European scissors are known as piranhas. The Tucuna Indians adorn their masks with piranha jaws. If one believes Paez's comments, piranha schools fulfill an important function in the flooded savannas of the Orinoco region, where they "prepare" the skeletons of the deceased. In these regions it is often impossible to bury bodies in the earth for months at a time; the piranha (known there as "caribe", e.g., cannibal) solves the problem very simply. In a few hours they eat the flesh off of corpses placed in the water. The skeletons are then dried, dyed and adorned, and are finally put on the high burial sites of these villages.

Piranhas are best characterized by their strong jaws with the sharp, firmly anchored teeth which form a continuous jagged saw. There is also a sawlike keeled belly and a spine in front of the dorsal fin which does not have the same origin as the similar structure in Stethaprioninae characins. The method of reproduction in piranhas is very interesting; apparently they care for the young, as is done in higher fishes (e.g., cichlids; see Volume V). The 30 cm long (and longer) *Serrasalmus* species are the most striking group in the family, but not the most numerous.

Myleinae, with over seven genera, forms a still more substantial part of the South American fishes. Most members have the same distinguishing characteristics as the Serrasalminae, including the spine on the belly. However a disc-shaped body, most prominent in the genera *Metynnis* and *Mylossoma*, is also present. Unlike the serrasalmids, this group has molars which are often similar to the cheek teeth, but are never like canines. The diet consists purely of plant material. Some species feed on fruits and blossoms and are thus dependent on the large trees on the river banks. The Indian name "pacu" refers to three different species, *Myleus pacu*, *Colossoma bidens* and *Colossoma oculus*. The two latter species reach the considerable size of 60–80 cm and a weight of 10 kg. They are an important source of food for the inhabitants of the great South American forests; the fish are eaten fresh or smoked. Some species are very popular in larger aquariums.

The third subfamily, Catoprioninae, has just one species, *Catoprion mento*. It maintains a parasitic scale-feeding life like the characins already mentioned (*Exodon* and the tetra *Probolodus*). This method of feeding probably developed as well from a predatory ancestor which was similar to the serrasalmids. However instead of using the scraping method of the serrasalmids another technique has been developed: its lower jaw is well-developed and has shovel-shaped teeth on the out-

◁

Characins: 1. *Thayeria obliqua;* 2. *Thayeria boehlkei;* 3. Swordtail characin (*Corynopoma riisei*); 4. *Poecilobrycon ocellatus;* 5. *Poecilobrycon eques;* 6. *Nannostomus beckfordi aripirangensis;* 7. *Nannostomus anomalus;* 8. Red tetra from Rio (*Hyphessobrycon flammeus*); 9. *Hyphessobrycon rubrostigma;* 10. Serpa tetra (*Hyphessobrycon callistus*); 11. *Hyphessobrycon heterorhabdus;* 12. Pristella (*Pristella riddlei*); 13. *Nematobrycon palmeri;* 14. Bloodfin (*Aphyocharax rubripinnis*); 15. Rummy nose (*Hemigrammus rhodostomus*); 16. Head-and-tail-light (*Hemigrammus ocellifer*); 17. *Hemigrammus pulcher.*

side. Werner Ladiges has observed "de-scaling" of a fish living in the same aquarium with a catoprionin. The stomachs of catoprionins caught in the wild contain numerous scales of fishes which are generally larger than themselves; some of the scales are partially digested.

The HATCHET FISHES (Gasteropelecidae), the only true flying fishes, have a heritage the equal of the piranha. There are only three genera and eight species of these small fishes, which are typically less than 6-8 cm long. They are completely adapted to a life near the water surface. Like modern speedboats, the front of which rises off the water at high speed, these fishes can also rise off the water surface. They literally fly several meters through the air, after a starting movement of one to several meters. The initiation movement has been seen in nature, but the actual flying has not yet been observed under these conditions, only in the laboratory. In both instances the process was accompanied by a humming sound.

Family: hatchet fish

Fig. 14-10. Distribution of the gasteropelecids (Gasteropelecidae).

The so-called FLYING FISHES (*Pantodon* in freshwater and Exocoetidae in the ocean) do not fly in the true sense of the term, but glide, even over long stretches. In contrast the gasteropelecids actually flap their pectoral fins somewhat like hummingbirds. The adaptation which gasteropelecids have is similar to that of birds. Although the gasteropelecids are otherwise closely related to the tetras, they have developed a sternum. It is formed from the coracoid bones (parts of the pectoral girdle), which become broadened, flattened and united to form a rounded keel. The large surface area of the bone enables powerful chest muscles to be attached to them, and these muscles activate the sickle-shaped winglike pectoral fins. This situation with a keeled chest and well-developed pectoral fins is not unusual among the characins. It has been found in Bryconinae, Rhaphiodontinae, Glandulocaudinae, Agoniatinae and Clupeacharacinae but is not as developed as in the gasteropelecids.

Fig. 14-11. Skeleton of a gasteropelecid (this fish has a sternum).

Thus the characins have made several attempts to glide, jump and finally fly over the water surface. This is probably of importance for the survival of these species; it enables them to catch flying insects and, of particular importance, to escape predators. Gasteropelecids can be easily kept in aquariums, but the containers must naturally be firmly covered.

The Hydrocynidae (also known as Alestidae) are the African counterpart of the tetras. What has been said about uniformity and life mode of the South American tetras also applies to them. They do not even differ from their New World relatives in their internal anatomy, although the separate development of the two groups has taken place over a period of 200–300 million years. In some respects they actually resemble the ancestral form more than the faster developing tetras; the genus *Chalceus* (which resembles the American *Brycon* genus), for

Family: Hydrocynidae

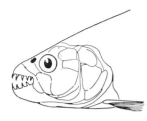

Fig. 14-12. The head of a hydrocinid.

Fig. 14-13. Distribution of Ctenoluciidae characins.

Fig. 14-14. A characidiid.

example, still has a number of these primitive features. The somewhat less than 100 species of African characins can be divided into two lines of development. The subfamily Alestinae, which according to some researchers consists of the single genus *Alestes,* or according to others has one or two genera, has thirty-four rather large species with molars; the species resemble the *Brycon* fishes. The largest *Alestes* fish has a L of 46 cm. The smaller and more diverse Petersinae have probably branched directly off the Alestinae, as the somewhat larger tetras have also done. According to Poll's recent work they contain fifteen genera with almost sixty species and are thus the most prominent African characin group. The most well-known species is *Phenacogrammus interruptus;* the magnificent, long-finned males reach a L of 10–20 cm in the aquarium. In nature the fins are not nearly as pronounced, because fin predators such as *Belanophago* do not permit the elongation of the fins.

The second line of African characins, the subfamily Hydrocininae, comprises the genus *Hydrocinus* with five species. They look like large Alestinae fishes, but they have well-developed canine teeth and are among the most important African predatory fishes. The largest species, *Hydrocinus goliath,* reaches a L of one meter and more and is an impressive fish. Zambia has a small schooling species, *Hydrocinus lineatus,* which hunts in groups like barracudas or piranha.

There are other resemblances between certain South American and African characin families. Thus the African family Hepsetidae with the single genus and species *Hepsetus odoe,* corresponds to the South American characin *Acestrorhynchus* in a rather striking way. Only some details of the skull structure reveal that these are two different lines of development which have a common origin but which have become separated. *Hepsetus odoe* is also one of Africa's prominent fish predators, even though it is much smaller than *Hydrocinus.* Behaviorally it is like pike.

The small Erythrinidae family with three genera is a South American predatory species in which the typical characin adipose fin is absent. This is probably a member of the oldest group of teleost (bony) fishes, because the erythrinids have very primitive features. One species, *Hoplias malabaricus,* from Guyana, attains a L of one meter. It is one of the best tasting fishes and probably owes its survival to the sole fact that Guyana is sparsely populated. Other species, such as *Erythrinus erythrinus* and *Hoplerythrinus unitaeniatus,* have special respiratory organs such as blood capillary nets on both of the gill covers; Lüling and others report that this enables them to move across land when their water dries up.

Another small family of predators is Ctenoluciidae. Although they are completely different structurally from the erythrinids, both groups probably have common ancestors. They are very elongated fishes with long, pointed jaws filled with sharp but tiny teeth and at most reach a L of less than 30 cm. They are not as ecologically significant as *Acestrorhynchus* or the piranhas and their African counterparts. There are only two genera with four species; they have a fleshy appendage of unknown function on either the upper or lower jaw.

The Characidiidae, unlike all other previously treated characins, are small bottom-dwelling fishes. Like the gudgeon of Europe, they are extremely well-adapted to this kind of life. Although they are of scientific importance as inhabitants of rapids and in general the river bottoms, very little is known about them. Presumably there are several genera with fifty or sixty species, of which some have not yet been discovered. Some are so highly adapted to living in rapids that they are comparable to catfishes or *Garra* carp. They have a flattened breast and spread pectoral fins, a mouth on the lower side of the body and eyes on the top of the skull, just like catfishes and the above-mentioned carp. Some *Elachocharax* species are among the smallest vertebrates in the world; they live between thin roots at the shores of narrow streams, such as the waters off the Rio Negro, together with the equally small gobies (*Gobiidae*; see Volume V). In Africa they have an astonishing counterpart in the dwarf characins of genus *Nannocharax*, which belong to a completely different subfamily.

The lebiasinids (Lebiasinidae) are another characin family which has very diverse subfamilies both in the anatomical and behavioral sense. Among them are some species which are very well known as aquarium fishes. *Lebiasina* is the only genus of the subfamily Lebiasininae with twelve species; this group looks like Bryconinae fishes. Some lebiasins also are losing the adipose fin. However Weitzmann found that the skull structure bears no similarity to the erythrinids', in which the adipose fin has completely disappeared. Since they are very resistant, these fishes are used to combat mosquitos in small standing bodies of water in which other fishes would not survive. Even though their native habitat is northwestern South America, they have been introduced into the Amazon region.

The second lebiasinid subfamily is Pyrrhulininae with the genera *Pyrrhulina, Copeina,* and *Copella.* There are more than twelve small species which inhabit most of South America and feed on small organisms; some feed on ants. Generally the fishes are found near the surface of shallow ponds which do not dry up. Many pyrrhulinins are popular aquarium fishes. One species, *Copella arnoldi,* which among aquarium hobbyists is still classed with the genus *Copeina,* spawns on leaves outside the water. The male sprays the eggs from time to time from the water surface. Other species also have developed a certain breeding method which is designed to protect the developing young. Caring for the young always means a lowered number of eggs in fishes; thus the majority of fishes are true "egg squanderers".

The third subfamily, PENCIL FISHES (Nannostominae), consists of both

Families:
Ctenoluciidae and
Characidiidae

Family: Lebiasinidae

▷

Characins: 1. *Carnegiella strigata;* 2. Black-winged hatchet fish *(Carnegiella marthae);* 3. Mexican tetra *(Astyanax fasciatus mexicanus);* 4. *Poecilobrycon espei;* 5. *Pyrrhulina vittata;* 6. *Pseudocorynopomadoriae;* 7. *Copella arnoldi;* 8. *Copeina guttata;* 9. *Roeboides guatemalensis;* 10. *Leporinus fasciatus;* 11. *Hemiodus semitaeniatus;* 12. *Prochilodus insignis;* 13. Headstander *(Chilodus punctatus);* 14. *Anastomus anastomus.*

into the rapids and spread poison under the cliffs. Thus it is no wonder that the swimming position of these fishes has not yet been determined.

The family Curimatidae also contains a collection of very disparate forms, each corresponding to its own mode of life. The subfamily *Chilodinae* also contains more upside down species, including the genera *Chilodus* and *Caenotropus* with three or four species. They lack teeth on the jaws, but some carry a few on the lips (these are very weak). These teeth are supplemented by numerous throat teeth and a special gill organ, the function of which is not yet understood.

Prochilodinae, with two genera and thirty-five to forty species, are rather large, bream-like fishes with a protrusible mouth that looks like a suction disc. L is up to 60 cm and teeth are present only on the lips. They feed primarily on organic waste in the mud. Their African counterparts are the genera *Citharinus* and *Citharidium* (see below). Prochilodins are found in Africa as well as South America. One species, *Prochilodus reticulatus magdalenae,* has commercial importance in Colombia while the species in Guyana are very rare.

Curimatinae, with several genera and over eighty species, contains mud feeders like the prochilodins. They have teeth only during the larval stage. These fishes fill the ecological niche left open by the absence of carp; they make the same demands as carp on the environment, the river bottom and their food base, and they resemble carp to a high degree. Only the carp barbs are absent. They do not grow as large as prochilodins and in spite of their prevalence have no commercial importance. In contrast to the three previous subfamilies the anodins (Anodinae) with one or two species are open water fishes. Thus they have a stick-shaped body, are good swimmers and filter their plankton diet with numerous gill filtration structures.

Family: Curimatidae

The few members of the family Crenuchidae differ from all other characins in terms of different anatomical features and behavioral aspects, so that they should perhaps be considered as a separate group. The family just contains three species; *Crenuchus spilurus, Poecilocharax bovallii* and *Poecilocharax weitzmani,* the last of which was just recently discovered. Their shape resembles that of toothed carp (see Chapter 19), including the sexual difference of males having much larger fins than females. The group is characterized by a small organ on the head which is in depressions in the skull and as far as is known occurs in no other fish group. The structure is composed of a web of sensory cells. Whether this is actually a sensory organ is not known. All similar organs in the animal kingdom, such as the heat detecting organ in snakes, give some idea of their function (like eye lenses or thin skin which would permit light or heat to have an effect). But these structures are covered by a thick, fibrous plug which would screen off practically any stimulus, or filter it.

Family: Crenuchidae

Fig. 14–16. Distribution of Crenuchidae species.

African characin
family: Citharinidae

The African family Citharinidae contains about twenty genera which, as Daget has shown, are characterized by the distinctive structure of the skull and jaws. They are less primitive anatomically than their American relatives. With the exception of the genus *Citharinus* they have ctenoid scales, which is not the case in the South American counterparts. Subfamily Citharininae has just two genera with eight species. These are bream-like fishes which feed in the mud and attain a great size. Specimens of *Citharinus distichodoides* 84 cm long and 18 kg in weight have been reported. Due to their prevalence they are important eating fish in tropical Africa. At inland ports they are the most popular fish; they are eaten smoked and are greatly enjoyed in spite of their many bones.

The subfamily Distichodontinae contains ten genera with more than fifty species. The largest genus is *Distichodus* with its rather stately fishes; their teeth are adapted for chewing large leaves. All other genera have few species and a less specialized diet. Some feed only on small organisms. Among them, the genus *Nannocharax*, with almost twenty species, corresponds fully to the South American characins *Characidium*. It is difficult to distinguish the African from the South American species without a magnifying glass.

Fig. 14-17. An ichthyo-borin (Ichthyoborinae).

The third citharinid subfamily, Ichthyoborinae, contains small or medium-sized predatory fishes. They are divided into nine genera, each typically with one to three species. Some species such as *Phagoborus* feed on fishes, while others like *Hemistichodus* have an insect diet. A third group has a semi-parasitic type of life; this contains *Phago*, *Belonophago*, and the 30 cm long *Eugnathichthys*. Matthes has shown that these fishes use their scissor-like jaws, jagged teeth and doubled lower jaw joint to bite off the fins of other fishes with a single bite. Daget found that *Ichthyoborus* has an intermediate position; it occupies the ecological niche of fin predators in those areas (e.g., in the Nile and Chari) where they do not occur. If the niche is occupied, as in Katanga, *Ichthyoborus* has a general diet consisting of crustaceans and fish. Both races of *Ichthyoborus* are considered to be systematically distinct subspecies.

Suborder: electric
and gymnotid eels

Distinguishing
characteristics

The electric eels (suborder Gymnotoidei), a highly developed off-shoot of the characins, are American freshwater fish which are active at night. The body is always very long, either rounded and naked as in electric eels (here in the narrower sense) or compressed and with scales as in the other species, the gymnotid eels. The gastric cavity and internal organs are immediately behind the head and only the swim bladder extends far to the rear. The main portion of the body is formed by a long "tail" or rear portion, generally lacking a caudal fin or having just a trace of one. The anus is well forward, located under the pectoral fins or on the throat. The dorsal fin is either greatly decreased in size or absent altogether; in some species without a caudal fin a fibrous structure replaces the dorsal fin. Pelvic fins are never present, and the pectoral fins are of various sizes. The anal fin is greatly elongated (one species has more than 500 rays on that fin!); it extends two-thirds or three-fourths of the entire L and in one species proceeds around the finless tail to the upper side of the body. The eyes are degenerate, and generally have an additional eyelid as an adaptation to the nocturnal life. The gill openings are small; several lower jawbones are either degenerate or absent. The swim bladder varies in shape but is usually in a bony capsule. The Weberian apparatus and rear of the skull are developed as in characins. There are four families: electric eels (Electrophoridae), gymnotid eels, Apteronotidae, and Rhamphichthyidae, with a total of fifteen genera and about forty species. Distribution is from Guatemala to the La Plata River, with greatest prevalence in the Amazon basin and the Guianas.

The prominent elongated anal fin in the electric eels and gymnotid eels aids in continual undulating motion and permits these fish to move backwards as well as forwards without also using the body and pectoral fins. This indicates that the fish must flee rapidly at times. The very long body is extremely important in the life of the electric eels. The anal fin rays can initiate such a powerful wave motion because of their supporting pterogiophores which have a rounded head permitting circular motion. This is generally not true of other teleost fishes.

The most striking feature which unites the electric eels with the mormyrids is the ability to produce electrical current. This is only found elsewhere in the rays and two bony fishes, the catfish *Malapterurus electricus* (Chapter 16) and the rare stargazer from the genus *Astroscopus* from the coast of the Virginias (see Volume V). The large ELECTRIC EEL *(Electrophorus electricus)* has a scientific name which doubly refers to this characteristic. This is among the most feared South American fish, after the freshwater stingray (Dasyatidae; see Chapter 4), the piranhas and the parasitic catfishes (Chapter 16). Since details on its natural history were published by Richter in 1729 it has been a notorious species. It was not the first electric eel known, for one of the few naturalists preceding Alexander von Humboldt into South America, namely Georg Marcgraf (1611–1644) described the gymnotid *Gymnotus carapo* in the first book dealing with South American fishes. As early as 1760 van Musschenbroek hypothesized that the emanations from electric eels were the same as those from Leyden jars. Then Alexander von Humboldt, on his famous research voyage in 1805, described the danger present in these shocks.

However only with the development of physics has it been possible to investigate the details of the electric eels shocks. Faraday was one of the pioneers in this study. He estimated the strength of the electric eel to be equal to fifteen fully-loaded Leyden jars, which corresponds

Electric organs

to a surface of about 2250 square meters. Carl Sachs, and more recently the Brazilian Couceiro, the Englishman Lissmann and the Frenchman Fessard have devoted study to this fish.

In the electric eels the current-producing organs are on both sides of the vertebral column in pairs and run almost the entire length of the body (with the exception of the forebody). They comprise about five sixths the L and three-eighths of its weight. The rest of the rear of the body is taken up by the muscles and the swim bladder. The tissue of the electric organs is a white, quivering gelatinous substance which is divided by numerous fibrous partitions. They can easily be detected simply by feeling the eel body. The chief electric organ is represented by three parts on each side of the body; it extends to three-fourths the tail length. Underneath and somewhat behind it is the narrow and smaller Hunters Organ. A third structure, the Sachs Organ, is more dorsal in the middle of the fish and like the others extends to the end of the tail. Each of these organs consists of a number of closely set, long prismatic lamellae; there are fifty to seventy of them on each side of the main electric organ. Each prism is then subdivided diagonally into a large number of electrical plates, of which there are 5000–6000. The total number of plates is about 500,000. Each plate has a tiny nerve coming from the group of nerves in the belly region. The entire system has its central portion in the brain just as is the case with mormyrids.

These electroplates are situated one behind the other and are in close rows side by side. They are wired in parallel. The power is summated and gives the discharges, which are released by the central nervous system, voltage as well as current intensity. Coates and Cox measured discharges of 550 volts and a little under two amps in an adult electric eel (which can reach a L of over two meters). This is about the strength of a kilowatt. The charge consists of four to eight separate charges and yet lasts only two- to three-thousandths of a second. As Sachs reports, these discharges can be repeated up to 150 times per hour without visible fatigue on the part of the eel. It is very likely that the shocks pose a danger even for large mammals when they are in the eel's electrical field; the field extends about the length of the body. The shocks are still worse when the eel is touched at two separate places on its body. Even though the shock is a defensive measure, it has been observed that a frog more than one meter away from an eel was initially disabled and then attacked by means of shocking. Thus the electrical organ also plays a role in obtaining prey.

A third role of this organ was also discovered in most other electric eels and mormyrids in the last twenty years. These organs develop much weaker shocks and have a structure unlike that of the powerful electric eels. Unlike the powerful electric plates of the dangerous species, they are comprised of just a small number of plates. They were

Fig. 14-18. Distribution of the Gymnotoidei electric eels.

considered to be electrically functionless for a long time, until their weak emanations could be detected.

Gymnotus carapo, a gymnotid, has an organ which is comparable to one in the mormyrid *Gymnarchus.* The tail contains four pairs of adjacent gelatinous tubes or bands which are filled with a sticky protein connective tissue; the tubes are wedged tight by a system of stoppers. In *Eigenmannia,* which belong to the rhamphichthyids (see below) there are six pairs of closely set electrical tubes. The electric plates consist of very narrow pocket structures. The comparable organ of *Apteronotus,* which is a member of the apteronotid eels (see below) is a paired structure in the rear of the body, but more toward the belly than in other species. Its two parts are united in the middle. The electrical elements themselves are very narrow, elongated tubes lacking partitions. In contrast to the structures in all other electric fishes these tubes are not considered to be modified muscles but organs which have developed from nerve fibers.

The rhamphichthyid eel *Steatogenys elegans* has another unusual organ. This consists of a pair of white bands which are about one centimeter long and are situated in a depression in the chin region. Each is divided into approximately forty microscopic discs. Like mormyrids electric eels release impulses of varying frequencies: *Gymnotus carapo* emanates thirty to sixty cycles, but more when it is excited; in *Apteronotus* the frequency is 470, and in *Eigenmannia* it is 250–300. However the maximum strength is one volt. Recent studies have shown that besides the strong shocks electric eels are able to give off impulses with slow but changing frequencies, and this is probably done with the Sachs Organ. These are comparable to the emanations of *Gymnotus* eels, but change according to breathing rate. All members of this suborder probably have their own electrical field, in which changes can be detected by small organs on the lateral line similar to those in mormyrids. Since the eels, like mormyrids, are nocturnal creatures, they thus have a mechanism for defense, locating prey and probably also for communication with other eels. This explains why the eels always swim with an outstretched body using chiefly the pectoral fins and the waves generated by the anal fin. They must constantly remain oriented to their electrical field, which is built up of organs along the body axis.

Another characteristic of electric eels is their ability to replace large parts of the rear body, including the scales, fins, muscles and pigmentation, when injured. Ellis found in a series of studies that these kinds of injuries, which are usually caused by predatory fishes, are usually not fatal; all the important organs are at the front end of the body. The greatly elongated tail, which greatly attracts enemies, can thus be subject to rather severe injuries without dire consequences for the eel. This situation is similar to one in some lizards. Many electric eels have stunted growths on their tails as evidence of such injuries. Only the

Fig. 14–19. *Apteronotus.*

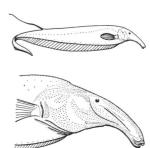

Fig. 14–20. *Sternarchorhynchus.*

▷
Serrasalmus nattereri, one of the feared South American piranhas.

▷▷
Above: the cyprinoid *Puntius schwanenfeldi* (see Chapter 15). Below: The cardinal tetra *Cheirodon axelrodi* may be distinguished from *Paracheidon innesi* by more intense red coloration; the former species is also a more sensitive one.

▷▷▷
A bitterling *(Rhodeus sericeus amarus)* male with "its" mussel. Only males in possession of mussels, into which the female lays her eggs, can have a mating companion. Therefore mussels are vigorously defended against intruders.

Fig. 14-21. The rhamphichthyid *Hypopomus*.

Family: electric eels

Family: gymnotid eels

Family: apteronotid eels

◁

Rasboras (*Rasbora heteromorpha*) set their eggs on the underside of leaves of aquatic plants. External fertilization takes place when the eggs are laid. The courting and spawning pair swim with their backs toward the floor.

electrophorid eels do not have such injuries, probably because of the great shocks they can generate.

Electric and gymnotid eels generally are found in lowlands and never in fast-flowing water. Like their fellow noctural mormyrids, they return during the day to their hiding places, which are shared by several eels. In their recesses they can orient themselves by means of their natural electrodes in the water, which are joined to an amplifier and loud speaker system. The noise which their current produces is readily detectable, and the frequency is specific for each species and thus permits identification of the various species on the basis of frequency. Nothing is known about reproductive behavior. In some species it is possible that they give birth to live young, although no definite evidence has been found for this.

The ELECTRIC EEL *Electrophorus electricus* is the single genus and species in the electrophorid (Electrophoridae) family. It is characterized by its size, the naked round body and the ability to fold the anal fin around the tail end so it looks like a caudal fin. Furthermore there are no skull bone seams. The species has a large throat with cylindrical teeth, and a powerful electric organ. This predatory eel lives on freshwater shrimp as a subadult, changing later to fishes.

All other electrophorid eels have a compressed body covered with small scales. The gymnotid genus *Gymnotus*, with just a few species, forms the second family, GYMNOTID EELS (Gymnotidae). They are also characterized by a flattened skull with no seams, a large mouth opening with a greatly degenerate upper jaw, the cylindrical teeth and the long cylindrical swim bladder. However there are also differences. The very prevalent gymnotid *Gymnotus carapo* and its subspecies have a body covered with scales, and the anal fin does not extend to the caudal fin. Their marbled coloration pattern provides them with effective camouflage. They have a maximum L of 50 cm and are predators as adults.

The third eel family, the apteronotids (Apteronotidae), contains nine genera, three with one species and five with just two or three species each. Only *Apteronotus* has six species. They all have a very small caudal fin on a thin tail shaft and generally a dorsal fold lacking rays, the function of which is unknown. Apteronotids and the *Porotergus* eels, which can reach a L of 50 cm, have a rather large mouth opening and can catch fish and shrimp. *Sternarchella* and *Oedemognatus* eels have a smaller mouth opening; they feed on small organisms as do the apteronotids *Sternarchogiton* and *Odontosternarchus* (which have relatively large mouths). A few species, such as *Sternarchorhampus*, which attains a L of 80 cm, *Sternarchorhynchus*, and *Odontosternarchus*, search the sandy bottom with their elongated snouts for tubificid worms (Tubificidae; see Volume I) and the larvae of non-biting midges (Chironomidae; see Volume II). They most strongly resemble the mormyrids. *Sternarchor-*

hynchus is a most unusual species. Its long, arched, trunklike snout is exactly like the same structure in the mormyrid *Campylomormyrus.*

The rhamphichthyids (Rhamphichthyidae) contains two lines of development: the subfamily Sternopyginae, consisting of predators with a short snout, and subfamily Rhamphichthyinae containing sand and mud burrowers with long snouts. The first subfamily contains the genera *Sternopygus, Eigenmannia* and *Rhabdalichops;* they have teeth in both jaws. *Sternopygus* furthermore has lidless eyes. *Eigenmannia* with five species is the largest genus; one of them, *Eigenmannia virescens,* attains a L of 40-50 cm and is prevalent throughout much of South America. The subfamily also contains *Hypopomus* and *Steatogenys;* they lack teeth and are also a prevalent South American species. They feed on cyclopids (Copepoda; see Volume I) and tubificid worms (Tubificidae).

The second subfamily, Rhamphichthyinae, has just two species, *Rhamphichthys rostratus* with several subspecies throughout South America and *Gymnorhamphichthys hypostomus* from the Guianas and the Amazon basin. Both have an elongated snout and search for sand-dwelling organisms which form their diet. While *Rhamphichthys* reaches a L of 1.80 m and has a good scale covering, *Gymnorhamphichthys* is considerably smaller and is naked up to the rear part of the body. The colorless species lives bured in the sand.

Family:
rhamphichthyid eels

15 Carp

Suborder: carp,
by P. Banarescu

Distinguishing
characteristics

Geographical
distribution

Family: carp

In contrast to the characins, etc., of the previous chapter, the carp suborder Cyprinoidei is characterized by the following characteristics: the jaw and other mouth bones lack teeth; a lower pharyngeal bone is sickle-shaped and has one to three rows of teeth; the mouth protrudes prominently; barbs are often present; the fins are well developed, but there is no adipose fin. Most species have scales, the head and gill cover being naked. All four gill cover bones are present. The swim bladder is divided by constrictions into two or three chambers but lacks an inner wall. The suborder has a vast diversity of species, often with very special environmental adaptations. There are six families: Minnows and carps, suckers, and four loach families.

While the *Ostariophysi* fishes (see Chapter 14), with their swim bladder-throat bones, are primarily distributed in southern latitudes, the carp are a more northern group. They are distributed throughout Eurasia including its islands, and east to the "Wallace Line," which runs between Borneo and Bali on one side and between Celebes and Lombok on the other, separating the Indo-Malayan faunal region from the Indo-Malayan-Australian transition region (see Vol. VII, Chapter 1). The distribution also includes Africa and the region from North America to Guatemala. Carp are not found in South America, western India, in much of Central America, in Malagasy or in Australia. The greatest diversity of species is in southeastern Asia (eastern India and southern China); this area is also considered to be the original habitat of the suborder. Four of the six families are restricted to this region; two others are primarily found here, while just a single family is primarily North American. Starting from southeastern Asia, carp invaded southern and western Asia and on to Africa in one direction, while the other direction of invasion took them to eastern Asia and Siberia, and from Siberia to North America.

The CARP group (Cyprinidae) is the basic group of the suborder and has more species than any other fish family. There is a low number of throat teeth arranged in one to three rows (with a maximum of seven in the major row). A masticating plate is present, consisting of a horny formation on the lower side of the occipital bone (on the rear of the skull); this is used together with the throat teeth to chew the food. The mouth is surrounded by the intermediate jawbones. Barbs, when present, consist at most of two pairs (rostral and maxillary); however the goby genus *Gobiobotia* has three pairs of chin barbs on the lower side of the throat. The swim bladder is usually large and is free; only a few east Asian species have degenerate swim bladders, which are enclosed in a capsule. The length and position of the dorsal and anal fins as well as the number of scales vary considerably. The body size and shape also lack uniformity within the suborder. The smallest species is probably *Rasbora maculata* from Malaysia and Sumatra; L is 16-20 mm; while the largest carp is *Catlocarpio siamensis* (L up to 3 m). In North America cyprinids are found only down to southern Mexico.

The division of the cyprinid family is still uncertain; at the present, ten subfamilies have been defined: danios, minnows and others, cultrins, xenocyprinins, bitterlings, gobies, barbins, schizothoracins, carp (in the narrower sense), and hypophthalmichthyins.

The DANIOS (Danioninae; also known as Bariliinae or Rasborinae) present the most difficult problem of defining and limiting. The genera grouped here are apparently related, but their various adaptations are so diverse that it is difficult to find characteristics common to all danios which distinguish them from other subfamilies.

Danio size varies tremendously, from 16 mm (in the dwarf danio) to 2 m (the sheltostshek). Most species are fairly elongated, while some are higher set but not as high as breams or carp. The body is always rather compact, the lower side being rounded; a sharp belly keel is found in just a few species. The mouth is at the end of the body or on the upper side. It is sometimes wide; the lips are thin and never fringed. There are one or two pairs of barbs in several genera. The throat teeth are almost always pointed and are generally arranged in three rows, less commonly in two or one row. The dorsal and anal fins are typically short. There is a noticeable tendency for the dorsal fin to be located toward the rear. The lateral line is either straight or greatly arched downward; some small species have incomplete lateral lines or none at all. The scale size varies greatly; one danio *(Swamba)* has no scales. Coloration also varies, most species being quite colorful. The basic color is a silvery hue on which darker colorful longitudinal or diagonal stripes are set. Instead of stripes, the markings may consist of spots. The fins are sometimes spotted.

There are also great behavioral differences among the danios. Most of them live in the upper water levels; some are found underneath plants, but there are no true bottom dwellers. Some genera are in the open water of large lakes (semipelagic life). Small African danios *(En-*

Distinguishing characteristics

Subfamily: danios

Distinguishing characteristics

Fig. 15-1. Distribution of the danios (Danioninae).

▷

Characins: 1. The hatchet fish *Gasteropelecus sternicla;* 2. *Metynnis schreitmuelleri;* 3. *Metynnis roosevelti;* 4. *Myloplus schultzei;* 5. *Hoplias malabaricus;* 6. *Boulengerella cuvieri.*

Life of the danio

graulicypris) inhabit the interior lakes of mainland Tanzania, Malagasy, etc., and are the chief prey of predatory fishes in those regions. Most danios feed on small organisms: some species eat zooplankton while others eat small invertebrates. However there are also true predators among danios, for example the sheltostshek and large danios in southern Asia and Africa. There is a greater tendency within this subfamily to develop predatory species than in any other within the carp group. The chief danio genus is *Barilius,* which contains both small and large species. All are more or less elongated; the lateral line is slightly arched, and the mouth is terminal. Some species have small barbs. Most species have distinct diagonal stripes which may be long or short. The mouth is large and has a wide opening. Distribution is from tropical and subtropical Asia to southern China; one species occurs in western Asia, and a few are in Africa. Most Asian species are small, while many of the African species are large predators. One species has been kept in an aquarium; it is the 10 cm-long *Barilius christyi,* from the Congo basin. Like many other members of the genus, *B. christyi* lives in schools. The species requires a large, amply planted aquarium with slightly acidic water and a temperature range of 22–24° C. The fish is fed living food; flying insects are taken from the surface of the water. The species has not yet spawned in captivity.

Danio genera and species

The danio genus *Danio,* the members of which are distributed from India to southern China, contains species with a rather high-set body, short barbs, a mouth on the upper side of the body, a complete lateral line and very long longitudinal stripes. The best-known species of this genus is the GIANT DANIO (*Danio malabaricus;* L up to 12 cm) from Ceylon and the Malabar Coast. Sexual maturity is attained at a L of 6–7 cm. The back is steel-blue with three or four lateral stripes running along the longitudinal axis. In a bright light the stripes take on an indigo-blue color. There is a golden spot on the gill cover. *Danio regina* from Thailand is similar to the previous species but has a still higher body and a black spot on the upper end of the gill slit. *Danio devario,* from northern India, is somewhat smaller, but higher-set, looking like a bitterling. Basic coloration is pale greenish-silver with irregular blue longitudinal stripes merging at the base of the tail.

The genus *Brachydanio* contains smaller, more slender danios with a short dorsal and anal fin and an incomplete lateral line. They have the same life habits as *Danio* fishes. Many species are well-known ornamental fishes, with the same breeding requirements as *Danio.* The ZEBRA FISH (*Brachydanio rerio*) is highlighted by its four blue longitudinal stripes. The species reaches a L of 4.5 cm and is distributed in central India. This is a popular, easy-to-breed aquarium fish. Its relatives, *Brachydanio kerri* and the SPOTTED DANIO (*Brachydanio nigrofasciatus*), are similar in anatomy and living habits. Both are from eastern India and are rarely kept in captivity. The recently introduced *Brachydanio frankei*

◁

Characins: 1. *Alestes longipinnis;* 2. *Neolebias ansorgei;* 3. *Neolebias landgrafi;* 4. *Alestopetersius caudalis;* 5. *Nannaethiops unitaeniatus;* 6. *Nannaethiops tritaeniatus;* 7. Feathertail tetra *(Phenacogrammus interruptus);* 8. *Distichodus sexfasciatus.*

has longitudinal stripes and round lateral spots arranged in rows. The zebra fish and *B. frankei* have been crossed successfully. The PEARL DANIO *(Branchydanio albolineatus)*, which is up to 5.5 cm long, is bright gray-green. With artificial upper lighting the body takes on a brilliant blue to shimmering violet color, while sunshine produces a grass-green color. The species is distributed in eastern India and on Sumatra.

The danios of genus *Esomus* are also small slender fishes with the mouth on the upper body side, very long upper-lip barbs, a short, rearward dorsal fin and one row of throat teeth. Distribution is restricted to India, the Malay Peninsula and Ceylon. These lively fishes behave much like the danios which have already been described. Four species are kept in aquariums. They require a long, large aquarium with a few plants and sufficient room for swimming freely. The most familiar species are the FLYING BARB *(Esomus danrica)* and *Esomus malayensis.*

The genus *Rasbora* contains the most species of any danio genus. Distribution is from India to southern China as well as the Indo-Australian islands from Lombok to Sumbava and the western Philippines. The fishes are small to very small and are typically elongated, although some are high-set. The dorsal fin is short, and is situated over the middle of the body. The lateral line is usually complete, the front part dipping sharply downward. The mouth is terminal and is pointed upward. The lower jaw has a button-shaped central protrusion; barbs are absent. There are three rows of throat teeth. Most species are silvery and often have very distinct dark longitudinal stripes extending from the eye to the caudal fin.

The rasbora danios are generally schooling fishes which inhabit the upper water levels in flowing or standing waters; they even spawn in groups to some extent. Many species have been brought to aquariums but few are popular. They need a free-swimming area edged by plants with soft, filtered water and a temperature of 24-25° C. They take both live and dead food. Many species, such as the SCISSORS-TAIL RASBORA *(Rasbora trilineata)*, *Rasbora lateristriata elegans,* and *Rasbora daniconius,* breed readily. They spawn between *Myriophyllum* bushes, over *Nitella* or onto artificial fibers.

The few species which evoke special attention vary greatly from the typical anatomy found in the genus. The most prevalent species, the RASBORA *(Rasbora heteromorpha)*, is one of these. It has a rather high body and an incomplete lateral line. The sides are silver-green with a dark red to violet matte sheen. A blue-black spot is on the rear of the body and the unpaired fins are carmine. The species spawns on the lower sides of aquatic plant leaves at a temperature of 24-28° C. (aquarium specimens have spawned at 20° C.); the young hatch after about twenty-four hours.

The Ceylonese species *Rasbora vaterifloris* is also a high-set species;

Fig. 15-2. Distribution of the rasbora *(Rasbora heteromorpha).*

the sides are grayish-green, the belly whitish or often orange; the unpaired fins are orange-yellow to orange-red. This fish also will spawn in captivity. *Rasbora maculata* is only 2.5 cm long and is the smallest species in the family. It is also one of the most beautiful, with its brick-red back, yellowish-red sides and yellow undersides. The fins are reddish-yellow and the dorsal fin has a black front edge and tip. Distribution is in southern Malaysia, Sumatra and Singapore.

Two species imported since 1938 are close relatives to the rasboras: *Aphyocypris pooni* and the WHITE CLOUD MOUNTAIN FISH *(Tanichthys albonubes)*. Both are small (L up to 4 cm), elongated, brilliantly-colored fishes with short dorsal- and anal fins and a diagonal mouth opening. The upper side of each is brown-olive or yellow-brown, while the underside is whitish. A band, which depending on the light hitting it is greenish or golden, extends from the head to the caudal fin; in juveniles the band is a shimmering blue-green. The brownish tones are darker in the white cloud mountain fish. Both species differ in the coloration of the fins.

These beautiful fishes are cared for in the same way as the danios; the favorable temperature range is 20-22° C. in summer and 16-18° C. during the winter. Both spawn in dense *Myrophyllum* growths and are so similar that not only their generic but even species distinction is questionable. Hybrids have frequently been produced in aquariums. In Hong Kong, for example, where they are bred in open tanks, one mainly finds hybrids.

Some danios are true predators. The largest species, distributed from the Amur basin to Vietnam is the SHELTOSTSHEK *(Elopichthys bambusa)*, which has a L of 2 m. This spindle-shaped fish has short unpaired fins; the dorsal fin is in the middle of the body. The mouth is wide; the lateral line is complete, and the fish has very small scales (up to 110 in the lateral line). It spawns in the Amur from late June to early July; spawning in China is earlier. The eggs float freely.

Subfamily: minnows

Distinguishing characteristics

The next carp subfamily, MINNOWS (Leuciscinae), is also difficult to distinguish and limit, even though its genera (or at least most of them) are truly closely related: natural, infertile hybrids have occurred. The genera can also be put into evolutionary order on the basis of anatomy. Some species reach sexual maturity at a L of 4–5 cm, while others do not become mature until they are one meter long. There are many elongated species as well as high-set ones (such as breams). The body and tail shaft are generally compressed along the sides; some North American goby-like species are round in cross-section. The mouth is terminal, partially toward the upper body side, or on the lower side; it is more or less horseshoe-shaped. Some genera have very short upper lip barbs. The lips are complete or interrupted and may be fleshy or thin. In some genera the lower lip is covered by a horny plate, but is never fringed or has warts as in many gobies or danios. The

Fig. 15-3. Distribution of minnows (Leuciscinae).

throat teeth are in one or two rows. The number of unpaired fin rays differs. The scales are medium-sized or small to very small; there are a few scaleless species and genera in arid regions. The lateral line, when it is complete, is always straight.

Biologically the various members of the subfamily are also quite diverse. Many species, usually with elongated bodies and flattened heads (e.g., chubs and minnows), live in flowing waters but are not so closely dependent on the floor as are other species (such as chondrostomes and several North American genera). This group, too, is characterized by an elongated body, but the mouth is on the lower side and in some species has barbs. The good swimmer *Alburnus alburnus*, with compressed sides, and with the mouth aimed upward, lives near the surface of slowly flowing or standing water, while the high-set species (breams and tench) are on the floor in standing water. Most species feed on small organisms which are found on the floor or between plants. *Alburnus alburnus* and a few related species are plankton-feeders and to a certain degree also insect predators. The ASP (*Aspius aspius*), as well as chubs as adults and the SQUAWFISH (*Ptychocheilus*) in western North America, are predators. However other species such as the rudd and especially the Chinese species feed almost exclusively on higher plants. All of these species spawn on plants over hard ground, with the exception of two Chinese species which simply set the eggs free in the water.

The EUROPEAN ROACH (*Rutilus rutilus*) can be considered to be the representative member of the subfamily. L in central Europe is at most 30 cm, while some eastern European subspecies are as much as 50 cm long. The body is compact and medium high-set with a complete lateral line having 39–48 scales. There is one row of throat teeth. The pectoral fins are under or a little in front of the dorsal fin. The back is olive-brown to blue-green; the sides are matte gray-silver to yellowish-silver; the unpaired fins are yellow to bright red, and the eyes are reddish.

Roaches live in standing or slowly flowing waters in the Baltic Sea and in salty or brackish waters of the Black Sea-Caspian Sea basin. They are found among plants and spawn from April to May on year-old plants or plant remains. Schools inhabiting desalted parts of the sea migrate for spawning into fresh-water. One female lays up to 150,000 eggs. Roaches feed on algae and plant remains as well as small invertebrates including mussels. They have different commercial importance in different regions and are of value in the Baltic Sea basin, in the Sea of Asov (where it is the subspecies TARAN [*Rutilus rutilus heckeli*]) and in the Volga (the VOBLA [*Rutilus rutilus caspicus*]). In these regions roaches are fished with nets and ring baskets. In central Europe they are often caught with hook-and-line using worms and flies as bait.

Other roach species inhabit southern and eastern Europe. Two

Life of minnows

European roach

Fig. 15–4. Distribution of European roaches (*Rutilus rutilus*) and its subspecies.

Fig. 15-5. Distribution of the rudd *(Scardinius erythrophthalmus).*

Rudds

Ctenopharyngodon

Fig. 15-6. Distribution of the minnow genus *Cteno-pharyngodon.*

Mylopharyngodon

Leuciscus minnows

species of the Black Sea-Caspian region are also found in the lakes of the Danube basin in Bavaria and Austria. They are: *Rutilus pigus virgo,* which differs from the European roach only in the somewhat smaller scales, the mouth which is partially on the lower body side, and the brilliant courtship coloration of the male; and *Rutilus frisii meidingeri,* an elongated, chub-like fish with still smaller scales (62–67 in the lateral line), which is caught with nets in great quantities during the spawning season.

The RUDD *(Scardinius erythrophthalmus)* is closely related to chubs. Its dorsal fin is more rearward than the pectoral fin and the throat teeth are in two rows and are jagged. The fins (excepting the dorsal fin) are bright red, while the eyes are yellow to yellow-red.

This fish inhabits almost exclusively standing waters rich with flora; it is also prevalent in small ponds and marshes. The eggs are laid on aquatic plants from April to June. Rudds feed chiefly on aquatic plants, and occasionally on invertebrates. It has little commercial value, but is sometimes caught with hook and line.

The minnow *Ctenopharyngodon idella* is closely related to rudds but looks completely different. L is up to 1 m, with a weight of 32 kg. The body is elongated with a flattened head and bright gray fins. The throat teeth are in two rows and are jagged as in rudd. Sexual maturity is attained in six to seven years.

The species spawns in summer in river bottoms. The free-floating eggs develop while they are carried downstream by the current. The young feed initially on invertebrates and as adults change to higher plants. In China and the Amur basin the species has great commercial importance. It is caught with various kinds of nets in rivers and lakes; the young are also bred in tanks until they have reached the size suitable for eating. Recently the species has been introduced to Japan, western Russia, Israel, Romania and other eastern and western nations, where it is set in ponds together with carp. Putting this fish into natural and artificial waters is very important, because that is not only an inexpensive means of preventing overdevelopment of certain aquatic plants, but these plants are also made industrially useful in the process.

Its near relative *Mylopharyngodon piceus* has a single row of molar teeth. Coloration is much darker than *Ctenopharyngodon,* being almost blackish. The species inhabits eastern Asia and feeds on snails and mussels. It is an important commercial species in China, and has recently been introduced into eastern European countries.

The chief genus of the subfamily, the MINNOWS *(Leuciscus),* is distributed through various subgenera in Eurasia and North America. They are medium-sized to large, elongated, with a slightly compressed body. The head is flattened and the dorsal and anal fins are rather short. The lateral line is complete. The two rows of throat teeth have hook-shaped points. Most species inhabit flowing waters.

Subgenus *Squalius* contains the CHUB *(Leuciscus cephalus)*. L is 60 cm. The head is large and the mouth is wide and at the end of the body. The scales are rather large with 38–47 in the lateral line. The central European subspecies has 44–47 scales in its lateral line. The dorsal and anal fins are rounded. The back is gray-brown with a greenish sheen and the sides are whitish-gray. Some older fishes have a golden shimmer. The dorsal and caudal fins are pale gray and the other fins are faintly reddish. Distribution is throughout Europe and Turkey.

Chubs inhabit flowing waters from the trout zone to the mouths of rivers; they have also been found in brackish water and occasionally in lakes and ponds. They are good swimmers. While young often form schools, the adults are solitary. The spawning period begins in April, lasting until June, or sometimes into July. Only the males develop distinct courtship coloration. Eggs are laid on stony, gravelly or loam ground, less often on plants. Chubs have a wide diet; the young feed on insect larvae, flying insects, crustaceans and similar organisms, but also eat plants, while adults are pure predators. The species is an important sport fish and is caught with rod and reel, taking almost any natural bait. Young chubs can easily be kept in an aquarium; they are quite active.

The DACE *(Leuciscus leuciscus)* is in the subgenus *Leuciscus* and resembles chubs except that it is smaller (L up to 30 cm). The mouth is partly on the lower side of the body; the anal fin is slightly indented, and the scales are smaller than in chubs (45–55 in the lateral line). Its habits are like those of chubs except that spawning is earlier in the year (March to April) and it is not a predatory species. Distribution is wider than for chubs, including all of Siberia; the species is not found in most of southern Europe and in Turkey. It is not a commercially valuable fish but is caught with hook and line and is used as bait.

The subgenus *Telestes* contains *Leuciscus souffia agassizi*. L is 24 cm. The fish looks like dace but has still smaller scales (50–61 in the lateral line) and a different coloration. The upper side has a dark gray sheen; the sides are gleaming silver with a broad dark violet stripe extending from the eye to the caudal fin (most clearly developed in spawning males). The fin bases and sides of the head are a bright orange-yellow.

The species inhabits the clear rivers and streams of the mountains, preferring gravel bottom. It is also found in a few cool lakes, where it is in schools at great depths. It feeds on small invertebrates and algae and spawns in spring. This is not a popular sport fish. Distribution is not great, including the Rhone basin, and the upper basin of the Rhine, Danube and Theiss rivers; other subspecies are in Spain, northern Italy and western Yugoslavia.

The subgenus *Idus* contains the IDE or ORFE *(Leuciscus idus)*. The ide has a higher-set body than its relatives. The mouth is terminal and the scales are small (55–61 in the lateral line). The anal fin is slightly indented. The sides are silvery, and the anal and paired fins are yellowish to reddish. The species inhabits lakes, streams and slowly flowing rivers in the plains. Small invertebrates form the main part of the diet. The fish spawns in spring or in the beginning of summer. Distribution includes all of Siberia, and eastern and central Europe to the Rhine. The fish is caught with dragnets. It can also be caught with rod and reel. Ides have been introduced to a number of dammed lakes in recent decades. A red-gold variety, *Leuciscus idus "forma orfus"*, is an ornamental fish in gardens and parks because of its striking beauty, and is also kept in aquariums.

A number of Japanese and northeast Asian species are grouped in the subgenus *Tribolodon.* They are close relatives of *Leuciscus souffia agassizi* but are larger. They are also the only carp which regularly spend part of their life in salt water (from the Amur region to Taiwan) but spawn in fresh water. This subgenus is rather important commercially.

Minnows

The MINNOW *(Phoxinus phoxinus)* is a small fish (L 9 cm, rarely 14 cm) with an elongated body, a small, terminal mouth, short, rounded fins, two rows of throat teeth, very small scales, and an incomplete interrupted lateral line.

Zoologist Karl von Frisch and his students used minnows for their classic investigations on the auditory capacity of fishes. In a short summary of his work in this area von Frisch wrote: "Many fish species have excellent hearing ability; this includes the carp family, to which most fresh-water fishes of the world belong. Within this family, the modest minnow is a very appropriate experimental subject. It is docile and can be kept easily in an aquarium. When it is fed a few times at the sound of a whistle it learns to come on command by the whistle. A well-trained fish will obey the faintest tone, and one is convinced that the minnow is no less sensitive to sound than the normal human is. It can also distinguish between several high tones. When fed regularly at the sound of a high tone, but punished by being hit slightly with a stick at a response to a low tone, it learns to come to the feeding tone and to flee when the warning tone is sounded. More precise studies have shown that minnows can distinguish half tones and even quarter tones. Minnows perceive sounds with the lower half of the auditory labyrinth, which corresponds to the cochlea in higher vertebrates, but the labyrinth lacks the tuned fibers. Thus, tone differentiation must be based in the sensory cells themselves and is simply refined in the cochlea of higher animals.

"There are fish genera with good hearing and with poor hearing. Thus it is of decisive significance whether there are assisting apparatus serving hearing and how they function. No fish has an eardrum. However in carp the swim bladder, which is continuously suspended in the water, has taken on the function of a sound receiver. The taut front

Fig. 15-7. Distribution of the minnow *(Phoxinus phoxinus)*.

chamber vibrates when acted upon by sound waves and transmits these vibrations through a chain of small bones which function just like the hammer, anvil and stapes of the inner ear in humans. Sound is transmitted through a small hole in the rear of the skull to the inner ear. Other supplementary structures are involved in the hearing process of other fishes. However, in many cases there is no means of amplification of sound (e.g., in trout, perch, pike and in most marine fishes). Those are the fishes which are hard of hearing and it takes loud tones to get them to react."

During the spawning season the sides of minnows take on an emerald-green hue; the corner of the mouth becomes carmine, the throat black, and the underside orange-red. This coloration disappears at once if the fish is frightened. Both sexes develop a distinct courtship coloration. Minnows inhabit mountain waters of streams, rivers and shore regions of the lakes, as well as ponds and pools. In northern Europe they descend as far as sea level. During the summer and early fall minnows are also found in the very warm river arms which have been transformed into pools. Clear, cool, flowing water, however, is required for spawning. Minnows almost always live in schools. The diet consists of plant matter, worms, crustaceans and insect larvae. The spawning season is long, extending from May to the end of July. The eggs are laid on gravel, stones or sand. In some areas minnows are eaten, for they have a good flavor; they are also very popular as bait for sport fishermen; and finally they are important prey of river trout. In spite of their preference for oxygen-rich water, minnows can be kept in aquariums, where they are lively, playful fish.

The BLEAK *(Alburnus alburnus)* is characterized by an elongated body which is quite compressed along the sides, a belly keel lacking scales, a mouth oriented upward, a long anal fin (with 16-20 branching rays) and a small dorsal fin somewhat toward the rear of the body. L is 10-13 cm or at most 20 cm.

This species has more schooling character than many other central European fishes. Schools varying widely in size are found just below the surface of all flowing and standing waters in Europe. In Italy and the southern Balkans there are several bleak subspecies. The species is also particularly prevalent in dammed lakes. The diet is primarily flying insects and plankton. During the spawning season (April to June) very large schools are formed. There is little development of courtship coloration. The eggs are laid on underwater plants or on hard ground; the female releases eggs in groups of three to six. The species has little commercial value, although its flesh is tasty and somewhat fatty; it is caught with nets and with hooks (flies are the best bait). The scales yield a substance which is used in making artificial pearls. Due to its prevalence in dammed lakes, the bleak will probably become more important to man than it has been. The attractive species has a

Fig. 15-8. Distribution of *Alburnus alburnus* and its subspecies.

▷

Eels: 1. Electric eel *(Electrophorus electricus)*; 2. The gymnotid *(Apteronotus albifrons)*; 3. The rhamphichthyid *(Eigenmannia virescens)*.

silver sheen and is quite lively. It lives well in large aquariums, and the young do well in small home aquariums. One should avoid sudden temperature changes and excessive heating. Bleaks take living and dried food.

Chalcalburnus

The genus *Chalcalburnus*, with the single species *Chalcalburnus chalcoides mento*, differs from the bleak in its thicker body, smaller scales (58–65 in the lateral line), shorter anal fin (with 14–17 branched rays) and keel, which is partly covered by scales and is situated toward the underside. L is up to 30 cm. The species inhabits streams, larger rivers, and especially lakes, including the brackish coastal lakes drained by the Black Sea. In central Europe the fish is found only in lakes off the upper Danube in Austria and Bavaria. Other subspecies are distributed in the basin of the Caspian Sea and the Aral Sea. The fish is rather valuable in southeastern Europe.

Leucaspius delineatus

Another minnow species, *Leucaspius delineatus*, resembles a young bleak; but it has an incomplete lateral line, 10–13 branching anal fin rays and no keel on the underside. L reaches 9 cm. The back has a blue sheen (in the bleak this is gray). Distribution is in eastern and central Europe from the Ural River to the Rhine. The species inhabits standing water and spawns from April to June; the eggs are laid in ring-shaped rows on plant stems and are guarded by the male. Sexual dimorphism is very distinct during the spawning period: males develop a prominent courtship coloration; the genital opening of the female is surrounded by two large rounded papillae. In Europe this is the most suitable aquarium fish because it is lively, social and resistant to higher temperatures; it takes all sorts of food and after a cold winter reproduces rapidly in large aquariums.

Alburnoides bipunctatus

The bleak *Alburnoides bipunctatus* resembles *Alburnus alburnus* but is smaller (L up to 14 cm), has a higher-set body, 12–17 intertwined rays in the anal fin and 42–53 scales in the lateral line. The sides do not gleam as much as in *Alburnus alburnus.* Wider and darker stripes extend from the gill cover to the base of the caudal fin, and the lateral line scales each have a pair of small black spots. Very narrow but distinct stripes are on the back from the end of the head to the base of the caudal fin. Paired fins are yellowish at their bases, turning orange during the spawning period. This bleak lives exclusively in flowing water, particularly in mountainous regions. It spawns from May to June on stones and feeds on aquatic invertebrates and flying insects. Distribution includes almost all of Europe with the exception of the southwest and part of Eurasia. It does not have commercial significance. The species thrives in larger aquariums but will not spawn there. It requires cool, oxygen-rich water.

◁

Cultrinae carp: 1. *Chela caeruleostigmata;* Danioninae carp: 2. *Esomus danrica;* 3. *Esomus malayensis;* 4. *Tanichthys albonubes;* 5. *Brachydanio albolineatus;* 6. *Brachydanio kerri;* 7. *Brachydanio nigrofasciatus;* 8. *Brachydanio rerio;* 9. *Danio malabaricus.*

The BREAMS *(Abramis brama)* share with the previous species a scale-covered keel between the pectoral and anal fins and a long anal fin. L is 50–70 cm, with a weight of 4–6 kg. The body is very high-set.

Tench also live exclusively in standing waters with muddy or boggy ground; thus the species is found in lakes, ponds and pools, less often in slowly flowing rivers between plants or near the bottom. During winter they burrow into the mud beneath deep water. Diet consists of invertebrates, flora and refuse. They spawn in groups on aquatic plants from May to July. The very small eggs are laid in groups of three or four. Tench is an important edible fish, with tender, tasty meat. It is fished in natural waters with dragnets and weir baskets and is bred together with carp. It is also a popular sport fish, caught with hook and line using worms, flies and other small organisms as bait.

Fig. 15-11. Distribution of tench *(Tinca tinca)*.

North America has more minnows than Europe: about 35 genera with some 220 species compared with the 130 species of Eurasia. Most of the American species are small, elongated fishes with a terminal mouth. The dace genus *Chrosomus* can be distinguished from the European-Siberian minnow genus *Phoxinus* solely on the basis of the single row of throat teeth in the former.

North American minnows

The SOUTH RED-BELLIED DACE *(Chrosomus erythrogaster)* has brilliant coloration: its back is dark brown to brownish-olive, sometimes with cloudy dark spots; two darker bands extend from the eyes to the caudal fin, enclosing a gleaming golden area. The belly is silver-white and the fins are yellow to brownish. During the spawning period the golden band becomes reddish, while the dark bands take on a pure black hue. This dace is distributed in the upper and central Mississippi basin. It feeds on all sorts of food and can be kept in a large aquarium. After a cool winter it spawns in the aquarium between plants and gets along well with other cold-water fishes. Sometimes another red-bellied dace species, *Chrosomus neogaeus*, is raised in aquariums.

The North American genus *Notropis* has the most species, including the SHINERS. These are small, fairly elongate, large-scaled fishes with a barbless, terminal mouth. Coloration varies; most species have shiny silver sides, dark backs and colorless fins. Some have longitudinal stripes or spots in various patterns. Some shiners resemble the European dace, chub, or roach. The most beautiful species is *Notropis welaka* (L 6-7 cm) from the southeastern Mississippi basin. This fish has a broad blackish longitudinal stripe; the dorsal and anal fins of the male become greatly elongated and rounded during the spawning period. The dorsal fin is completely black and the snout and head are sky-blue during this period. Despite its beauty, this fish is not kept in aquariums.

Two other shiners, however, are ornamental fishes in Europe: *Notropis hypselopterus* and *Notropis lutrensis*. The former has a rather high-set body which is basically brownish-green with a black longitudinal stripe below a parallel red band. The fins are partially red at their bases. The species is distributed in Florida, Georgia and Alabama. It is an unpretentious, active and harmless schooling fish which can be kept in the aquarium successfully if given enough room to swim. It takes both natural and dried food. Spawning in the aquarium takes place between plants at a temperature of 21-23° C.; during the winter 12-16° C. will suffice.

Notropis lutrensis resembles the European bitterling quite closely; L is 8 cm. The whitish body has reddish, blue and green tones; occasionally males have a vivid red head; the fins are yellowish to red. Males develop courtship coloration in May. The species is kept under the same conditions in aquariums as the previous shiner. This fish, distributed in the southeastern United States and northern Mexico, requires fresh, clear, well-ventilated water.

A few North American species are adapted to bottom-living in flowing waters. These are elongated, barely compressed fishes with a mouth partially or completely on the lower body side. The dorsal and anal fins are short; the lateral line is complete; the scales are small to medium-sized. Many of these species have short upper lip barbs. The lips, thin or fleshy, never have warts or fringes. Most species also have a row of lateral spots or a longitudinal stripe. In shape, coloration and life habits these American species, which may variously be called chubs, daces or minnows, particularly resemble the European gobies. They differ among themselves in the shape of the body and tail shaft, in the size of the eyes, coloration, length of the intestinal tract and other characteristics. All these are adaptations to their respective habitat conditions, such as different water velocities, feeding patterns, etc.

The CHUB genus, *Hybopsis*, contains the most species and is very much like the European gobies. DACES *(Rhinichthys)* have smaller scales and dark coloration and most closely resemble the southeastern European barb *Barbus meriodionalis petenyi*.

The BLACKNOSE DACE *(Rhinichthys atronasus)*, L 12 cm, is characterized by its handsome coloration: the upper side is olive-green to blackish and is often spotted. A wide black band runs on the sides from the snout to the base of the tail; above it is a golden stripe, beneath it a yellow stripe. The fins are yellowish. During the spawning period the longitudinal band becomes black-red. Blacknose daces can be kept in cold-water aquariums and will breed there. After winter conditions of 4-8° C. they spawn on stones almost daily for several weeks.

Some particularly primitive species live west of the Rocky Mountains. Most are small elongate fishes with small to very small scales; they look like European minnows in shape and coloration. The genus *Meda* from the Colorado basin has lost all its scales and the last ray of its dorsal fin has been modified into a spine.

Western North America is dry country; many rivers are very low during the summer, while others disappear completely. These conditions are very unfavorable for fishes, yet many species have adapted to living in unusual habitats. For example the genera *Meda, Moapa* and

Eremichthys, each with a single species, are very small and are limited to the springs of certain rivers. In the struggle for existence these few primitive species are always at a disadvantage to other fishes. The introduction of several new species from eastern North America or Europe will not only result in a sharp decrease in the numbers of these carp but can also mean their disappearance.

The cultrin subfamily (Cultrinae) are either high-set or elongate species. The bodies of all species are greatly compressed. The belly keel, always present, has no scales. It extends from the anal fin to the pectoral fins or to the throat. The barbless mouth is terminal or on the upper body side and is wide in some species. The pointed throat teeth are in three, rarely two, rows. They generally have hooked points. Many genera have a toothlike process on the lower jaw. The gill structures are longer and more numerous than in most other carp species; there are 9–106 gill rakers on the first gill arch. The short dorsal fin almost always has seven entwined rays; in most east Asian species the last unbranched ray is transformed into a spine. The dorsal fin is located behind the middle of the body and may be far to the back in many genera, in which it is behind the base of the medium to long anal fin. Except in a few dwarf forms, the lateral line is complete; in many species the front of the lateral line arches downward sharply. The intestinal tract is short and the belly is silver, brownish or blackish; the well-developed swim bladder is divided into two or three chambers. Body size varies considerably, from 28 mm in the smallest species, *Chela dadyburjori,* to 100 cm in *Erythroculter illishaeformis.* There are 21 genera with about 77 species.

These are white, silvery fishes. Only a few southeast Asian species have dark spots on the body or the fins. Most cultrins are good swimmers, having adapted to life near the surface of the water. Some species even live partially in open water. The deep-set Chinese breams *Parabramis* and *Megalobrama* are found at great depths, however, just like the European breams, but without being true bottom-dwelling species. Some species feed on plankton, others on flying insects, and still others on fauna or floor invertebrates; on the other hand, a few large species are pure predators. Eggs of all east Asian cultrins drift freely in the water and develop while they are carried downstream. Some south Asian species also spawn in between plants. The subfamily gets its name from the genus *Culter,* the best-known species of which is *Culter alburnus.* Distribution of this fish is throughout eastern Asia. L is 35 cm; the mouth is on the upper body side; there are 64–72 scales in the lateral line; the belly keel extends to the throat. *Hemiculter leucisculus,* also distributed throughout eastern Asia, has a L of 22 cm, a short anal fin and a lateral line arched downward. Its life habits and requirements are about the same as the European bleak. It feeds on plankton and insects.

The much larger members of *Erythroculter* are predators with wide

Subfamily:
cultrin carp

Distinguishing
characteristics

Fig. 15–12. Distribution of cultrin carp (Cultrinae).

mouths. The lateral line is straight. Unlike the two previous species, the belly keel is present only between the pectoral and anal fin. *Erythroculter illishaeformis*, distributed throughout the east Asian mainland, has its mouth on the upper side of the body. It is commercially important and in the Amur basin, where it is called verkhoglyad, is caught with weir baskets at the rate of 200–300 metric tons per year. The somewhat smaller *Erythroculter mongolicus* differs from the previous species by the terminal mouth. The south Chinese and north Vietnamese *Rasborinus lineatus* is a small spineless fish with a high-set body, a dorsal fin located well to the rear, and indistinct dark longitudinal stripes along the sides. In Singapore, where the species has probably been introduced, it is known as *Rasborichthys altior* and is sold under this name as an aquarium fish.

The south Asian cultrins are typically small and, unlike most other east Asian species, do not have a dorsal fin spine. The smallest species are in the genus *Chela*, and two of them, *Chela laubuca* and *Chela cachius*, are kept as aquarium fishes. These are high-set, beautiful, schooling species with a long anal fin, a dorsal fin located to the rear, an elongated first pectoral fin ray and a sharply arched lateral line. They prefer the upper water levels. Both species spawn in captivity; courtship begins with onset of dusk. The young hatch at 25–26° C. after 20–24 hours; after three to four days they can be fed rotifera (see Vol. I), and later crustacean larvae (see Vol. I).

Pelecus cultratus is a European species which resembles the southeast Asian cultrins not only in shape but also in its life habits and skeletal structure to such a degree that it is doubtless a member of this subfamily. L is 50 cm, rarely 60 cm, with a weight of 2 kg. The body is greatly compressed; the small mouth is on the top side of the body. The dorsal fin is very short and has been pushed to the rear of the body. The anal and particularly the pectoral fins are very long. The lateral line has a distinctive undulating shape. The keel is sharp, extending to the throat. The fish lives in lakes, including brackish coastal lakes, and in larger rivers. The diet includes plankton, crustaceans, flying insects and fishes (even bottom-dwelling fishes). As in all cultrins, the eggs float freely. In south Russian lakes and in Lake Balaton, Hungary, this species is rather important commercially, but is of no interest to sport fishermen.

Subfamily: Xenocyprininae

The Chinese subfamily Xenocyprininae contains small to medium-sized minnows which resemble the European chondrostome anatomically. The small mouth has a diagonal opening and is on the lower body side or almost terminal. The lower lip is stunted. There are one, two or three rows of hookless throat teeth with very long chewing surfaces. The short dorsal fin has a smooth spine and is slightly shorter than the anal fin. In some species the keel is between the pectoral and anal fins. The intestine is very long. The underside is black.

The life habits of xenocyprinins are not well known. Those species

Ichthyologists have found a number of biological reasons for this curious adaptation. In 1933 Wunder showed in the European bitterling that the development of the egg-laying tube in females and the display coloration of the male are released by a stimulus emanating from the mussel. Males and females kept in an aquarium devoid of mussels do not develop the courtship coloration or the tube. However, as soon as mussels are put into the aquarium, these structures appear rapidly; females even spawn into the mussels in the absence of males.

The bitterling subfamily (Acheilognathinae) has five or six genera with about forty species which differ only in rather insignificant features: a complete or incomplete lateral line, smooth or furrowed throat teeth and the presence or absence of a spine in the dorsal fin.

The EUROPEAN BITTERLING *(Rhodeus sericeus amarus)* exists as a subspecies distinct from the CHINESE BITTERLING *(Rhodeus sericeus sericeus)* of the Amur basin and northern China. The great gap between these two is a consequence of the Ice Age; before its onset bitterlings were also prevalent in Siberia. Distribution of the European bitterling includes central, eastern and southeastern Europe from northern France to the Ural River and Macedonia, as well as northern Turkey and the Transcaucasus. The fish lives mainly in ponds and pools, along lake shores and in slowly flowing plains rivers. In scattered places it is also found in stony rivers of hilly country. L is 7–8 cm. The lateral line is incomplete; barbs are absent, and the dorsal and anal fins are short. Males are characterized by a striking display coloration. The spawning period lasts several months, from April or May to August, and a single female spawns several times during this interval. The larvae stay within the mantle cavity of river mussels *(Unio* and *Anodonta;* see Vol. III) until they can swim. In spite of their beauty and the feasibility of keeping them in aquariums, bitterlings are not popular aquarium fishes.

The GUDGEONS (Gobioninae) are small to medium-sized carp with either elongate, low, or medium high-set bodies. The underside is flattened or rounded. The mouth is small and is terminal or on the lower body side. Many genera have a pair of upper lip barbs, but snout barbs are never present. *Gobiobotia* gudgeons still have three pairs of gill barbs on the underside of the body. Some genera have thin lips, while others have fleshy or fringed lips with warts; still other genera have horny plates on the lips. There are either one or two rows of throat teeth. Both dorsal and anal fins are short. The length of the intestinal tract varies considerably. Belly coloration is silvery or brown to black. The swim bladder is developed in most genera but is degenerate in highly specialized genera inhabiting flowing water (e.g., *Microphysogobio* and *Gobiobotia);* in these species the swim bladder is enclosed by a bony or fibrous capsule. Coloration varies considerably; there is a row of round or elongated lateral spots or a longitudinal stripe in many genera. Others have large, indistinct spots.

The European bitterling

Subfamily: gudgeons

Some gudgeon genera (e.g., *Sarcocheilichthys, Pseudorasbora, Gnatho-pogon* and others) prefer standing or slowly flowing waters; the BIWA LAKE GUDGEON *(Gnathopogon coerulescens)* which is found only in Lake Biwa in Japan, is partly free-swimming. However, most genera, particularly the highly developed ones, are found in rivers. As an adaptation to this kind of life they have a low-set body, the mouth on the lower body side, long barbs and warts. The dorsal fin is located to the front and the body has become heavier due to the degeneration of the air bladder. Gudgeons feed on bottom-dwelling invertebrates or algae; some species chew off fauna growing on stones with the help of horny plates on the jaws.

Many species, including the half-pelagic Lake Biwa gudgeon, spawn on the bottom or on plants. Some care for the young, as for example *Abbottina rivularis,* which builds a saucer-shaped depression at shallow sites as a nest. The male protects the larvae and drives off animals approaching the nest. The STONE MOROKO *(Pseudorasbora parva)* male protects the larvae which are on stones. *Sarcocheilichthys* gudgeons, with the single exception of *Sarcocheilichthys sinensis,* lay their eggs in the mantle cavities of living river mussels just as the bitterlings do. The females also have a long egg-laying tube. Several other genera are highly specialized bottom-dwellers *(Gobiobotia, Saurogobio,* and others, including the species *Sarcocheilichthys sinensis)* and have freely floating eggs which develop while being carried downstream by the river. This particular adaptation appears to be a relatively recent one. Spawning display coloration appears in several species but is not well developed in European gudgeons, while some eastern Asian fishes have greatly developed coloration.

The gudgeon subfamily contains eighteen genera with seventy-three species. With the exception of the COMMON GUDGEON *(Gobio gobio),* distributed from England to central China, all genera are limited to eastern Asia. The stone moroko is a small fish with a medium high-set body, and the barbless mouth on the upper body side. The sides are silvery, most scales having a dark spot. Sexual maturity is reached in one year. The moroko inhabits all ponds, lakes and slowly flowing rivers from eastern Asia to southern China. In 1960 it was introduced into Romania and probably into other European nations along with several other immature Chinese commercial fishes. The following year spawning took place and the fishes were transferred from breeding ponds into natural waters. In a few decades the "colorful goby" (as the Romanian fishermen call it) will be a typical representative of the European animal kingdom along with sunfishes and catfishes.

Saurogobio gudgeons, L 40 cm, are distributed in most rivers of the east Asian mainland, but not in Japan. Their elongated body, long and slender tail shaft, warty lips and the forward dorsal fin all indicate that this is the carp best adapted to life at the bottom of flowing waters.

Fig. 15-14. Distribution of common gudgeon *(Gobio gobio).*

A typical representative is *Saurogobio dabryi*. *Gobiobotia* contains the most species, which, unlike all other carp, have four pairs of barbs. Some authors group them with loaches.

The GUDGEON genus *Gobio* is the only one prevalent in Europe. It contains thirteen species, five found in Europe. Four, including the common gudgeon, inhabit central Europe including the Danube basin. This fish is distributed from Ireland, England and the Pyrenees to central China. It has a moderately elongate body with a laterally compressed tail shaft and eyes pointed outward. It is typically found in rivers, streams, and sometimes in ponds, lakes, and marshes, but spawns only in flowing waters. In western and central Europe it is in all flowing waters on sandy, gravel, or loam bottoms, but it is being forced out of the central and lower Danube basin, where other species of the same genus occur, into the smaller loamy or muddy tributary streams. The species is a popular dish in France.

The gudgeon *Gobio uranoscopus* is distributed solely in the Danube basin. It is more slender than the common gudgeon and the tail shaft is long and slender. Its barbs are also longer and the eyes are pointed upward. The general coloration is darker, but its fins have almost no spots. Although this species was described in Munich in 1828, little more is yet known about it. Unlike common gudgeons, this fish occurs only in fast-flowing rivers on stony bottoms.

Gobio kessleri, like the previous species, has a slender body and slender tail shaft, but it has larger eyes, shorter barbs and brighter coloration. Distribution is in the sandy rivers of the central and lower Danube basin beneath the zone of *G. uranoscopus*. It appears in large schools which sometimes contain hundreds of individuals.

Gobio albipinnatus has a shape between that of the two above species. It has a rather thick and only slightly compact tail shaft, large eyes and bright coloration. Distribution is from the Volga to the Danube. It is the only representative of its genus in the Danube below Vienna.

More than half of all carp species belong to the subfamily BARBINS *(Barbinae)*, but the variations within this subfamily are not as great as in, for example, danios and minnows. Most barbins are adapted to a bottom-dwelling life in rivers; they have elongate, slightly compressed bodies with the mouth on the lower body side or almost terminal. In other species living in standing waters between plants and other obstacles the body is high-set (although never so high as in breams or common carps) and laterally compressed. The mouth is terminal and a belly keel is found only in the southeast Asian genus *Rohtee*. These are good swimmers of the upper water levels; there are no true predators or pelagic species among the barbins.

Barbins have one or two pairs of barbs; the throat teeth are typically arranged in three rows, but some genera have one or two rows. The shape of these teeth is quite diverse, and all tooth forms which appear

Gudgeons of genus *Gobio*

▷
In one carp *(Cyprinus carpio)* the scales are totally absent.

▷▷
Breeds of goldfishes, from top to bottom: Lion's head, a variety with thickened processes on the head and a double caudal fin. The black telescoped fog-tail has protruding eyes and a double caudal fin, which is particularly large here. This variety has protruding eyes from the side of the head which are oriented upward. None of these remarkable breeds could live in nature; they require human care to survive.

Subfamily: barbins

in the carp family are represented in the barbins, including grasping teeth (the primordial type), molars of various shapes, long and compressed cutting teeth and others. In many species the last unbranching ray of the dorsal fin has ossified and has teeth, while in others it is thin. Even closely related species can be quite different in this respect. The dorsal fin is located in the middle of the body or somewhat to the front. In most species the anal fin is short. The size of the scales varies greatly; a few species have no scales at all. The lateral line, when it is complete, is straight and ends below the middle of the caudal fin (except in one species). The length of the intestinal tract also varies; the belly coloration is silvery, brown or blackish. Degeneration of the swim bladder is an adaptation in some barbins (genus *Garra*) to bottom-dwelling in rivers. The lips can be thin (as in *Puntius*) or fleshy, folded, warty or covered by a horny layer. Many barbins have sucking discs (for example the sucking barbins). There are many genera and species.

Barbins inhabit all sorts of inland waters from fast-flowing mountain streams to pools. Almost all feed on the bottom or on plants; barbin feed on small organisms and algae. Eggs are laid on the hard ground or on plants. Spawning display coloration appears in many genera. The barbins are the only carp group with blind species. Some examples are the cave species *Caecobarbus* in the Congo basin, *Phreatichthys* in Somalia, *Typhlogarra* from Iraq, and *Iranocypris* from Iran.

Barbs

The BARBS *(Barbus)* are medium-sized to large elongate fishes with flattened underside, two pairs of barbs, the mouth on the lower side of the body, and a complete lateral line with 40–100 scales. The dorsal fin has eight entwined rays, while the anal fin has five, rarely six. The throat teeth are hook-shaped. The intestinal tract is short. Belly coloration is whitish or brownish. Many species exist, mostly in western Asia; several species occur in southern Europe and there are two in central Europe.

The best-known species is the BARB *(Barbus barbus)* or BARBEL. L is up to 50 cm, rarely 90 cm, with a weight of 8.5 kg. The last unbranching dorsal ray fin is spine-shaped and is jagged on its rear edge. There are 56–62 scales in the lateral line.

⊲⊲

The well-known goldfishes differ from the wild eastern Asiatic form *(Carassius auratus)* only in color.

⊲

Like many another fish, tench *(Tinca tinca)* is good eating. They breed well in warm, muddy, even oxygen-poor ponds.

Barbels inhabit only streams and rivers with clear water and sandy or gravel bottoms. Young barbels are also found in smaller streams. During the day the fish lives on the bottom of the river, feeding at dusk, primarily on insect larvae, especially of May flies *(Palingenia* and *Polymytarcis;* see Vol. II) in larger rivers and also crustaceans, worms, rarely plants and fish spawn. Wintering sites are in deep tranquil inlets and pools. During the spawning period (May and June, rarely to July) large groups composed mostly of older females ascend the rivers and sometimes penetrate as far as mountain streams, laying the eggs on stones in rivers. Sexual maturity is attained in three to five years.

Barbels are often considered good eating, for their meat is tasty, and they are caught with all sorts of nets and hooks. They are also popular sport fish, putting up a good fight when hooked. The best bait includes earthworms, and in hot months cheese and "Leberkäse" (a south German meat product). The hook must be well concealed, for barbels bite carefully.

The barb *Barbus meridionalis petenyi* differs from the barbel by its thin, non-jagged dorsal fin ray; the anal fin is much longer, extending behind the caudal fin base. L is at most 28 cm. The back is brownish with darker and brighter spots; the sides are gold-yellow and spotted; the dorsal and caudal fins are spotted, while the other fins are yellowish; the barbs are yellow and lack a red axis.

The species is non-migratory, inhabiting mountain rivers and streams with gravel bottoms, penetrating as far as the trout zone. It spawns rather late in the summer. Diet consists of invertebrates. The species has no commercial value. Distribution is in the Danube basin west to Moravia or even Austria, and also in the upper basin of the Dnyestr, Oder, Weichsel and Wardar Rivers; subspecies are found in the southern Balkans, Italy, southern France, Spain and Portugal, while closely related species occur in northwestern Africa and the Caucasus countries.

Fig. 15-15. Distribution of the barbins (Barbinae).

The much larger species *Barbus brachycephalus caspius* and *Barbus capito*, from the Caspian-Aral basin, are migratory fishes living in brackish water and migrating up rivers during the spawning season. Both have commercial significance.

In southern Asia the barbs are represented by the closely related genus *Tor*, in which the scales are much larger. Distinguishing both genera is difficult since some west Asian species take an intermediate position. The large-scaled barbs are known as "mahseers" in the Anglo-Indian literature. Many of them are large, commercially important fishes. *Tor tor* (L up to 2 m) has 25-27 scales and a strong spine in the dorsal fin. Distribution is throughout India, where it is one of the most popular sport fishes. During the spawning period it migrates upstream like the European barbs.

Large-scaled barbs

Most Asian species belong to the genus *Puntius*. These are small, high-set and laterally flattened fishes; the snout is generally short and the mouth is terminal. Many species have four barbs; some have just two and others have no barbs. The last unbranching dorsal fin ray is either thin or spine-shaped. Most of these barbs look quite different from the *Barbus* species; but certain west Asian and African species occupy transitional states making it difficult to draw distinctions between the two groups. Some researchers thus unite *Puntius* and *Tor* with the *Barbus* barbs. Only a few *Puntius* fishes are commercially important; the most important of them is *Puntius javanicus*, raised in breeding ponds in tropical countries.

Ornamental barbs

Fig. 15-16. Distribution of the barb *(Barbus barbus)* and its subspecies.

Many *Puntius* species, however, are kept in aquariums; at least 31 Asian species are bred and raised in captivity. These species, from India, Ceylon, Thailand, Malaysia and Indonesia, often live in great schools in very diverse aquatic environments. Their activity requires very large aquariums (with a capacity of no less than 50 l). It is important that the ground not be too soft and that there be an ample floral covering, although the plants should not be so thick that there is not enough swimming area. Most species inhabit the lower water levels. *Puntius* barbs are less sensitive to low temperatures than other tropical fishes, and can withstand a temporary drop to 17° C. *Puntius* are docile and can generally be kept together with other small fishes. They take living and dried food of all kinds. Breeding *Puntius* is easy and is very simple in some species. They usually spawn in large glass aquariums without an earthen bottom; they spawn on fine-leaved plants which do not stand too close to each other. Almost all species prefer soft water. The parents should be put in the breeding tank in the evening; they can usually be seen spawning the next morning as the first rays of sunshine appear. The partners press together at the plants and release eggs and sperm. The fertilized eggs adhere to the plants or fall to the bottom. When well fed, the fishes spawn several times a year.

The most important ornamental *Puntius* species kept in aquariums are: 1. *Puntius conchonius*, a very beautiful and popular fish from northern India; 2. *Puntius everetti* from Borneo and Singapore; 3. *Puntius lateristriga* from Malaysia and Indonesia; 4. *Puntius nigrofasciatus* from southern Ceylon; 5. *Puntius titteya* from Ceylon. There are also four other *Puntius* fishes which have four to six vertical dark stripes on a reddish-silver basic color. Recent studies have revealed that these are two species, each with two subspecies: *Puntius pentazona pentazona* from Borneo, *Puntius pentazona hexazona* from Sumatra and Malaysia, *Puntius tetrazona tetrazona* from Sumatra and Borneo, and *Puntius tetrazona partipentazona* from Thailand and Malaysia. These are not as docile as the other species and should not be kept together with long-finned fishes, since they tend to bite off the ends of the fins.

Many African barbs are also classified with the genus *Puntius* and about fifteen of them are kept in aquariums. In most African species the edge of the dorsal fin is slightly concave, while Asian species have a slightly convex fin edge. Among the African ornamental barbs, *Puntius holotaenia* has a narrow stripe extending from the snout end to the caudal fin. The Angolian species *Puntius fasciolatus* is silver with a blue-green sheen and twelve narrow blackish diagonal stripes. The species *Puntius hulstaerti*, from the lower Congo basin, is distinct from the other species. Its anal fin begins somewhat beneath the middle of the dorsal fin base. *Puntius viviparus*, from Natal (South Africa), has mistakenly been characterized as giving birth to live young. About 250 barb species in tropical and southern Africa have been described and an examination of these species is greatly needed. Most are small with large or medium-sized scales. The mouth may be on the lower side of the head, partially on the lower side, or terminal. The body is either elongate or rather high-set; the lateral line may be complete or incomplete.

The barb genus *Capoeta* contains small to middle-sized fishes which are closely related to the previous barbs. However, their mouth opening is always diagonal, both jaws are cartilaginous, and the lower jaw has in addition a thin horny covering. The pointed lower lip chews off portions from plants, as in chondrostomes. The belly coloration is black; the intestinal tract is very long, up to ten times the length of the body. These fishes primarily inhabit the drier parts of western Asia; they are also found in the upper courses of rivers in mountains as well as lower courses which sometimes end in marshes or deserts. Some *Capoeta* barbs inhabit lakes. All species feed chiefly on plants and most spawn on stony ground in rivers. Some species, like the Transcaucasian barbs (*Capoeta capoeta* and *Capoeta heratensis natio steindachneri*) are commercially important. The small *Capoeta damascinus* is also kept in aquariums, reproducing readily in captivity. The African genus *Varicorhinus* resembles the west Asian *Capoeta* species.

The barbin genus *Garra* contains small, large-scaled gudgeon-like or roach-like barbins with a flattened underside and the mouth on the lower side of the body. Part of the lip has been transformed into a sucking disc. This disc is only slightly developed in the primitive species, but the higher species have very well-developed discs. Some species have a high degree of sexual dimorphism with a modified head structure in males. The intestinal tract is long. The peritoneum is black. As with most barbins, the throat teeth are arranged in three rows. In some *Garra* barbins the swim bladder has degenerated. The elongated body, presence of a sucking disc, and degeneration of the swim bladder are adaptations to living in fast-flowing water; indeed, most sucking barbins inhabit mountain streams and rivers of southern and western Asia and Africa; they even occur in some rapids. Some west Asian fishes live in regions which in recent geological times have become rather unfavorable environments because the rivers are drying up. In this area a small sucker barbin, assigned to a separate genus, *Typhlogarra*, has adapted to subterranean life.

The genus of the subfamily with the most species, after *Barbus* and *Puntius*, is *Labeo*, its members known as LABEOS. This contains both small and large fishes; the body shape and height are quite diverse. In many species the edge of the dorsal fin is greatly indented, while in others it arches outward. The mouth opening is on the lower side of the body. The lips are always thick, sometimes folded, and the lower lip is covered by a pointed horny layer. Both lips together serve as a sucking organ. The throat teeth are in three rows and are compressed

Capoeta barbs

Sucking barbs
(Garra)

▷

Leuciscus (minnow) species: 1. Bleak (*Alburnus alburnus*); 2. Chub (*Leuciscus cephalus*); 3. Ide or orfe (*Leuciscus idus*); 4. *Leuciaspius delineatus*; 5. Bream (*Abramis brama*); 6. Rudd (*Scardinius erythrophthalmus*); 7. Tench (*Tinca tinca*); 8. Gudgeon (*Gobio gobio*); 9. Barb (*Barbus barbus*).

Labeos

with their crowns very near each other. This transforms the entire upper surface into a chewing area.

Labeos are distributed in the flowing and standing waters of south-eastern Asia (as far north as the Yangtze) and Africa. Most of them are herbivorous. Some large species are commercially important in central and southern China and in the large African lakes. Four small species (two south Asian and two African) are aquarium fishes, the best-known being *Labeo bicolor*. It is velvet-black with a sharply separate orange-red to blue-red caudal fin. The second popular species is *Labeo erythura*. As Hediger reports (see Vol. XIII), *Labeo velifer* visits and "cleans" hippopotamuses. Labeos are more difficult to keep in captivity than barbins. They require peaty water which must regularly be renewed; the aquarium should not be in too much light, and the fish should have places for concealing themselves. These are reports from South Africa of breeding in captivity.

The giant barb

The largest carp species, *Catlocarpio siamensis,* with a L of up to 3 m, belongs to the barbins. It looks like a monster; the head is very large (comprising almost one third of the L); the mouth is deeply incised and pointed upward slightly, and there are no barbs. This giant barbin is caught with nets and with hooks. A closely related species, the INDIAN CARP *(Catla catla)*, was formerly widely distributed in India. It is a valuable eating fish with few bones. Today the species is uncommon and has disappeared from areas because of a lack of breeding sites.

Schizothoracins

The schizothoracin subfamily Schizothoracinae contains medium-sized to large, barbin-like fishes. The mouth is on the lower side of the body; the body and tail shaft are somewhat compact. The lower side is flattened and the paired fins are horizontal. The subfamily is distinguished from barbins by a series of larger scales around the anal opening and anal fin. The other scales are small or absent altogether. There are two or three rows of throat teeth; the peritoneum is black and the intestinal tract is long. The subfamily is considered to be a branch of the barbins. There are twelve genera with about forty species.

Fig. 15-17. Distribution of the schizothoracins (Schizothoracinae).

Schizothoracins inhabit fast-flowing mountain rivers and streams. Some species live in lakes, and others are found in plains rivers. They spawn on stones, usually in rivers. The spawn of the chief genus, *Schizothorax*, is poisonous and is dangerous to mammals (including man) and birds but not to other fishes or amphibians. Some species are herbivorous while others feed on microscopic organisms, and still others are predators. With the exception of one south Indian genus, the schizothoracins live in upper Asian elevations. A few are found as far west as Iran, and one species is even in Turkey. Unlike most other fresh-water fishes, a few species of schizothoracins are found in the upper courses of rivers including the Syr-Darya and Amu-Darya as well as the Indus and the Yangtze basin.

◁
The asp *(Aspius aspius)*, a member of the minnow group.

found even in brackish water in the Bermudas, although they are fresh-water species.

In China, with a tradition of developing surprising forms, many goldfish races have been developed; there are about 120 in all, not counting many transitional forms. Today Japanese and Chinese breeds are being hybridized. Some of the most popular modern breeds are briefly described here: there are "harlequins" with a colorful tuft, or with double tails which make them look like small peacocks. The "comets" have a tail which is three times the normal length. Fog-tails, the fins of which actually look like fog, have been bred in China since the 15th Century. The "Moors," with their velvet-black bodies, indicate clearly when the aquarium water becomes unfavorable; "egg fishes" have greatly rounded bellies; others have telescoped eyes sticking out of the head like tubes, pointing upward in some varieties. The "lion heads" have large warty growths like manes on their heads; "pompons," first bred in Hiroshima, have a protrusion on the nose which expands when the fish inhales. The "sleepers," which usually rested on the aquarium floor, died out in 1772 and breeders have not been able to produce them again. In the "pearlfishes" each scale (and a goldfish may have 650 scales on the body) is arched outward like a ball and has a dark edge. If such a fish loses a "pearl," a normal scale grows back to replace it; these unusual scales grow only once in a lifetime.

The flood of tropical fish species is pushing the goldfish from the indoor aquarium into open ponds. Many fish hobbyists consider the bizarre goldfish forms to be malformed or cripples. By the same logic any dog breeds which do not correspond to wolves or jackals should be done away with. Actually all these extreme forms belong to the most interesting creatures in the animal kingdom; they tell us much more than do many wild animals. Why shouldn't the many domesticated tropical fishes also be considered cripples? How else would one designate the various guppies and great variety of live-bearing characins and other fishes? Goldfish have long been a part of man's culture and few other domestic animals are understood as well. Opinion may vary about the taste of Chinese breeders; the ancient Chinese merely saw the possibility of developing new varieties from the original wild goldfish and acted upon this idea. No variety has been developed by artificial force; they all have a careful genetic (e.g., breeding) history.

In 1500 the goldfish entered Japan through Korea, and in Japan breeds correspond to Japanese tastes: dainty, fine fishes like swimming flowers which are meant to be observed only from above. The first Japanese breeds were reported between 1700 and 1710. In Europe goldfish were known only through reports until the first specimens arrived in Portugal in 1691. Madame de Pompadour, the extravagant mistress of King Louis XV at the court of Versailles, received them from China as a gift. In England they apparently appeared during the reign of James I (1603-1625). In 1711 the Duke of Richmond secretly received a large Chinese vase containing goldfish. It was only because of their stable character that these fish could endure such a long journey.

All these comments deal strictly with the common goldfish. The first Japanese fog-tailed fish reached Paris in 1872, creating a sensation. In Germany the famed Berlin fish breeder Paul Matte received the first usable fish from Japan in 1885. This "Matte tribe" still lives, although European interest in goldfish has declined.

The Oriental goldfish was sensationally received in Europe. In 1780 the first book about it arrived in Europe, shortly after Count von Heyden, the Prussian ambassador in Holland, brought the first goldfish to Berlin. In St. Petersburg, ten years later, Prince Potemkin decorated the tables with goldfish from the Far East in a banquet prepared for Czarina Katharina in his magnificent winter garden. It seems surprising that the United States did not receive its first goldfish until Rear Admiral Daniel Ammen brought them in 1878. However the country's first goldfish farm appeared ten years later.

Today, according to a Japanese newspaper article, the region of Koriyama alone has 12-14 million goldfish in breeders' tanks. The Asian goldfish breeders are concerned with varieties completely different from those which can be bought in pet shops for a small amount of money in Europe. They breed just the rare and expensive varieties. In Peking's Park of the Forbidden City goldfish breeding has recently begun in the old breeding tanks of the Chinese emporers, but the fish they produce now are commercial items. European common goldfish are bred by the hundreds of thousands in Italy, primarily around Bologna. In England there are goldfish breeding clubs which have large annual exhibitions. Two English goldfish hobbyists, Hervey and Hems, have written a substantial book about these Chinese fish, filled with fascinating details about them. The title picture is a colored reproduction of an oil painting of a goldfish, by former Prime Minister Winston Churchill.

Goldfish have a remarkably good memory. They can easily be trained in an aquarium or pond to come to the feeding place at the sound of a tone like a bell. Like carp, they have good hearing. Thus it is not surprising that just before eating, goldfish murmur, making various sounds. The goldfish's ability to draw relationships was made known to a factory owner in England. Water warmed while cooling the factory machines passed into the fish tank through a tube. The goldfish liked that and always stayed near the mouth of the tube. On Sunday the machines did not operate, so there was no warm current running through the tube. As soon as the machines started up Monday morning, though, the goldfish gathered expectantly at the mouth of the tube, even though it took some time before the warm water came through. A goldfish can become quite tame and will take food from the hand, but it seems unable to differentiate its owner from other people.

Hundreds and even thousands of dollars are spent for some goldfish. The Chinese have developed forms which have letters on their sides, although it has been said that they help somewhat by tattooing and etching off certain scales. New goldfish varieties become popular like women's clothing. Goldfish entered in competition are prepared in advance for the exhibition. They are kept for weeks in the owner's living room so they lose their fear of people; they are fed small amounts so they stay lively, and they are made accustomed to electric lights. During the time of Napoleon III it was faddish to carry tiny live goldfish inside of glass balls worn on earrings. One of the most famous goldfish was "Old Black Joe" of World War II (named for Josef Stalin, who enjoyed great popularity at that time), a coal-black fog-tailed goldfish used in advertisements for Liberty Bonds. It belonged to Otto Gneidig, of New Jersey, and lived to be over twenty years old. The fish supposedly changed color several times from black to red to blue.

Goldfish are also extremely hardy. In the late 1930's a London house-wife, while cleaning her fireplace, found something on the grillwork which she took at first for a rust-covered piece of fat. It turned out to be a live but very dirty goldfish. How it came through the chimney into the fireplace (perhaps dropped by a bird) was never established. The fish recovered quite rapidly when it was put into water. These resistant animals have even been kept in wet moss for half a day.

Another wild goldfish, the GIBEL *(Carassius auratus gibelio),* is found in the Aral Sea basin and in Europe, particularly in southern and southeastern Europe. It differs from the east Asian variety only in the body measurements. Its origin is unknown; it is perhaps the descendant of east Asian goldfish which were introduced to the Aral Sea basin by the Chinese in early times; they may have been brought to Europe by the Tartars. The fact that in some regions, such as in the lower and middle Danube basin, gibels appeared in the last few decades and multiplied rapidly, pushing the common carps out, indicates that they were not the original inhabitants of these areas.

Gibel reproduction is interesting: many populations contain only females, and in others there are just a few males. The eggs are fertilized by males of other species, including other carp. This, however, is not true fertilization; the cell nuclei do not unite. Instead, the sperm nuclei stay in one corner of the egg and later disappear; they only serve to stimulate part of the egg nucleus. This method of reproduction, gynogenesis, is one form of parthenogenesis; in addition to gibels it is found in groups of Central American live-bearing toothed carp which contain only females. Gibels have important commercial value, and are fished with all sorts of nets. They are also important accompanying fish when breeding carp.

Other carp-like species, by P. Banarescu

The gibel

▷

Ornamental barbs:
1. *Puntius cumingii;*
2. *Puntius oligolepsis;*
3. *Puntius eugrammus;*
4. *Puntius nigrofasciatus;*
5. *Puntius semifasciolatus;*
6. *Puntius gelius;* 7. *Puntius conchonius;* 8. *Puntius pentazona pentazona;* 9. *Puntius tetrazona tetrazona;*
10. *Puntius filamentosus;*
11. *Puntius vittatus;*
12. *Puntius schwanefeldi;*
13. *Puntius arulius;*
14. *Puntius titteya.*

Subfamily:
Hypophthalmich-
thyinae

Fig. 15-20. Distribution of the hypophthalmichth-yins (Hypophthalmichth-yinae).

◁

Barbs: 1. *Barilius christyi;* 2. The ornamental barbs *Puntius hulstaerti;* Labeos: 3. *Labeo wecksi.*

The HYPOPHTHALMICHTHYINS (subfamily Hypophthalmichthyinae) are a highly developed line of carp which probably evolved from the barbins. These are large, high-set fishes with laterally flattened sides. The large head and broad forehead with the upward-pointed mouth give it a peculiar appearance. The eye is small and below the axis of the head. The lateral line is complete and the scales are very small. The dorsal fin is short, while the anal fin is long. One peculiarity in the group is the merging of the gill folds with each other; they do not join with the partitioning walls as in other carps. This characteristic enables these fishes to feed on the smallest plankton. The intestinal tract is very long, as much as fifteen times the fish's length. The throat teeth are in one row and are greatly flattened.

The hypophthalmichthyins inhabit rivers and lakes. Their eggs float freely. In eastern Asia there are three species belonging to two closely related genera. *Hypophthalmichthys molitrix* (L up to 1 m, weight up to 8 kg) lives in the Amur basin and throughout China. Its scaleless keel extends from the throat to the anal fin base. The sexually mature fish migrate upstream; the spawning sites typically lie behind sandy spits of land and in islands on the edges of rivers. Up to 500,000 eggs are laid, and are carried downstream by the water. After consuming the yolk sac, the young feed on the plankton in the rich tributary waters. Adults feed exclusively on phytoplankton. One peculiarity of this species is that it jumps out of the water when a motor approaches or a shadow is on the water surface; this leap can be as much as 2 m above the water level.

H. molitrix is an important commercial fish, with tasty meat. In the Amur region of the Soviet Union about 1000 tons are caught annually; in pre-World War II China about 32,000 tons were caught every year. In the Amur it is caught in long weir baskets. In China it is also bred in fish ponds. Its larvae are caught in rivers or lakes together with those of *Ctenopharyngodon* carp, and put into ponds. Since about 1958 the species has been introduced into other countries, including the western Soviet Union, Israel and Romania. It is kept in ponds and reproduction takes place by artificial fertilization. Since keeping carp and *Hypophthalmichthys* together has a number of economic advantages, this practice will probably increase in the future. *Ctenopharyngodon* prevents over-development of the aquatic plants; the hypophthalmichthyins prevent excessive algae growth. Both species are ready for eating after the second year. Since the hypophthalmichthyin feeds on flora, it increases the energy exchange in the water. Carp appear to grow more satisfactorily in this situation than when they are bred alone.

Family: suckers

The SUCKERS (Catostomidae) are similar to cyprinid carps. The family includes small and large species which on the basis of appearance belong to two major groups. Some are high-set with long, thin lower throat bones and small, densely packed throat teeth and a long concave or convex dorsal fin. Species in this group inhabit rivers and lakes of the plains. The second group inhabiting mountain streams and fast-flowing rivers, contains elongate, spindle-shaped fishes with shorter lower-throat bones, strong throat teeth, short, light gill structures and a short dorsal fin. Suckers differ from cyprinids in their very numerous throat teeth which are in a single row. In all species the mouth is small, inferior and greatly protrusible. The very fleshy lips have small, hair-like tufts and form a sucking disc. There are no barbs. The scale covering is complete and the lateral line is straight. The large, free swim bladder consists of two or three chambers.

Suckers primarily inhabit flowing water; some species live in lakes but spawn mostly in the tributaries. Many migrate in spring into very fast-flowing sites in rivers. Others, particularly lake inhabitants, spawn on plants in shallow water. Males are generally smaller than females. During the spawning period, males and females assume a bright coloration which includes the head and fin rays. Species in flowing water feed on invertebrate bottom-dwelling organisms, while those in standing water feed additionally on plants. Suckers probably originated in southern or southeastern Asia, but most of the 100 present-day species are found in North America. One genus, containing a single species, lives in central China (and is a close relative of the North American *Carpiodes*), while one subspecies of sucker *(Catostomus catostomus)* inhabits northeastern Siberia. Due to their size they are more important commercially than the small resident carp.

The GYRINOCHEILIDS (Gyrinocheilidae) are small, elongate, slightly compressed, barbin-like fishes with medium-sized scales, a complete lateral line and an inferior mouth which forms a sucking disc enabling the fish to adhere to rocks. Thus they can withstand strong currents while scraping algae off the river bottom. The chief characteristic of gyrinocheilids is the presence of a supplementary slit above the actual gill slits, which is for sucking in respiratory water; unlike other fishes, the gyrinocheilids do not take in water through their mouths. This enables the gyrinocheilids to fulfill their oxygen needs even when their sucking disc is strongly adhered to some object. There are no throat teeth. The paired fins are horizontal. The intestinal tract is very long, and the swim bladder is small. Gyrinocheilids live exclusively in flowing waters, usually in mountain streams, but also in plains rivers. They feed on plants and adhering algae. Their reproductive process is still unknown, but they probably spawn on or under stones. The family contains just one genus, *Gyrinocheilus*, with three species, two in Thailand and the third in Borneo.

The Siamese gyrinocheilid *Gyrinocheilus aymonieri* (L about 25 cm) was introduced about fifteen years ago as an aquarium fish. It is very useful since it feeds on algae and other aquatic plant growths from the aquarium plants, stones and walls without injuring higher plants. If

Fig. 15-21. Distribution of suckers (Catostomidae).

Family: gyrinochelids

Fig. 15-22. Distribution of 1) Psilorhynchidae and 2) Gyrinocheilidae.

Fig. 15-23. Distribution of hillstream loaches (Homalopteridae).

algae are not present in sufficient quantities it will take artificial food and wilted lettuce. One can readily observe its fast gill cover movements. The fish is very lively and jumps about; it requires oxygen-rich water and temperatures around 25° C.

Family: Psilorhynchidae

The psilorhynchids (Psilorhynchidae) are a family with one genus, *Psilorhynchus,* and three species, inhabiting mountain rivers and streams in northwestern India, Assam and upper Burma. These are small, spindle-shaped fishes with horizontal paired fins and a short dorsal and anal fin. The inferior mouth has a very distinctive feature: it is separated on the left and right from the lower side of the head by an indentation. There are no barbs. The lower throat bone is thin and there is just one row of throat teeth. As in homalopterids (see below) and unlike minnows, they have a number of unentwined fin rays. The basypterigium (a bone at the base of the skull) is well developed, as in homalopterids. The swim bladder is free, having no bony parts. On the basis of these characteristics the psilorhynchids could be considered closer to the homalopterids or to minnows, depending on which characteristics were considered. All three spindle-shaped psilorhynchids inhabit waters with a strong current. Their life habits are not well known. They have no commercial value and are unsuitable as aquarium fishes.

Family: hillstream loaches

The HILLSTREAM LOACHES (Homalopteridae) are small fishes which have adapted to living in fast-flowing water and have a series of distinctive features. In all these loaches, the forebody and head are very much flattened and the underside is uniformly smooth. The small mouth, on the lower half of the head, has warts and two or three pairs of barbs. The throat teeth are in one row; paired fins are horizontal, greatly widened and with long bases. The hillstream loaches have a number of unbranching fin rays. The tail shaft is rather long and is either thick or thin. The swim bladder has degenerated, and its front portion is in a bony capsule. The most primitive genera (e.g., *Homaloptera*) are like cobitid loaches (see below) and are not very flattened. Genera including the SUCKERBELLY LOACHES *(Gastromyzon),* with the greater anatomical development on one side of the body, have very broad and ray-like bodies, and the paired fins form a sort of sucking disc with which the fishes hold firmly onto rocks in the rapid current. In some genera both pectoral fins have merged completely in the rear, and a number of highly specialized respiratory structures have developed. Water which does not flow into the mouth flows along the sides of the head and not underneath the body. Little water passes through the small gill openings. Hillstream loaches feed on plants growing on the rocks.

Family: loaches, by T. Nalbant

The last carp family is the LOACH (Cobitidae) family. These are small to medium-sized fishes. L is between a minimum of 30–60 mm and a maximum of 200–300 mm. The largest European member is the subspecies *Cobitis elongata bilseli.* The body is either short and compact or elongate and somewhat cylindrical; it can be quite slender in some species. The head is usually small, especially in the elongated species. The small eyes are completely covered by a layer of skin. The mouth is on the lower part of the head and is arched, surrounded by fleshy lips. In all species there are three pairs of barbs; the fourth or fifth pair found in some species are actually greatly developed chin flaps of the lower lip. The location of the barbs differs; cobitin loaches have a pair of snout barbs and two upper lip barbs. The chin flaps in some genera (e.g., *Misgurnus, Lepidocephalus*) are well developed; the only cobitid without barbs is *Neoeucirrichthys maydelli.* In *Niwaella* the barbs are very short while the lips are well developed, forming a kind of sucking disc. The number of loach species can only be estimated since a few which have been regarded as species are actually only subspecies. There are three subfamilies: 1. Botiins (Botiinae); 2. Noemacheilin loaches (Noemacheilinae); and 3. Cobitin loaches (Cobitinae).

In most loaches the dorsal and pectoral fins are short and are in about the middle of the body. The pectoral fins are a little behind the base of the dorsal fin (except in a few cobitins). The dorsal fin typically has six to ten, rarely up to fourteen, rays (although some species have thirty, sixty or more). The rear edge of the caudal fin is straight or rounded. In botiin loaches, many noemacheilins, and a few cobitins (*Lepidocephalus* and others) this fin is indented and has pointed lobes. The body is usually covered by small cycloid scales; a few noemacheilins lack scales altogether. The botiin genus *Leptobotia*, the cobitin *Lepidocephalus*, and some others also have scales on the sides of the head. *Lepidocephalus* loaches even have scales on the top of the head. The lateral line is straight and either complete (in botiins and many noemacheilins) or incomplete, in which case it may be short and not extend as far as the tip of the pectoral fin (e.g., in pond loaches, cobitins and *Sabanejewia*); or it may extend to the middle of the body and beyond as in many noemacheilins.

Sexual dimorphism appears, although not in all loaches. In males the innermost ray of the pectoral fins can be ossified and spine-like (as in *Lepidocephalus*); in a number of noemacheilins the outer pectoral fin rays (or all of them) may be ossified, widened and covered with spawning coloration. The outermost ray of the pectoral fins can be still more thickened and elongated, as is the case in pond loaches and a few cobitins. This outermost ray can also have a horny plate (the lamina circularis or canestrinis scale) at its base, as in most cobitins and the pond loach *Misgurnus erikssoni.* Sometimes the body has two thickenings somewhat in front of the dorsal fin; this is the case in all pond loaches and all *Sabanejewia* loaches. Males in *Cobitis elongata* (the large European loach), *Niwaella* species, and *Lepidocephalus*, are much smaller than females. Furthermore in several species the two sexes are colored differently.

Fig. 15-24. Distribution of loaches (Cobitidae).

Loaches may be divided into two major types based on coloration, although there is great variation within each of these types. Most botiins and noemacheilins, as well as *Acanthophthalmus* and a few cobitins, have diagonal stripes. Most cobitins and a few noemacheilins have one or more longitudinal rows of spots as well as a dark or black spot on the upper half of the caudal fin base. The basic coloration is whitish or yellowish; the stripes and spots are gray, gray-blue, brownish or blackish. *Sabanejewia aurata bulgarica* has a beautiful brownish-violet basic color. The tropical species often have striking coloration (red or orange and black); this is found in the botiin genus *Botia,* in a few noemacheilins, and *Acanthophthalmus.*

The loach swim bladder has developed in an unusual way; it consists of two chambers. The front chamber is always enclosed by a capsule which is either fibrous (in most botiins) or has ossified (in all noemacheilins and cobitins). This fore-capsule is related to the Weberian apparatus. The rear chamber is always free and is well developed in botiins and a few noemacheilins, but has degenerated in most noemacheilins and all cobitins. The forechamber and surrounding capsule is widened on the sides in noemacheilins; in cobitins it is uniform and either conical or wider than its length. The digestive system is also different among different loach species. Botiins and noemacheilins have specialized stomachs, a feature absent in cobitins. In some noemacheilins the intestines are as long as the body or longer and have several convolutions; they are also long in botiins but short and almost straight in all cobitins.

Most loaches inhabit flowing waters; a few noemacheilins have even adapted to life in extremely rapidly flowing waters in mountain streams. However, there are also species, especially in the botiins and noemacheilins, which inhabit lakes, ponds, and even marshy pools. Species in temperate zones usually spawn from April to June; the spawning period of tropical species is unknown. The central Japanese species *Niwaella delicata* spawns in January, the middle of winter. The number of eggs laid varies between 100–300 in *Niwaella* and 700–3000 in botiins, noemacheilins, the pond loach, cobitins and *Sabanejewia.* Loaches are ground and plant spawners. Their eggs have a diameter of at most 2.5 mm. The young hatch after 1–8 days. The larger species, such as botiins, Chinese botiins, pond loaches and *Lepidocephalus,* feed on insects or insect larvae, worms and crustaceans; many small species feed on algae and other food from the muddy river bottoms. With the exception of a few Asian species from mountainous regions and central Asia, loaches have no commercial value; a few species are used for bait and others are kept in aquariums.

Subfamily: botiins

The botiins subfamily Botiinae contains loaches found only in eastern and southern Asia. They are small to medium-sized fishes; the body is either short or elongated and is always laterally compressed. There are two pairs of closely set snout barbs and a pair of upper-lip barbs. The spine beneath the eye is short and slightly bent; its two branches lie one on the other. The body always has scales and the small cycloid scales are either rounded or long and sometimes are on the side of the head. The lateral line is complete and is in the middle of the body. The caudal fin is deeply indented and its lobes are usually pointed. There are two genera: 1. *Botia* and 2. *Leptobotia.*

The *Botia* loaches live primarily in southern Asia but have also reached eastern Asia as far north as the Yangtze basin. About twenty species are known, and are divided into three subgenera: A. Botiins (in the narrower sense) *(Botia)* with the better-known species *Botia macracanthus* and *Botia lohachata,* all in southern Asia; B. *Hymenophysa,* including *Botia hymenophysa, Botia berdmorei, Botia horae,* and *Botia sidthimunki,* the last being from southern China while all others are from southern Asia; C. *Sinibotia,* exclusively distributed in eastern Asia.

Fig. 15-25. Distribution of botiins (Botiinae).

All botiins are group-living. In aquariums individual botiins will become aggressive, but when six or seven of them are kept together they immediately become docile. The dwarf botiin has been observed to attack the larger *Botia macracanthus.* The fishes need clean, soft, oxygen-rich water and sandy or gravel bottoms. Often they dig small depressions or pits for concealment. Breeding them in aquariums is difficult.

The larger, usually elongated loaches with scales on the sides of the head belong to genus *Leptobotia.* These east Asian loaches extend northward to the Amur basin and Japan.

The noemacheilin subfamily Noemacheilinae differs from both others particularly in the absence of a spine beneath the eye. The body is generally elongate and is more or less cylindrical; the body is short and compact in only a few species. The head is usually flattened. There are two pairs of snout barbs and a pair of upper lip barbs. Chin folds are never present. The dorsal fin in most species is short (with 7–10, rarely more, entwined fin rays), except in *Vaillantella.* The body either lacks scales or has a partial or complete covering. The caudal fin may be rounded, clipped, or indented, but never to the extent of the botiins. Many species have a well-developed adipose process on the upper edge of the tail shaft. With about 120 species, this is the largest loach subfamily. Most are in the genus *Noemacheilus;* there are also two to four small genera.

Subfamily: Noemacheilinae

Most of these loaches inhabit fast-flowing rivers and streams. A few, especially those in mountainous Asia, have adapted to living in ponds and lakes. In these species the swim bladder (unlike that of most loaches) is well developed and has two or three chambers. The rheophilic ("current-loving") species prefer pure, oxygen-rich water and sandy or gravel bottoms. They feed mostly on insect larvae and other bottom-dwelling small organisms or algae. The genus *Vaillantella* is completely divergent; its members have a very long dorsal fin and about sixty entwined fin rays.

Fig. 15-26. Distribution of *Noemacheilus* loaches.

The major genus *Noemacheilus* will probably be subdivided into several genera in the future. It includes the STONE LOACH (*Noemacheilus barbatulus,* maximum L 16 cm). This elongated loach dwells on the firm bottom of flowing, clear waters. It is also found in regions where trout, graylings and chondrostomes live. Its diet consists of insect larvae and crustaceans. The spawning period is in spring (March and April). The second species, *Noemacheilus angorae bureschi,* is distributed in the basin of the Struma in Bulgaria and Macedonia. It is the typical central European loach. These and other subspecies also inhabit Turkey, the Transcaucasus, and adjacent regions.

Subfamily: Cobitinae

The cobitin subfamily (Cobitinae) is characterized by an elongated, more or less laterally compressed body, with one pair of snout barbs and two pairs of upper-lip barbs, a front and a rear barb (on the edge of the mouth). Chin folds are always present, and are sometimes quite well developed. The spine below the eye is short and its two branches are horizontal. There are almost always small cycloid scales, which in some species are also on the sides of the head. The lateral line is usually incomplete, extending at most to the middle of the body. In most pond loaches (*Misgurnus*) and a few other species (e.g., *Niwaella*) the dorsal and pectoral fins are behind the middle of the body. In *Acanthophthalmus* the dorsal fin is far to the rear, almost above the anal fin, while the pectoral fins remain in the middle of the body. The caudal fin is rounded or clipped off, rarely being somewhat indented. There are fourteen genera and about fifty species distributed throughout most of Eurasia as well as Japan, Borneo, Java, Ceylon and Morocco.

Fig. 15-27. European distribution of pond loaches (*Misgurnus*).

Most cobitins inhabit flowing and very rapidly flowing waters; some live in lakes and ponds (such as the pond loaches and species which dig into sand or mud; the best example of the mud-diggers is *Cobitophis*). The life habits of many Asian species are unknown. Cobitins do have special adaptations, for example a very elongated, spindle-shaped body, as in *Acanthophthalmus* and *Cobitophis,* as well as doubled upper lips and the lack of barbs in *Neoeucirrichthys maydelli.*

The European POND LOACH (*Misgurnus fossilis;* L up to 30 cm) is a well-known fish of standing waters. Its body is almost perfectly round but toward the back the sides are flattened. The skin is slimy. The spine beneath the eye is degenerate and immovable. Pond loaches live in muddy ponds which are rich in plants or even peat. They are also found in marshy ponds and pools. During winter and periods in which there is a shortage of water they burrow into the mud. The ability to respire through the intestinal tract is better developed in pond loaches than in other members of the family. The spawning period is from April to July; pond loaches spawn on plants. They feed chiefly on river-bottom invertebrates. Related species occur in eastern Asia from the Amur basin to northern Burma.

Fig. 15-28. Asian distribution of the pond loach (*Misgurnus*).

The European cobitin *Cobitis taenia* is much smaller, with a L of up

to 12 cm (males do not exceed 9 cm). The body and especially the head are compressed laterally, and the spine beneath the eye is well developed. There are several subspecies, of which a few are now considered to be distinct species. The fish lives in flowing and standing water but prefers clear water with hard or sandy bottoms. It feeds nocturnally on invertebrates. In spring and summer it spawns on the river bottoms.

Sabanejewia aurata is similar to *Cobitis taenia,* but it has just two rows of longitudinal spots and no black spot in the caudal fin. With a wealth of subspecies, it inhabits the flowing and rapidly flowing waters of eastern Europe, penetrating the Danube as far as Austria. It is also found in parts of western Asia.

Several south Asian species belong to the genus *Acanthophthalmus*, which contains a number of species which are very difficult to differentiate, including: *Acanthophthalmus kuhli kuhli, A. myersi, A. semicinctus,* and *A. shelfordi.* Like *Lepidocephalus guntea*, they have been bred in aquariums.

Fig. 15-29. Distribution of the cobitid loach *(Cobitis).*

▷
Goldfish breeds: 1. Little Red Riding Hood; 2. Orchid dragoneye; 3. Red dragoneye, with rotated gill cover and nose lobes; 4. Skygazer; 5. Bouquet head; 6. Pearl-scaled goldfish; 7. Lion's head; 8. Spotted dragoneye; 9. Opal fish; 10. Bugeye.

16 Catfishes

Order: catfishes,
by D. Vogt

Distinguishing
characteristics

◁
Barb-like fishes: 1. *Garra taeniata,* a sucker; 2. *Epalzeorhynchus siamensis;* 3. *Epalzeorhynchus kallopterus,* closely related to suckers; 4. The gyrinocheilid *Gyrinocheilus aymonieri* from the side and (a) from below, showing the sucking mouth; 5. The homalopterid *Gastromyzon borneensis.*

One usually thinks of a catfish as being a plump, barbed fish which lies on the bottom and becomes active only when seeking food. However, there are many catfishes which live in open water, and some of them do not look at all like typical catfishes. Catfishes are distributed throughout the world, but primarily in South America, where 1200 of the 2000 catfish species are found. Even though they are typically fresh-water inhabitants, a few species are found in oceans.

The CATFISHES (order Siluriformes) are generally bottom-dwelling fishes with barbs on the head. There are no scales; the body is either naked or covered with bony plates arranged in rows or forming an armored covering. The upper jaw has typically degenerated and functions as a point of attachment for the barbs. Sometimes the barbs have supportive cartilage; others have a support mechanism which permits the barbs to relax as soon as the fish dies. The barbs contain taste organs and thus in a sense are an extension of the tongue. This is logical since the majority of catfishes are active at dusk and at night and need a supplementary organ to detect food. The number of catfish families varies according to different authorities; we divide the order into 32 families, of which we will discuss the 21 most important.

Various structures for gaining a suction hold on objects have developed within individual catfish families. These structures are usually found in species which inhabit fast-flowing water or rapids. Catfishes also differ considerably in the degree of parental care. Many of them spawn in open water or between plants; others have true parental care, usually paternal; and still others are mouth-breeders, laying eggs which can be up to 1 cm in diameter.

Because of their secretive life habits and dismal appearance it is no wonder that catfishes have given rise throughout history to all sorts of tales and fables. We know very little about their life; and a few, primarily very small catfishes, are better understood only because they have been kept and bred in aquariums. Catfishes are nonetheless,

are very colorful and have been put into aquariums throughout the world. *Leiocassis siamensis* (L up to 20 cm) has a deeply forked caudal fin. This catfish also prefers concealed recesses in cavities or under protuberances and may assume the unusual position of lying on its back with its belly up, pressed firmly against some object.

The rivers of Borneo and Sumatra contain a bagrid catfish, *Bagrichthys hypselopterus* (L about 40 cm), with a very unusual feature: its first hardened dorsal fin ray may be 30 cm long. The purpose of this structure is unknown.

Bagrids are represented in Africa by *Auchenoglanis occidentalis* (L up to 50 cm), which is primarily found in central Africa and the Nile region. The juvenile coloration is flesh-colored to pink, with dark spots in various numbers and shapes on this basic coloration. The long, moderately inclined upper side of the head is a prominent feature. The caudal fin is not deeply forked. Only smaller specimens of this species are kept in home aquariums, since the large ones require roomy tanks. Even intense studies have revealed little about the life habits of this catfish. It is only known that it prefers concealment and is active at dusk.

The EURASIAN CATFISHES (Siluridae) have given the entire order its scientific name. The various species differ considerably in body size. The European sheat-fish may be several m long like some southeast Asian species. However, there are also dwarfs among the European catfishes which are just a few cm long. The skin is always naked; there are two or three barb pairs which can be quite long. With a few exceptions these are bottom-dwelling species which spend most of the day lying on the floor or concealed in recesses. The pure free-swimming varieties live in schools and small swarms and are active by day.

Family: Eurasian catfishes

The largest European catfish is the EUROPEAN WELS (*Silurus glanis*; L 2.5 m, occasionally over 3 m). The largest are in the Danube, according to reports by fishermen and travelers. Even 30 cm-long welses are predatory and feed on everything they can handle; in addition to fishes, this includes frogs and small water birds. The welses are not particularly lively. Even young ones just over 10 cm long lie motionless on the floor during daytime and seek concealment in depressions and recesses. They do not become active until offered some food. Werner Ladiges observes: "As soon as I arrived in Leutstetten, in Bavaria, people told me about catches of very large welses. Of course, all kinds of fishing stories are told and one can't remember all of them. For a whole year I was not much impressed by these welses until one day an American officer brought one to me that he had caught; it was 70 cm long. From then on I became interested in them and even caught one in a small basket. Then a very large wels was sighted in the castle pond in Leutstetten. Prince Rasso of Bavaria, who is very much interested in fishes, and I spent an afternoon and half the night waiting

The European wels

to see this wels just once. We did not see the big one but spotted one which was 50 cm long under the protruding edge of the marshy shore. The fish undulated but stood quietly in the water and turned around from time to time when the current pushed it out of its recess. From the edge of the boat we had a look into the strange, romantic underwater world of the wels. A large, branched tree stump lay in a deep pool, and giant old carp and very large perch stood under; we knew it was likely that a large wels would be there, but there was none.

"Then came the Summer and Fall of 1947 with the great dry period. The water in the Würm River in Leutstetten sank to an unheard-of level; one day I put on my long rubber boots and set out to find welses. I looked all along the shore, through every overhanging root and every collection of floating wood, and finally I found some. In a very short time I caught nine of them, the largest being about 65 cm long. Right at the beginning I lost a huge one, which on the basis of its pull must have weighed at least 12.5 kg. I explored with a large hand-net and then felt around with a piece of wood and even my hand in there. When I did this by a deep root, a black body suddenly rushed up with a powerful lunge, and pulled the net out of my hand; I had difficulty holding up the large wels. I saw it fighting for a moment in shallow water; then it disappeared in the mass of wood it had stirred up. Later I found three welses after the first frosts had come in fall; they were together with barbs and large chubs under the roots of a large beech tree. They were so entwined in the roots that it took almost forty-five minutes for me to get them into my net. I got my last wels in 1948; the prince brought it to me in his hat. It was all of 5 cm long and thus the smallest I had ever seen."

The elongated wels body consists of a powerful forebody and a laterally greatly compressed tail shaft; the prominent anal fin merges with the caudal fin. This fish, with its calm, undulating tail movements, normally has its long pair of upper jaw barbs pointed straight forward, while the four smaller barbs of the lower lip hang down. The dorsal fin, consisting of just four rays, seems small for such a powerful animal. This mysterious fish is still mistaken for a pike or eel in many stories.

For reproduction males clean a site in shallow water and surround it with a wall of plant material, building a sort of nest. One or several females lay their eggs here and the male guards the spawn. After hatching, the young tadpole-like fishes stay in the nest for a while; but they soon become so active that the guardian father cannot keep them together. Depending on the food supply, the young grow at different rates and reach sexual maturity after about four years.

Many stories have developed about the age which European welses can attain; it is not uncommonly reported that specimens over 100 years old have been found. These claims are certainly untrue, but there is no doubt that in suitable waters the wels can indeed reach a ripe old age.

A second European wels, *Silurus aristotelis,* lives in southern Greece and Eurasia. Unlike the case for most European fishes, we know practically nothing about its natural life. Its method of reproduction was known to the ancient Greeks. The male cleans a site in shallow water by means of fin and body motions and guards the eggs. This species looks like the other wels, but its dorsal fin is smaller, and it has just two pairs of barbs. The caudal fin is distinct from the anal fin.

Silurus aristotelis

One of the largest welses is *Wallagonia attu,* distributed from Indonesia to India and Ceylon. Specimens have been found which were over 2 m long, but generally L is 60–100 cm. It inhabits large rivers almost exclusively, and its presence is clearly evident since it seeks prey during the day just under the surface of the water. In hunting it often leaps out of the water and falls back with a loud splash. *Wallagonia* vigorously attacks schools of small fishes on their annual upstream migrations. With their powerful teeth these predators not only feed on small fishes but larger ones as well. It is also a popular fish for human consumption.

Wallagonia attu

New species of the catfish genus *Kryptopterus* are continually being introduced as aquarium fishes. The most popular member of this group in that respect is *Kryptopterus bicirrhis* (see Color plates, pp. 388 and 393), which has a L rarely over 15 cm. This species does not belong to the schilbeid catfish family (Schilbeidae), which will be described next, but to the silurids; in German it is called the Indian glass wels because it is almost transparent. The vertebral column and bones are quite distinctly visible, and only the body cavity is not transparent. The species is primarily distributed in Indonesia but also is in eastern India, where it is not as prevalent. The single pair of barbs is very long and always extends forward. The dorsal fin consists of a single ray; however, the anal fin is very long, with up to 70 rays. The fish moves about in open water with the help of the barbs, gathering in large schools in the shade. The fin makes undulating movements but does so in coordination with the entire body. Often this species stands with its head against the current and can thus feed on floating prey quite easily.

In the SCHILBEID CATFISHES (Schilbeidae) there are two to four pairs of barbs. These are generally medium-sized species from Africa and Asia, the majority of which have long anal fins. Almost all are edible and are commercially exploited, particularly in southern Asia.

Eutropiella debauwi (L up to 8 cm), from the Congo River, has been introduced in recent years as an aquarium fish in Europe. In coloration it is exceptional among the schilbeids: this almost transparent fish has a silvery shimmer and, according to the nature of light hitting it, shows all colors of the rainbow. Furthermore it has three black, slightly violet-blue, shiny longitudinal bands. Males are considerably more slender than females in this schooling, open-water fish. Nothing is known about its reproduction. Its swimming position in the water is also distinctive: it usually has its head oriented against the current and steers with the body and the long anal fin in undulating movements, with the three short barb pairs extended forward.

The African schilbeid *Physailia pellucida* (L up to about 10 cm) has a disc-shaped, flattened, high-set, slender body which is still more transparent than that of *Kryptopterus bicirrhis*. However, the African species differs from its silurid counterpart in that it has no dorsal fin, having instead an adipose fin, and that it has four almost equally long barbs. The African species is also a schooling fish.

Giant catfishes

Until very recently the GIANT CATFISHES (Pangasiidae) were included in the schilbeid family. Although most of the giant catfishes are barely more than 50 cm long, there are some species with a L of over 1 m, and in some cases 2.5 m. This family includes what is perhaps the largest freshwater catfish of all, if the measurements given by fishermen are used as an indication: the GIANT CATFISH (⊹ *Pangasianodon gigas;* L 2.5 m). An important food source in eastern India, it differs from most other catfishes in its life habits. In spite of its great size, this catfish feeds exclusively on plants. In older specimens the teeth are missing, although they are present in the young. Interestingly, no extremely small or juvenile giant catfishes have ever been described. The fish inhabits the large east Indian rivers and has regular migrations which coincide with the rainy period. At that time the giant catfishes move upstream into Lake Tali, China, where they spawn. Although the species is heavily fished with nets and baskets during the migrating period, sufficient numbers still complete the migration. Although more detailed investigations are underway, our knowledge of the species is limited to what you have read here.

Mouth-breeding catfishes

As their name implies, the MOUTH-BREEDING CATFISHES (Tachysuridae) incubate their eggs in the mouth cavity, this being done by the males. The development of barbs in this family varies. Some have very small ones *(Batrachocephalus)*, while others have three long pairs of barbs extending behind the head (e.g., *Tachysurus*). Mouth-breeding catfishes inhabit the coastal regions from India to Indonesia. Their average L is 30–50 cm, although some species reach 1 m. A few characteristics related to sexual differences and having to do with reproduction are found in the group.

One of the most prevalent mouth-breeding catfishes along the Gulf of Siam is *Osteogeneiosus militaris* (L about 30 cm), an edible species although not a popular one. The swim bladder is used in producing gelatin. The species has a single pair of long, stiff, partially ossified barbs which extend to the pectoral fins. Although the species is prevalent and often is caught by fishermen along the coast and lower courses of rivers, almost nothing is known about its life habits. It is only known that the species reproduces throughout the year and is a mouth breeder. The male takes the yellow eggs, with a diameter of about 1 cm, and incubates them in its mouth. Only about 10–15 eggs are laid.

The mouth-breeding genus *Tachysurus* contains many species, of which a few are good eating while others are hardly known or are rarely caught. Once again the males do the incubating. The two sexes can be differentiated by the anatomy of the pectoral fin: females have a widened part on the rear of this fin which is used as an adhering surface for the eggs. During spawning only one egg appears at a time, which is then taken by the male into its mouth. It was once observed that a male *Tachysurus sagor* not only had 39 eggs in its mouth but also held 4 freshly hatched young, 4 cm long, all in its mouth! The male had taken up new eggs before releasing the young from the previous spawning. This 50 cm-long fish possesses silver-white to blue-green diagonal bands on the back and sides.

The SEA CATFISHES (Ariidae) have recently been grouped together with the mouth breeders. Many of these species inhabit salt water. The dorsal and pectoral fins have powerful thorns; the nasal openings are very closely set, and the rear ones are closed by flaps. These are generally open-water, often schooling, fishes, which have received the popular name of CRUCIFIX CATFISHES. Werner Ladiges reports on the history of this name:

Sea catfishes

In South American ports, such as Georgetown, Guyana, where all sorts of living or mummified animals are offered for sale, one can find very strange whitish ones which have crosses and all sorts of Christian symbols painted on them. If one examines such a creature closely it becomes clear that this is actually a skeleton of the CRUCIFIX CATFISH *Arius proops* (see Color plate, p. 386) which is the basic material for this unusual memento. Sea catfishes are popular eating along the South and Central American coasts and rivers. Their scientific name, *Arius*, is derived from the Greek word Αρειος (pronounced Araios), which means "sanctified by the god of war," and refers to the bony structure extending from the skull covering to the powerful thorn of the first dorsal fin ray.

"With the skull covering turned around," Ladiges adds, "the underside has a bony structure which can be seen when all the flesh and soft parts are removed. This structure looks roughly like a crucifix or like a painting of Christ crucified, and is surrounded by a Weberian bone in the form of a halo. The rough upper surface of the skull roof, which is also clearly visible on the living fish, looks like a monk with a cowl and hood and arms outstretched in prayer. Other people see in this structure a Roman mercenary with armor on his chest. The prominent dorsal fin spine is the lance point with which the warrior opened Christ's side. However, the auditory ossicles, which rattle

▷

Minnow relatives;
1. European minnow *(Phoxinus phoxinus)*; 2. European bitterling *(Rhodeus sericeus amarus)*; Loaches:
3. Pond loach *(Misgurnus fossilis)*; 4. Stone loach *(Noemacheilus barbatulus)*;
5. European cobitin loach *Cobitis taenia.*

when the skull is shaken, are the rolling dice which the mercenaries tossed for the vestments. Thus if the individual skull bones are carefully separated, each individual one represents some part of the Passion of the Savior." These crucifix fishes were reported for the first time in a travel report written in 1789.

Amblycipitidae catfishes

The catfish family Amblycipitidae is distributed in India and eastward as well as on a few southeast Asian islands. L is about 15 cm; the body is slender, and the caudal fin is large and indented to form two tips. The dorsal and anal fins are very small. Four pairs of barbs surround the mouth, and one pair of nasal barbs is between the eyes.

These small catfishes, which are all in a single genus (*Amblyceps*), are distributed through a wide region but have been observed very little in nature. Their main habitats are fast-flowing mountain streams and cliff areas. For respiration they have developed a curious fold under the end of the head in front of the pectoral fins, which enables them to breathe even in very fast-flowing streams. Local fishermen report that they can live outside the water for long periods of time. Although these catfishes are found almost exclusively in mountain waters, a few are known off smaller coastal islands, where they live in brackish water. *Amblyceps mangois* looks like a loach in coloration and shape, with the exception of the long barbs extending behind the head.

Sucker catfishes

A few genera of SUCKER CATFISHES (Sisoridae) have developed a sucking apparatus in two different ways; they use these in fast-flowing mountain streams to gain a hold on objects. The other species in this family do not have such a mechanism. The dorsal and pectoral fins have powerful spines. Six barbs are present, differing in length among members of the genera but always protruding just slightly beyond the head.

The largest sucking catfish is *Bagarius bagarius*, distributed from India to Indonesia. It attains a L of 2 m, but on the average is 1 m long when brought to market. This active predator feeds not only on small fishes but also on frogs, fresh-water shrimp and other animals.

The sucking apparatus is found in genus *Glyptothorax*, among others. These species prefer the fast-flowing mountain streams of southern and southeastern Asia. They have adapted to this life by a surface of skin folds on the lower side of the forebody and head. With this folded structure they can hold on to stones, roots or other firm objects. A typical example of the group is *Glyptothorax trilineatus*, distributed from Nepal across India and Burma to Thailand. In some areas it attains a L of 30 cm. The basic coloration is blackish-dark brown, brightening toward the belly. The three bands across the body facilitate recognition of the species: one yellowish to dirty white longitudinal band extends on the edge of the back from the rear of the head to the caudal fin; a second is along the middle of the body and lateral line, and a third,

◁
Loaches: 1. *Botia sidthimunki;* 2. *Botia maracanthus;* 3. *Botia hymenophysa;* 4. *Botia horae;* 5. *Acanthophthalmus semicinctus;* 6. *Acanthophthalmus kuhli;* 7. *Lepidocephalus thermalis,* with scales on its head; 8. *Noemacheilus fasciatus kuiperi;* 9. *Acanthopsis choirorhynchus.*

the shortest, begins at the pelvic fins and extends to the lower caudal fin. The adipose fin is quite large and the caudal fin is deeply indented.

The sucker catfishes of genus *Oreoglanis* have developed another sucking mechanism. The first rays of the pelvic and pectoral fins are greatly broadened and diagonally striped. The lips are also wide and rather flattened, forming a supporting sucking organ. Smith, who discovered the Siamese species *Oreoglanis siamensis,* wrote that the fish was caught shortly before dusk and was observed the following morning. Its upper side was uniformly olive-green and the underside was flesh-colored. Two cream-colored oval spots were located at the base of the dorsal fin on each side of the fin. When resting, the fish adhered to any surfaces and the sucking lips were supported by folds of the front and sides of the head.

As in a few other fish orders, catfishes also contains species which inhabit small, standing bodies of water where they must breathe atmospheric oxygen. These waters would otherwise be uninhabitable. The catfish group living this way is the LABYRINTHIC CATFISHES (Clariidae), distinguished by a supplementary respiratory apparatus in the upper part of the gill cavity, used for breathing oxygen from the air. Their gills are small, and the fishes die if they cannot get to the air. The dorsal and anal fins are very long and in some species form a continuous seam which includes the caudal fin. Labyrinthic catfishes are distributed from Africa to southeast Asia, where they are found in larger bodies of flowing water and in small, standing waters.

Labyrinthic catfishes

The species in the main genus *Clarias* (see Color plate, p. 399) are commercially important in their home distribution. For this reason they have been introduced elsewhere. Almost all species resemble each other in basic coloration. Their body is typically gray-black to dirty brown without any striking pattern. All are predators which attack whatever they can handle. Werner Ladiges reported from his Angola expedition that *Clarias* catfishes jumped out of the water after flying insects when they could find no more prey in the water.

With their additional breathing apparatus, the labyrinthic catfishes can also migrate over land when they are in arid regions and must move from one water hole to another. In most species the maximum L is 40 cm, but some giant forms exceed 1 m. Of the four pairs of barbs, one is oriented approximately vertically on the nasal openings, and two are very long and are almost always outstretched to the front, and a shorter pair points downward. The long barbs make these fishes look like insects with feelers. If any fish disturbs these barbs, the otherwise tranquil catfish reacts instantly. Its mouth is so large that when fully opened it seems clear that prey is usually swallowed whole in a single gulp. Labyrinthic catfishes are often cave-dwellers and often are albinos.

Gymnallabes catfishes differ from the *Clarias* species in their anatomy. As their German name, meaning eel and worm catfishes, indicates, they look like worms or eels. *Gymnallabes typus* (L 70 cm) has a long dorsal fin and an almost equally long anal fin. Its barbs are short but powerful. These west African catfishes inhabit very murky water. Aquarium observations have shown that they prefer concealment and are active at dusk. Once they find a recess, they return to it again and again when danger threatens. They have a striking yellowish-brown to golden-brown coloration. They are easily kept in large tanks and are predators, as are the other members of the family.

The stinging catfish

The INDIAN CATFISH or STINGING CATFISH (*Heteropneustes fossilis*; see Color plate, p. 393; L about 30 cm), from southeastern Asia, is the single member of the family Heteropneustidae. Some authors place it with *Clarias*. The Indian catfish is distinguished by the fact that it has a long, hollow, cylindrical canal extending from the gill cavity to the tail shaft on each side of the body. This organ is used for respiring atmospheric oxygen. There are four pairs of barbs, the longest extending to the tail shaft when the fish lays it back. Distribution is in various bodies of waters in southern and southeastern Asia.

Indian catfish are good food sources in Thailand, according to H. M. Smith, but fishermen do not like to cope with this species because the pectoral fin rays can inflict painful wounds. Furthermore, the spines have, incorrectly, been called poisonous. This bottom-dwelling fish is not particularly distinctive in its life habits. It often uses objects and recesses as hiding or resting sites; during the hours of dusk the fish swims quite skillfully, with undulating motions, just above the floor, rising to the surface from time to time for atmospheric oxygen. If there is no source of danger in the area the fish can stay at the surface of the water breathing air for several seconds. Its breathing rate increases, however, when it is threatened or feels disturbed. In this situation it breathes air with short, quickly turning body motions.

The Indian catfish is a predator which may include large fishes in its diet. Its greatly distensible mouth enables it to swallow fishes half its own length in a single piece. When that happens the body shape around the stomach takes on very unusual proportions. Occasionally Indian catfishes are also used as aquarium fishes, but only at a size of 10 cm. They have even been bred, but the precise course of courtship has not been reported.

Chacidae catfishes

Members of the catfish family Chacidae (L about 30 cm), from India, Indonesia, and other areas east of India, have the most divergent shape of all catfishes. The head is greatly flattened and widened, giving these fishes a most unusual impression. Their barbs are small, in some cases hardly noticeable. A few have small, branching processes along the sides of the head, for reasons still unknown to science.

In recent years the best-known species, *Chaca chaca* (see Color plate, p. 393), has often been sent from Singapore to public aquariums.

Unfortunately, only scanty observations have been made thus far, because of the tranquil, almost stoic life of this fish. It generally rests motionless on the bottom showing signs of life only at feeding time. The caudal fin is shaped like a fin seam and extends from the rear of the back to behind the anal fin. The tail shaft is very thin relative to the body. This mysterious fish feeds on whatever it can overcome, including small fishes.

The only species of the family of ELECTRIC CATFISH (Malapteruridae) is the ELECTRIC CATFISH *(Malapterurus electricus)*, from Africa. L extends to 120 cm, with a weight of about 25 kg. The body is long and cylindrical, with a short tail shaft. The electric organ is just under the skin and extends from the head through the body to the front of the tail shaft. There are three pairs of barbs.

Electric catfishes

An electric catfish looks at first glance like a sausage, for its plump body is dirty flesh-colored or gray-white, and the eyes, very small in adults, give the impression that the animal is blind. Electric catfish occasionally is a food source for man, but it is generally feared because of its discharges, which in adults can be as much as 350 volts. The young do not have such a strong current.

It was not known until 1958 how the fishes generated electricity. Then Lissmann and Machin, studying of this phenomenon primarily in mormyrids, which produce electricity of much lower voltage than the electric catfish, found (as we have already described in Chapter II) that they from an electrical field about themselves which serves to inform them about the environment. It is probably a similar case with the electric catfish. Lissmann could follow them in African rivers with the aid of apparatus to detect the electrical discharges.

Aquarium observations have shown that the electric catfish stuns or kills prey with strong shocks. Its electric organ, consisting of glands and dovetail-shaped, toothed, electroplate tissue, also serves as a defensive mechanism. A single large nerve cell, originating on the spinal-column gray layer, supplies the organ with nervous impulses.

Anyone who has grabbed even a young 20 cm-long electric catfish knows how powerful their electric shocks can be. In zoological gardens it has been learned that adults can prove dangerous if they are not treated with caution. Of course, such a striking fish is known to commercial fishers. Even the ancient Egyptians knew the species and often drew it in their pictorial displays.

The African UPSIDE-DOWN CATFISHES (Mochocidae) are characterized by the comb-like paired lip barbs. The lower-jaw barb pair is also distinct in most species, while in some there is less of a comb. Catfishes normally spend most of the day on the bottom. The upside-down catfishes also do this, but in a surprising way; they adhere to objects within the water regardless of whether the objects are vertical or horizontal. Thus they assume positions which are very unusual for fishes.

Upside-down catfishes

They can rest with the head up or down; they may be at an angle or have the belly-side on top. This feature is especially developed in one species, the BACK-SWIMMING CONGO CATFISH *(Synodontis nigriventris)*, which typically swims with the belly up, using its pectoral and pelvic fins together with the anal fin to maintain this unusual position.

While there are very inconspicuously colored species which are usually a bluish-black-gray, there are also very colorful upside-down catfishes. These can have white and yellow-gold colors with reddish-violet shimmering dark points and spots in a fairly regular pattern. An example is *Synodontis angelicus* (see Color plate, p. 399), of which the juveniles, with their blue-violet body with many bright spots, are a colorful addition to an aquarium. These upside-down catfishes spawn on dark parts of the aquarium, laying the eggs and sperm on smooth stones or on the glass of the tank. These otherwise peaceful fish become quite agitated during courtship and the partners chase each other and drive off other inhabitants of the tank, which is not their normal behavior.

Considering the beauty and striking behavior of all upside-down catfishes, it is no surprise that they have been brought into the aquarium, even though this is done only with the small species. Some species emit audible sounds, probably by moving the hard fin rays within their compartments. Whether the sounds are accidental or have a biological meaning, possibly to scare off enemies, has not yet been clarified. The large species are eaten by man in their African habitat, although not in the same quantities as other fishes because of their strong thorns. The ancient Egyptians knew the upside-down catfishes and depicted them in their drawings. One of the large species, *Synodontis schall,* is the main food of the African sea eagle (see Vol. VII), according to Wolfgang Fischer.

Pinter has described spawning in the back-swimming Congo catfish: "Reproduction occurs in water at a hardness rating of six (German scale) and temperature of 24–27° C. Small piles are left on the dark parts of the aquarium edges. I did not actually observe copulation, unfortunately. The light yellowish eggs have a diameter of 2.5–3 mm. They do not adhere to the glass very strongly and a few fall to the floor during the course of development. The young hatch after seven to eight days, and live four more days from the yolk sac, after which they feed on crustacean larvae. During the first weeks of life they swim normally, that is, with the belly down. Swimming on the back begins at an age of seven to ten weeks."

No other upside-down catfish has been as well studied as the back-swimming Congo catfish, although breeding reports have been written by fish hobbyists. These catfishes often are found in schools; however, since they do not become active until dusk, they are practically impossible to observe. They greatly prefer anything which conceals them, and they seek such recesses as soon as they are disturbed. In spite of their plump shape, all upside-down catfishes are skillful swimmers and can attain considerable speeds for short distances.

The DORADID ARMORED CATFISHES (Doradidae; L up to about 20 cm) include species which have bony plates along the sides of the body, on which there usually are small teeth pointed toward the rear. The first ray of both the dorsal fin and the pectoral fin are very powerful, and both rays can have a thorn-like process. It has been reported that if grabbed by someone, one of these fishes can inflict injuries by the combined effect of fins and body teeth. There are two or three pairs of barbs, very long in some species.

Most doradid armored catfishes are very quiet, mysterious fishes. The better-known species include *Amblydoras hancocki* and the TALKING CATFISH *(Acanthodoras spinosissimus)*, both from tropical South America. Their dark basic coloration varies from brown to blue-black, which can be covered by an irregular pattern of dark spots. The barbs have alternating light and dark rings. The thorny process along the body is generally lighter than the body itself. *Amblydoras* and *Acanthodoras* emit clear, grating noises. The pectoral fin rays make movements within their housing which probably produce these sounds. It is not known whether these sounds have specific purposes in the wild. The tones are generally heard in aquariums when the catfishes are lifted out with a net, and the motion of the pectoral fins is quite obvious. Males and females are hardly distinguishable on the basis of coloration but vary in size, males being visibly smaller. In courtship the male rides on the female's head between the eyes; the manner in which copulation ensues is not known. Like all catfishes, these species prefer to spend their time in soft ground during the day, burrowing so deeply that only their high-set eyes stick out of the ground.

One of the most beautiful doradids is *Doras costatus;* its dark brown to brownish-black body has three yellow to yellowish-white longitudinal bands. It inhabits the upper Amazon basin. *Doras* contains the famed "singing catfishes of the Ukayali," of which the Polish traveler Arkady Fiedler has written: "Toward evening one begins to hear wonderful sounds in the water as if bells were ringing. Those are the Ukayali catfishes 'singing.' I heard them for the first time near our hut during a brilliant sunset and as a thunderstorm was approaching. At first a rare quietness reigned on the river and elsewhere about us; then suddenly the sound of a bell was clearly heard from the water; another bell sound followed at another spot, and these were followed by more. This lyrical sound was so impressive that I had a heightened feeling which one sometimes gets in the concert hall." It is uncertain whether the sound he heard was from the catfishes or from frogs. However, this single report on the "singing catfishes" is not as improbable as it appears. Fishes commonly utter sounds; it has only recently been

Doradid armored catfishes

shown, with finer apparatus, that these sounds have a rather wide range. The swim bladder always plays an important role as a resonating base for tone production, while the tones are usually produced in specialized fin rays.

Bunocephalid catfishes

Catfishes which are compressed from top to bottom, with the forebody having an almost disc-shaped roundedness and the long, thin tail shaft like a pan handle, belong to the family Bunocephalidae. The German name for this group translates to FRYING PAN CATFISHES. There is just one row of plates along the sides of the body; the pectoral fin has a jagged first fin ray. There are one long and two short pairs of barbs. Distribution is in tropical and subtropical South America.

The better-known members of this family include *Bunocephalus bicolor* (L about 20 cm), which occurs in the Amazon basin to La Plata. This bottom-dwelling fish lies buried in the sand during the day with just its high-set eyes protruding. It also likes sites where it can conceal itself, leaving them only at dusk. Since its eyes are weak, the barbs play an important role in finding food. Not much is known about the natural life of these catfish, since in aquariums they usually withdraw when bright lights are present. They feed chiefly on bottom-dwelling worms and insects but also take refuse. Interestingly, they swim just above the bottom, and rather clumsily at that, their barbs touching the ground in search of something edible. If something is found, the fish can turn quite suddenly and bore into the sand head first to get at it.

Plotosid sea catfishes

Among the few marine catfishes are the PLOTOSID SEA CATFISHES (Plotosidae), found only rarely in fresh water. These fishes have an eel-shaped elongate body; the dorsal fin extends from the front of the back over the entire body, and the anal fin is not much shorter. The dorsal, caudal, and anal fins form a continuous seam with a pointed end. These fishes have eight barbs. Plotosid sea catfishes are often found in very large schools; Klausewitz has described the schools as looking like a sea cucumber moving through the water. They often move in cuneate (wedge-shaped) schools. Most of those over 20 cm long are poisonous and can be extremely dangerous; they have poison glands on a few spines located on the dorsal and pectoral fins.

A few years ago, the sea catfish *Plotosus lineatus* (L about 30 cm) was introduced into aquariums. This is one species which thrives in captivity. As Rössel has shown in his studies, it is light-sensitive. It is smaller than *Plotosus canius,* with a L of over 75 cm. *P. canius* also penetrates river mouths as far as the lower courses of rivers; *P. lineatus* rarely does this. Apparently *P. canius* is more tolerant of fresh water.

Among all the catfishes no group gives rise to more diverse opinion about their flavor than do the plotosid sea catfishes. In South Africa they are considered a delicacy, while people in the Indo-Pacific region do not eat them at all. Plotosids have a curious growth behind the anus, a branching organ, the function of which is unknown. During the spawning period the fishes are found in rocky reef and cliff areas, where they lay their eggs in the crevices.

The South American family with the greatest number of genera is the PIMELODID CATFISH group (Pimelodidae). They can make very fast motions with their long antennae, especially when feeding. Some species are found in open water. The adipose fin is often large. Some of the most beautiful and striking members of this family are those in the genera *Sorubim* and *Pseudoplatystoma*.

Pseudoplatystoma fasciatum (see Color plate, p. 387), like its relatives, has a long head which is flattened in front and looks like a duck's bill. The three long pairs of barbs are generally pointed to the front and are quite stiff. The basic coloration of the long, slender body, which is laterally only slightly compressed, is olive-brown with lighter portions. This is covered by saddle-shaped black to dark brown bands and a few spots, which sometimes are arranged in rows. These strikingly colored catfish are very mysterious in their life habits, preferring to stay between plants, among trees which have fallen in the water, or under shore projections. In the aquarium they resemble the pike behaviorally. In their home distribution they are caught with bow and arrow or with hooks, and are very popular eating.

Sorubim lima is a still more beautiful pimelodid catfish (see Color plate, p. 400). Since the upper jaw is longer than the lower jaw, the mouth appears to be on the lower part of the head.

A blind species, *Typhlobagrus kronei*, is found in the Caverna des Areias, a cave near Saõ Paulo. Its body has a fleshy-whitish hue and in adults the eyes are absent. No reports have been published on the life of this fish.

One of the larger pimelodids is *Rhamdia sebae* (L up to 40 cm). Its long first barb pair extends all the way to the caudal fin when it is laid back. The dorsal fin is short relative to the adipose fin, and the caudal fin is forked. A narrow black longitudinal band extends from the eye to the caudal fin; the body has a reddish-brown basic coloration. The species is another dusk-and-nocturnal animal which lies almost motionless during the day on the bottom or in some recess. It is allegedly a predator, searching not only in open water, but also burrowing through the floor for food.

One of the smallest species of the family is *Microglanis parahybae* (L at most 7 cm), distributed from southeastern Brazil to Argentina. Its short barbs just reach the rear of the head. The faintly reddish-brown to yellowish basic coloration is covered by irregular brownish to black diagonal stripes which look a little like a lobster's rings. *Microglanis* is typically found between plant roots, but can be active during the day, especially when highly desirable food is available.

The PARASITIC CATFISHES (Trichomycteridae), widely distributed in South America, are quite slender and sometimes so thin that they can

Pimelodid catfishes

Parasitic catfishes

be considered to be worm-shaped or even fiber-shaped. Most species have a dorsal fin consisting of a single, rather soft ray. In addition to nonparasitic species which are found in open water, the family also has the only true parasites in the entire fish group; they feed on the blood of the host animal. They are known to parasitize mammals—man included—when they have the opportunity. Generally the parasitic species inhabit the gill cavities of larger fish species; they have spines aimed toward the rear of the body which aid them in holding on to the host. They always swim against the current so that they get into the water which is squeezed through the gills in the large fish which they parasitize.

Not surprisingly, little is known about the life of the parasitic catfishes. They swim in open water but are sometimes found in soft ground as well. The barbs are short. Most parasitic catfishes do not exceed a L of 6 cm. According to South Americans, members of the genus *Vandellia* are called "candiru"; in English it is known as the URETHRA FISH. Other parasitic catfishes have no common names but use names of other fishes. In this one case, however, there is a good, if not happy, reason for using a common name. The candirus penetrate the urethras of bathers who urinate in the water. Since the fishes have gill cover spines pointed to the rear, it is already too late once their presence is noticed since they cannot simply be pulled out. This has repeatedly caused death. If the afflicted individual does not want to have blood poisoning he must undergo an amputation. For this reason some Indian tribes put some covering about the lower torso when they go swimming in these waters, something which they do not normally do. Since the candirus have very small eyes and live in water which transmits light down to a depth of 1 m, they apparently do not find the urethras by vision but by scent.

Callichthyid armored catfishes

The open-water forms feed on aquatic organisms such as insect larvae, worms, insects and small crustaceans. Since the parasitic catfishes feed on the gill elements and probably the blood of the host as well, they can quite correctly be designated as parasites, living at the expense of another animal without killing that animal.

The CALLICHTHYID ARMORED CATFISHES (Callichthyidae) are, to most aquarium hobbyists, the best-known members of the order. L is 3.5–25 cm. The upper body has two rows of V-shaped bony plates of armor; the belly side lacks this covering. There are four to six short barbs which in most species are laid back into depressions in the head while the fish is swimming, reducing the resistance. The downward-pointed barbs are not the lower jaw barbs but those of the upper jaw; the two sets of barbs cross at the corners of the mouth. In larger species the barbs are supported by cartilaginous masses, while smaller callichthyids have a turgor mechanism. The armor in some species is extremely strong and up to 3 mm thick. Distribution is primarily across central

South America in the Amazon basin, but also in the northern and southern parts of the subcontinent.

The armored catfishes, poor swimmers, usually "hop" about on the bottom, where they are usually in large schools; they also feed off the bottom. The barbs are put to good use at this time: in most of their native waters light penetrates very poorly, so vision is of little use in finding food. The barbs, which, as in other catfishes, have taste organs on them, provide a means of finding food in this murky environment. Smaller species in the genera *Corydoras* (see Color plate, p. 411) and *Brochis* (see Color plate, p. 411) can even use the barbs like hands in picking up food; the fish does not have to change its body orientation while doing this.

Armored catfishes are eaten occasionally but are not important food sources. They are caught in wire baskets, and roasted over the fire like chestnuts. Since the armor is very hard, the fishes have to be cracked open, again like chestnuts!

Individual callichthyid armored catfishes are found in diverse environments. Many are found in quiet inlets, along shores and behind sandbanks, while others occur in muddy, oxygen-poor water. Since fishes in such muddy waters do not get enough oxygen from the water, they have evolved a supplementary respiratory organ in the intestinal tract. Rising to the surface, they swallow atmospheric air, which is then expelled through the anus.

Until recently the reproduction of this family was incorrectly described. Since the first breedings in captivity, the claim was made for decades that the female adhered to the genital region or the genital papillae of the male, sucking out the sperm and swallowing it, producing a different sort of "internal fertilization." There were already anatomical reasons for rejecting such an assertion, but the belief was not corrected until Knaak did so in 1955. He observed and photographed the armored catfishes during copulation and found that females do not gain this suction hold on the males. The male lies in front of the female, surrounds the now-prodding female with its powerful pectoral fin rays under the barbs and thus holds fast to her body. Then the male simply releases its sperm into the water while the female releases one or more eggs and gathers them. At that point the male releases its hold, and the female swims through the cloud of sperm carrying the eggs in a pocket formed by the pectoral fins. So internal fertilization does not take place in this catfish. Most armored catfishes prepare for egg-laying by cleaning a site on the ground and removing all dirt from the site. This cannot truly be called cleaning behavior, because later the fishes often put the eggs on surfaces which have not been cleaned.

True incubation of the eggs is found in the genera *Callichthys* and *Hoplosternum* (see Color plate, p. 411); males usually assume this chore.

Fig. 16-1. Spawning of callichthyid armored catfishes (from top to bottom): spawning is initiated by great activity among the fishes. Females are often approached by several stimulated males. ..."Sniffing" and touching by the male. ...The male succeeds in standing perpendicular to the female, holding her barbs firmly with his pectoral fins. Several eggs are released into a pocket formed by the pectoral fins of the female, and they are fertilized in the pocket. ...The female seeks a site to lay eggs; it superficially cleans the site.

The male makes a nest at the water surface with foamy air bubbles and plant material. The female often participates in nest-building. However, when spawning is complete the male cares for the developing eggs. Spawning proceeds as in the small catfishes of genus *Corydoras* except that the eggs are not stuck to the nest but are brought to it by the female. The male then receives the nest and protects it from fishes and water-insect larvae. These nests have been observed in nature, but the process of spawning has only been observed in aquariums. Since caring for the developing young is paternal, the brood does not stay together very long after hatching; the young even separate within the nest.

For some, if not all, species, the freshly hatched young have a characteristic behavior pattern: the 5 mm-long young disappear for the first few days into the ground and are thus protected from enemies and from drifting away with the current. They reappear after a few days and move in a group over the nutrient-rich mud or even in the mud. Their first food consists of all sorts of refuse and small organisms. The barbs are already present at this stage of growth. Aquarium observations have shown that most armored catfishes maintain a distinct spawning period even after many generations; it extends from November or December to April or May.

Some of the largest callichthyid armored catfishes belong to the genera *Callichthys* and *Hoplosternum* (L about 25 cm). In appearance the two genera are quite similar; both have long barbs, somewhat laterally compressed bodies, and similar coloration. The key to distinguishing them is the caudal fin. *Callichthys* has a rounded fin while in *Hoplosternum* the caudal fin is slightly indented with two tips, or is straight. *Callichthys callichthys* (see Color plate, p. 411) is the best-known member of its genus. It is distributed widely in South America, its coloration varying in different regions. *Callichthys* requires a large aquarium when kept in captivity; those from southern parts of South America can be kept in European outdoor ponds during the summer months. The two well-known *Hoplosternum* species, the HOPLO CATFISH (*H. thoracatum*; see Color plate, p. 411) and *H. littorale,* have the same basic coloration; metallic hues often make them appear greenish-blue. Both *Callichthys* and *Hoplosternum* have elongated foreheads which give the head a flattened appearance. The barbs are usually aimed toward the front. They are laid back on the body while the fishes are swimming. The barbs are supported by cartilage and have rings.

Small ARMORED CATFISHES often kept in aquariums belong to the genera *Corydoras* and *Brochis*. L is generally about 8 cm, with the exception of the MICRO CATFISH (*Corydoras hastatus;* see Color plate, p. 411) and *Corydoras pygmaeus.* There are four to six short barbs. The third pair of barbs forms from the lower lip beneath the mouth opening, which in these species is located on the inferior side of the body. A rather uniformly arching line extends from the upper lip along the back to the dorsal fin and from there runs in a straight line to the caudal fin. It is sometimes interrupted by the adipose fin. The pectoral fins can be spread like wings. The dorsal and pectoral fins each have a pointed, hardened ray. When upright and stiffened, these are powerful defensive weapons.

The *Corydoras* species can be grouped into three major classes on the basis of coloration; this has nothing to do with their systematic classification, however. The first group has an olive-green to brownish basic color with irregular, often cloud-like spots of various sizes. Since the members of this color group are widely distributed, their appearance differs considerably. Three of the best-known examples of the group are *Corydoras palleatus* (see Color plate, p. 411), the BRONZE CATFISH *(Corydoras aeneus)*, and *Corydoras schultzei.*

The members of the second color group have a whitish-gray basic coloration, covered by black spots arranged in irregular patterns or in lines. The dorsal fin or head often have black bands. Examples of the group are the GUIANA CATFISH *(Corydoras melanistius;* see Color plate, p. 411), the LEOPARD CORYDORAS *(Corydoras julii),* and *Corydoras punctatus.* Since the resemblance in these corydoras catfishes often goes beyond external appearance, their systematic relationship has recently been more closely investigated. It is quite possible that new species will be added to the list or old ones discarded in the near future. This group contains one of the most beautiful corydoras of all, *Corydoras schwartzi.* Two color types of the species are known, both apparently from the Rio Negro. In the more beautiful one the front edge of the dorsal fin is bright red, and reddish tones appear in the other fins as well. The second color type does not have these reddish hues. Whether the red color is maintained in aquarium conditions remains to be seen; with tropical fishes it has been repeatedly found that certain colors, which probably are dependent upon the nature of the diet, often disappear in time or are not found in succeeding generations.

The third corydoras color group has a yellowish to dirty orange basic coloration with various black longitudinal stripes. Thus the BOWLINE CATFISH *(Corydoras arcuatus;* see Color plate, p. 411) has a black longitudinal stripe extending from the mouth along the back to the lower part of the tail base. Another form of the black stripe is found in *Corydoras metae.* Here a short band extends from the dorsal fin to the caudal fin. The systematic relationships have not been clarified within this color group.

Exceptional species of the corydoras group are the dwarfs, *Corydoras pygmaeus,* and the MICRO CAT *(Corydoras hastatus).* They spend most of their time in open water, and when seen briefly they look more like tetras or some other open-water species than catfishes. These schooling fishes move rapidly, and while swimming lay their barbs back. The

▷
The sea catfish *(Plotosus lineatus)* of the Indian and Pacific Oceans is a schooling fish. In spite of its small size (L 25–30 cm), its poisonous fin rays make it one of the dangerous fishes of the tropical coral reefs.

▷▷
The skull of the "crucifix fish" *(Arius proops)* has the shape of a crucifix on its lower side. The upper side, shown here, resembles a praying bishop. In western India and South America this impression is emphasized by painting the skull. It is sold as a holy amulet.

▷▷▷
The pimelodid catfish *Pseudoplatystoma fasciatum* has the same flattened skull of the "crucifix fish".

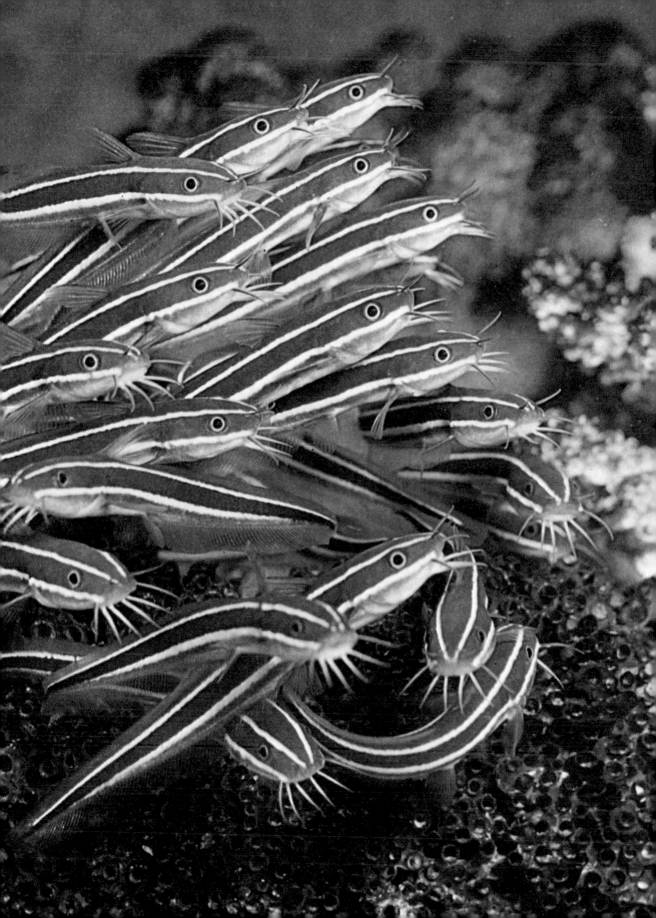

the incoming light, the smaller the pupil, and the more this structure can distend over the eye.

Reproduction in loricariids has been closely studied in just two genera, the WHIPTAIL LORICARIA *(Loricaria)* and *Ancistrus. Otocinclus* has also been bred in captivity, but only accidentally, and the process was not closely observed. Copulation and care of the eggs is perhaps similar in all loricariids, since the reproductive sequences in those species already studied are in principle the same. Generally the fishes spawn in a cavity, such as in rocks, hollow logs or other suitable recesses. Bamboo shoots or glass tubes serve this purpose in the aquarium. The recesses which loricariids choose are not very roomy, making one wonder where they find room in such a narrow enclosure. After spawning, the male cares for the eggs until the young disperse and live independently. Even before they disperse, the young do not have a close relation to the father, remaining only among themselves.

Fig. 16-2. Spawning in loricariid armored catfishes.

The largest loricariids are the *Plecostomus* species (see Color plate, p. 412); L in some individuals exceeds 1 m. Most plecostomids inhabit fast-flowing water, and they migrate into the clear Andes streams at an altitude of about 4000 m. They use the sucking mouth for propulsion while migrating, gaining a sucking hold in very rapidly flowing spots, such as in waterfalls, and pushing forward, centimeter by centimeter, with short jerky movements, all the while maintaining the suction hold. By this means they can get through places where the water is roaring past with an extremely powerful current. They prefer concealed sites when resting near the shore or behind stones. They also adhere to tree trunks which have fallen into the water. Such a loricariid will hold on to the stone even when lifted out of the water.

Fig. 16-3. A young loricariid armored catfish.

Almost all loricariids have a brownish basic coloration with some gray or reddish colors. The body has irregular black or dark brown spots arranged in different ways, resulting in rather beautiful patterns. The dorsal fin is higher than the body itself and forms an impressive sail when erect.

The dorsal fin of *Pterygoblichthys* is still higher. Beside the large pectoral fins, which are horizontal and protrude like wings, the large caudal fin is also striking. It rests on the powerful tail shaft and is tapered on the lower side. Nothing is known about reproduction in this genus. The local populace eat petrygoblichthyids.

The loricariids in genus *Ancistrus* are of similar shape, with flattened belly and broadened forebody; in aquariums they are often known under the generic name *Xenocara.* Their German name translates to ANTENNA ARMORED CATFISHES, and is derived from the antenna-like protruding structures which males develop during the spawning season and which older specimens carry about continuously in varying numbers on the head. Some species, such as *Ancistrus bufonia*, even have branching antennae which look like antlers of deer. These antennae look very menacing but actually they are soft and cannot inflict injury. Their natural function is unknown.

Ancistrus dolichopterus (see Color plate, p. 412; L is about 15 cm) is known in German as the BLUE ANTENNA CATFISH since during the spawning period males have a bluish coloration; otherwise they are a rather inconspicuous dirty gray-brown. During the spawning period the males have a blue-black color with greatly contrasting white spots. The species has been bred repeatedly in captivity for many years. When spawning, the partners seek narrow recesses, often laying the eggs on the ceiling of the enclosure. The male guards the spawn, removes unfertilized eggs, and supplies the eggs with oxygen by fanning the water with its fins. After hatching occurs, the care behavior of the male ceases quickly; the young become very active. When these catfish rest quietly on the floor they support the forebody with the large wing-like pectoral fins so that the gill openings are lifted off the ground. They do not like muddy floors and if there is too much mud in the tank they find objects in the water to which they adhere.

The LORICARIA *(Loricaria)* are much more delicate, even seeming fragile. The forebody is also somewhat widened but not to the same extent as in relatives of this genus. Some loricaria are as much as 30 cm long, but many species remain smaller. The tail shaft is the most striking structure, being extremely elongated, and tapering practically to a thin fiber. The hard, bony plates which overlap like roof tiles make the tail shaft quite resistant to injury. Loricaria are distributed throughout South America from Argentina northward into Central America, from lowland waters to mountains. The group has clear sexual dimorphism. At the spawning period the cheeks and pectoral fins of males develop a short, stubby beard like a brush, which disappears when spawning is over. Males are also more uniformly slender than females, which are thicker at the belly.

The systematic position of the individual loricaria species is still uncertain. *Loricaria parva* (see Color plate, p. 412) is often kept in aquariums and has been bred in captivity. Its caudal fin, with elongated upper and lower rays, often quite long, has an impressive appearance. The species reproduces in the same manner as the antennae-bearing armored catfishes, although Froesch observed that the former also lay their eggs on plants. Depending on the species, the eggs are orange, reddish or greenish. Loricaria often burrow into the ground, shaking the fins and body laterally and thus disappearing in the sand. The only parts of the body remaining visible are the eyes, set high on top of the head. Loricaria are often found in schools.

Some of the most unusual catfishes are in the genus *Farlowella* (see Color plate, p. 412). The front of the head has a long, snout-like protuberance, the function of which is not known. Often fishes with such structures inhabit soft ground, through which they burrow with the protruding snouts. *Farlowella* catfishes do not live in muddy habitats, but are found where there is soft sand. Apparently they are very sensitive to the oxygen content of the water. In nature they have varying life habits. They feed chiefly on algae and other plants, but also take bottom-dwelling organisms. They have spoon-shaped, two-pointed teeth which are used for scraping. These shy fishes, which have not been seen much in captivity, lead a life which is still largely unknown to us.

The *Otocinclus* members (see Color plate, p. 412), with a L between 3.5 and 8 cm, are among the smallest loricariids. Their basic coloration is generally a dirty silver-white with a brownish coating. Some species also have dark to black longitudinal bands or spots. Since they are very small, these are popular aquarium fishes. They are useful ones since they scrape their food, algae and other plants, off the walls or ornaments within the aquarium. Generally *Otocinclus* loricariids are schooling species, but unlike more typical schooling fishes, they are quiet, and do not have the great activity more commonly associated with schools of fishes.

▷
Eurasian catfishes:
1. *Kryptopterus bicirrhis;*
Labyrinthic catfishes:
2. Stinging catfish
(Heteropneustes fossilis);
3. *Chaca chaca;* Bagrid
catfishes: 4. *Leiocassis
siamensis;* 5. The striped
dwarf catfish *(Mystus
vittatus).*

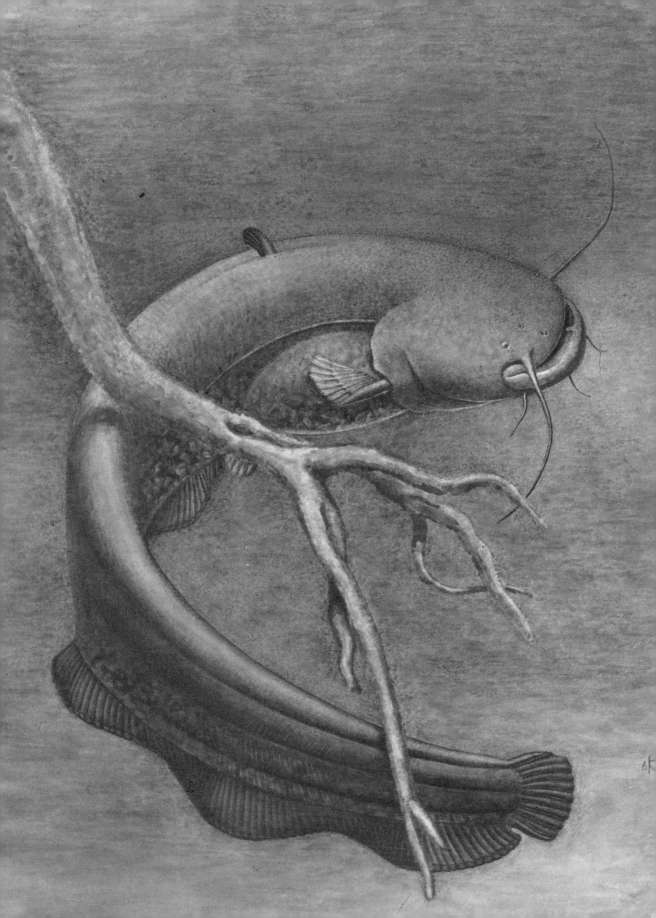

17 Trout-perches, Toadfishes, Clingfishes and Anglerfishes

The five orders, trout-perches, toadfishes, clingfishes, anglerfishes and cod (see Chapter 18), share a few striking developments of the jaw muscles and infraorbital bones (i.e., those beneath the eyes); furthermore the caudal portion of the skeleton has just two or three hyperalia, the fan-shaped bones behind the last caudal vertebra, which are sometimes very prominent structures in these fishes. A few ichthyologists therefore class all five orders as primitive spiny-finned fishes (Paracanthopterygii). Like the other spiny-finned fishes (see Vol. V), they are characterized by spined rays of the fins.

Order: trout-perch,
by W. Ladiges

In all trout-perches (Percopsiformes) the pelvic fins are behind the pectoral fins, and some species have an adipose fin. There are three suborders: 1. Cavefishes; 2. Pirate perches; and 3. Trout-perches.

Suborder: cavefishes

The CAVEFISHES (Amblyopsoidei), with the single family Amblyopsidae, have the anus in the throat region and more or less degenerate eyes. This eye development, often associated with a greater development of the organs of touch, characterizes the cavefishes. Another characteristic feature is the uniform body coloration, due sometimes to a lack of skin pigmentation, while others share a dark hue. As the American ichthyologist Eigenmann reported in 1917, those fish groups in which the eyes are less significant than the organs of touch tend to occupy niches in semi-darkness (like catfishes or eels); and in the most extreme form of this development they inhabit caves. In America (for example, in Cuba, Yucatan, the Mississippi basin and Brazil) many caves have been found with resident fishes. These are not all related forms, but fishes which in most cases are related to above-ground species in the immediate vicinity. The same is true for the cave-dwelling fishes in Africa and Asia. Thus it is possible for fishes living near caves to occupy these habitats relatively successfully, to the extent that the necessary behavioral and anatomical modifications—all under genetic control—are made to adapt to the lack of light, lower temperature, etc.

◁

The European wels
(Silurus glanis), a member
of the Eurasian catfish
family.

Thus, in a work on Yucatan caves, C. L. Hubbs writes that in addition to true blind cavefishes (*Typhlias pearsei* and *Pluto infernalis*) there were also fishes which deviated little or not at all externally from above-ground species, including a few subspecies of the cichlid *Cichlasoma urophthalmus* (see Vol. V). The various subspecies of the pimelodid catfish *Rhamdia guatemalensis* show every phase from completely developed eyes to the complete absence of eyes. It is not surprising that a catfish, which is naturally shy of bright light, should assume a cave-dwelling existence, but to find cave dwellers among the light-loving group of cichlids is indeed astonishing. According to Hubbs, the cave-dwelling cichlids have slightly larger eyes than their above-ground relatives. In the chapter on characins it was shown that the blind characin *Anoptichthys jordani,* a true cave fish, developed from *Astyanax fasciatus mexicanus,* which has large eyes but also many tactile sensory bodies. All this contradicts Eigenmann's statements and indicates that his definition of cavefishes is too narrow. Due to the superficial similarities between various groups, we have pictured a number of different cavefishes together on p. 417.

Many other cavefishes developed from groups which in general do not show indications of a development toward living in the dark, including the blind fishes of the suborder Amblyopsoidei. There are four species of cavefishes in limestone caves in Kentucky, Missouri, Arkansas, southern Illinois, southern Indiana and Tennessee, belonging to the genera *Amblyopsis, Typhlichthys* and *Chologaster;* these fully or partially blind fishes with degenerate eyes externally resemble carp (see Chapter 19). They are closely related to the non-blind species *Chologaster cornutus,* from marshes in Georgia, South Carolina and the Virginias, a species which shows the typical development line to subterranean life since it has the well-developed tactile organs on its head and body.

Habitats of the cavefishes in the Mississippi Valley are primarily determined by completely or partially subterranean rivers which have worn through the soft masses of stone. Individual subterranean species are separated from each other by obstacles which no fish could penetrate, so presumably each species has undergone its own development. All cavefishes in Amblyopsidae apparently are oviparous. Eigenmann first described raising the young in the MAMMOTH CAVE BLINDFISH (*Amblyopsis spelaeus;* see Color plate, p. 417). The sexually mature female's genital opening protrudes so far that the eggs are laid in the gill cavity. They are carried between the gills until the young hatch. The young remain there until they are about 6 mm long. All the cavefishes are small species, 4–5 cm long, which feed on small invertebrates, especially crustaceans.

The PIRATE PERCH (*Aphredoderus sayanus*), the only member of the pirate perch suborder (Aphredoderoidei), inhabits the warm parts of

Suborder: pirate perch

the central Mississippi basin and the coast from New York to Texas. The inconspicuous, 12 cm-long fish is olive-brown with a dark pattern; like cavefishes, it has its anus in the throat region but lacks an adipose fin. The pirate perch builds a nest and cares for the young; both parents guard the nest. It feeds on all sorts of small organisms.

Suborder: trout-perches

The TROUT-PERCHES (Percopsoidei, family Percopsidae), with two genera with one species in each, occur in North America. They have a perch-like shape and an adipose fin like that of a salmon. *Percopsis omiscomaycus* (L up to 20 cm) occurs in the Great Lakes region while the smaller *Columbia transmontana* lives in the Columbia River. In the spring these fishes from cold habitats migrate short distances from their lakes into rivers to spawn between rocks. Both trout-perches are carnivorous.

Order: toadfish, by W. Ladiges

The toadfish order (Batrachoidiformes) contains one family with four particularly interesting genera and about thirty species. The head is quite large relative to the body, and is flattened, with a large mouth opening. The body tapers considerably toward the rear. The jaw has many teeth. There are two dorsal fins: the first has just two to four fairly powerful spiny rays; the second is long and broad-rayed. The pelvic fins are in front of the large pectoral fins. There are just three pairs of gills and a very narrow gill opening.

The TOADFISHES (genera *Thalassophryne*; see Color plate, p. 418; and *Thalossotia*) inhabit the warm and tropical coasts of the Americas. They are also found in brackish water and occasionally in fresh water. There are shallow-water species as well as those which alternate between great depths and the shallow coastal area. They undertake migrations which are probably not associated with spawning. In shallow water these inactive fishes are found lying on the ground, hidden between stones and algae. Coastal people fear them greatly, because they can inflict great injury to someone wading through the water who steps on one. The first two rays of the first dorsal fin and the spines of the gill cover are hollow, with a poison vesicle at each base. Like most fish poisons, this one acts as a shock with a strong pain. Both effects subside with time. No fatalities have been reported from these injuries. Toadfishes emit a grunting, frog-like tone, and do so especially when attacked.

The MIDSHIPMEN (*Porichthys*) even whistle like their namesake. Two common North American species include the Pacific *Porichthys notatus* (L 30–40 cm) and the smaller Atlantic *Porichthys porosissimus* (L 15–20 cm), distributed from South Carolina to Argentina. Midshipmen are characterized by the many pores on the head and body; they also have luminous organs on the belly arranged in rows or spots. It has been found that males can produce a considerable amount of light during the spawning period. With strong physical or chemical stimulation one also gets a strong but short-lasting burst of light from these flashes.

The best-known species of the genus *Opsanus* is the OYSTER TOADFISH (*Opsanus tau;* see Color plate, p. 418), distributed along the Atlantic Coast from Maine to Florida on to Cuba. It is the single fish species known to have reacted to man's influence by developing the curious habit of seeking, during the spawning period, only water which has all sorts of refuse such as tin cans, old shoes, and other little recesses in which to spawn. One of the parents, usually the male, or sometimes both partners, guards the eggs; the young hatch after 20-26 days. During this period these normally aggressive fish are even more prone to attack. All toadfishes are predators which easily chew up mollusks, crabs and fishes with their strong teeth.

The CLINGFISHES (Gobiesociformes; family Gobiesocidae) have long, slender, naked bodies. The head is large and flattened. L rarely exceeds 10 cm. There is a large, often two-part sucking disc behind the base of the pectoral fins on the underside of the forebody. The widely separated pectoral fins form just a small part of the suction disc, which consists mostly of skin and muscle (unlike lumpsuckers and sea snails; see Vol. V). The upper side of the suction disc is covered by a thick, sole-like epidermis. The dorsal fin is solid, with no spiny rays; the dorsal and anal fins are well to the rear. There are fifteen genera with fifty species, most of which live between stones in tidal zones and are generally limited to temperate seas.

These small fishes feed on a wide variety of small organisms. The sucking disc gives them a very firm hold on stones even when water is quite agitated. A somewhat larger species is the NORTHERN CLINGFISH (*Caularchus maeandricus;* L about 16 cm) from the Pacific Coast of the U.S.A. It is found in pools along coastal cliffs. *Bryssetaeres pinniger* is usually found in pools along California's coast, along with sea urchins. *Chorisochismus dentex,* one of the three South African species, is among this genus's larger species. This fish willingly takes bait on a hook but adheres so strongly to the fishing line that the line breaks before the fish can be taken out of the water. When a sucker fish just 15 cm long is pressed to a large rock until it gains a hold, the fish and stone can be lifted up together. Sucker fishes can also glide smoothly along smooth rocks and stones with a distinctive movement of the sucking disc. Some species are also found in fresh water on the Galapagos Islands. One sucker fish, *Lepadogaster bimaculatus* (BL 8 cm), lives in the North Sea.

All members of the anglerfish order (Lophiiformes) have a modified single movable dorsal fin ray which carries a kind of "bait" at its end; this is the basis of the name anglerfish. The "bait" is known technically as the illicium. Other unusual modifications in this group include the pectoral fins, which enable the fishes to crawl along the ground. All anglerfishes are slow-moving, almost motionless fishes, attracting their prey by means of their natural bait, at which time they suck in

▷
Upside-down catfishes:
1. *Synodontis angelicus;*
2. *Synodontis schall;* Schilbeid catfishes: 3. *Pareutropius mandevillei;*
Labyrinthic catfishes:
4. *Clarias lazera.*

Order: clingfish, by W. Ladiges

Order: anglerfish, by W. Villwock

in six-sided compartments in a sort of honeycomb arrangement. Eventually the structure tears and the eggs float individually in the water until the young hatch. The 2 mm-long larvae swimming near the surface of the water do not resemble their parents at all; they look much more like typical fish larvae. After a four-month larval period, with growth up to a L of 10–15 cm, and a complex change in body shape, the juvenile goosefishes have the adult appearance and behavior.

In spite of their unusual appearance, goosefishes are popular as food. The head and leathery skin are removed, however, so that little remains of the "frying pan" look of the whole fish. In Europe the fishes are sold fresh or smoked as "trout sturgeon." Insulin is extracted from the islands of Langerhans in the pancreas.

The FROGFISH suborder (Antennarioidei) contains four families, usually of small fishes. L does not exceed 30 cm. The suborder contains the FROGFISHES (Antennariidae in the narrower sense) and BATFISHES (Ogcocephalidae). These primarily inhabit tropical seas. A few of them, such as Histrio histrio, look uncannily like the algal bundles floating in the Sargasso Sea towards the north of the Gulf Stream. Besides occupying dense algal growths, frogfishes are also found between coral in warm seas.

Suborder: frogfish

Fig. 17-2. A frogfish (Antennarius).

In some frogfishes, including the Atlantic species Antennarius scaber or Histrio histrio, the first fin ray is moveable and has been transformed into a natural baiting device as in goosefishes. The frogfish Antennarius moluccensis, from the South Seas, still has the "fishing line" in the form of a moveable dorsal fin ray, but the "bait" at the end is absent. It has not been established yet whether this fish can "fish" as well as its Atlantic relative does.

All frogfishes have a peculiar body shape, enabling them to resemble their surroundings to such a high degree that they blend into the algae "forests" like the Sargasso frogfish. This mimicry of the surroundings is created by many skin folds and various skin formations, supplemented by the ability to adapt to some extent in color and pattern to the background. Frogfishes which do not swim well are prevented from falling through the Sargasso seaweed or other objects by their pectoral fins, which function much like human hands in grabbing and holding objects. The reproduction and other life habits of this suborder are practically a complete mystery. Female frogfishes observed in aquariums have produced spawning tubes similar to those in goosefishes.

Among the remaining families of the suborder, the families Brachyonichthyidae and Chaunacidae are less well known while the BATFISHES (Ogcocephalidae) are more notorious; there are about thirty species of them distributed in warm and temperate seas from shallow waters to great depths. L reaches 35 cm, averaging 15–20 cm. Representative species include *Ogcocephalus nasutus, Ogcocephalus vespertilio* (see Color plate, p. 421) from the Atlantic, and *Halieutaea retifera,* one of the smallest species in the group (L 5 cm), from the Hawaii Island region.

Herald writes, "The peculiar batfishes 'roll' across the bottom like tanks, pushing themselves forward with their large, arm-like pectoral fins and small, leg-like pelvic fins." Batfishes may occasionally be seen swimming; the nature of their swimming motions indicates that they resist changing sites and do so infrequently. The "fishing line" of batfishes is short and is in a pit directly above the mouth. It can be extended and has a worm-like bait. The mouth opening is unusually small for an anglerfish. They can perhaps "afford" the luxury of a small mouth because of the special structure and situation of the fishing line and bait. The deepsea batfish has a luminous organ at the end of the bait.

Suborder: deepsea anglerfish

The third suborder of anglerfishes consists of the DEEPSEA ANGLER-FISHES (Ceratioidei) which originally contained eleven families but today has ten families since Laevoceratiidae and Aeschynichthyidae were combined to form the diceratiid family Diceratiidae. Deepsea anglerfishes are characterized by the absence of pectoral fins and the fact that only females have a fishing organ. While goosefishes and frogfishes live primarily in shallow water and may migrate into deeper water only for spawning, the approximately 120 species of deepsea anglerfishes are found at substantial oceanic depths varying from 300 to 4000 m. Any typical bait on the end of a fishing line would no longer be recognizeable at these depths, so deepsea anglerfishes have a luminous organ at the end of the line. It has not yet been established how they produce light in the bait structure, but it is probably due to luminous bacteria.

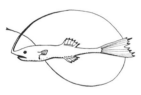

Fig. 17-3. The deepsea anglerfish *(Gigantactis).*

Some deepsea anglerfishes, including the DEEPSEA ANGLER *(Linophryne arborifer;* see Color plate, p. 421), have a luminous structure with tree-like branches hanging from its chin. In this case, too, the mechanism producing light has not been determined. Most deepsea anglers have high-set backs which when seen in sunlight make an astonishing impression. Most of them are small; only the females in a few species in the genus *Ceratias* reach a L of 1 m. Sexual dimorphism is very highly developed in deepsea anglerfishes; males and females of the same species differ in size and shape.

Fig. 17-4. A deepsea anglerfish *(Caulophryne).*

The ceratiid anglers (Ceratiidae), photocorynid anglers (Photocorynidae) and linophrynid anglers (Linophrynidae) are species with very small males which parasitize the females and stay with them (see Color plate, p. 421). The males do have their own gill respiratory system and the necessary vessels to supply their organs with oxygen, but their food is obtained from the bloodstream of their female hosts, with certain blood vessels in the males' heads in a dependent relationship with vessels in the females. Close examination of the dwarf males shows that several systems are more or less degenerated in them, particularly the teeth and the intestinal tract. They also lack the fishing line, as do all male deepsea anglerfishes.

The biological significance of parasitic males is related to the method of propagation of the species. Despite numerous signaling systems, including the luminous organs, it remains difficult for the different sexes to find each other in the deep sea, where all sunlight is absent. This dangerous disadvantage for the species is offset by having just one partner seek food while the other parasitizes the "breadwinner." The differences in size are based on figures supplied by Nikolski. He describes a female *Ceratias hollbolli* (see Color plate, p. 421) 103 cm long with two males 8.5 and 8.8 cm long, respectively, parasitizing her. Three males, each 1.8 cm long, were attached to the belly of one 7 cm-long *Edriolychnus schmidti*. In the other families, Melanocetidae, Himantolophidae, and Oneirodidae, the males are also smaller than the females, but they feed independently and do not cling to the females.

The female deepsea anglerfishes are very active predators which feed on large organisms. Their jaws have powerful teeth and their stomachs are so greatly distensible that they can swallow prey which is larger than they are. The stomach of an 8 cm-long JOHNSON'S BLACK ANGLERFISH (*Melanocetus johnsoni*; see Color plate, p. 421) contained a lantern fish (*Lampanyctus crocodilus*) nearly twice the size of the angler. The non-parasitic males are open-sea swimmers and feed on small organisms. Marshall reports on the anglerfishes' development:

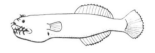

Fig. 17-5. A deepsea anglerfish *(Neoceratias)*.

"It is quite possible that maturity and releasing of eggs takes place in the great depths. When that occurs, the tiny eggs apparently float toward the top while developing, because they are quite prevalent at the surface. The young initially develop there, feeding on plankton such as copepods. The sexes can be differentiated even in the young larvae, since the females already have the beginnings of illicia on their heads. When the larvae are 3 mm long the fins are already well developed and the eyes appear. Deepsea anglerfish larvae are characterized by their inflated appearance, which is due to the development of a gelatinous tissue under the skin. Since the larvae live in the warmer, less viscous parts of the ocean, and have no swim bladder, the enlargement of the body from this tissue is perhaps a buoyancy mechanism. As soon as the larvae are 8 mm long the body of the male becomes more slender and the female more spherical. These changes precede metamorphosis."

Deepsea anglerfishes primarily inhabit warmer parts of the Atlantic, Pacific and Indian Oceans; their numbers decrease greatly in the northern and southern temperate latitudes. Only three species have been found in the North Atlantic: the ceratiid anglers *Ceratias hollbolli*, *Cryptosaras conesi* and *Himantolophus groenlandicus*.

18   Codfishes

Order: codfishes,
by H.-J. Messtorff

Distinguishing
characteristics

The members of the following fish order live almost exclusively in salt water. The order includes the well-known common codfish which gives the order its name, CODFISHES (Gadiformes).

The more primitive cod species are characterized by a uniform unpaired fin seam which, as in eels (see Chapter 9), combines the dorsal, caudal, and anal fins. In higher cod species the caudal fin is clearly distinct, but the long fin seam is still maintained by a long uniform dorsal fin and anal fin. Some groups have other structural peculiarities of the fin seam: the dorsal or anal fins may have indentations or even be split into at most three dorsal and two anal fins. The position of the paired fins is also an important characteristic: pelvic fins are usually present but in some species are not situated in the normal position. They are always well to the front and are almost always clearly in front of the pectoral fins. Codfishes are also called soft-finned fishes because as a rule their fins are composed entirely of soft rays. A hardened or spiny ray in the first dorsal fin is found only in the grenadier fish. The swim bladder lacks an air duct (the ductus pneumaticus) to the intestinal tract. There are five suborders: 1. Eel cods; 2. Codfishes; 3. Cusk eels; 4. Viviparous blennies; and 5. Grenadiers; there are ten families with at least 150 genera and over 200 species.

The oldest fossil finds of codfishes are from Lower Tertiary levels (the Eocene and Oligocene, about 25–55 million years ago); presumably this fish group began developing into the present-day forms toward the end of the Mesozoic era in the Cretaceous period (120 million years ago). Codfishes are believed to have developed in the Tethys Sea, which was a tropical central sea uniting all oceans in an east-west direction from the earliest period of the earth's history into the Lower Tertiary.

Today, codfishes are distributed worldwide. Their richest period of development and diversification occurred after the Tethys Sea was united with the northern seas in the Arctic region; thus cod underwent a second period of growth in the northern seas, and today most species and the largest populations of cod are found in the northern hemisphere, primarily in the northern Atlantic. Penetration into the Pacific Ocean presumably occurred later (in the Pliocene, 2–12 million years ago) from the Arctic Sea when the land bridge between Asia and North America was broken by the Bering Strait. Thus, the relatively young northern Pacific cod species differ little from the Arctic forms.

The little-known EEL CODS (suborder Muraenolepioidei; the single family is Muraenolepidae and the only genus *Muraenolepis*) form the smallest and simplest codfish group. The three known species live exclusively in subantarctic and antarctic seas. Their name is derived from their eel-shaped bodies and eel-like scales. With the exception of a single isolated forefin ray, the dorsal fin is undivided and forms a uniform seam with the caudal and anal fins. Hardly anything is known about the life habits of this fish.

Suborder: eel cod

Fig. 18-1. Eel cod.

The largest, most diverse and most important cod group is the CODFISHES *(Gadoidei).* There are four families with different distribution and life habits: 1. Deepsea codfishes; 2. Bregmacerotid codfishes; 3. Codfishes (in the narrower sense); and 4. Hake.

Suborder: cod (gadoids)

The DEEPSEA CODFISHES (Moridae), with about seventeen genera and over seventy species, live primarily at depths of over 500 m. The great uniformity of conditions throughout the world at those depths has enabled the family to be distributed worldwide while maintaining a low degree of development. Deepsea codfishes are found in all oceans in both hemispheres. Basically they resemble eel codfishes in body shape, but the unpaired fins are better developed. The dorsal fin is clearly divided into two parts while the caudal fin is indented and the anal fin is in one or two parts.

Family: deep-sea codfish

The great number of genera and species of deepsea codfishes is not evidence for great overall numbers of the family; since their habitat is generally outside the commercially fished zones, there are no accurate population estimates available. Deepsea codfishes are infrequently caught. Practically nothing is known about their life habits. Some species attain a L of 70 cm, but apparently most are medium-sized, not exceeding 50 cm in length. One of the larger species is the northern Atlantic BLUE HAKE *(Antimora rostrata),* which is occasionally caught together with rosefishes (see Vol. V). The species has an unusual violet coloration with striking blue-black fin edges. The genus is found not only in the northern Atlantic, but also in every other ocean. The Australian *Antimora viola* is perhaps the same species. The genus *Laemonema* is distributed in west African seas from Madeira to the Cape of Good Hope. Other widespread genera are *Mora, Physiculus, Lepidion* and *Lotella.*

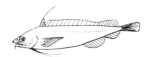

Fig. 18-2. Blue hake.

The BREGMACEROTID CODFISHES *(Bregmacerotidae)* are found exclusively in the tropical and subtropical parts of the Atlantic, Indian and Pacific

Family: bregmacerotid codfish

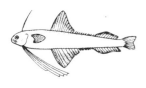

Fig. 18-3. Bregmacerotid cod.

Family:
codfish (Gadidae)

Fig. 18-4. Distribution of codfishes (Gadidae). Note the presence of species of genus *Gaidropsarus* off South Africa and New Zealand.

The cusk

Oceans. These delicate pelagic fishes are found near the surface and at substantial depths. There is only one genus, *Bregmaceros* (L about 10 cm), with a few species, all characterized by a horn-like protruding structure on the rear edge of the head. It is the remains of the first dorsal fin, consisting of a single long ray. The second and third dorsal fins, associated with each other by a row of short rays, not only resemble the anal fins but practically form a mirror image of them and give the fish the same tapering arrow-shaped appearance whether viewed from top or bottom; this is emphasized by the greatly elongated pelvic fins beneath the throat. Little is known of their life habits.

The CODFISH family (Gadidae) has at least twenty-two genera with over twice that many species. The life habits vary considerably. A great diversity of forms developed as a result of adapting to different environmental conditions. Codfish habitat is generally limited to the shelf regions off the coasts. A few species also inhabit the shelf itself as far down as 1000 m, but they avoid great oceanic depths.

The annual world codfish catch (6.5 million tons in 1967) is second only to that of herring. In terms of human consumption, codfishes are in first place, since most herring are processed into fish meal. In the northern Atlantic shelf region codfishes probably exceed all other fish families in prevalence and regular occurrence. Regardless of where and when a fisherman sets his net he will surely catch at least one cod species in it.

For centuries the ample codfish populations in the northern Atlantic have been the basis for the development of a constantly increasing deep-sea fishing industry. Catching advances and increased demand have led in the last few decades to an intense increase in the catch. From 1958 to 1967 the catch of codfish species rose from 4.1 to 6.5 million tons (in the northern Atlantic from 3.6 to 4.7 and in the northern Pacific from 0.5 to 1.8 million tons per year). That represents, for some of the most important codfish populations in the northern Atlantic, the highest possible catch rate. To ensure that there will still be cod available in the future, to meet the increasing needs of the growing human population, international regulations already in effect will be broadened and made more strict.

The very primitive genus *Brosme*, with its only member, the CUSK (*Brosme brosme*; see Color plate, p. 431; L about 90 cm), is found on the shelf as far down as 1000 m. The rounded caudal fin is united with the long, undivided dorsal fin and the similar but shorter anal fin, even though deep indentations at the points of merging serve to separate the fins. The fin seam is emphasized by a black-and-white edge. The barb on the lower jaw, found in most codfish species, is particularly well developed. At an age of eight to ten years the cusk is 40–50 cm long and sexually mature. During the spawning period in spring, the females lay up to 2 million eggs, which float in the water and have a diameter of 1.5 mm. The 4 mm larvae hatch after about 10 days. The adults, primarily bottom dwellers, feed chiefly on larger crustaceans but also on other bottom-dwelling fishes. Their firm white meat tastes something like lobster. The species is regularly caught in modern fishing equipment but never in great numbers and is among the lesser known commercial species. The amount caught is about one per cent of the total codfish catch from the northern Atlantic.

Fig. 18-5. Distribution of the cusk (Brosme brosme).

Several other codfish genera (Gaidropsarus, Onogadus, Enchelyopus and Ciliata) all resemble each other closely; they have the usual barb on the lower jaw and also two to four more above the mouth. L is about 60 cm. The dorsal fin is in two parts, but the front one is greatly degenerated and has partially sunk into a longitudinal groove in the back. The caudal fin is clearly distinct from the dorsal fin and the one-part anal fin.

These codfishes live primarily in shallow water of the algal zone near the shore, but they are occasionally found in deeper water. The chief distribution is in the northern Atlantic, but the three-barbed varieties (e.g., the genera Gaidropsarus and Onogadus), which have the widest distribution, are found off Japan (Gaidropsarus pacificus), New Zealand (G. novaezealandiae), and South Africa (G. capensis), and in the Mediterranean and Black Sea (G. mediterraneus). Onogadus argentatus is found only in the northern Atlantic. In the northwestern Atlantic, O. ensis is distributed from Cape Hatteras to Greenland.

Of the four-barbed species (Enchelyopus), only one is found on both sides of the northern Atlantic: Enchelyopus cimbrius. It is also in the western Baltic Sea.

The five-barbed varieties (Ciliata) of these codfishes have two species, C. mustela and C. septentrionalis, restricted to the northern Atlantic including Iceland.

These commercially insignificant species are bottom dwellers, and their characteristic barbs, which are around the tip of the snout and have tactile and taste senses, are used in seeking prey, primarily small bottom-dwelling fishes, crustaceans and worms. They spawn in the first half of the year, and, as with most codfishes, the eggs float freely in the water.

These four genera resemble the genus Raniceps in fin shape and arrangement as well as life habits, but Raniceps lacks the characteristic barbs above the mouth. Other divergent external and internal features separate this genus from the previous four. This commercially unimportant genus has a single species, Raniceps raninus (see Color plate, p. 432; L 30 cm), found exclusively on the European Atlantic coast from the Bay of Biscay to Trondheim Fjord and in the Baltic Sea to the coast of Mecklenburg, an East German state. It prefers rocky bottoms, 20–30 m deep, with a rich plant growth. The widened, flattened frog-like skull and the laterally compressed tail tapering toward the rear give

Raniceps codfish

Fig. 18-6. Gaidropsarus.

the uniformly brown-black fish a tadpole-like appearance. Indeed, its common German name translates to FROG CODFISH. They are solitary and do not undertake extensive migrations. Their diet consists chiefly of invertebrate bottom dwellers (e.g. crustaceans, worms, mollusks and echinoderms). The spawning season is from the summer months into fall, when they move into somewhat greater depths, about 50–75 m. Each of the floating eggs, less than 1 mm in size, has a small oil globule.

Forkbeards

Another distinct codfish group, the FORKBEARDS, contains the genera *Phycis* in the eastern Atlantic and Mediterranean, and the genus *Urophycis* in the western Atlantic. These codfishes are characterized by long, forked pectoral fins (from which the German common name meaning FORKED CODFISH is derived), with just two very long rays and one shorter ray. The first dorsal fin is triangular with a fiber-like elongated third ray, the length of which depends on the species. The second dorsal fin, extending to the tail, is generally lower and elongated. The anal fin is shaped like the second dorsal fin, but is about one-third shorter than it and is undivided. The caudal fin is relatively small, rounded in the rear, and is clearly distinct from the dorsal and anal fins by a developed tail shaft.

Fig. 18-7. Distribution of the forkbeards (1, 2 *Phycis*; 3, 4, 5; *Urophycis*).

These codfishes live just above the floor of the sea, primarily on the shelf at depths of 200–1000 m; they prefer soft, muddy ground. Their diet consists chiefly of small crustaceans and fishes. Rather sluggish swimmers, they offer little resistance when hooked by a fisherman, even though they are among the larger codfishes and can be up to 120 cm long. Their eggs develop while floating in the water and the young fishes feed at the surface or in shallow water. Little more is known about the life habits of this group.

The east Atlantic species are commercially important only in the Mediterranean. In the northeastern Atlantic *Phycis blennoides* is found only at great depths and is caught infrequently, among other fishes in deep-sea nets.

Fig. 18-8. East Atlantic forkbeard.

The more populous groups of forkbeards in the northwestern Atlantic, however, are commercially more important. The important species are the CODLING or HAKE (*Urophycis tenuis*) and the MUD HAKE (*U. chuss*), which resemble each other and are distributed off the North American coast from Cape Hatteras to southern Newfoundland. Their combined annual catch is 100,000 tons, which is about two per cent of the total catch of codfish from the northern Atlantic.

A few other species in the same genus, distributed further south into the Gulf of Mexico, are less prevalent and commercially insignificant. The prevalence of the forkbeard *Urophycis brasiliensis* is noteworthy, since it is south of the equator along the Brazilian coast, while all other forkbeards on both sides of the northern Atlantic are found only around the Gulf Stream and the bordering Mediterranean region.

Fig. 18-9. West Atlantic forkbeard.

Practically identical external and internal features between BURBOT *(Lota)* and LING *(Molva)* point to their close relationship. Both species have very elongate, almost eel-shaped bodies. The arrangement and division of the unpaired fins is the same as in forkbeards. The only difference is that the first dorsal fin of both the burbot and the ling is rounded and does not have as greatly elongated a fin ray. However, in spite of their close relationship, the two genera are found in entirely different habitats. Burbot is the sole fresh-water codfish species, while ling cod are found only in oceans.

BURBOT *(Lota lota;* see Color plates, pp. 256 and 476-477), with a L of up to 1 m and a weight of 25 kg, is distributed around the North Pole and is found in rivers and lakes of central and northern Europe, northern Asia and North America. Three subspecies are distinguished within this distribution: *Lota lota lota* in middle and northern Europe and northern Asia, *Lota lota leptura* in eastern Siberia, Alaska and northwestern Canada, and the ALEKEY TROUT or EEL-POUT *(Lota lota maculosa)* in the rest of Canada and down to its southern distribution border in the U.S.A.

Burbot prefer clear water and coarse or rocky ground which offers sites for concealment. While the body has a predominant brownish, marbled pattern, there are many local variations. In the mouths of rivers they rarely penetrate as far as brackish water. Unlike many other fishes, burbot are most active in winter, spending the summer burrowed in the ground. While the young feed mainly on small crustaceans and insect larvae, the chief diet of adults is fishes and fish eggs. Burbot reach sexual maturity in three to four years. Large, older females can lay over 3 million eggs in one spawning season. They spawn in the late winter months, chiefly in sites in the upper courses of rivers. The 1 mm-large eggs lie on the floor or float just above the bottom. Burbot are of commercial importance only locally, particularly along Siberian rivers such as the Ob.

Three ling species are found in the northern Atlantic, inhabiting the same regions to some extent but largely separated. They all avoid the currents from cold, polar waters. They are not found near the North American coast.

The LING *(Molva molva;* L up to 2 m) occurs primarily near the floor of the upper part of the continental shelf between 200 and 400 m. It occasionally forms schools. The coloration is dark gray-brownish on the head and back, lighter on the sides of the body, and white on the underside. Weighing over 50 kg, this is the largest member of the codfish family. The species preys on any fishes it can seize, but does not feed on crustaceans and echinoderms. Sexual maturity is attained at an age of eight to ten years, at which time males are already 80 cm long, while the consistently larger females are 90–100 cm long. In the spawning season from April to June, large females can lay up to 60

Burbot and ling

▷
Small callichthyid armored catfishes: 1. The micro catfish *(Corydoras hastatus);* 2. *Corydoras barbatus;* 3. *Corydoras macropterus;* 4. *Corydoras caudimaculatus;* 5. Guiana catfish *(Corydoras melanistius);* 6. The bowline catfish *(Corydoras arcuatus);* 7. *Corydoras reticulatus;* 8. *Corydoras paleatus;* 9. Hoplo catfish *(Hoplosternum thoracatum);* 10. Callichthys *(Callichthys lichthys);* 11. *Brochis coeruleus;* 12. *Dianema longibarbis.*

of the pack ice, while in fall large schools of polar cod move toward the shores, where they spawn in winter at an ambient temperature below freezing. The eggs float; and polar cod are pelagic fishes as adults. The diet consists chiefly of plankton. Sexual maturity is attained after four years, and the maximum lifetime is seven years. Most polar cod spawn just once in their life. The species has minor commercial significance, mainly around the polar coasts of Russia. However, as an important prey for arctic toothed whales and seals, it plays a vital role in the biology of the region.

The EAST SIBERIAN COD (*Arctogadus borisovi)* and the ARCTIC GREENLAND COD (*A. glacialis)* are the northernmost codfishes in the world. They look very much alike but live in completely separate areas in low-salinity waters close to the northeastern Siberian and northern Greenland polar waters, respectively. Their life habits have not been studied.

Codfishes in the genera *Eleginus* and *Microgadus* can barely be distinguished externally, differing only in the lateral lines. Their life habits and reproduction sequences are also quite similar. Because of this, their dissimilar distributions are surprising. The distribution of the WACHNA COD or NAVAGA (*Eleginus navaga;* L at most 32 cm) includes the Murmansk coast, the White Sea and the coast of the Kara Lake to the mouth of the Ob. These small and very prevalent coastal codfish penetrate into the lower courses of rivers, but spawn in December outside the brackish-water zone.

The somewhat larger FAR EASTERN NAVAGA (*Eleginus gracilis;* L reaches 52 cm) is distributed in the northern Pacific from the Korean coast to the coasts of the Bering Sea and Alaska. On the basis of the present distribution, these codfish must have come into the Pacific from the Siberian polar seas.

Microgadus codfishes probably came into the Bering Sea via another route, from the northwestern Atlantic along the North American polar coast. The ATLANTIC TOMCOD or FROSTFISH (*Microgadus tomcod;* L at most 35 cm) is quite prevalent on the North American Atlantic coast from Virginia to southern Labrador, living just off the coast and in river mouths; it is even found in fresh water. The PACIFIC TOMCOD (*Microgadus proximus)*, distributed from Alaska to San Francisco, lives the same way. The commercial importance of these small but flavorful fishes is only of local significance.

The ALASKA POLLACK (*Theragra chalcogramma;* L 40–60 cm) has the same distribution as the Pacific cod. However, the pollack is less dependent on being near the coast and may be found at depths as great as 300 m. Externally the pollack resembles polar cod, although the pollack is considerably larger and has a slightly less forked caudal fin. Like polar cod, the Alaska pollack is also found in open water (which is, however, warmer). Its diet consists of plankton and small schooling fishes. In the western Pacific Ocean isolated populations have devel-

Fig. 18-14. Distribution of polar cod (*Boreogadus saida).*

Fig. 18-15. Polar cod.

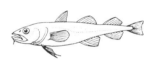

Fig. 18-16. East Siberian cod.

Fig. 18-17. Far eastern navaga.

▷

Blind cavefishes from various orders and suborders: Trout-perches: 1. Mammoth Cave blind-fish (*Amblyopsis spelaeus);* Carp: 2. The Congo blind barb (*Caecobarbus geertsi),* 3. The blind characin *Anoptichthys jordani;* Cod: 4. *Stygicola dentatus* (see Chapter 18).

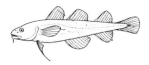

Fig. 18-18. Atlantic tomcod or frostfish.

Haddock

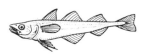

Fig. 18-19. Alaska pollack.

Fig. 18-20. Distribution of haddock (*Melanogrammus aeglefinus*).

The pollock

◁

Toadfishes: 1. The Atlantic midshipman (*Porichthys porosissimus*); 2. Toadfish (*Thalassophryne maculosa*); 3. Oyster toadfish (*Opsanus tau*), a shallow-water species found in refuse from "civilization."

oped differences in growth and the spawning season (October-December vs. spring). The free-floating eggs rise to the surface and develop there. The Alaska pollack has recently become more important commercially. From 1958 to 1967 the annual catch increased five times, to 1.7 million tons. Thus the Alaska pollack is now commercially the second most important codfish species, after Atlantic cod.

The other codfish genera inhabit the northern Atlantic and the Mediterranean, with the exception of a single South Atlantic species. They are quite similar to the northern species, and their distributions even overlap somewhat; although the northern border of their distribution is more southern than that of the northern species. Three important commercial species belong to the group. Among them HADDOCK (*Melanogrammus aeglefinus;* see Color plate, p. 431; L up to 100 cm), with an annual catch in 1967 of 480,000 tons, is the most important. Its distribution in the northern Atlantic is the same as that of Atlantic cod. However, haddock is rarely found off Greenland, and in the northwestern Atlantic it is found only off Newfoundland's southern coast, off Nova Scotia and in the Gulf of Maine. Haddock differs from all other codfishes by a black spot above the pectoral fin. The lower jaw is very short and the barb is small. Its chief diet consists of invertebrate bottom-dwelling organisms and herring spawn. As a pure shelf inhabitant, haddock is rarely found below 200 m. The highest age of this medium-sized fish is about fourteen years. Like Atlantic cod, haddock undertake periodic migrations between the feeding grounds and the spawning sites. In the northeastern Atlantic the chief spawning sites are in the northern North Sea and off the Norwegian coast. Haddock reach sexual maturity after just three to four years, spawning in the spring, somewhat later than Atlantic cod. It has lower fertility, but the development period of the eggs is shorter than in Atlantic cod. The initially pelagic young are often found under the umbrellas of large medusas (see Vol. I) before changing to bottom dwelling in the fall of their first year of life; their length at that time is 11 cm. Unlike other codfishes, which migrate toward the shallow coastal waters, haddock spend their first years in the open sea.

The POLLOCK, SAITHE OR BLISTER-BACK (*Pollachius virens;* see Color plate, p. 431; L about 120 cm) has about the same commercial significance as the haddock. With a 1967 catch of 410,000 tons, it occupies third place among North Atlantic commercial species, after the North Atlantic cod and haddock. In body shape, fin position and size it resembles Atlantic cod, but other characteristics clearly distinguish the two species. The lower jaw protrudes somewhat and has a very small barb. The dark-colored mouth is also a distinctive feature. The pollock is a pelagic predator which, as a fast, skilled swimmer, feeds chiefly on schooling fishes, especially herring. However, its extensive migrations follow less well-established routes than those of other codfishes. Its inconstant, changing life habits make following its growth dynamics difficult, which does not ease matters for the pollock industry. The fish can reach an age of eighteen to twenty years. It attains sexual maturity in four to five years and thereafter spawns each spring in practically the same region as haddock. The egg and larvae drift with the current and the young migrate into coastal waters after their pelagic development period; they spend their first year in coastal waters.

The second member of the genus, the POLLACK (*Pollachius pollachius*), is not very prevalent and is unimportant commercially. It differs from the pollock only in lacking a barb, in the distinct protrusion of the lateral line above the pectoral fins, and by its brownish-colored back and brass-colored sides. It is generally found above stony floors.

The smaller and short-lived WHITING (*Merlangius merlangus*; L 60-70 cm) takes fourth place among codfishes with a 1967 catch of 189,000 tons. Its meat is quite popular in Great Britain and France, but less so in Germany. Its far-flung distribution extends across the entire northern and western European shelf from the north polar cap to the Atlantic coast of Spain and the southern coast of Iceland. One subspecies, the MEDITERRANEAN COD or MOLO, extends the distribution into the Mediterranean and Black Seas. Whitings are not found in the northwestern Atlantic. It is most prevalent in the North Sea and off the western coast of England. Whiting resemble haddock somewhat in coloration, and also have a lateral black spot, which, however, is much smaller than the haddock's and is located at the base of the pectoral fin. As pure shelf inhabitants, whiting are found in even shallower water than haddock and are prevalent in regions with soft, muddy ground at depths between 60 and 150 cm. Their diet consists mostly of small crustaceans and fishes. Young whiting maintain the same relationship with medusas as do haddock.

Three other genera of small, usually pelagic codfish species will be discussed here. Some of them form very large groups and are important prey for their larger relatives. Only a few of these species are important to humans, as sources of fish meal.

COUCH'S WHITING or the POUTASSOU (*Micromesistius poutassou*; L 50 cm) lives in large schools at 80-300 m depths, generally above great depressions on the edge of the continental shelf. It bears some resemblance to the whiting (*Merlangius merlangus*), but has a bluish shimmering coloration and much larger eyes. Distribution extends from Bear Island in the north via Iceland to the western Mediterranean. Apparently the poutassou is not found around Greenland. Recently the presence of this species was confirmed in North American waters for the first time, and it is possible that the poutassou will spread to the northwestern Atlantic. A second species, *Micromesistius australis*, is found under similar environmental conditions in the southwestern Atlantic along the Patagonian coast and around the Falkland Islands.

The pollack

Whiting

Fig. 18-21. Distribution of 1. Pollack (*Pollachius pollachius*); 2. Pollock (*Pollachius virens*).

Fig. 18-22. Whiting.

The poutassou and other codfish

▷

Anglerfishes: 1. Deepsea angler (*Linophryne arborifer*); 2. The ceratiid angler *Ceratias hollbolli*; a male with two females; 3. Johnson's black anglerfish (*Melanocetus johnsoni*), shown before (a) and after (b) swallowing the much longer lantern fish *Lampanyctus crocodilus*; 4. The batfish (*Ogcocephalus vespertilio*).

Fig. 18-23. Poutassou.

Fig. 18-24. *Trisopterus esmarkii.*

Family: hake

Fig. 18-25. Silvery pout.

Fig. 18-26. Distribution of hake (Merluccidae): 1. Hake *(Merluccius merluccius)*; 2. American hake *(Merluccius bilinearis)*; 3. *Merluccius productus*; 4. *Merluccius gayi*; 5. *Merluccius hubbsi*; 6. Stockfish *(Merluccius capensis)*; 7. *Merluccius australis.*

◁

Like other anglerfishes, the frogfish *Antennarius moluccensis* has a "fishing line and bait" on its head, with which it attracts prey. The "line" is a modified first dorsal fin ray.

Among the three species in genus *Trisopterus* the majority are *Trisopterus minutus,* a dwarf cod just 25 cm long at most, and *Trisopterus esmarkii.* The former is distributed from the Norwegian coast to the Bay of Biscay, while the latter is found in the same area but also extends to Iceland. The distribution of *Trisopterus luscus* (L 30 cm) is more southerly and closer to the coast, from the northern North Sea into the Mediterranean.

The SILVERY POUTS *(Gadiculus)* are the smallest codfish species (L 15 cm). They are found at greater depths, 400–1000 m, on the continental shelf. The northern species, *Gadiculus thori,* is distributed from Norway and southern Iceland to the Bay of Biscay, while the southern species, *Gadiculus argenteus,* is found from the southernmost part of the distribution of *G. thori* to northern Africa and the western Mediterranean.

The HAKES *(Merlucciidae)* are closely related to the codfish family, but have a special systematic position due to their unusual distribution. The family has just one genus, *Merluccius.* Its slender body, skull structure, and the large-toothed mouth give this carnivorous fish a gar-like appearance. There are two dorsal fins and one long anal fin, which is almost the mirror image of the second dorsal fin in shape, size, and position.

The HAKE *(Merluccius merluccius)* is found in the northeastern Atlantic off the western and southwestern coasts of Europe, along the continental shelf. The northern border of the distribution is formed where branches of the Gulf Stream meet masses of polar waters. This is also the northern limit of the AMERICAN HAKE *(Merluccius bilinearis;* see Color plate, p. 431). Living in deep water has enabled the hakes to penetrate the tropical Atlantic and inhabit oceanic regions in the southern hemisphere with temperate to subtropical conditions. This accounts for the large South Atlantic populations of STOCKFISH *(Merluccius capensis)* off southwestern Africa and *Merluccius hubbsi* from the coasts of southern Brazil and Argentina. There are also Pacific species, *Merluccius gayi* and *Merluccius productus,* off the western coast of North and South America. Their presence has been explained by a presumed migration around Cape Horn. The New Zealand species *Merluccius australis* may also have come by this route.

Hakes can be over 1 m long, but there are small and medium-sized species as well. They are predators, feeding chiefly on herring and other schooling fishes. The European hake seeks its prey at night in the upper water levels; during the day it is less active and is near the floor, at which time it can be caught very easily, even with a dragnet. This species spawns in spring, apparently without preferred spawning sites. The floating eggs then drift within the hake distribution region. The commercial importance of hakes has increased sharply since 1960. The total catch was 374,000 tons in 1958 and by 1967 reached 1,600,000 tons, more than a fourfold increase.

The CUSK EEL suborder (Ophidioidei), with two families, may or may not belong to the codfish group. Both share some external features, including the pectoral fins at the throat and the absence of hardened rays in the fins; they also share certain internal features. The elongated, eel-shaped body with an undivided fin seam is a sign that the cusk eels descended from primitive codfishes.

The CUSK EEL family (Ophidiidae), with many genera, is distributed world wide. These are generally small and not very prevalent fishes, scaleless or with very small scales. The pectoral fins, always found at the throat, have only a few rays. Cusk eels are usually found at greater depths.

Due to their good-tasting meat, cusk eels in genus *Genypterus*, which reach a size of 1.5 m in the cooler waters of the southern hemisphere, are fished a great deal. The larger species in genus *Brotula*, from the Indo-Pacific, are commercially fished in Japanese waters. The family is represented in the northwestern Atlantic by *Lepophidium cervinum*. The species *Ophidium barbatum* is found in the eastern Atlantic to the English Channel and in the Mediterranean and Black Seas. Several cusk eels now found in caves, such as the Cuban cavefishes *Lucifuga subterranea* and *Stygicola dentatus* (see Color plate, p. 417), developed from marine species.

The PEARLFISHES (Carapidae, sometimes called Fierasferidae; see Color plate, p. 432) are primarily distributed in warmer seas but scattered specimens are found in the northern Atlantic. These small fishes (L up to 20 cm) have adapted to their environment by developing a slender, tapering body lacking scales and almost without pigmentation. It lacks pectoral fins, and the anal opening is in the throat region. The unpaired fin seam tapers to a point in the rear. These fishes live inside sea cucumbers (see Vol. III, Chapter 12), into which they slip skillfully, with their tapered tails first. They only leave the host when seeking food and reproducing. Pearlfishes lay small elliptical eggs which rest on the floor, surrounded by a gelatinous envelope. The larvae swim freely and have a long, thread-like process on the head.

The VIVIPAROUS BLENNIES (Zoarcoidei) also have eel-shaped bodies and a fin seam joining the unpaired fins. The small pelvic fins, when present, are in front of the pectoral fins. The single family (Zoarcidae) has several genera and many species distributed in the northern Pacific, northern Atlantic and in arctic and antarctic waters. The antarctic species developed from deep-sea species in the northern hemisphere. All viviparous blennies are bottom dwellers which feed chiefly on invertebrates inhabiting the floor.

Their German name, meaning EEL MOTHERS, is appropriate, for the members of *Zoarces* give birth to living young (e.g., they are viviparous). The European species, the EEL POUT or VIVIPAROUS BLENNY (*Zoarces viviparus;* see Color plate, p. 432; L up to 60 cm), inhabits coastal waters

Suborder: pearlfish

Fig. 18-27. A cusk eel from genus *Genypterus*.

Fig. 18-28. Penetration of a cusk eel into a sea cucumber: the fish initially probes headfirst at the anal opening of the host; then the fish bends the tail up to the head of the sea cucumber and leads it inside; after straightening its body the fish wriggles backwards into the host animal. Occasionally several cusk eels are found in one host. Cusk eels have also been found in sea stars and mussels.

from the White Sea to the English Channel, including the Baltic Sea and river mouths. During the spawning season in summer an internal fertilization takes place, and in winter females give birth to as many as 400 fully developed young, 35–55 mm long, which live immediately as bottom-dwelling fish. They become sexually mature after two years. Blennies rarely live over three or four years.

The closely related, somewhat larger MUTTON FISH (*Macrozoarces americanus*), from the Atlantic coast of North America, is not viviparous, but lays its eggs on the floor and guards them.

The eel pouts *Lycodes* and *Lycenchelys* live primarily at substantial depths in the North Atlantic shelf-edge regions.

The genus *Gymnelis* is characterized by the absence of pelvic fins. The small northern species *Gymnelis viridis* is distributed in the arctic, generally where the ambient temperature is below 0° C.

Suborder: grenadiers

The GRENADIER suborder (*Macrouroidei*) is a group distributed world-wide, with many genera and species, primarily deep-sea fishes. Their relationship to codfishes is not easily discernible externally. Generally they have two dorsal fins, sometimes with a spiny ray. The second dorsal fin forms a fin seam with the anal fin, which extends well forward; the seam terminates in a point on the extremely thin tail. There is no caudal fin. The large head often is shaped quite differently in different species. The rear of the body tapers sharply behind the short body cavity.

Fig. 18-29. A grenadier.

Grenadiers look rather ugly, an impression strengthened by the armored skin which often has large, spiny scales. Fishermen also call them armor grenadiers or armored rats, since they look something like chimaeras, which in German are called sea rats (see Chapter 4). Grenadiers such as the SMOOTH-SPINED RAT-TAIL (*Macrourus berglax*), the SOLDIER FISH (*Coelorhynchus carminatus*), the ROCK GRENADIER (*Coryphaenoides rupestris*) and others, all found in large numbers, inhabit the northern Atlantic continental shelf beginning at 300 m and proceeding deeper. They are often inadvertently caught together with rosefishes. In some places, particularly in the northwestern Atlantic south of Labrador, they occur in such dense schools that fishing nets meant for rosefishes are instead filled with 10 to 20 tons of grenadiers. Even though they are apparently tasty, grenadier fishes are rarely for human consumption; their hard skin makes them impractical for producing fish meal.

19 Flyingfishes, Toothed Carp and Silversides

The SILVERSIDE order (Atheriniformes) is an extremely diverse group. It contains pure marine forms and those which temporarily penetrate river mouths of tidal zones characterized by brackish or fresh water, as well as fishes which have evolved from marine species but spend their entire lives in fresh water, and thus are secondary fresh-water fishes. The order contains three suborders: 1. Flyingfishes; 2. Toothed carp; and 3. Silversides.

The FLYINGFISH suborder (Exocoetoidei) actually does not contain any true "flying" species; it does, however, include such diverse fishes as half-beaks, needlefishes, and sauries. However, the most striking members of the suborder, for specialist and layman alike, are the FLYING FISHES (Exocoetidae; see Color plate, p. 449). They live primarily in tropical to warm seas and bordering temperate regions. Where they are prevalent they are found in schools. It is quite an impressive sight to stand at the railing of a ship and see schools of flyingfishes suddenly leaping out of the water. Generally the individual fishes jump in all directions, occasionally landing on the deck of the ship; they are said to be able to jump 5 m high.

Flyingfishes in genus *Exocoetus* are quite easy to recognize. They roughly resemble herring both in appearance and size. The back is a brilliant blue, as in many pelagic fishes, while the sides and underside are silvery. The location and shape of the fins are characteristic. The well-rearward dorsal fin indicates that flyingfishes are active near the surface. In all flyingfishes, the caudal fin is distinctively asymmetrical: the lower half is larger and longer than the upper half and is important in "flying" since it acts as a "motor" when preparing for "flight." The wing-like enlarged pectoral fins are particularly striking.

For a long time it was disputed whether these fishes, with their large pectoral fins, carry out bird-like or butterfly-like flying motions. Sophisticated electronic strobe photographs have recently resolved the

Order: silversides, by W. Villwock

Suborder: flyingfish

Family: flyingfish

problem, showing that the pectoral fins are used like airplane wings and are not flapped like birds' wings. In some species the pectoral fins are considerably broadened and outspread, improving their "airplane wing" function. Thus, flyingfishes are only "gliders," as are dragons (*Draco volitans,* see Vol. VI), flying squirrels (see Vol. XI), and flying phalangers (see Vol. X). The gasteropelecid characins, or hatchetfishes (see Chapter 14), which are distributed solely in the Amazon region, have flying organs in their pectoral fins which are similar to birds' wings and enable their fins to flap; this is also true, to a limited degree, in the butterflyfish (*Pantodon buchholzi;* see Chapter 11).

When flyingfishes and their relatives prepare to "fly," they swim with increasing speed beneath the water surface until they have sufficient speed to glide through the air. The length and time of the flight are variable; stretches of 45–50 m have been observed, as well as flying times of about 3 seconds. The average flight speed is 55 km per hour. At the conclusion of a flight, the fishes sink back in the water, at which time the function of the caudal fin lobes becomes apparent. If the fishes prefer to not dive and swim further, they rapidly shake their fin lobes to the side, thereby increasing their speed sufficiently to begin a new gliding flight. These lobes can beat at a rate of up to 50 strokes per second. The fishes can repeat the process 3 or 4 times. This has resulted in flights lasting a total of 13 seconds and a distance in excess of 200 m. The average height above the water is 1 m, but this depends on wind strength and direction. Occasionally, under favorable up-wind conditions along the hulls of ships, flyingfishes can glide considerably higher.

Like many other fishes, flyingfishes lay their eggs on floating seaweed and other algae in the water, including the famed sargassum seaweed east of the Gulf of Mexico. The freshly hatched juvenile flyingfishes, L 1–2 cm, differ from the adults by their fin shape: the wing surface has not yet developed, and the fins look like those of any typical fish. In the following stages of development, 2–5 cm-long juveniles develop broadened, almost circular pectoral fins. They live at the surface of the water during this period. They do not begin leaping out of the water until they are 5 to 8 cm long; their first leaps are 1–10 m long. From then on, development toward the adult flying form proceeds rapidly.

The development of the individual animal (or ontogeny) gives us an insight into the evolutionary development (or phylogeny) of the species: the attainment of flying ability has proceeded in steps, according to the various developmental stages of today's flyingfishes. The oldest *Exocoetus* species developed in the Triassic Period (about 220 million years ago) in the Alps and southern Italy. These fishes are known as *Thoracopterus niederristi. Thoracopterus* already had the striking broadened pectoral fins and a highly developed lower caudal fin lobe. However, the distribution of flyingfishes and the development of present species has not yet been explained.

The most primitive of present-day flyingfishes are the genera *Fodiator* (see Color plate, p. 449) and *Parexocoetus*; the latter is found in giant schools off western India. Among modern flyingfishes, two basic forms are differentiated, those with two and those with four "wings." The "two-winged" species have only the well-developed pectoral fins for the wing surface; "four-wingers" also have enlarged pelvic fins which occasionally are as large as the pectoral fins. The flyingfish genus *Exocoetus*, with *Exocoetus volitans* (see Color plate, p. 449) in all warm seas and *E. obtusirostris* restricted to the Atlantic Ocean (both being 20-25 cm long), contains "two-winged" species. The largest flyingfish, *Cypselurus californicus* (L up to 45 cm), found on the Californian coast, is also a two-winged species. *C. heterurus,* however, belongs with the four-wingers. The juvenile *Cypselurus* species differ from *Exocoetus* juveniles by the chin barbs in the former; these barbs regress and eventually disappear while the young approach the adult stage.

The FLYING HALFBEAKS (*Oxyporhamphus*) are characterized by the greater number of vertebrae and the appearance of an elongated lower jaw in the young fishes. In these and other characteristics, the flying halfbeaks resemble the halfbeaks, which are described later. On the other hand, their fin development and flight ability place them close to flyingfishes. For this reason the flyingfishes, flying halfbeaks and halfbeaks are all put into a single family.

The HALFBEAKS (*Euleptorhamphus, Hyporhamphus, Hemirhamphus, Dermogenys* and *Nomorhamphus*) are shaped somewhat like pike. Their pectoral fins are short and small, permitting them to make only short leaps into the air. They have the least flight ability in the family. According to the present viewpoint, the halfbeaks form the starting point and the flyingfishes the culmination of the development of flight ability within the group. They are called halfbeaks because of the arrangement of the upper and lower jaws. The upper jaw (varying gradually from species to species in this respect) is more or less short, while the lower jaw protrudes considerably.

The halfbeaks include some 70 species. Their average L is 30 cm. Most of them live in the oceans, primarily in the Pacific Ocean. This area contains the largest halfbeak, *Euleptorhamphus viridis* (L up to 45 cm). This species also has the best-developed pectoral fins and the characteristic elongated lower caudal fin lobe. *Hypohamphus unifasciatus* is distributed in both the Pacific and Atlantic Oceans. *Hemirhamphus brasiliensis* is a pure Atlantic species, inhabiting the warmer and temperate zones.

Most halfbeaks lay eggs. There are, however, a few species in which the eggs mature within the body of the mother. In such cases the larvae leave the protective egg case when they are laid, and thus in an exaggerated sense are born live. This sort of live bearing is designated as ovoviviparous, in contrast to the true viviparous birth, which is also found in silversides. In order to fertilize the eggs within the female, a transmitting mechanism is required for the male's sperm. The halfbeaks have such an apparatus in their gonopodium comparable to a similar structure in the toothed carps, but which in the case of the carps is formed from the front part of the well-rearward anal fin.

The best-known ovoviviparous halfbeak is *Dermogenys pusillus* (see Color plate, p. 451), which grows to a length of just 7 cm. It is found in fresh water off eastern India and Indochina. After the Siamese fighting fish *Betta splendens,* it is the most popular tournament fish in the Indo-Malayan region. The males, bred to be especially aggressive, take part in prolonged "tournament" contests according to specific rules, in which there is usually no resulting injury. *Dermogenys* has been introduced in Europe and has been kept and bred there in aquariums. As a surface fish, the species feeds on flying insects, especially on those which are at the water surface. The lower beak, resembling a small piece of driftwood, seems to be used as an attractant on which insects might land. A bright red spot on its tip strengthens the impression of this structure as a fly trap. The lower jaw is immobile, while the upper jaw can be moved.

Other halfbeaks are the two Celebes species of genus *Nomorhamphus,* which externally do not resemble halfbeaks; their jaws are equally long, and the characteristic lower jaw elongation is found only in juveniles.

Family: needlefish

The second family of flyingfishes contains the NEEDLEFISHES or GARFISHES (Belonidae). This includes two closely related groups which formerly were known as Tylosuridae and Petalichthyidae. Needlefishes resemble halfbeaks in external shape and fin position, but, unlike halfbeaks, both needlefish beaks are elongated in the adults. The larvae and young juveniles of the GARFISH *(Belone bellone)* still have the halfbeak-like unequal jaws. Both upper and lower jaws of needlefishes have many needle-shaped teeth. These predators seek schools of surface fishes. When pursured themselves, they can escape by making long, repeated leaps out of the water.

The family contains about 60 species. They inhabit all warm and temperate seas; a few move through brackish water into fresh water. One example of a fresh-water species is *Potamorrhaphis guianensis,* from the Amazon basin. A number of needlefishes reach a L of 100–120 cm, such as the Indo-Pacific *Strongylura crocodila* and its Atlantic relative *Strongylura marina.* The GARFISH *(Belone bellone;* L up to 100 cm) is distributed in the eastern Atlantic to the North Sea and Baltic Sea; it is also prevalent in the Mediterranean and Black Seas. Body size varies among different species but averages about 50 cm. All needlefishes have one striking feature: their skeleton (rarely also the musculature) has a turquoise-green hue. Although not dangerous for human consumption, this feature has led to the fact that needlefishes do not have the position they deserve among eating fish.

The third family of flyingfishes is the SAURY family (Scomberesocidae). Their scientific name is derived from *scomber* (mackerel) and *esox* (pike, which the sauries resemble in shape and life habits. Their German name, in fact, means mackerel pike). Externally they resemble elongated mackerels with a distinctive pike-like snout. This resemblance, which has nothing to do with any evolutionary relationship, is largely caused by the five to seven small fins standing behind the dorsal and anal fins. Like pike, sauries are very active predators. Only four pelagic species are known, but all are quite prevalent, and the largest measure 40 cm. The ATLANTIC SAURY *(Scomberesox saurus)* is distributed in the Atlantic, and the Mediterranean and Black Seas, and occasionally in the North Sea. Sauries also inhabit the other warm and temperate seas. Like needlefishes, they have juvenile forms in which the jaws are unequally long as in halfbeaks. All this is evidence for including all these species with the flyingfish suborder.

The second suborder of silversides, the TOOTHED CARPS (Cyprinodontoidei) is one of the biologically most diverse groups of fishes in the world. The name toothed carp is based in part on the resemblance these fishes have to carp, many of them looking like miniature editions of carp, but also to a major difference between the two groups, for the toothed carps, unlike true carps, have teeth in their jaws. Distribution is worldwide excepting the cold zones in the northern and southern hemispheres. The group has adapted to widely different climatic conditions.

A few toothed carps are marine species living near the shore; examples are the toothed carps in the eastern Mediterranean, the Persian Gulf, and the Gulf of Mexico. Other species inhabit pure salt-water biotopes. Still other toothed carps are found in warm springs in the western Orient, northern Africa, and in the midwestern U.S.A. Other species of toothed carps (entire families, in fact) are restricted to very small distributions and are known as endemites. This is the case of the North American family Empetrichthyidae or the South American group Orestiidae. Those in genus *Orestias* are particularly important scientifically in terms of their well-developed speciation (development to various distinct species). Today *Orestias* toothed carps are found in the Lake Titicaca region in the Peruvian and Bolivian highlands or altiplano, which averages 3800–4000 m above sea level and which is bordered by the western and eastern Andes mountains. Where did these fishes come from? How have they developed such a diversity of species? These questions cannot be answered completely. The highland Indians once ate these fishes, but, since the first trout were

Family: sauries

Suborder:
toothed carp

▷

Codfishes: 1. The pollock *(Pollachius virens)*; 2. Atlantic cod *(Gadus morhua)*; 3. Haddock *(Melanogrammus aeglefinus)*; 4. American hake *(Merluccius bilinearus)*; 5. Cusk *(Brosme brosme)*.

introduced into Lake Titicaca in 1937, the largest toothed carp (◊ *Orestias cuvieri*), like trout a predator, has steadily declined in numbers. In 1960 extensive searches were made by Kosswig, Villwock and Luecken, yet they found no *Orestias cuvieri* at all. All these highland toothed carps share the characteristic absence of pelvic fins.

Family: killifish

The most diverse toothed carp family is the KILLIFISHES (Cyprinodontidae), which formerly was divided into two families, Cyprinodontidae and Fundulidae. The cyprinodontid fishes are primarily distributed in western North America, the Mediterranean region and western Asia to Iran and the Pakistani coast; the fundulids, on the other hand, are predominantly in the tropics of Central and South America, Africa and India.

A highly interesting group, genus *Aphanius*, is found concentrated in western Asia, with a few species around the Mediterranean into western Europe. As secondary fresh-water fishes (i.e., descendants of marine species), they have maintained the ability to withstand water with great salinity, a sign of their ancestry. In the past, droughts and the dry steppe climate often concentrated the salt content in what were once large fresh-water bodies. What is noteworthy is that these salts are not the same ones found in ocean water, but are dissolved bitter salts, primarily of magnesium and sodium. The toothed carps in these areas were able to adapt to these unusual environmental pressures over a period of thousands of years.

One of the inhabitants of inland brackish or salt waters is *Aphanius sophiae*, a popular aquarium fish from the Niriz lake region and thermal springs of Shiraz, Iran. L, as in all *Aphanius* species, is 5–7 cm. Females are usually inconspicuously colored, as are most females in the genus, and are silvery-brown or gray with many spots. Generally the fins are colorless and transparent. Males are bright yellow on the belly and olive-colored on the back, with many diagonal stripes extending to the base of the caudal fin.

Two species restricted to Turkey, *Aphanius chantrei* and *Aphanius anatoliae*, are so similar in appearance that to the uninitiated they look like the same species. However, hybridization experiments have shown that they are truly distinct species. Male hybrids cannot reproduce. Thus there are fertility limitations present for preserving the two groups as independent entities; externally the two species are identical. *Aphanius anatoliae* also contains groups which, probably because of the composition of their water, have lost some of their scales; they also tend to have fin modifications and other anatomical changes; interestingly, crossing these with the more typical *Aphanius anatoliae* killifishes does produce fertile young.

Some species prefer pure fresh water, such as *Aphanius mento* (see Color plate, p. 450), with pitch-black, aggressive males, and the recently rediscovered Algerian killifish *Aphanius apodus*. Still other

◁

1. *Raniceps raninus;* 2. Viviparous blenny *(Zoarces viviparus);* 3. Rock grenadier *(Coryphaenoides rupestris);* 4. A pearlfish *(Carapus acus);*

Aphanius killifishes prefer oceanic water or water with the same salt composition as the ocean. They live either in the sea (e.g., *Aphanius dispar*, most of which inhabit the coasts of the Arabian peninsula, the Persian Gulf, the Red Sea and the eastern Mediterranean) or in regions which historically contained sea water. Because of salt deposits remaining in the lakes and rivers, these bodies of water have a marine-like composition. The Tunesian salt marshes are probably old marine regions which were flooded from the Mediterranean Sea between the Ice Ages and after the most recent Ice Age. Even the Siwa oasis in Egypt, 300 km from the present coast of the Mediterranean, once must have had a connection with the sea via the Nile River, as indicated by remains of marine toothed carps and shells of marine mollusks found at the oasis.

The chief species of northern Africa and the southern coast of Europe east to France is *Aphanius fasciatus.* The male of this species has brilliant saffron fins. In the west we find *Aphanius iberus,* primarily in saline waters on Spain's Mediterranean coast. It can live in a salinity of six per cent. In the 19th Century the species was also on the other side of the Strait of Gibraltar in the highlands of Oran (Algeria), but land renewal has exterminated the species in this region. During courtship males develop a blue-black coloration tending toward golden yellow on the undersides with many silvery diagonal stripes terminating at the tail shaft in irregular rows of spots.

The North American killifishes of genus *Cyprinodon* live very much like the Asian killifishes. They inhabit the Californian coast or the edges of large salt lakes in Nevada. Well-known species include *Cyprinodon variegatus* and *Cyprinodon nevadensis.* The Atlantic side of the continent has the FLAG FISH (*Jordanella floridae;* see Color plate, p. 450; L 5 cm) in the Florida Everglades. This species, with brilliantly colored males, distinguishes itself from all other members of the family because it is the only species which cares for the young. The male guards the eggs which the female lays, fanning water toward them with the pectoral fins to supply them with new oxygen. This is rather important in the marshy everglades.

While the North American representatives of the family feed primarily on plants, the Old World forms are predators. Both groups share the same shape and arrangement of teeth. These typically three-pointed teeth stand in one row in the upper and lower jaws. They are suitable for chewing plants and prey animals, such as worms, threadhorn larvae (see Vol. II), crustaceans, water fleas and cyclopids (for the last two see Vol. I). Cannibalism has not been observed as it has in some fundulids, but it may exist.

The cyprinodontids are all characterized by spawning using an adhering substance. The females use a sticky glandular secretion to cement the eggs to a solid substrate, such as on fine plant parts. The development period proceeds rapidly, usually taking one week, rarely two weeks, for the young to hatch. The rapid sequence of generations, high number of young, and their easy care make the cyprinodontids popular for aquarium hobbyists and interesting for studying processes of speciation and other evolutionary questions.

The fundulids are more difficult to keep but are therefore more fascinating for many aquarium hobbyists. They are primarily distributed in tropical South America, Africa and Asia. The group can be organized into two main subgroups. One includes annual fishes which occur in seasonally dry lakes and river in South America and Africa. The second group contains species from waters which do not dry out. The African genus of KILLIES (*Aphysemion;* see Color plates, pp. 450 and 454) is particularly interesting since it contains species found both in permanent and seasonal bodies of water. Their life expectancy as adults is between two drought periods. With the onset of a drought, or at the very latest with the drying out of their home waters at the high point of a drought, the adults perish.

The secret of their adaptation to the periodically returning drought of their habitat is that the fertilized, thick-walled eggs can withstand unfavorable periods while they are buried in the ground. A muddy pool has sufficient moisture for the eggs to develop. And now comes the second surprise: not only are the eggs capable of surviving during these periods, but the embryos develop within them! It is no surprise that the embryonic development in these species is quite unlike that of typical bony fishes, on the basis of their unusual environmental conditions. Development does not proceed continuously, but in stages. One such interruption is known as a diapause. Peters, who extensively studied these fishes, reports:

"There are obligatory and facultative diapauses, the latter being determined primarily by the changing oxygen content of the water caused by rainy and dry periods. Thus, in an early embryonic stage, before the embryonic shield is present, development will cease until the water dries up and the atmospheric oxygen can penetrate into the ground. On the other hand, the developmental pause in the embryo now capable of independent life is interrupted when the oxygen supply is below some threshold quantity. This happens when the first rains come after an extended drought period; they deplete the oxygen supply in the ground."

These conditions can be imitated in the aquarium and thus the actual developmental processes can be studied. Amazingly, a large proportion of the fully-developed embryos hatches 2–4 hours after the ground is moistened. The process is probably equally rapid in nature. The growth rate to the adult form proceeds very rapidly as well. One can realize how inhabitants of such regions would believe in fish rains if they did not understand how these fishes develop.

As previously stated, the killies are not only bottom dwellers adapted to seasonally dried out waters, but a larger number of them are found in permanent waters where the spawn adheres to fine water plants.

The ground-spawning species include *Aphyosemion arnoldi*, the GOLDEN PHEASANT KILLY (*Aphyosemion sjostedti*; see Color plate, p. 450). the LYRETAIL (*Aphyosemion australe*; see Color plate, pp. 450 and 454), which attaches its eggs to objects, the TWO-STRIPED KILLY (*Aphyosemion bivittatum*), and the RED-SPECKLED KILLY (*Aphyosemion cognatum*). One species, *Aphyosemion calliurum*, has been described by Peters as a transitional form between ground spawners and those which attach the eggs. It is impossible to describe the brilliant colors and variety of shapes in these fishes; the color drawings and photographs (pages 450 and 454–455) do this much more effectively.

The killies are primarily distributed in western and southwestern Africa; some, like the golden pheasant killy, are now put in the genus *Roloffia*. Other killies such as the *Nothobranchius* killies (see Color plates, pp. 450 and 455) live in central and eastern Africa and on Zanzibar. Most *Nothobranchius* killies behave like ground spawners such as the *Aphyosemion* killies, while others are found in permanently saline waters near the coast. The best-known species are *Nothobranchius guentheri* (see Color plate, p. 455) from East Africa including Zanzibar, and the FIRE KILLY (*Nothobranchius rachovi*; see Color plates, pp. 450 and 455), the latter being the most brilliant *Nothobranchius* species. Its males are as colorful as hummingbirds, but the females are inconspicuously yellowish-brown with colorless fins.

All African seasonal fishes in the *Aphyosemion* and *Nothobranchius* genera have three diapauses (or interruptions) during their embryonic development. The South American genera of PEARL FISHES (*Cynolebias*; see Color plates, pp. 450, 454 and 455) and LONGFINS (*Pterolebias*) have basically the same adaptations to seasonally dry waters as their African relatives, but the pearl fishes lack the intermediate diapause. Since these toothed carps have developed independently in Africa and South America, it must be presumed that these coinciding adaptations developed independently.

The best-known South American species are *Cynolebias nigripinnis* (see Color plate, p. 455), the ARGENTINE PEARL FISH (*Cynolebias belotti*; see Color plate, p. 455), and *Cynolebias ladigesi* (see Color plate, p. 450). The LONGFINS (e.g., *Pterolebias longipinnis*; see Color plate, p. 450 and the PERUVIAN LONGFIN [*Pterolebias peruensis*]), with a L of 10–12 cm, are about twice as large as the pearl fishes. The pearl fishes have very interesting sexual differences: there are different numbers of fin rays in the dorsal and anal fins of males and females. Because of that, the males and females were originally assigned different names until their true relationship was discovered. The South American RIVULUS (*Rivulus*; see

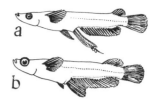

Fig. 19-1. (a) Male and (b) female glass carp.

Fig. 19-2. Goodeid top-minnow distribution.

Fig. 19-3. Head of a four-eyed fish. The eyes are divided by a band of skin into upper and lower elements; this enables the fish to see above and below the water surface.

Color plates, pp. 450 and 455) fishes have an almost amphibian life. They live in very shallow, heavily overgrown waters in marshes and even on wet rocks near waterfalls.

Fundulids are more difficult to keep in captivity than other killifishes not only because of their peculiar characteristics as ground spawners but also because of the unique composition of the water they inhabit. This so-called "black water" is yellowish to dark brown and contains a great deal of humus and very little calcium. If this kind of water is provided (which can be done by using rainwater or other calcium-poor water), most of the fundulid "problem fishes" can be kept successfully and even bred.

The African killies *Aplocheilichthys* (see Color plate, p. 450) usually do not live more than one year. However, they can be bred easily, which compensates for the short time they will live. As surface fishes, they often jump out of the water, something which an aquarium hobbyist should remember. Other surface-dwelling species include the western and central African genus *Epiplatys,* of which the most widely kept species in aquariums is *Epiplatys dageti* (see Color plate, p. 450). The genus *Aplocheilus* (not to be confused with the African *Aplocheilichthys*) is found only in Asia (see Color plates, pp. 450 and 455). *Aplocheilus* killifishes are more resistant and less sensitive to aquarium keeping than any other fundulid.

The fundulids also contains an interesting species found in the temperate to subtropical waters of the southern U.S.A., the BLUEFIN KILLY *(Chriopeops goodei);* and *Valencia hispanica,* from the mouth of the Guadalquivir River in southern Spain and perhaps from Corfu as well.

Two somewhat divergent families are also included with the toothed carp suborder. The Celebes family Adrianichthyidae, with just three species from fresh-water lakes in Celebes (with a L of 8–20 cm they are the largest toothed carp), contains *Xenopoecilus poptae,* from the Posso Lake, which is fished only during winter. These fishes also spawn during winter and have a kind of ovoviviparity. For this reason some researchers include them with the livebearer family, Poeciliidae.

The JAPANESE KILLIFISHES (Oryziatidae) also are striking in terms of reproduction because the females do not lay eggs but carry them about in a droplet. At a later stage of development the eggs are actually laid. All Japanese killifishes have laterally compressed bodies, large, high-set pectoral fins and in other respects have typical surface-active-fish features. They have also been domesticated like the fighting fishes from Malaya and the guppies in North America and Europe.

The other silversides, by M. Dzwillo

In 1940 a new Indian species, *Horaichthys setnai,* was discovered; it is the only member of the family Horaichthyidae, the GLASS CARP. This small, transparent fish reaches sexual maturity at a L of just 2 cm. The male has a complex copulatory organ, with which it sticks the sperm packet (or spermatophore) at the genital opening of the female. In spite of this internal fertilization, females lay the eggs at an early part of their development.

The GOODEID TOPMINNOW family (Goodeidae), found in rocky streams and lakes in central Mexico, also has internal fertilization. However, males do not have specialized copulatory organs. The first six rays of the anal fin are shortened and separated from the rest of the fin by a partition. There are about two dozen species of goodeid topminnows, most of which are a few cm long. They all are viviparous. The eggs are poor in yolk supply, and the embryos are fed by cord-like gill processes from the mother's body.

The FOUR-EYED FISH (*Anableps anableps;* see Color plates, pp. 456–457; L 10–20 cm) is the single representative of a family with the same name (Anablepidae). In males the front of the anal fin is tubular and is a copulatory organ surrounded by skin and scales. The egg envelopes of the four or five young born at a time burst immediately at the moment of birth. The four-eyed fish is a shallow-water and surface-active species distributed in fresh water in Central America and northern South America. The eye of the fish is divided in two by a band of skin; the upper portion of each eye protrudes out of the water, while the lower one remains submerged. The single eye lens is oval and shaped in such a way that the upper eye can focus out of water and the lower eye is adapted for underwater vision; the upper eye is flatter than the lower eye. Although four-eyed fish are almost always found in a horizontal position at the water surface, they always feed underwater.

The JENYNSIID TOPMINNOWS (Jenynsiidae, with the single species *Jenynsia lineata*), also has a copulatory organ which is a tube covered by skin. The embryos are also nourished from the mother's body. This small toothed carp is distributed in southern Brazil and northern Argentina.

The copulatory organ of some individual four-eyed fishes and jenynsiid topminnows can only be moved to the right, and in others only to the left. The genital opening of the females is also accessible from just one side, so that a female with a right-sided genital opening can only copulate with a male with a copulatory organ movable to the left. Left-sided and right-sided individuals are present in about equal numbers in the species.

Perhaps the best-known silverside family is the LIVEBEARERS (Poeciliidae). Their natural distribution encompasses the warmer and temperate zones of the New World. Man has introduced them throughout North America as well. Members of this family are not only popular and easy-to-keep aquarium fishes; various livebearers have also been used to combat mosquitos in waters throughout the warm regions of the world, and the livebearers have often multiplied rapidly under these conditions. These are small fishes (L at most 20 cm) which occupy various habitats. In general they avoid large, open waters. The vast majority are found in fresh water; some species also live in brackish or salt water near river mouths.

The livebearers are distinguished by the male copulatory organ, the gonopodium. This structure develops from the anal fin with the onset of sexual maturity. The third, fourth and fifth rays of this fin become greatly elongated and form a groove through which the sperm can be transmitted into the female's genital opening. At the same time the front rays become shortened. Generally the individual sperm cells combine in great numbers to form oval or rounded structures. The tip of the copulatory fin often has fine hooks and toothed edges as well as a spoon-shaped elongation of the fin skin. The varying structure of the copulatory fin tip from species to species is significant in determining the interrelationship of the species; however, it serves no special function. Even males in which the tip of the copulatory fin was cut off and grew back in an atypical manner were capable of copulating. In order for this organ to function in copulation, it can be moved forward and to the side, but it can also swing out to the right or left. To provide a firm hold for the organ, certain internal skeletal parts have been developed which join the copulatory fin with the vertebral column.

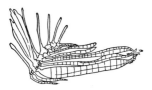

Fig. 19-4. The gonopodium (copulatory fin) of the male platy (*Xiphophorus maculatus*).

A single intromission of sperm usually suffices for fertilization of several groups of young. The excess sperm are stored in folds of the ovarian cavity and are viable for a long time. One female guppy was put in isolation after copulating and still produced eleven successive litters without copulating again. The eggs of most livebearers have a rich yolk supply. They develop within the ovarian cavity but are generally not nourished from the body of the mother. At the moment of birth the protective envelope of the embryo bursts. There are also relatives of livebearers in which the embryos do receive nourishment from the mother, as has been shown in the four-eyed fish and jenynsiid topminnows. The period of development varies from species to species and is particularly dependent on the ambient water temperature. In most species, litters are produced every four weeks at a temperature of 20-25° C. Many members of the family reproduce almost throughout the year. The young are well developed when born and can feed independently. Many livebearers feed on their young if the young do not find shelter very quickly between plants. Some aquarium hobbyists have noted that the male waits under the female when she is giving birth, in order to eat the young as soon as they are born. This form of cannibalism is dependent upon the food supply and the population density of the fishes.

Subfamily: Tomeurinae

One subfamily, Tomeurinae, contains the single species *Tomeurus gracilis*, distributed in fresh water in eastern Venezuela, Guyana, and Surinam. The species lacks the chief characteristic of the family in that it lays eggs. Each female lays one hard-shelled egg onto aquatic plants; the egg has fibrous processes on it. A well-developed fish soon hatches. The copulatory fin of the male is located well forward in the throat region. Even though the egg-laying livebearer thus diverges from the typical Poeciliidae structure, it belongs with them because it has all the other characteristic features.

Most livebearers are in the subfamily Poeciliinae. The most important genera are: 1. LIVEBEARERS (Poecilia), with the familiar guppy and molly; 2. PLATYFISHES and SWORDTAILS (Xiphophorus); 3. Phalloceros; 4. GAMBUSIAS (Gambusia); 5. LUCINOS (Belonesox); 6. MOSQUITO FISHES (Heterandia); 7. Poeciliopsis; and 8. MERRY WIDOWS (Phallichthys).

Subfamily:
livebearers
(Poeciliinae)

The GUPPY (Poecilia reticulata; see Color plates, pp. 451 and 460) has been introduced into aquariums throughout the world. American hobbyists have often called this easily kept species the "missionary of aquarium fishes." Today, more than sixty years after its initial introduction, probably more guppies are kept than any other aquarium fish. Its original distribution can no longer be determined, since guppy populations are now found in many warm parts of the Americas and other parts of the world; they are descended from fishes which were introduced to combat insects. Guppies have also been taken from captivity and introduced to wild waters, which has also contributed to their spread. At any rate, guppies originally inhabited Venezuela, Barbados, Trinidad and many other Central American islands.

Male guppies reach a L of 3 cm, while females can be twice as long. While females are a uniform, plain yellow-gray and rarely have an indistinct dark or yellow color in the caudal fin, males vary considerably in body and caudal fin coloration as well as in the length and shape of the caudal and dorsal fins. Black or red spots on the sides and the caudal fin are accompanied by bluish or greenish shimmering patterns. The caudal fin can have a fan-like elongation or be extended on the upper and lower margins like a sword. Decades of artificial breeding by hobbyists have produced guppy varieties which far exceed the wild forms in their brilliant colors.

The great diversity in guppy males is of scientific as well as popular interest. The Danish biologist Johannes Schmidt—who made well-known studies of eel migrations—and Ölaf Winge were the first to establish that the genes for coloration in males are all in the sex chromosomes. In a single paper published in 1927 Winge described eighteen genes for various color patterns and caudal fin shapes of the guppy. The distinction among specific color patterns of guppies prevalent in certain geographic areas was responsible for the fact that some varieties were mistakenly considered to be new species and even new genera. However, hybridization experiments showed that all these fish belong to one species, and this eliminated the profusion of scientific names for them.

If you look at a school of guppies in an aquarium it is at once apparent that the very lively males almost constantly display around the females, and copulation apparently takes place quite often. The males touch the genital region of the females with their forward-protruded copulatory fin. The American biologists Clark and Aronson, however, found, on the basis of extensive observations, that most of these contacts do not lead to fertilization of the female. Successful copulation with sperm transmission occurs relatively rarely and includes a longer union with the female than occurs in these pseudo-copulations. A fertilized guppy female produces young every four weeks in litters of from 10 to 100 young.

The guppy is a fresh-water fish only seldomly encountered in brackish water, even though it can become accustomed to salt water in the aquarium. However, one live-bearer, *Poecilia branneri* (see Color plate, p, 451), is found usually in brackish water. This species, in which the males are at most 2.5 cm long, lives in the lower Amazon region. One species which is very similar in size, shape, and life habits is *Poecilia parae* (see Color plate, p, 451), distributed along the coasts of northeastern South America. *Poecilia melanogaster* (see Color plate, p, 451) and *Poecilia nigrofasciata* (see Color plate, p, 451), from Haiti, were once popular aquarium fishes. The back of the male of the latter species becomes increasingly high-set with age.

One series of species in the genus *Poecilia* was formerly known under the generic name of *Mollienesia* among aquarium hobbyists. Two members of this group, *Poecilia latipinna* and *Poecilia velifera* (see Color plate, p. 452), are characterized by sail-like dorsal fins. Both reach a L of 12–15 cm. The former lives off the Atlantic coast of North America from North Carolina to Yucatan, while the latter is restricted to Yucatan. While most live-bearers, including the guppy, feed on animals (insect larvae, small crustaceans, etc.), the members of the former genus *Mollienesia* feed on flora, particularly on algae. Another member of this group is *Poecilia sphenops,* distributed from Mexico to northern South America. Many hybrid varieties of this species have been introduced into aquariums. The best-known of them is the BLACK MOLLY (see Color plate, p. 51). In 1962 the American ichthyologists M. Gordon and D. E. Rosen described a cave-dwelling variety of *Poecilia sphenops* in a limestone cave in Mexico. These fish show clear signs of acquiring cave-dwelling characteristics: their eyes are much smaller than their above-ground relatives, and the body coloration is poorly developed.

The molly genus *"Mollienesia"* also contains the AMAZON MOLLY *(Poecilia formosa).* It is from North America, and is named not for the river but for its unusual life habits. This Amazon of fishes occurs only as a female. The Amazon molly inhabits southern Texas together with *Poecilia latipinna,* and northeastern Mexico with *Poecilia sphenops.* The males of both other species (and, in captivity, males of other livebearers as well) copulate with the Amazon molly females. However, no hybridization occurs; the foreign sperm cells only activate the development of the eggs, and the resulting young are identical to the mother. This unusual form of reproduction is known as gynogenesis and is a special case of parthenogenesis.

The PLATYFISHES and SWORDTAILS (Xiphophorus), from Central America, are closely related to the livebearers. The MOONFISH or PLATY (Xiphophorus maculatus; see Color plates, pp. 452 and 454) attains a L of about 6 cm and even in its natural distribution has many varieties. While its body shape is uniform, many different black and red patterns appear on the sides of the body and on the fins. However, the brilliant red platys often seen in aquariums have been bred artificially by crosses with other Xiphophorus species; they do not exist in nature. The VARIATUS (Xiphophorus variatus; see Color plate, p. 452), a close relative, often has an orange-yellow coloration. While platy species have a fully rounded caudal fin which is identical on top and bottom, male Xiphophorus xiphidium platys have an elongated lower caudal fin edge. This tip on the caudal fin can be interpreted as the first indication of the same characteristic which is more fully developed in male swordtails.

Almost every aquarium hobbyist knows the SWORDTAILS or HELLERIS (Xiphophorus helleri; see Color plates, pp. 51 and 452). The adults attain a L of 7–12 cm (not including the sword), the females being somewhat larger than males, as is found in most livebearers. The male's sword, a prominent, black-edged, almost body-length process on the lower side of the caudal fin, is colored brilliant green to orange-yellow. The sword is used to stimulate females during courtship. Swordtails are very fertile, and a female can produce up to 200 young in a single litter. Aquarium newsletters often carry stories on sex transormation in swordtails. Adult females allegedly become fully transformed to males; the characteristic sword develops on the caudal fin, and the anal fin is modified to form the copulatory organ. However, recent studies have shown that some swordtail males take a long time to reach maturity and the secondary male sexual characteristics do not develop until the males reach the same size as adult females. Late-maturing males are more compact (as are females) than the early-maturing males from the same litter. So this is not a case of sexual transformation.

As was the case with the platy, many artificially-bred swordtails have been produced for aquariums. The popular red swordtail hybrids were produced by crosses with platys. The color-determining genes, which produce only small red spots in the platy, produce a red hue on the entire body when a cross is made with swordtails. In the same way, genes producing black patterns on parts of the body and fins in platys produce black coloration of the entire body in hybrid crosses with swordtails. The hybrids often have pigmented swollen areas, which become more marked by re-crossing the hybrids with sword-tails. These swollen areas are derived from normal platy coloration features. The study of coloration was initiated by the American geneticist M. Gordon, and by C. Kosswig in Germany. Since these first studies were made in the 1920s, these platy and swordtail species have become important species for genetic research.

The first livebearer kept as an aquarium fish was the BARRIGUDINYO (*Phalloceros caudimaculatus*; L 2.5 cm in males and up to 6 cm in females). In 1898 this small, easily kept species was first introduced from eastern South America. Hobbyists became greatly enthusiastic about the species because they had previously been familiar only with egg-laying fishes. Today the only variety which is kept in aquariums is *Phalloceros caudimaculatus,* variety *retriculatus* (see Color plate, p. 451).

The GAMBUSIAS (*Gambusia*; see Color plates, pp. 51 and 452) are distributed in the eastern U.S.A., on a few West Indian islands, and in Mexico and elsewhere in Central America. They are not very popular aquarium species because of their plain coloration and aggressiveness. A few gambusia species have been used to fight mosquitos and have been introduced throughout the world as a means of combating malaria. In southern Europe *Gambusia affinis* has been placed into many pools, lakes and streams. This has often resulted in many problems, since the gambusias feed not only on mosquito larvae but also the brood of other fishes, often eliminating part of the resident population.

The LUCINO (*Belonesox belizanus;* see Color plate, p. 452) is a pure predator. With a maximum L of 20 cm in females, this eastern Central American species is the largest livebearer. Its body form, functionally related to its predatory life, resembles that of pike. Lucinos feed on smaller fishes.

At the opposite end of the size scale, the MOSQUITO FISH (*Heterandria formosa;* see Color plate, p. 452), in which females reach a L of 3 cm and males of at most 2 cm, is definitely the smallest member of the family. For a long time this livebearer from Florida was considered to be the smallest vertebrate in the world until smaller gobies were discovered on the Philippines (see Vol. V). Reproduction in mosquito fish is fascinating. While the young of most livebearers are born in one litter, female mosquito fish always have, within their bodies, several litters of young at various stages of development. Thus the time of birth extends over a period of one to two weeks. The female bears several young daily or at greater intervals during this period. This overlap of developing embryos is known as superfetation; it is also found in *Poeciliopsis* livebearers and in a somewhat different form in hares (see Vol. XII). Some *Poeciliopsis* also reproduce by gynogenesis, like the Amazon molly. Gynogenesis is otherwise not found in any vertebrate.

The MERRY WIDOW (*Phallichthys amates;* see Color plate, p. 451) also has a long and prominent copulatory fin. The size of this popular aquarium fish, from the Atlantic side of Guatemala, varies between 3 and 5 cm. One female produces up to forty young per litter.

The SILVERSIDE suborder (Atherinoidei) was once placed close to mullets (Mugilidae; see Vol. V) on the basis of many shared characteristics. Recent studies of internal anatomy by the American ichthyologist D. E. Rosen, however, have confirmed their closer relationship to the toothed carps. Silversides differ from toothed carps in that they possess two dorsal fins, the front one having hard rays, and the rear one soft rays.

The RAINBOW FISHES (Melanotaeniidae) consist of a few Australian fresh-water fishes which are very popular for aquariums because of their bright, shimmering coloration and the ease with which they can be kept. *Melanotaenia maccullochi* (see Color plate, p. 453; L up to 7 cm) only occurs in northern Australia. The AUSTRALIAN RAINBOW FISH (*Melanotaenia nigrans*; see Color plates, pp. 453 and 456; L 10 cm) is distributed in eastern Australia including far southward as well as the Sydney region. In some regions every pool or stream contains this species. In one spawning period a single female can produce up to 200 eggs. They have short fibers which are used to adhere to aquatic plants. Interestingly, males have a fast, sideward head movement which they execute during courtship; this behavior is characteristic of many egg-laying toothed carps. The American researcher Foster also believes that this is a characteristic which relates the toothed carps and the silversides.

The SILVERSIDES (Atherinidae), unlike the higher-set rainbow fishes, are primarily elongated fishes. Their common name is derived from the silvery sheen in most species. Most of them live in large schools near the coasts of tropical and temperate seas. Since they are generally relatively small, they are not important as food despite their great prevalence.

European waters contain several *Atherina* silversides. Schools of *Atherina hepsetus* (L up to 14 cm) are found throughout the Mediterranean and in the Black Sea. They are primarily found in the open sea, approaching the shore only during the very extended spawning season. *Atherina mochon* (L up to 12 cm) is quite the opposite, being found throughout the year along the shore. In spite of their small size, Mediterranean silversides are often caught with nets, since they are very flavorful fishes. A third silverside, *Atherina presbyter*, is distributed along the western coast of Europe from the western Meditersnean to Denmark. Some European silversides penetrate into fresh water; *Atherina mochon* is found in rivers and fresh-water lakes in southern Europe.

In tropical regions there are silversides which are exclusively fresh-water fishes. One of them which is known to aquarium keepers

Suborder: silversides

Rainbow fishes

Family: silversides

is *Telmatherina ladigesi* (see Color plate, p. 453; L up to 7 cm). The rays of the second dorsal fin and the anal fin are greatly elongated in males and extend toward the rear. Females lack these elongated fin rays, and they have plainer coloration. Spawning is preceded by a striking courtship display. The yellow eggs are attached to fine-leaved aquatic plants, and hatch after eight to ten days.

Mexican lakes also have fresh-water silversides, which must have migrated in earlier times into these lake regions from the Pacific Ocean. Today there are no routes for fishes to travel between the ocean and these lakes. These Mexican species are of some commercial importance for the local population. One relatively large member of these silversides, the JACKSMELT SILVERSIDE *(Atherinopsis californiensis)*, attains a L of 50 cm. These fishes spawn on eel-grass. When this suitable substrate cannot be found the eggs can hang on the female for over a week; they are carried about like a long flag.

One of the most unusual silversides is the GRUNION *(Leuresthes tenuis;* L about 15 cm). It has the typical silverside appearance and is found only in a very limited region off shallow, sandy coasts of southern California. However, thousands of grunion appear there during the spawning season in massive schools. While silversides typically lay their eggs on aquatic plants, grunion deviate from this pattern and in many respects has an unusual means of reproduction. These fish spawn at night during the spring high tides moving with the waves onto the beaches and laying their eggs in the sand. Since these spawning periods are predictable, many tourists and local residents gather in groups by the full moon or when the new moon is out, bringing lanterns to admire this natural spectacle or to catch grunion, using, according to the law, only their bare hands. They may drop their catch into buckets, but may use no tools for the actual capture.

The fish appear on the beach in the first four nights after the spring tide from March to August. Although many people watch the grunion schools, very few get to see the actual spawning process, which takes place on the beach for twenty to thirty seconds between two incoming waves. Females burrow into the sand with their tail and fins until they are in a vertical position up to their pectoral fins in the soft, moist sand. Then a male, sometimes several, wraps around the female which moves rapidly to and fro. Finally the female frees itself from the sand. During this time the eggs are laid and fertilized, and when the next wave comes the grunion return to the water. As soon as the grunion schools have departed, there is no remaining sign of them on the beach. However, if one digs into the sand, the eggs can be found about 5 cm deep, where they are often in piles of two thousand eggs. For a few days after the spring tide the surface of the sand is still somewhat moist, but it dries out and becomes very hot under the Californian sun. However, at 5 cm depth, where the eggs are developing, the sand is

Fig. 19-5. Spawning grunions.

still moist and only moderately warm. The next tide, two weeks later, washes the eggs out of the sand, and the young hatch and pass into the ocean. During their first year grunion grow very rapidly so that they attain adult size in the following spring and soon after that begin spawning on the beaches. They live three to four years. The SARDINE SILVERSIDE *(Hubbsiella sardina)*, also from the west coast of North America, reproduces in a similar manner.

As different as the various spawning processes may be in the individual silversides, the entire family is characterized by rather large schools. The American zoologist Evelyn Shaw studied the development of schooling in the young of the Atlantic silverside (genus *Menidia*). She found that the larvae, which are 4.5 mm long when they hatch, began forming small groups of up to ten individuals when they were 11-12 mm long. By the time the fishes were 14 mm long, their schooling was so well developed that they swam constantly and orderly in a school within the test aquarium. Schooling, which occurs among many fishes from very different orders, is of fundamental importance. When the individual fishes form groups, this greatly reduces the problem of finding a mating partner. Food searching is also often easier in groups than alone. Furthermore this serves a protective function and thereby betters chances for survival.

Isonidae, a family which was formerly part of the silverside family, contains two genera of fishes from the Indo-Pacific region: *Iso,* with five species, and *Notocheirus,* with the single species *Notocheirus hubbsi.* The American D. E. Rosen has determined that for anatomical reasons these genera belong to a distinct family. They are characterized by the pectoral fins, which are located very high and close to the back.

Family: Isonidae

A closer relative of the silversides is the PHALLOSTETHID superfamily, Phallostethoidea. This contains two closely related families, the PHALLOSTETHID PRIAPRIUMFISHES (Phallostethidae) and the NEOSTETHID PRIAPRIUMFISHES (Neostethidae), with 10 genera and about 20 species. As typical members of the silverside suborder, the phallostethids are slender, have a relatively deeply forked caudal fin, and generally have two dorsal fins. The front dorsal fin, however, has just one or two hardened rays, and in some species has disappeared completely. Pectoral fins are absent on all phallostethids. The anal and genital openings are very far forward, located in the throat region.

Superfamily: Phallostethids

The most striking and unusual feature of the phallostethids is the bizarre, asymmetrical copulatory organ of the males, the priapum, a strange, bony, muscular structure on the lower side of the head and throat of males. The priapum develops from part of the shoulder and pelvic girdles, the pectoral fins, and the first rib pair. The long, arched bones on the priapum enable the male to grasp the female firmly when fertilizing the eggs. The eggs are fertilized internally, in the ovarian cavity, and are laid when still in an early stage of development. They have fibrous processes similar to those of silversides.

Close studies have been made of these fishes by the American ichthyologist Albert W. C. T. Herre. The phallostethids are distributed in southeastern Asia from India to the Philippines. The phallostethids and neostethids can be organized into three groups on the basis of their habitats. The most widely distributed inhabit the brackish and salt water of the coastal region of Malaysia, Singapore, Thailand, Borneo, and the Philippines. The second group inhabits streams and rivers in the mountains of Luzon; there are also a few species in this group found in fresh-water streams in the lowlands of Thailand and Luzon. The third group is the species from Philippine fresh-water lakes. In spite of their prevalence, these fishes are not well known to the local populace, because they are very small—just a few cm long—and almost invisible. Except for the eyes, one can see almost nothing of these transparent fishes when they are swimming in small schools near the water surface.

Often the salt-water species are found in great quantities in tidal pools, into which they were brought by the tidal floods. These coastal fishes, prefer to spawn in the mangrove zone. Huge schools of juveniles can be found around the mangroves. The species in mountain streams avoid the strong current, staying in quieter locations. The marine species are found primarily near the shore and in quiet bays, where they find protective concealment among the aquatic plants. Phallostethids feed like other silversides on plankton and small bottom-dwelling organisms. Those in mountain streams feed particularly on insects.

Fig. 19-6. A male phallostethid.

A. SILVERSIDES. THE SILVERSIDES CONTAIN A LARGE NUMBER OF SPECIES WHICH ARE VERY POPULAR AQUARIUM FISHES. THEY INCLUDE THE TOOTHED CARP AND THE LIVEBEARERS.

First Picture Page
Flyingfishes: 1. *Cypselurus furcatus*; 2. *Fodiator acutus*, a primitive small flyingfish; 3. *Exocoetus volitans*, an Atlantic flyingfish.

Second Picture Page
Egg-laying toothed carp:
1. *Aphanius mento*; 2. The pearl fish *Cynolebias ladigesi*; 3. The flag fish *(Jordanella floridae)*; 4. A fundulid, the goldear killy *(Fundulus chrysotus)*; 5. A killifish, *Epiplatys annulatus*; 6. The African killy, *Epiplatys dageti sheljuzkoi*; 7. *Epiplatys dageti monroviae*; 8. Lyretail *(Aphyosemion australe)*; 9. *Aphyosemion filamentosum*; 10. *Aphyosemion nigerianum*; 11. Golden pheasant killy *(Aphyosemion sjoestedti)*; 12. Rough-back killy *(Pachypanchax playfairi)*; 13. Fire killy *(Notobranchius rachovi)*; 14. A rivulus *(Rivulus cylindraceus)*; 15. An African killy *(Aplocheilichthys myersi)*; 16. Malabar killy *(Aplocheilus lineatus)*; 17. Longfin *(Pterolebias longipinnis)*.

Third Picture Page
Freshwater silversides:
1. *Bedotia geayi* from Malagasy; 2. *Pseudomugil signifer* from Australia; 3. *Telmatherina ladigesi* from Celebes; 4. A rainbow fish *(Melanotaenia macculochi)* from Australia; 5. The Australian rainbow fish *(Melanotaenia nigrans)*.

Fourth Picture Page
Egg-laying toothed carp:
1. A male *Aphyosemion filamentosum*; 2. *Aphyosemion exiguum* male; 3. *Aphyosemion fallax*; 4. *Aphyosemion occidentalis toddi* male; 5. *Aphyosemion nigerianum* male; 6. Lyretail *(Aphyosemion australe)* male; 7 and 8. A pearl fish *(Cynolebias elongatus)*; the female stimulates the male in courtship (8); after spawning the pair comes out of the bottom, in which the eggs rest (7); 9. *Austrofundulus transilis* spawning. Livebearers: 10. "Fighting" platy males *(Xiphophorus maculatus)*.

Fifth Picture Page
Egg-laying Toothed Carp:
1. Fire killy *(Nothobranchius rachovi)* male;
2. *Nothobranchius neumanni*; 3. A pearl fish *(Cynolebias belotti)* before spawning: the male seeks a spawning site while the smaller female follows;
5. *Epiplatys annulatus* male;
6. *Aplocheilus dayi* male;
7. The rivulus *Rivulus milesi*; 8 and 9. *Nothobranchius guentheri*: two males are shown in a ritualized fight. Part of this fight consists of "mouthpulling"; 10. *Nothobranchius melanostilus* spawning.

Sixth and Seventh Picture Pages
Silversides (upper pictures): Australian rainbow fish *(Melanotaenia nigrans*, left; *Bedotia geayi*, right). Four-eyed fishes (lower picture): The four-eyed fish *(Anableps anableps)*, a shallow-water fish, in which the upper eye-half is different from the lower half so that the fish can see both above and below the water.

Eleventh Picture Page
Flyingfishes: 1. An ovoviviparous halfbeak, *Dermogenys pusillus*, a non-flying species. Livebearers:
2. *Poecilia parae*; 3. *Poecilia branneri*; 4. The guppy *(Poecilia reticulata)*; a female and six males of different breeds are shown; 5. *Poecilia nigrofasciata*; 6. *Poecilia melanogaster*; 7. Barrigudinyo *(Phalloceros caudomaculatus forma reticulatus)*;
8. Merry widow *(Phallichthys amates)*.

Twelfth Picture Page
Livebearers: 1. Variatus *(Xiphophorus variatus)*;
2. Platy *(Xiphophorus maculatus)*; 3. Swordtail *(Xiphophorus helleri)*; 4. Lucino *(Belonesox belizanus)*;
5. Mosquito fish *(Heterandria formosa)*; 6. *Poecilia velifera*; 7. A gambusia *(Gambusia affinis holbrooki)*.

B. CODFISHES (SEE CHAPTER 18)

Eighth and Ninth Picture Pages
The burbot *(Lota lota)* is the single freshwater codfish distributed in Europe and North America.

Tenth Picture Page
Male guppies: the guppy *(Poecilia reticulata)* is one of the livebearing toothed carps. As a popular aquarium fish it is a truly domesticated animal and is found in a great variety of colors and shapes. The males have particularly striking fin shapes and color patterns.

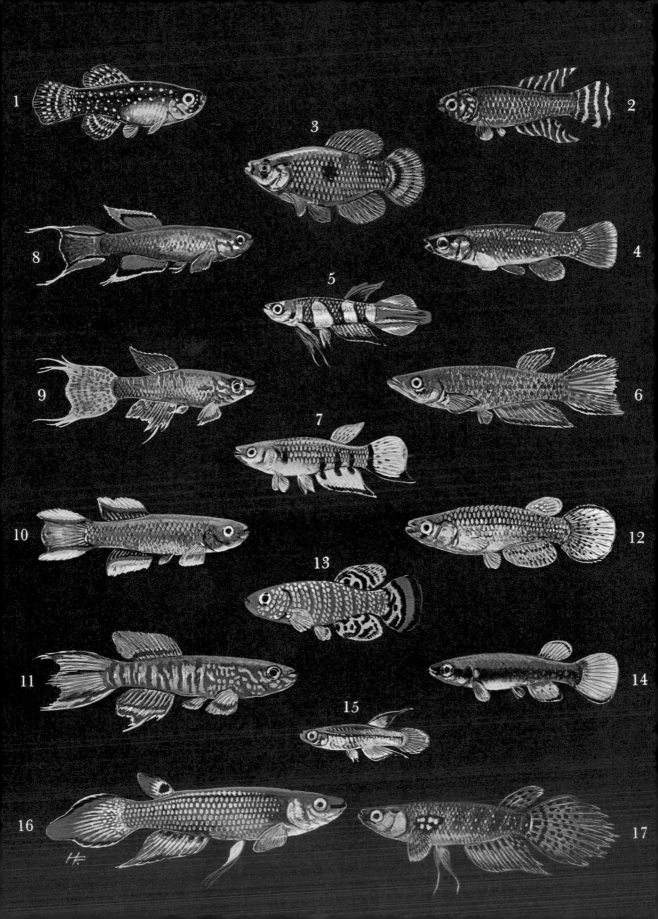

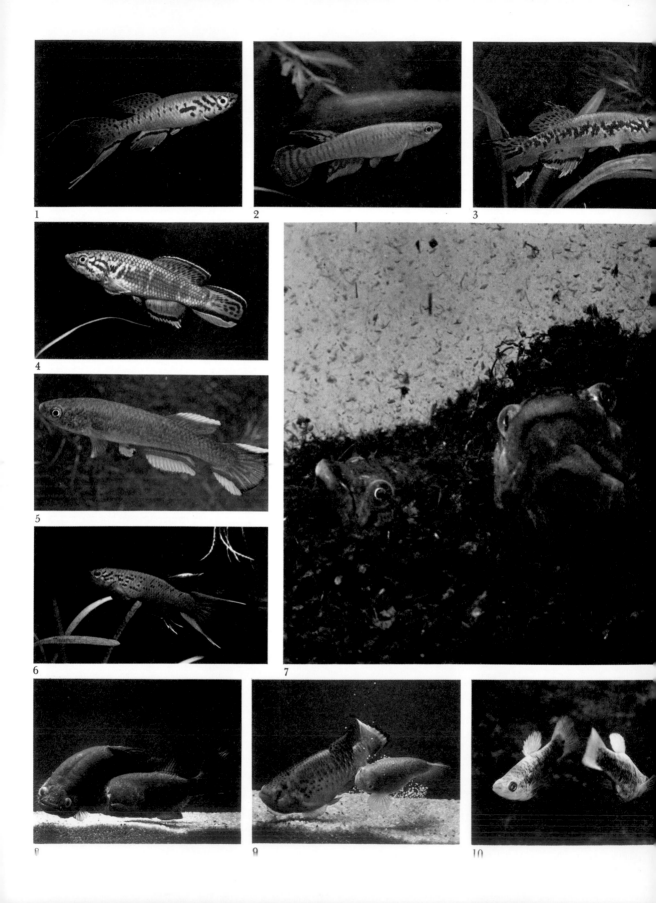

1   2   3

4

5

6   7

8   9   10

2

3

5

6

7

9

10

Systematic Classification

Fossil species have not been included in this survey. + in front of the scientific name indicates that the species or group is extinct, and ⟡ means that the species or group is endangered.

SUBPHYLUM VERTEBRATES (VERTEBRATA)

Superclass Jawless Fishes (Agnatha)

Class Ostracoderms (+Ostracodermata)

Class Cyclostomes (Cyclostomata)

Subclass Hagfishes (Myxini)

A few groups in this extinct class of vertebrates are described in the text or pictured therein (Chapter 2); however these species are not part of this survey.

Family Myxinidae	31	**Subfamily Bdellostomatinae**	31
Subfamily Myxininae	31	Genus *Bdellostoma*	33
Genus *Myxine*	31	Japanese Hagfish, *B. burgeri* Girard, 1854	33
North Atlantic Hagfish, *M. glutinosa* Linné, 1758	31		

Subclass Lampreys (Petromyzones)

Family Petromyzonidae	31	1758)	36
Subfamily Petromyzoninae	34	Brook Lamprey, *L. planeri* (Bloch, 1784)	35
Genus Australian Lampreys (*Geotria*)	34	Genus *Petromyzon*	35
Genus *Ichthyomyzon*	35	Sea Lamprey, *P. marinus* Linné, 1758	37
American Brook Lamprey, *I. fossor* (Reighard and Cumming, 1916)	35	American Sea Lamprey, *P. marinus dorsatus* Wilder, 1883	38
Genus *Lampetra*	35		
Freshwater Lamprey, *L. fluviatilis* (Linné,		**Subfamily Mordacinae**	34

Superclass Fishes (Pisces)

Class Cartilaginous Fishes (Chondrichthyes)

Subclass Elasmobranchs (Elasmobranchii)

Order Sharks (Selachii)

Suborder Comb-toothed Sharks (Notidanoidei)

Family Hexanchidae	90	1880	90
Genus *Hexanchus*	47	Genus *Heptotranchias*	90
Cow Shark, *H. griseus* (Bonnaterre, 1788)	90	Narrow-headed Seven-gilled Shark, *H. perlo* (Bonnaterre, 1788)	90
Six-gilled Shark, *H. corinus* Jordan & Gilbert,			

Suborder Chlamydoselachoidei

Family Frilled Sharks (Chlamydoselachidae)	90	Frilled Shark, *C. anguineus* (Garman, 1884)	90
Genus *Chlamydoselachus*	47		

Suborder Horn Sharks (Heterodontoidei)

Family Horn Sharks (Heterodontidae) 90
  Genus *Heterodontus* 91

Japanese Horn Shark, *H. japanicus* (Dumeril, 1865) 91

Suborder Sharks (Galeoidei)

Family Sand Sharks (Carchariidae) 91
  Genus *Carcharias* 91
    Fierce Shark, *C. ferox* (Risso, 1810) 91
    Tiger Shark, *C. taurus* (Müller & Henle, 1837) 91

Family Goblin Sharks (Scapanorhynchidae) 92
  Genus *Scapanorhynchus* 92
    Japanese Goblin Shark, *S. owstoni* (Jordan, 1898) 92

Family Mackeral Sharks (Isuridae) 92
  Genus *Carcharodon* 95
    Great White Shark, *C. carcharias* (Linné, 1758) 95
  Genus *Lamma* 88
    Porbeagle Shark, *L. nasus* (Bonnaterre, 1788) 95
  Genus *Isurus* 95
    Mako Shark, *I. oxyrhynchus* Rafinesque, 1810 96

Family Basking Sharks (Cetorhinidae) 96
  Genus *Cetorhinus* 96
    Basking Shark, *C. maximum* (Gunner, 1765) 96

Family Thresher Sharks (Alopiidae) 97
  Genus *Alopias* 97
    Thresher Shark, *A. vulpinus* (Bonnaterre, 1788) 97

Family Whale Sharks (Rhincodontidae) 98
  Genus *Rhincodon* 98
    Whale Shark, *R. typus* Smith, 1829 98

Family Carpet Sharks (Orectolobidae) 98
  Genus *Stegostoma* 99
    Zebra Shark, *S. fasciatum* Müller & Henle, 1837 99
  Genus *Orectolobus* 99
    Wobbegong, *O. maculatus* (Bonnaterre, 1788) 99

Family Dogfishes (Scyliorhinidae) 99

  Genus *Scyliorhinus* 100
    Smaller Spotted Dogfish, *S. caniculus* (Linné, 1758) 100
    Larger Spotted Dogfish, *S. stellaris* (Linné, 1758) 100
  Genus *Galeus* 100
    Black-mouthed Dogfish, *G. melanostomus* Rafinesque, 1841 100

Family Smooth Hounds (Triakidae) 101
  Genus *Mustelus* 101
    Smooth Hound, *M. mustelus* (Linné, 1758) 101
    Stellate Smooth Hound, *M. asterias* Cloquet, 1821 101
  Genus *Triaenodon* 101
    Smooth Dogfish, *T. obesus* (Rüppell, 1835) 101

Family False Cat Sharks (Pseudotriakidae) 102
  Genus *Pseudotriakis* 102
    Atlantic False Cat Shark, *P. microdon* Capello, 1868 102

Family Grey Sharks (Carcharhinidae) 102
  Genus *Prionace* 102
    Great Blue Shark, *P. glauca* (Linné, 1758) 102
  Genus *Galeocerdo* 105
    Tiger Shark, *G. cuvieri* (Le Sueur, 1822) 105
  Genus *Galeorhinus* 105
    Tope, *G. galeus* (Linné, 1758) 105
  Genus *Carcharhinus* 106
    Atlantic Grey Shark, *C. plumbeus* Nardo, 1827 106
    Black-tip Reef Shark, *C. melanopterus* (Quoy & Gaimand, 1824) 106
    C. menisorrah Müller & Henle, 1837 107

Family Hammerhead Sharks (Sphyrnidae) 106
  Genus *Sphyrna* 106
    Great Hammerhead, *S. mokkaran* Rüppell, 1835 106
    Common Hammerhead, *S. zygaena* (Linné, 1758) 115

Suborder Spiny Dogfishes (Squaloidei)

Family Angular Rough Sharks (Oxynotidae) 115
  Genus *Oxynotus* 115
    Humantin, *O. centrina* (Linné, 1758) 115

Family Spiny Dogfishes (Squalidae) 115
  Genus *Squalus* 115

Piked Dogfish, *S. acanthias* (Linné, 1758) 115
  Genus *Etmopterus* 116
    Lantern Shark, *E. spinax* (Linné, 1758) 116

Family Spineless Dogfishes (Dalatiidae) 116
  Genus *Somniosus* 116

Greenland Shark, *S. microcephalus* (Bloch & Schneider, 1801) — 116
Sleeper Shark, *S. rostratus* (Macri, 1819) — 116
Genus *Dalatius* — 116
Darkie Charlie, *D. licha* (Bonnaterre, 1788) — 116

Family Bramble Sharks (Echinorhinidae) — 116
Genus *Echinorhinus* — 116
Bramble Shark, *E. brucus* (Bonnaterre, 1788) — 116

Suborder Saw Sharks (Pristiphoroidei)

Family Pristiphoridae — 119
Genus Saw Sharks *(Pristiophorus)* — 119

Suborder Squatinoids (Squatinoidei)

Family Monk Fishes (Squatinidae) — 119
Genus *Squatina* — 119
Angel Shark, *S. squatina* (Linné, 1758) — 119

Order Rays and Skates (Rajiformes)

Suborder Sawfishes (Pristioidei)

Family Sawfishes (Pristidae) — 120
Genus *Pristis* — 120
Sawfish, *P. pectinatus* Latham, 1794 — 120
Sawfish, *P. pristis* (Linné, 1758) — 120

Suborder Guitar Fishes (Rhinobatoidei)

Family Rhinobatidae — 121
Genus Guitar Fishes *(Rhinobatos)* — 121
Guitar Fish, *R. rhinobatos* (Linné, 1758) — 121
R. productus Ayres, 1854 — 117
R. cemiculus (St. Hilaire, 1817) — 121

Suborder Electric Rays (Torpedinoidei)

Family Electric Rays (Torpedinidae) — 121
Genus *Torpedo* — 121
Eyed Electric Ray, *T. torpedo* (Gmelin, 1788) — 122
Marbled Electric Ray, *T. marmorata* Risso, 1810 — 122
Black Electric ray, *T. nobiliana* — 122

Family Narkidae — 121

Family Temeridae — 121

Suborder Rays (Rajoidei)

Family Skates (Rajidae) — 120
Genus *Raja* — 88
Thornback Ray, *R. clavata* Linné, 1758 — 123
Flapper Skate, *R. batis* Linné, 1758 — 123
Undulate Ray, *R. undulata* Lacépède, 1798 — 128

Family Anachantobatidae — 122

Family Arynchobatidae — 122

Suborder Eagle Rays (Myliobatoidei)

Family Stingrays (Dasyatidae) — 123
Genus *Dasyatis* — 123
Common Stingray, *D. pastinaca* (Linné, 1758) — 123
Stingray, *D. centroura* (Mitchill, 1815) — 123
Blue Stingray, *D. violacea* (Bonaparte, 1832) — 124
Genus *Taeniura* — 124
T. lymna (Cuvier, 1829) — 124

Family Urolophidae — —

Family Potamotrygonidae — —

Family Butterfly Rays (Gymnuridae) — 124
Genus *Gymnura* — 124
Butterfly Ray, *G. altavela* (Linné, 1758) — 124

Family Eagle Rays (Myliobatidae) — 124
Genus *Myliobatis* — 124
Spotted Eagle Ray, *M. aquila* (Linné, 1758) — 124
Genus *Pteromylaeus* — 124
Bull Ray, *P. bovinus* (St. Hilaire, 1809) — 124

Family Manta Rays (Mobulidae) — 125

Genus *Mobula*   125
Devil-fish, *M. mobular (Bonaterre, 1788)*   125

Genus *Manta*   125
Manta Ray, *M. birostris* (Walbaum, 1792)   125

Subclass Chimaeras (Holocephali)

Family Chimaeridae   126
Genus *Chimaera*   126
Rabbit Fish, *C. monstrosa* (Linné, 1758)   126
Genus *Hydrolagus*   128
Ratfish, *H. colliei* (Lay & Bennett, 1839)   128

Family Rhinochimaeridae   126

Genus *Harriotta*   129
H. raleighana Goode & Bean, 1894   129

Family Callorhynchidae   126
Genus *Callorhynchus*   129
Monkeyfish, *C. capensis* Dumeril, 1865   129

Class Teleost (Bony) Fishes (Osteichthyes)

Subclass Spiny-rayed Fishes (Actinopterygii)

Superorder Bichirs (Polypteri)

Order Polypterids (Polypteriformes)

Family Polypteridae   132
Genus Bichirs *(Polypterus)*   132
P. senegalus Cuvier, 1829   132

P. ornatipinnis Boulenger, 1902   132
Genus Reedfishes *(Calamoichthys)*   132
C. calabaricus Smith, 1865   132

Superorder Chondrostei

Order Sturgeon (Acipenseriformes)

Family Sturgeon (Acipenseridae)   134
Subfamily Sturgeon (Acipenserinae)   134
Genus *Huso*   135
European Sturgeon, *H. huso* (Linné, 1758)   135
Kaluga, ◊*H. dauricus* (Georgi, 1775)   136
Genus *Acipenser*   135
Common Atlantic Sturgeon, *A. sturio* Linné, 1758   136
Sterlet, *A. ruthenus* Linné, 1758   136
Adriatic Sturgeon, *A. naccari* Bonaparte, 1836   137
A. stellatus Pallas, 1771   137
A. gueldenstaedti Brandt, 1833   137
A. gueldenstaedti gueldenstaedti Brandt, 1833   137
A. gueldenstaedti persicus Borodin, 1897   137
A. gueldenstaedti colchicus v. Marti, 1940   137
A. nudiventris Lovetzky, 1828   137
Siberian Sturgeon, *A. baeri* Brandt, 1869   137
Amur Sturgeon, ◊ *A. schrencki* Brandt, 1869   137
A. Sinensis Gray, 1834   138
A. dabryanus Duméril, 1868   138
A. kikuchii Jordan & Snyder, 1908   138
A. multiscutatus Tanaka, 1908   138
A. medirostris Ayres, 1854   138
White Sturgeon, *A. transmontanus* Richardson, 1836   138
Atlantic Sturgeon, ◊ *A. oxyrhynchus* Mitchill,

1814   138
A. brevirostrum Le Sueur, 1818   138
Lake Sturgeon, ◊ *A. fulvescens* Rafinesque, 1817   138

Subfamily Shovel-nosed Sturgeon (Scaphirhynchinae)   141
Genus American Shovel-nosed Sturgeon *(Scaphirhynchus)*   141
Shovel-nosed Sturgeon, *S. platorhynchus* (Rafinesque, 1820)   141
S. albus (Forbes & Richardson, 1905)   141
S. mexicanus (Giltay, 1929)   141
Genus Asiatic Shovel-nosed Sturgeon *(Pseudoscaphirhynchus)*   141
P. kaufmanni (Bogdanow, 1874)   141
P. hermanni (Kessler, 1877)   141
P. fedtschenkoi (Kessler, 1872)   141

Family Paddlefishes (Polyodontidae)   142
Genus *Polyodon*   142
American Paddlefish, *P. spathula* (Walbaum, 1792)   142
Genus *Psephurus*   143
Chinese Sturgeon, *P. gladius* (Martens, 1826)   143

Superorder Lower Bony Fishes (Holostei)

Order Gars (Lepisosteiformes)

Family Gars (Lepisosteidae) 144
Genus *Lepisosteus* 147
Giant Gar, *L. tristoechus* (Bloch & Schneider, 1801) —
L. tropicus (Gill, 1863) 147
Longnose Gar, *L. osseus* (Linné, 1758) 147

Gar, *L. productus* (Cope, 1865) —
L. platyrhincus (De Kay, 1842) 147
L. platostomus (Rafinesque, 1820) 147
Mississippi Alligator Gar, *L. spatula* (Lacépède, 1803) 147

Order Bowfins (Amiiformes)

Family Bowfins (Amiidae) —
Genus *Amia* 149

Bowfin, *A. calva* Linné, 1766 149

Superorder Bony (Teleost) Fishes (Teleostei)

Order Tarpons (Elopiformes)

Suborder Tarpons (Elopoidei)

Family Elopidae 152
Genus *Elops* 152
Lady fish, *E. saurus* Linné, 1766 152

Family Megalopidae 153

Genus *Megalops* 153
Atlantic Tarpon, *M. atlanticus* (Cuvier & Valenciennes, 1846) 153
Ox-eye Herring, *M. cyprinoides* (Broussonet, 1782) 153

Suborder Bonefishes (Albuloidei)

Family Albulidae 156
Genus *Albula* 156
Bonefish, *A. vulpes* (Linné, 1758) 156
Genus *Dixonina* 156
D. nemoptera Fowler, 1911 156

Family Deepsea Bonefishes (Pterothrissidae) 156
Genus *Pterothrissus* 156
Japanese Deepsea Bonefish, *P. gissu* Hilgendorf, 1877 156

Order Eels (Anguilliformes)

Suborder Anguilloidei

Family Freshwater Eels (Anguillidae) 160
Genus *Anguilla* 160
European Eel, *A. anguilla* (Linné, 1758) 160
American Eel, *A. rostrata* (Le Sueur, 1821) 160
Japanese Eel, *A. japonica* Schlegel, 1850 163

Family Moringuidae 163
Genus *Moringua* 163
M. macrochir Bleeker, 1855 163
M. bicolor Kaup, 1856 163
M. javanica (Kaup, 1856) 163
Genus *Stilbiscus* 163
S. edwardsi Jordan & Bollmann, 1888 164

Family Myrocongridae 164
Genus *Myroconger* 164

M. compressus Günther, 1870 164

Family Morays (Muraenidae) 164
Genus *Echidna* 164
Zebra Moray, *E. zebra* (Shaw, 1797) 164
E. nebulosa (Ahl, 1789) 168
Genus *Gymnothorax* —
G. flavimarginata (Rüppell, 1828) —
Genus Morays *(Muraena)* 165
Moray Eel, *M. helena* Linné, 1758 165
M. pardalis Schlegel, 1847 164
Genus *Thyrsoidea* 165
T. macrurus (Bleeker, 1854) 165

Family Pike Eels (Muraenesocidae) 165
Genus *Muraenesox* 165

Batavia Putyekanipa, *M. cinereus* (Forskal, 1775) ... 165
Larger Putyekanipa, *M. talabon* (Cantor, 1850) ... 165
Indian Putyekanipa, *M. talabonoides* (Bleeker, 1853) ... 165

Family Conger Eels (Congridae) ... 165
Genus *Conger* ... 165
Conger Eel, *C. conger* (Linné, 1758) ... 165
Genus *Ariosoma* ... 166
A. balearica (De la Roche, 1809) ... 166
Genus Deepsea Conger Eels *(Bathycongrus)* ... 166
B. mystax (De la Roche, 1809) ... 166
Genus *Promyllantor* ... 166
P. latedorsalis (Roule, 1916) ... 166

Family Garden Eels (Hetercongridae) ... 166
Genus *Gorgasia* ... 173
Garden Eel, *G. maculata* Klausewitz & Eibl-Eibesfeldt, 1959 ... 173

Family Snake Eels (Ophichthyidae) ... 166
Genus *Ophichthys* ... 169
O. gomesii (Castelnau, 1855) ... 169
O. ophis (Linné, 1758) ... 169
Genus *Pisoodonophis* ... 169
P. cruentifer Goode & Bean, 1895 ... 169

Family Cutthroat Eels (Synaphobranchidae) ... 169
Genus *Synaphobranchus* ... 169
Gray's Cutthroat Eel, *S. pinnatus* (Gronov, 1854) ... 169

Family Ilyophidae ... 169
Genus *Ilyophis* ... 169
I. brummeri Gilbert, 1891 ... 169

Family Snubnose Eels (Simenchelyidae) ... 169
Genus *Simenchelys* ... 169
Slime Eel, *S. parasiticus* Gill, 1879 ... 169

Family Snipe Eels (Nemichthyidae) ... 169
Genus *Nemichthys* ... 169
N. scolopaceus Richardson, 1848 ... 169
Genus *Cercomitus* ... 169
C. flagellifer Weber, 1913 ... 169

Family Serrivomeridae ... 170
Genus *Serrivomer* ... 170
S. sector Garman, 1899 ... 170
Genus *Spinivomer* ... 170
S. goodei Gill & Ryder, 1883 ... 170

Family Cyemidae ... 170
Genus *Cyema* ... 170
Deepwater Eel, *C. atrum* Günther, 1878 ... 170

Suborder Swallowers (Saccopharyngoidei)

Family Swallowers (Saccopharyngidae) ... 170
Genus *Saccopharynx* ... 170
Pelican-fish, *S. ampullaceus* (Harwood, 1827) ... 170

Family Gulpers (Eupharyngidae) ... 170

Genus *Eupharynx* ... 171
E. pelecanoides Vaillant, 1882* ... 171

Family Monognathidae ... 171
Genus *Monognathus* ... —

Order Notacanthiformes

Family Spiny Eels (Notacanthidae) ... 171

Family Halosauridae ... —

Genus *Halosauropsis* ... 176
H. macrochir (Günther 1878) ... 176

Order Herring (Clupeiformes)

Suborder Denticipitoids (Denticipitoidei)

Family Denticipitidae ... 172
Genus *Denticeps* ... 172

D. clupeoides Clausen, 1959 ... 172

Suborder Herring (Clupeoidei)

Family Herring (Clupeidae) ... 177
Genus Herring *(Clupea)* ... 177
Atlantic Herring, *C. harengus* Linné, 1758 ... 178
C. harengus membras Pallas, 1811 ... 72
Pacific Herring, *C. pallasii* (Cuvier &

Valenciennes, 1847) ... 178
Genus Sprats *(Sprattus)* ... 189
Sprat, *S. sprattus* (Linné, 1758) ... 190
Falkland Sprat, *S. fuegensis* Jenyns, 1842 (= *Clupea fuegensis*) ... —

Genus *Harengula* 190
  H. zunasi Bleeker, 1854 191
Genus *Clupeonella* 191
  C. delicatula (Nordmann, 1840) —
Genus *Sardinella* 191
  S. aurita (Cuvier & Valenciennes, 1847) 191
Genus *Sardinops* 191
  Pacific Sardine, *S. caerula* (Girard, 1854) 191
  South American Sardine, *S. sagax* (Jenyns, 1842) 191
  Japanese Sardine, *S. melanosticta* (Temminck & Schlegel, 1846) 191
  Australian Sardine, *S. neopilchardus* (Steindachner, 1879) 191
  South African Sardine, *S. ocellata* (Poppe, 1853) 191
Genus *Sardina* 192
  Pilchard, *S. pilchardus* (Walbaum, 1792) 192
Genus Shad *(Alosa)* 193
  Shad, *A. alosa* Linné, 1758 193
  American Shad, *A. sapidissima* (Wilson, 1811) 194
  Twaite Shad, *A. fallax* (Lacépède, 1800) 194
  A. alabamae Jordan & Evermann, 1896 195
Other important genera:

Opisthonema, Lile, Rhinosardina, Kowala, Clupeoides, Pellonula, Poecilothrissa, Potamalosa, Hyperlophus, Ethmalosa, Hilsa, Dorosoma, Nematalosa, Ilishu, Caspialosa, Pomolobus, Spratelloides, Jenkinsia, Dussumieria, Etrumeus —

Family Anchovies (Engraulidae) 196
Genus *Engraulis* 196
  Anchovy, *E. encrasicholus* Linné, 1758 196
  Anchoveta, *E. ringens* Jenyns, 1842 (= *E. anchoveta*) 200
  Japanese Anchovy, *E. japonicus* Temminck & Schlegel, 1842 200
  South African Anchovy, *E. capensis* Gilchrist, 1913 200
  North American Anchovy, *E. mordax* Girard, 1854 200
Other Genera:
Cetengraulis, Anchovia, Anchoa, Anchoviella, Pterengraulis, Hildebrandichthys —

Family Wolf Herring (Chirocentridae) 200
Genus *Chirocentrus* 200
  Wolf Herring, *C. dorab* (Forskal, 1775) 200

Order Osteoglossids (Osteoglossiformes)

Suborder Bonytongues (Osteoglossoidei)

Family Bonytongues (Osteoglossidae) 202
Genus *Arapaima* 202
  Giant Arapaima, *A. gigas* (Cuvier, 1817) 202
Genus *Osteoglossum* 203
  Arowana, *O. bicirrhosum* Vandelli, 1829 203
  O. ferrairai Kanazawa, 1966 204
Genus *Scleropages*
  Malayan Bonytongue, ♦ *S. formosus* (Müller & Schlegel, 1844) —

  Australian Bonytongue, *S. leichhardti* Günther, 1864 —
Genus *Clupisudis* 202
  African Bonytongue, *C. niloticus* (Cuvier, 1829) 204

Family Butterfly Fishes (Pantodontidae) 204
Genus *Pantodon* 288
  Butterfly Fish, *P. buchholzi* Peters, 1876 204

Suborder Featherbacks (Notopteroidei)

Family Mooneyes (Hiodontidae) 205
Genus *Hiodon* 205
  Mooneye, *H. tergisus* Le Sueur, 1818 205
  H. alosoides (Rafinesque, 1819) 205
  H. salenops Jordan & Bean, 1877 205

Family Featherbacks (Notopteridae) 205
Genus *Notopterus* 205
  Featherback, *N. chitala* (Hamilton & Buchanan, 1822) 205
Genus *Xenomystus* 205
  X. nigri (Günther, 1868) 205

Order Mormyrids (Mormyriformes)

Family Mormyrids (Mormyridae) 210
Genus *Petrocephalus* 210
Genus *Marcusenius* 210
  M. stanleyanus (Boulenger, 1897) 210
  M. monteiri (Günther, 1873) —
  M. sphecodes (Sauvage, 1878) 210
  M. isidori (Cuvier & Valenciennes, 1846) 210

Genus *Stomatorhinus* 210
Genus *Mormyrops* 209
  M. boulengeri (Pellegrin, 1900) 211
Genus *Mormyrus* 211
  M. kannume Forskal, 1775 209
Genus *Boulengeromyrus* 211
  B. knoepffleri Taverne & Gery, 1968 211

Genus *Hyperopisus* 211
  H. bebe (Lacépède, 1803) 211
Genus *Myomyrus* 211
Genus *Hippopotamyrus* 211
Genus *Isichthys* 211
Genus *Campylomormyrus* 210
Genus *Genyomyrus* 211
  G. donnyi Boulenger 1903 211

Genus *Gnathonemus* 211
  G. petersi (Günther, 1862) 212
  G. tamandua (Günther, 1864) 223

Family Gymnarchids (Gymnarchidae) 212
Genus *Gymnarchus* 212
  Gymnarchid, *G. niloticus* Cuvier, 1830 212

Order Salmon (Salmoniformes)

Suborder Salmon (Salmonoidei)

Family Salmon (Salmonidae) 213
Subfamily Salmoninae 213
Genus *Salmo* 213
  Atlantic Salmon, *S. salar* Linné, 1758 214
  Brown Trout, *S. trutta* Linné, 1758 218
  Oceanic Trout, *S. trutta trutta* Linné, 1758 218
  River Trout, *S. trutta fario* Linné, 1758 218
  Lake Trout, *S. trutta lacustris* Linné, 1758 218
  Rainbow Trout, *S. gairdneri* Richardson, 1836 225
Genus Pacific Salmon (*Oncorhynchus*) 227
  Pink Salmon, *O. gorbuscha* (Walbaum, 1792) 227
  Sockeye Salmon, *O. nerka* (Walbaum, 1792) 227
  Chinook Salmon, *O. tschawytscha* (Walbaum, 1792) 227
  Dog Salmon, *O. keta* (Walbaum, 1792) 227
  Silver Salmon, *O. kisutch* (Walbaum, 1792) 227
  Masu, *O. masou* (Brevoort, 1856) 227
Genus Charrs (*Salvelinus*) 230
  European Charr, *S. alpinus* (Linné, 1758) 231
  Brook Trout, *S. fontinalis* (Mitchill, 1815) 232
  Lake Trout, *S. namaycushi* (Walbaum, 1792) 232
Genus *Hucho* 214
  Danube Salmon, *H. hucho* (Linné, 1758) 233
  H. taimen (Pallas, 1773) 233
Genus *Brachymystax* 214
  Lenok, *B. lenok* (Pallas, 1776) 233

Subfamily Whitefishes (Coregoninae) 234
Genus *Stenodus* —
  Inconnu, ⚥ *S. leucichthys* (Güldenstädt, 1772) —
Genus *Coregonus* 235
  Tungun, *C. tugun* (Pallas, 1811 239

Tschirr, *C. nasus* (Pallas, 1776) 239
C. albula Linné, 1758 240
Shallowwater Cisco, *C. albula artedi* Le Sueur, 1818 241
Siberian Whitefish, *C. albula sardinella* Valenciennes, 1848 241
Houting, *C. larvaretus* (Linné, 1758) 241
C. oxyrhynchus (Linné, 1758) 243
C. fera Jurine, 1825 243
C. pidschian (Gmelin, 1788) 242

Subfamily Graylings (Thymallinae) 244
Genus Graylings (*Thymallus*) 244
  European Grayling, *T. thymallus* Linné, 1758 244
  T. arcticus (Pallas, 1776) 244

Family Ayus (Plecoglossidae) 245
Genus *Plecoglossus* 245
  Ayu, *P. altivelis* (Schlegel, 1850) 245

Family Smelt (Osmeridae) 246
Genus *Hypomesus* 246
Genus *Taleichthys* 246
Genus Smelt, *Osmerus* 246
  Smelt, *O. eperlanus* (Linné, 1758) 246
  European Smelt, *O. eperlanus eperlanus* (Linné, 1758) 246
  American Smelt, *O. eperlanus mordax* (Mitchill, 1815) 247
Genus Capelin (*Mallotus*) 246
  Capeling, *M. villosus* (Müller, 1776) 247

Suborder Galaxiids (Galaxioidei)

Family Icefishes (Salangidae) 248
Genus *Salanx* —

Family New Zealand Trout (Retropinnidae) 248

Family Galaxiids (Galaxiidae) 248
Genus *Galaxias* 249
  G. alepidotus (Forster, 1844) 249
  G. maculatus (Jenyns, 1842) 249

  G. attenuatus (Jenyns, 1842) 249
Genus *Brachygalaxias* 249
  B. bullocki (Regan, 1908) 249
Genus *Neochanna* 249
  N. apoda Günther, 1867 249

Family Aplochitonids (Aplochitonidae) 249
Genus *Aplochiton* 250
Genus *Prototroctes* 250

Suborder Pike (Esocoidei)

Family Pike (Esocidae)   250
  Genus Pike *(Esox)*   251
    Northern Pike, *E. lucius* Linné, 1758   250
    Amur Pike, *E. reicherti* Dybowski, 1869   251
    Muskellunge, *E. masquinongyi* Mitchill, 1824   251
    Chain Pickerel, *E. niger* Le Sueur, 1818   251
    Pickerel, *E. americanus* Gmelin, 1788   251
    Grass Pickerel, *E. vermiculatus* Le Sueur, 1846   251

Family Mudminnows (Umbridae)   251
  Genus Umbra   251

European Mudminnow, *U. krameri* Walbaum, 1792   251
Central Mudminnow, *U. limi* (Kirtland, 1840)   252
U. pygmaea (De Kay, 1842)   252
Genus Novumbra   252
Olympic Mudminnow, ⚥ *N. hubbsi* Schultz, 1929   252
Genus Dallia   252
Alaska Blackfish, *D. pectoralis* Bean, 1879   252

Suborder Argentines (Argentinoidei)

Family Argentines (Argentinidae)   252

Family Deepsea Smelts (Bathylagidae)   252

Family Barreleyes (Opisthoproctidae)   252
  Genus Opisthoproctus   252
  Barreleye, *O. grimaldii* Zugmayer, 1911   252

Suborder Stomiatoids (Stomiatoidei)

Family Deepsea Scaleless Dragonfishes (Melanostomiatidae)   253
  Genus Bathophilus   253

Family Deepsea Scaly Dragonfishes (Stomiatidae)   253

Family Deepsea Stalkeye Fishes (Idiacanthidae)   253
  Genus Idiacanthus   257
  Deepsea Stalkeye Fish, *I. fasciola* Peters, 1876   266

Family Deepsea Snaggletooths (Astronesthidae)   253

Family Deepsea Loosejaws (Malacosteidae)   253
  Genus Ultistomias   253
  Deepsea loosejaw, *U. mirabilis* Beebe, 1933   253

Family Deepsea Bristlemouths (Gonostomatidae)   253

Genus Cyclothone   254
  C. signata Garman, 1899   266
Genus Maurolicus   254
  M. muelleri (Gmelin, 1788)   266
Genus Diaphus   266
  D. elucens Brauer, 1904   266
Genus Vinciguerra   254
  V. lucetia (Garman, 1890)   254

Family Deepsea Hatchet Fishes (Sternoptychidae)   254
  Genus Argyropelecus   254
  Deepsea Hatchet Fish, *A. affinis* Garman, 1899   266

Family Deepsea Viperfishes (Chauliodontidae)   254
  Genus Chauliodus   254
  Viperfish, *C. sloani* Schneider, 1801   254

Suborder Deepsea Slickheads (Alepocephaloidei)

Family Platyproctidae   258
  Genus Dolichopteryx   258

Family Searsiidae   258

Suborder Bathylaconoidei

Suborder Lanternfishes (Myctophoidei)

Family Lanternfishes (Myctophidae)   258
  Genus Lampanyctus   259
    L. leucopsarus Eigenmann & Eigenmann, 1890   259
    L. crocodilus (Risso, 1810)   404

Genus Myctophum   259
  M. punctatum Rafinesque, 1810   259
  M. affine (Lütken, 1892)   259
  M. cocco (Cocco, 1829)   —

M. phengodes (Lütken, 1892)   266

Family Neoscopelidae   260
   Genus *Neoscopelus*   260
     N. macrolepidotus Johnson, 1863   260

Family Aulopodidae   260
   Genus *Aulopus*   260
     Aulopid, *A. japonicus* Günther, 1880   260
     Sergeant Baker, *A. purpurrissatus* Richardson, 1843   260
     A. filamentosus (Bloch, 1791)   260

Family Lizard Fishes (Synodontidae)   260
   Genus *Synodus*   260
     S. foetens (Linné, 1766)   260
     S. synodus (Linné, 1758)   261
     S. lucioceps (Ayres, 1855)   261
   Genus *Trachynocephalus*   261
     T. myops (Schneider, 1801)   261
   Genus *Bathysaurus*   260
   Genus *Saurida*   261

Family Bombay Ducks (Harpodontidae)   261
   Genus *Harpodon*   261

Family Chlorophthalmidae   261
   Genus *Chlorophthalmus*   —

Family Bathypteroidae   261
   Genus *Bathypterois*   261
     B. ventralis Garman, 1899   266

Family Ignopidae   —

Family Paralepidid Lanternfishes (Paralepididae)   262
   Genus *Paralepis*   262
     P. atlantica Kroyer, 1891   —
   Genus *Lestidium*   262

Family Anotopteridae   262
   Genus *Anotopterus*   262
     A. pharao Zugmayer, 1911   262

Family Pearleyes (Scopelarchidae)   262

Family Evermannellidae   263

Family Omosudidae   263
   Genus *Omosudis*   266
     O. lowii Günther, 1887   266

Family Lancetfishes (Alepisauridae)   263

Family Scopelosauridae   263

Order Cetomimiformes

Suborder Ateleopids (Ateleopodoidei)

Family Ateleopids (Ateleopodidae)   264

Suborder Mirapinnatoids (Mirapinnatoidei)

Family Mirapinnidae   264   **Family Eutaeniophoridae**   264

Family Kasidoridae   —

Suborder Cetomimoidei

Family Cetomimidae   264   **Family Rondeletiidae**   267
     Genus *Rondeletia*   267

Family Barbourisiidae   267

Suborder Giganturids (Giganturoidei)

Family Giganturids (Giganturidae)   268   **Family Rosaurids (Rosauridae)**   268
   Genus *Gigantura*   268
     G. vorax Regan, 1925   268

Order Ctenothrissiformes

Family Macristiidae   268   *M. chavesi* Regan, 1903   268
   Genus *Macristium*   268

Order Gonorynchiformes

Suborder Milkfishes (Chanoidei)

Family Milkfishes (Chanidae) 269
  Genus *Chanos* 269
    Milkfish, *C. chanos* Forskal, 1755 269

Family Kneriids (Kneriidae) 271
  Genus *Kneria* 271
    K. polli Trewavas, 1936 285
  Genus *Parakneria* 271

Family Cromeriids (Cromeriidae) 273

  Genus *Cromeria* 273
    C. nilotica Boulenger, 1901 —

Family Grassé Fishes (Grasseichthyidae) 274
  Genus *Grasseichthys* 274
    Grassé Fish, *G. gabonensis* Géry, 1964 274

Family Phractolaemiids (Phractolaemiidae) 274
  Genus *Phractolaemus* 274
    P. ansorgei Boulenger, 1901 274

Suborder Gonorynchoidei

Family Gonorynchidae 274
  Genus *Gonorynchus* 274

    G. gonorynchus Ridewood, 1905 274

Order Carp (Cypriniformes)

Suborder Characins (Characoidei)

Family Characins (Characidae) 278
Subfamily Agoniatinae 277
  Genus *Agoniates* 277

Subfamily Rhaphiodontinae 277
  Genus *Rhaphiodon* 278
    R. gibbus Agassiz, 1829 278
  Genus *Hydrolycus* 278

Subfamily Characins (Characinae) 277
  Genus *Charax* 278
  Genus *Cynopotamos* 278
  Genus *Acestrocephalus* 278
  Genus *Gilbertolus* 278
  Genus *Gnathocharax* 278
  Genus *Hoplocharax* 278
    H. goethei Géry, 1968 278
  Genus *Exodon* 278
  Genus *Roeboides* 278
    R. guatemalensis (Günther, 1864) 278
  Genus *Roeboexodon* 278
  Genus *Oligosarchus* 278
  Genus *Acestrorhynchus* 278

Subfamily Bryconinae 279
  Genus *Brycon* 279
  Genus *Chalceus* 279
  Genus *Salminus* 279
  Genus *Triportheus* 279

Subfamily Clupeacharacinae 279
  Genus *Clupeacharax* 279
    C. anchoveoides Pearson, 1924 279

Subfamily Iguanodectinae 279

Subfamily Paragoniatinae 279

Subfamily Aphyocharacinae 279
  Genus *Aphyocharax* 286
    A. rubripinnis Pappenheim, 1921 286

Subfamily Glandulocaudinae 279
  Genus *Tyttocharax* 279
  Genus *Xenurobrycon* 279
  Genus *Corynopoma* 279
    C. riisei Gill, 1858 279
  Genus *Pseudocorynopoma* 291
    P. doriae Perugia, 1891 291

Subfamily Stethaprioninae 279
  Genera *Stethaprion, Brachychalcinus, Poptella, Ephippicharax* —

Subfamily Tetras (Tetragonopterinae) 279
  Genus *Hemigrammus* 280
    Head-and-tail-Light, *H. ocellifer* (Steindachner, 1883) 280
    H. pulcher Ladiges, 1938 286
    Rummy Nose, *H. rhodostomus* E. Ahl, 1924 286
  Genus *Moenkhausia* 280
  Genus *Hyphessobrycon* 280
    Rosy Tetra, *H. ornatus* E. Ahl, 1934 280
    Tetra Perez, *H. erythrostigma* (Fowler, 1943) 282
    False Tetra, *H. simulans* Géry, 1963 283
    Red Tetra, from Rio, *H. flammeus* Myers, 1924 280

H. *rubrostigma* Hoedemann, 1956   286
Serpa Tetra, *H. callistus* (Boulenger, 1900)   286
H. heterorhabdus (Ulrey, 1895)   286
Genus *Hemibrycon*   –
Genus *Probolodus*   287
Genus *Anoptichthys*   280
   A. jordani Hubbs & Innes, 1936   287
Genus *Astyanax*   280
   A. fasciatus mexicanus (Filippi, 1853)   280
Genus *Stygichthys*   281
   S. typhlops Britton & Böhlke, 1965   281
Genus *Gymnocharacinus*   282
Genus *Thayeria*   286
   T. obliqua Eigenmann, 1908   286
   T. boehlkei Weitzmann, 1957   286
Genus *Nematobrycon*   286
   N. palmeri Eigenmann, 1911   286

Subfamily Cheirodontinae   282
Genus *Megalamphodus*   282
Genus *Pristella*   282
   Pristella, *P. riddlei* Meek, 1907   282
Genus *Cheirodon*   282
   Red neon, *C. axelrodi* Schultz, 1956   282
Genus *Odontostilbe*   282
Genus *Paracheirodon*   283
   Neon Tetra, *P. innesi* Myers, 1936   283

Subfamily Rhoadsinae   283

Family Serrasalmidae   283

Subfamily Serrasalminae   –
Genus Piranhas *(Serrasalmus)*   284
   Piranha, *S. piraya* Cuvier, 1819   284
   Piranha, *S. nattereri* (Kner, 1859)   284

Subfamily Myleinae   287
Genus *Metynnis*   287
   M. schreitmuelleri E. Ahl, 1922   307
   M. roosevelti Eigenmann, 1915   307
Genus *Mylossoma*   287
Genus *Myleus*   287
   M. pacu (Schomburgk, 1841)   287
Genus *Colossoma*   287
   C. bidens (Spix, 1829)   287
   C. oculum (Cope, 1872)   287
Genus *Myloplus*   307
   M. schultzei E. Ahl, 1938   307

Subfamily Catoprioninae   287
Genus *Catoprion*   287
   C. mento (Cuvier, 1819)   287

Family Hatchet Fishes (Gastropelecidae)   288
Genus *Gastropelecus*   307
   Hatchet Fish, *G. sternicla* (Linné, 1758)   307

Genus *Carnegiella*   291
   Black-Winged Hatchet Fet, *C. marthae*
   (Myers, 1927)   291
   C. strigata (Günther, 1864)   291

Family Hydrocynidae   288

Subfamily Alestinae   289
Genus *Alestes*   289
   A. longipinnis (Günther, 1864)   289

Subfamily Petersinae   289
Genus *Phenacogrammus*   289
   Feathertail Tetra *P. interruptus* (Boulenger,
   1899)   289
Genus *Alestopetersius*   308
   A. caudalis (Boulenger, 1899)   308

Subfamily Hydrocininae   289
Genus *Hydrocinus*   289
   H. goliath Boulenger, 1898   289
   H. lineatus Bleeker, 1862   289

Family Hepsetidae   289
Genus *Hepsetus*   289
   H. odoe (Bloch, 1794)   289

Family Erythrinidae   289
Genus *Hoplias*   289
   H. malabaricus (Bloch, 1794)   289
Genus *Erythrinus*   289
   E. erythrinus (Schneider, 1801)   289
Genus *Hoplerythrinus*   289
   H. unitaeniatus (Spix, 1829)   289

Family Ctenoluciidae   289
Genus *Ctenolucius*   –
Genus *Boulengerella*   307
   B. cuvieri (Spix, 1829)   307

Family Characidiidae   290
Genus *Elachocharax*   290
Genus *Characidium*   –

Family Lebiasinids (Lebiasinidae)   290
Subfamily Lebiasininae   290
Genus *Lebiasina*   290

Subfamily Pyrrhulininae   290
Genus *Pyrrhulina*   290
   P. vittata Regan, 1912   291
Genus *Copeina*   290
   C. guttata (Steindachner, 1875)   291
Genus *Copella*   290
   C. arnoldi (Regan, 1912)   290

Subfamily Pencil Fishes, (Nannostominae) 291
  Genus *Nannostomus* 293
    N. beckfordi Günther, 1872 286
    N. beckfordi aripirangensis Meinken, 1931 286
    N. beckfordi anomalus Steindachner, 1876 —
    N. beckfordi beckfordi Günther, 1872 —
  Genus *Poecilobrycon* 286
    P. eques (Steindachner, 1876) 293
    P. unifasciatus (Steindachner, 1876) 293
    P. espei Meinken, 1956 291
    P. ocellatus Eigenmann, 1909 286

Family Hemiodontidae 293
Subfamily Hemiodontinae 293
  Genus *Hemiodus* 291
    H. semitaeniatus Kner, 1859 291

Subfamily Bivibranchiinae 293

Subfamily Paradontinae 293

Family Anostomidae 293
  Genus *Leporinus* 293
    L. fasciatus (Bloch, 1795) 291
  Genus *Anostomus* 293
    A. anostomus (Linné, 1758) 291
  Genus *Abramites* 293
  Genus *Schizodon* 293
  Genus *Sator* 293
    S. respectus Myers & Cavalho, 1959 293
  Genus *Gnatholemus* 293
  Genus *Synaptolemus* 293

Family Curimatidae 294
Subfamily Chilodinae 294
  Genus *Chilodus* 294
    Headstander, *C. punctatus* Müller & Troschel, 1845 294
  Genus *Caenotropus* 294

Subfamily Prochilodinae 294
  Genus *Prochilodus* 291
    P. reticulatus magdelenae Steindachner, 1878 294
    P. insignis Schomburgk, 1841 291

Subfamily Curimatinae 294

Subfamily Anodinae 294
  Genus *Anodus* —

Family Crenuchidae 294
  Genus *Crenuchus* 294
    C. spilurus Günther, 1863 294
  Genus *Poecilocharax* 294
    P. bovallii Eigenmann, 1909 294
    P. weitzmani Géry, 1965 294

Family Citharinidae 295
Subfamily Citharininae 295
  Genus *Citharinus* 294
    C. distichodoides Pellegrin, 1919 295

Subfamily Distichodontinae 295
  Genus *Distichodus* 295
    D. sexfasciatus Boulenger, 1897 295
  Genus *Nannocharax* 290
  Genus *Neolebias* 308
    N. ansorgei Boulenger, 1912 308
    N. landgrafi E. Ahl, 1928 308
  Genus *Nannaethiops* 308
    N. unitaeniatus Günther, 1871 308
    N. tritaeniatus Boulenger, 1913 308

Subfamily Ichthyoborinae 295
  Genus *Phago* 295
  Genus *Belonophago* 295
  Genus *Eugnathichthys* 295
  Genus *Ichthyoborus* 295
  Genus *Phagoborus* 295
  Genus *Hemistichodus* 295

Suborder Electric Eels (Gymnotoidei)

Family Electric Eels (Electrophoridae) 303
  Genus *Electrophorus* 296
    Electric Eel, *E. electricus* (Linné, 1766) 296

Family Gymnotid Eels (Gymnotidae) 303
  Genus Gymnotus 296
    Gymnotid Eel, *G. carapo* Linné, 1758 296

Family Apteronotidae 303
  Genus *Apteronotus* 298
    Apteronotid Eel, *A. albifrons* (Linné, 1758) 298
  Genus *Porotergus* 303
  Genus *Sternarchella* 303

  Genus *Ordemognathus* —
  Genus *Sternarchogiton* —
  Genus *Odontosternarchus* 303
  Genus *Sternarchorhamphus* 303
  Genus *Sternarchorhynchus* 303

Family Rhamphichthyid Eels (Rhamphichthyidae) 304
Subfamily Sternopyginae 304
  Genus *Sternopygus* 304
  Genus *Eigenmannia* 304
    E. virescens (Valenciennes, 1847) 304
  Genus *Rhabdalichops* 304

Genus *Hypopomus*	304
Genus *Steatogenys*	304
S. elegans (Steindachner, 1880)	304
Subfamily Rhamphichthyinae	304

Genus *Rhamphichthys*	304
R. rostratus (Linné, 1768)	304
Genus *Gymnorhamphichthys*	304
G. hypostomus Ellis, 1913	304

Suborder Carp (Cyprinoidei)

Family Carp (Cyprinidae)	305
Subfamily Danios (Danioninae)	306
The most important genera and species:	
Genus *Swamba*	—
Genus *Engraulicypris*	306
Genus *Barilius*	309
B. christyi Boulenger, 1920	309
Genus *Danios (Danio)*	309
Giant Danio, *D. malabaricus* (Jordan, 1844)	309
D. regina (Fowler, 1934)	309
D. devario (Hamilton & Buchanan, 1822)	309
Genus *Brachydanio*	309
Zebra Fish, *B. rerio* (Hamilton & Buchanan, 1822)	309
B. kerri (H. M. Smith, 1931)	309
Spotted Danio, *B. nigrofasciatus* (Day, 1896)	309
B. frankei (Meinken, 1963)	309
Pearl Danio, *B. albolineatus* (Blyth, 1860)	310
Genus *Esomus*	310
Flying Barb, *E. danrica* (Hamilton & Buchanan, 1822)	310
E. malayensis (Mandie, 1909)	310
Genus *Rasbora*	310
Scissors-Tail Rasbora, *R. trilineata* Steindachner, 1870	310
R. lateristriata elegans Volz, 1903	310
R. daniconius (Hamilton & Buchanan, 1822)	310
Rasbora, *R. heteromorpha* Duncker, 1904	310
R. vateriforis Deraniyagala, 1930	310
R. maculata Duncker, 1904	311
Genus *Aphyocypris*	311
A. pooni Lin, 1939	311
Genus *Tanichthys*	311
White Cloud Mountain Fish, *T. albonubes* Lin, 1932	311
Genus *Elopichthys*	311
Sheltostchek, *E. bambusa* (Richardson, 1844)	311
Subfamily Minnows (Leuciscinae)	311
Most important genera and species:	
Genus Squawfishes *(Ptychocheilus)*	312
Genus *Rutilus*	313
European Roach, *R. rutilus* (Linné, 1758)	312
Taran, *R. rutilus heckeli* (Nordmann, 1840)	312
Vobla, *R. rutilus caspicus* (Yakovlev, 1873)	312
R. pigus virgo (Heckel, 1852)	313
R. frisii meidingeri (Heckel, 1852)	313
Genus *Scardinius*	313

Rudd, *S. erythrophthalmus* (Linné, 1758)	313
Genus *Ctenopharyngodon*	313
C. idella (Valenciennes, 1844)	313
Genus *Mylopharyngodon*	313
M. piceus (Richardson, 1844)	313
Genus Minnows *(Leuciscus)*	313
Subgenus *Squalius*	314
Chubb, *L. (Squalius) cephalus* (Linné, 1758)	314
Subgenus *Leuciscus*	314
Dace, *L. (Leuciscus) leuciscus* (Linné, 1758)	314
Subgenus *Telestes*	314
L. (Telestes) souffia agassizi Cuvier & Valenciennes, 1844	314
Subgenus *Idus*	314
Ide, *L. (Idus) idus* (Linné, 1758)	314
L. (Idus) idus forma orfus Linné, 1758	315
Subgenus or Genus *Trilobodon*	315
Genus *Phoxinus*	315
Minnow, *P. phoxinus* (Linné, 1758)	315
Genus *Alburnus*	316
Bleak, *A. alburnus* (Linné, 1758)	316
Genus *Chalcalburnus*	319
C. chaloides mento (Agassiz, 1832)	319
Genus *Leucaspius*	319
L. delineatus (Heckel, 1843)	319
Genus *Alburnoides*	319
A. bipunctatus (Bloch, 1728)	319
Genus *Abramis*	319
Bream, *A. brama* (Linné, 1758)	319
A. sapa (Pallas, 1811)	320
A. ballerus (Linné, 1758)	320
Genus *Blicca*	320
Silver Bream, *B. bjoerkna* (Linné, 1758)	320
Genus *Vimba*	320
V. vimba (Linné, 1758)	320
V. vimba elongata (Valenciennes, 1844)	321
Genus *Chondrostoma*	321
C. nasus (Linné, 1758)	321
Genus *Aspius*	321
Asp, *A. aspius* (Linné, 1758)	321
Genus *Tinca*	321
Tench, *T. tinca* (Linné, 1758)	321
Genus *Chrosomus*	322
South Red-Bellied Dace, *C. erythrogaster* Rafinesqu, 1820	322
C. neogaeus (Cope, 1866)	322
Genus Shiners *(Notropis)*	322
N. welaka Evermann & Kendall, 1897	322
N. hypselopterus (Günther, 1868)	322

N. lutrensis (Baird & Girard, 1853) 322
Genus Daces (Rhinichthys) 323
   Blacknose Dace, R. stronasus (Mitchill, 1815) 323
Genus Chubs (Hybopsis) 323
Genus Meda 323
Genus Moapa 323
Genus Eremichthys 324

Subfamily Cultrins (Cultrinae) 324
Genus Culter 324
   C. alburnus (Basilevsky, 1855) 324
Genus Hemiculter 324
   H. leucisculus (Basilevsky, 1855) 324
Genus Erythroculter 324
   E. mongolicus (Basilevsky, 1855) 325
   E. illishaeformis (Bleeker, 1871) 324
Genus Rasborinus 325
   R. lineatus (Pellegrin, 1907) 325
Genus Parabramis 324
Genus Megalobrama 324
Genus Chela 325
   C. dadyburjori (Menon, 1952) 324
   C. laubuca (Hamilton & Buchanan, 1822) 325
   C. cachius Hamilton & Buchanan, 1822 325
   C. caerulostigmata (Smith, 1931) 318
Genus Pelecus 325
   P. cultratus (Linné, 1758) 325

Subfamily Xenocypridinae 325
Genus Xenocypris 329

Subfamily Bitterlings (Acheilognathinae) 329
Genus Rhodeus 329
   Bitterling, R. sericeus (Pallas, 1811) 329
   Chinese Bitterling, R. sericeus sericeus (Pallas, 1811) 330
   European Bitterling, R. sericeus amarus Bloch, 1782 329
Genus Acanthorhodeus 329

Subfamily Gobies (Gobioninae) 330
Genus Sarcocheilichthys 329
   S. sinensis Bleeker, 1860 331
Genus Gnathopogon 331
   Biwa Lake Goby, G. coerulescens (Sauvage, 1883) 331
Genus Abbottina 331
   A. rivularis (Basilevsky, 1855) 331
Genus Pseudorasbora 331
   Stone Moroko, P. parva (Schlegel, 1842) 331
Genus Gobiobotia 330
Genus Microphysogobio 330
Genus Saurogobio 331
   S. dabryi (Bleeker, 1871) 332
Genus Gobies (Gobio) 331
   Common Goby, G. gobio (Linné, 1758) 331
   G. uranoscopus Agassiz, 1828 332

G. kessleri Dybowski, 1862 332
G. albipinnatus Lukasch, 1933 332

Subfamily Barbins (Barbinae) 332
Genus Barbs (Barbus) 337
   Barb, B. barbus (Linné, 1758) 337
   B. meridionalis petenyi Heckel, 1847 338
   B. brachycephalus caspius Berg, 1914 338
   B. capito (Güldenstädt, 1775) 338
   B. mosal (Hamilton & Buchanan, 1822) = Tor tor (?) 55
Genus Tor 338
   T. tor (Hamilton & Buchanan, 1822) 338
Genus Rohtee —
Genus Puntius 337
   P. javanicus (Bleeker, 1850) 338
   P. cumingii Günther, 1868 351
   P. oligolepis Bleeker, 1893 351
   P. semifasciolatus Günther, 1868 351
   P. gelius Hamilton & Buchanan, 1822 351
   P. filamentosus Cuvier & Valenciennes, 1844 351
   P. schwanefeldi Bleeker, 1853 351
   P. arulius Jerdon, 1849 351
   P. eugrammus Silas, 1956 351
   P. vittatus Day, 1865 351
   P. conchonius (Hamilton & Buchanan, 1822) 339
   P. everetti Boulenger, 1894 339
   P. lateristriga Cuvier & Valenciennes, 1842 339
   P. nigrofasciatus Günther, 1868 339
   P. titteya Deranijgala, 1929 339
   P. pentazona Boulenger, 1894 339
   P. pentazona pentazona Boulenger, 1894 339
   P. pentazona hexazona Weber & Beaufort, 1912 —
   P. tetrazona Bleeker, 1855 339
   P. tetrazona tetrazona Bleeker, 1855 339
   P. tetrazona partipentazona Fowler, 1934 339
   P. holotaenia Boulenger, 1904 339
   P. fasciolatus Günther, 1868 339
   P. hulstaerti Poll, 1945 339
   P. viviparus Weber, 1897 339
Genus Caecobarbus 337
   C. geertsi Boulenger, 1921 417
Genus Phreatichthys 337
Genus Iranocypris 337
Genus Capoeta 340
   C. capoeta (Güldenstädt, 1773) 340
   C. heratensis natio steindachneri (Kessler, 1872) —
   C. damascinus (Valenciennes, 1842) —
Genus Varicorhinus 340
Genus Garra 340
   G. taeniata Smith, 1931 362
Genus Typhlogarra 340
Genus Labeos (Labeo) 340
   L. wecksi Boulenger, 1909 352
   L. erythrura Fowler, 1937 343
   L. velifer Boulenger, 1898 343

L. bicolor Smith, 1931 — 343
Genus *Catlocarpio* — 343
  C. siamensis Boulenger, 1898 — 343
Genus *Catla* — 343
  Indian Carp, *C. catla* (Hamilton & Buchanan, 1849) — 343
Genus *Epalzeorhynchus* — 362
  E. siamensis Smith, 1931 — 362
  E. kallopterus Bleeker, 1850 — 362

Subfamily Schizothoracinae — 343
Genus *Schizothorax* — 343
Genus *Schizopygopsis* — —
  S. stoliczkai Steindachner, 1866 — —
Genus *Diptychus* — —

Subfamily Carp (Cyprininae) — 344
Genus *Cyprinus* — 344
  Carp, *C. carpio* (Linné, 1758) — 344
Genus *Carassius* — 345
  Crucian Carp, *C. carassius* (Linné, 1758) — 345
  Goldfish, *C. auratus* (Linné, 1758) — 345
  Gibel, *C. auratus gibelio* (Bloch, 1783) — 350

Subfamily Hypophthalmichthyinae — 353
Genus *Hypophthalmichthys* — 353
  H. molitrix (Valenciennes, 1844) — 353

Family Suckers (Catostomidae) — 353
Genus *Carpiodes* — —
Genus *Catostomus* — 354
  Sucker, *C. catostomus* (Forster, 1773) — 354

Family Gyrinocheilids (Gyrinocheilidae) — 354
Genus *Gyrinocheilus* — 354
  G. aymonieri (Tirant, 1883) — 354

Family Psilorhynchids (Psilorhynchidae) — 355
Genus *Psilorhynchus* — 355

Family Hillstream Loaches (Homalopteridae) — 355
Genus *Homaloptera* — 355
Genus *Gastromyzon* — —
  Suckerbelly Loach, *G. borneensis* Günther, 1874 — —

Family Loaches (Cobitidae) — 355

Subfamily Botiins (Botiinae) — 357
Genus *Botia* — 357
Subgenus *Botia* — 358
  B. (Botia) macracanthus (Bleeker, 1852) — 358
  B. (Botia) lohachata Chaudhuri, 1912 — 358
Subgenus *Hymenophysa* — 358
  B. (Hymenophysa) hymenophysa (Bleeker, 1852) — 358
  B. (Hymenophysa) berdmorei (Blyth, 1860) — 358
  B. (Hymenophysa) horae Smith, 1931 — 358
  B. (Hymenophysa) modesta (Bleeker, 1865) — —
  B. (Hymenophysa) sidthimunki Klausewitz, 1959 — 358
Subgenus *Sinibotia* — 358
Genus *Leptobotia* — 358

Subfamily Noemacheilins (Noemacheilinae) — 358
Genus *Noemacheilus* — 358
  Stone Loach, *N. barbatulus* (Linné, 1758) — 359
  N. angorae bureschi Drensky, 1928 — 359
  N. fasciatus (Cuvier & Valenciennes, 1846) — 372
  N. fasciatus kuiperi De Beaufort, 1939 — 372
Genus *Vaillantella* — 358

Subfamily Cobitins (Cobitinae) — 359
Genus *Misgurnus* — 356
  Pond Loach, *M. fossilis* (Linné, 1758) — 359
  M. erikssoni Rendahl, 1930 — 356
Genus *Cobitophis* — 359
Genus *Neoeucirrichthys* — 356
  N. maydelli Banarescu & Nalbant, 1968 — 356
Genus *Niwaella* — 356
Genus *Cobitis* — 360
  C. taenia (Linné, 1758) — 359
  C. elongata Heckel & Kner, 1858 — 356
  C. elongata bilseli Battalgil, 1942 — 356
Genus *Sabanejewia* — 356
  S. aurata (de Filippi, 1856) — 360
  S. aurata bulgarica Drensky, 1928 — 357
Genus *Acanthophthalmus* — 359
  A. myersi Harry, 1949 — 360
  A. kuhli kuhli (Cuvier & Valenciennes, 1846) — 360
  A. semicinctus (Fraser-Brunner, 1940) — 360
  A. shelfordi Popta, 1901 — 360
Genus *Lepidocephalus* — 356
  L. guntea (Hamilton & Buchanan, 1822) — 360
  L. thermalis (Valenciennes, 1844) — 372
Genus *Acanthopsis* — 372
  A. choiorhynchus Bleeker, 1894 — 372

Order Catfishes (Siluriformes)

Family Diplomystid Catfishes (Diplomystidae) — —

Family North American Freshwater Catfishes (Ictaluridae) — 364
Genus *Ictalurus* — 364
  Brown Bullhead, *I. nebulosus* (Le Sueur, 1819) — 364
  I. furcatus (Le Sueur, 1840) — 364
  Channel Catfish, *I. punctatus* (Rafinesque, 1818) — 364
Genus *Trogloglanis* — 364
  ◊ *T. pattersoni* Eigenmann, 1919 — 365
Genus *Satan* — 364
  ◊ *S. eurystomus* Hubbs & Baily, 1947 — 364
Genus *Schilbeodes* — —

Family Bagrid Catfishes (Bagridae) 365
  Genus Dwarf Catfishes *(Mystus)* 365
    M. nemurus (Cuvier & Valenciennes, 1839) —
    Striped Dwarf Catfish, *M. vittatus* (Bloch,
    1797) 365
    M. tengara (Hamilton, 1822) 365
  Genus *Leiocassis* 365
    L. siamensis Regan, 1913 366
  Genus *Bagrichthys* 366
    B. hypselopterus (Bleeker, 1852) 366
  Genus *Auchenoglanis* 366
    A. occidentalis Cuvier & Valenciennes, 1840 366

Family Eurasian Catfishes (Siluridae) 366
  Genus *Silurus* 368
    European Wels, *S. glanis* Linné, 1758 366
    S. aristotelis Agassiz, 1856 368
  Genus *Wallagonia* 368
    W. attu (Bloch, 1801) 368
  Genus *Kryptopterus* 368
    K. bicirrhis (Cuvier & Valenciennes, 1839) 368

Family Schilbeid Catfishes (Schilbeidae) 368
  Genus *Eutropiella* 368
    E. debauvi (Boulenger, 1901) 368
  Genus *Pareutropius* 399
    P. mandevillei Poll, 1959 399
  Genus *Physailia* 369
    P. pellucida Boulenger, 1901 369

Family Giant Catfishes (Pangasiidae) 369
  Genus *Pangasianodon* 369
    Giant Catfish, ⟁ *P. gigas* Chevey, 1930 369

**Family Mouth-Breeding Catfishes
(Tachysuridae)** 369
  Genus *Batrachocephalus* 369
  Genus *Osteogenoiosus* 369
    O. militaris (Linné, 1758) 369
  Genus *Tachysurus* 369
    T. sagor (Hamilton, 1822) 370

Family Sea Catfishes (Ariidae) 370
  Genus *Arius* 370
    A. proops (Cuvier & Valenciennes, 1839) 370

Family Amblycipitidae 373
  Genus *Amblyceps* 373
    A. mangois (Hamilton, 1822) 373

Family Sucker Catfishes (Sisoridae) 373
  Genus *Bagarius* 373
    B. bagarius (Hamilton, 1822) 373
  Genus *Glyptothorax* 373
    G. trilineatus Blyth, 1860 373
  Genus *Oreoglanis* 374
    O. siamensis Smith, 1933 374

Family Labyrinthic Catfishes (Clariidae) 374
  Genus *Clarias* 374
    C. lazera Cuvier & Valenciennes, 1840 399
  Genus *Gymnallabes* 374
    G. typus (Günther, 1867) 375

Family Heteropneustidae 375
  Genus *Heteropneustes* 375
    Stinging Catfish, *H. fossilis* (Bloch, 1797) 375

Family Chacidae 375
  Genus *Chaca* 375
    C. chaca (Hamilton & Buchanan, 1822) 375

Family Olyridae —

Family Electric Catfishes (Malapteruridae) 376
  Genus *Malapterurus* 376
    Electric Catfish, *M. electricus* (Gmelin, 1789) 376

Family Upside-Down Catfishes (Mochocidae) 376
  Genus *Synodontis* 377
    Back-Swimming Congo Catfish,
    S. nigriventris David, 1936 377
    S. angelicus (Schilthuis, 1891) 377
    S. schall (Bloch & Schneider, 1801) 377

Family Doradid Armored Catfishes (Doradidae) 378
  Genus *Amblydoras* 378
    A. hancocki Cuvier & Valenciennes, 1888 378
  Genus *Acanthodoras* 377
    Talking Catfish, *A. spinosissimus* Eigenmann
    & Eigenmann, 1888 377
  Genus *Doras* 378
    D. costatus (Linné, 1766) 378

Family Bunocephalidae 379
  Genus *Bunocephalus* 379
    B. bicolor Steindachner, 1882 379

Family Plotosid Sea Catfishes (Plotosidae) 379
  Genus *Plotosus* 379
    P. lineatus (Thunberg, 1791) 379
    P. canius Hamilton, 1882 379

Family Pimelodid Catfishes (Pimelodidae) 380
  Genus *Pseudoplatystoma* 380
    P. fasciatum (Linné, 1766) 380
  Genus *Sorubim* 380
    S. lima (Bloch & Schnieder, 1806) 380
  Genus *Typhlobagrus* 380
    T. kronei Ribeiro, 1912 380
  Genus *Rhamdia* 380
    R. sebae (Cuvier & Valenciennes, 1840) 380
    R. guatemalensis (Günther, 1864) 396
  Genus *Microglanis* 380
    M. parahybae (Steindachner, 1876) 380

Family Parasitic Catfishes (Trichomycteridae) 380
Genus *Vandellia* 381

Family Callichthyid Armored Catfishes (Callichthyidae) 381
Genus *Callichthys* 382
Callichthys, *C. callichthys* (Linné, 1758) 383
Genus *Hoplosternum* 382
Hoplo Catfish, *H. thoracatum* (Cuvier & Valenciennes, 1840) 383
H. littorale (Hancock, 1828) 383
Genus *Corydoras* 382
C. paleatus (Jenyns, 1842) 384
Bronze Catfish, *C. aeneus* (Gill, 1858) 384
C. schultzei Holly, 1940 384
Guiana Catfish, *C. melanistius* Regan, 1912 384
C. punctatus Bloch, 1794 384
C. punctatus punctatus Bloch, 1794 —
Leopard Corydoras, *C. punctatus julii* Steindachner, 1906 —
C. barbatus (Quoy & Gaimard, 1840 411
C. macropterus Regan, 1913 411
C. caudimaculatus Rössel, 1961 411
C. reticulatus Frazer-Brunner, 1938 411
C. schwartzi Rössel, 1963 384

Bowline Catfish, *C. arcuatus* Elwin, 1939 384
C. metae Eigenmann, 1914 384
Micro Catfish, *C. hastatus* Eigenmann & Eigenmann, 1888 383
C. pygmaeus Knaack, 1966 383
Genus *Brochis* 382
B. coeruleus (Cope, 1872) 411
Genus *Dianema* 411
D. longibarbis Cope, 1872 411

Family Loricariid Armored Catfishes (Loricariidae) 389
Genus *Ancistrus* 390
A. bufonius (Valenciennes, 1840) 390
A. dolichopterus (Kner, 1854) 391
Genus *Plecostomus* 390
P. punctatus (Cuvier & Valenciennes, 1840) 412
Genus *Pterygoblichthys* 390
Genus *Loricaria* 390
Whiptail Loricaria, *L. parva* Boulenger, 1895 391
Genus *Otocinclus* 392
O. vittatus Regan, 1904 412
Genus *Farlowella* 391
F. acus Kner, 1853 412

Order Trout-Perches (Percopsiformes)

Suborder Cavefishes (Amblyopsoidei)

Family Cavefishes (Amblyopsidae) 395
Genus *Amblyopsis* 396
Mammoth Cave Blindfish, *A. spelaeus* De Kay, 1842 396

Genus *Typhlichthys* 396
Genus *Chologaster* 396
C. cornutus Agassiz, 1853 396

Suborder Pirate Perches (Aphredoderoidei)

Family Aphredoderidae 396
Genus *Aphredoderus* 396

Pirate Perch, *A. sayanus* (Gilliams, 1824) 396

Suborder Trout-Perches (Percopsoidei)

Family Trout-Perches (Percopsidae) 397
Genus *Percopsis* 397
P. omiscomaycus (Walbaum, 1792) 397

Genus *Columbia* 397
C. transmontana Eigenmann & Eigenmann, 1892 397

Order Toadfishes (Batrachoidiformes)

Family Toadfishes (Batrachoididae) 397
Genus *Thalassophryne* 397
Toadfish, *T. maculosa* Günther, 1861 418
Genus *Thalossotia* 397
Genus *Porichthys* 397

P. notatus Girard, 1854 397
Atlantic Midshipman, *P. porosissimus* Cuvier & Valenciennes, 1837 397
Genus *Opsanus* 398
Oyster Toadfish, *O. tau* (Linné, 1766) 398

Order Clingfishes (Gobiesociformes)

Family Clingfishes (Gobiesocidae)	398
Genus *Caularchus*	398
Suck Fish, *C. maeandricus* (Girard, 1858)	398
Genus *Bryssetaeres*	398
B. pinniger (Gilbert, 1890)	398

Genus *Chorisochismus*	398
C. dentex (Pallas, 1767)	398
Genus *Lepadogaster*	398
L. bimaculatus (Pennant, 1812)	398

Order Anglerfishes (Lophiiformes)

Suborder Goosefishes (Lophioidei)

Family Goosefishes (Lophiidae)	401
Genus *Lophius*	401
Allmouth, *L. piscatorius* Linné, 1758	401

Suborder Frogfishes (Antennarioidei)

Family Frogfishes (Antennariidae)	402
Genus *Antennarius*	402
A. scaber (Cuvier, 1817)	402
A. moluccensis Bleeker, 1855	402
Genus *Histrio*	402
H. histrio (Linné, 1758)	402

O. nasutus (Cuvier & Valenciennes, 1837)	403
O. vespertilio (Linné, 1758)	403
Genus *Halieutaea*	403
H. retifera Gilbert, 1903	403

Family Batfishes (Ogcocephalidae)	402
Genus *Ogcocephalus*	403

Family Brachyonichthyidae	402
Family Chaunacidae	402

Suborder Deepsea Anglerfishes (Ceratioidei)

Family Diceratiidae	403
Family Photocorynidae	403
Family Linophrynidae	403
Genus *Linophryne*	403
Deepsea Angler, *L. arborifer* Regan, 1925	403
Family Ceratiidae	403
Genus *Ceratias*	403
C. hollbolli Kroyer, 1844	404
Genus *Edriolychnus*	—
E. schmidtii Regan, 1925	—
Genus *Cryptosaras*	404
C. conesi Gill, 1883	404

Members of other Families:	
Genera Cigantactis, Caulophryne and	
Neoceratias	—
Family Melanocetidae	404
Genus *Melanocetus*	404
Johnson's Black Anglerfish, *M. johnsoni* Günther, 1864	404
Family Himantolophidae	404
Genus *Himantolophus*	404
H. groenlandicus Regan, 1926	404
Family Oneirodidae	404
Genus *Galatheathauma*	56

Order Codfishes (Gadiformes)

Suborder Eel Cods (Muraenolepioidei)

Family Eel Cods (Muraenolepidae)	406
Genus *Muraenolepis*	406

Suborder Codfishes (Gadoidei)

Family Deepsea Codfishes (Moridae)	406
Genus *Antimora*	406
Blue Hake, *A. rostrata* Günther, 1876	406
A. viola Goode & Bean, 1879	406

Genus *Mora*	406
Genus *Laemonema*	406
Genus *Physiculus*	—
Genus *Lepidion*	406
Genus *Lotella*	406

Family Bregmacerotid Codfish (Bregmacerotidae) 406
Genus *Bregmaceros* 406

Family Codfish (Gadidae) 407
Genus *Brosme* 407
Cusk, *B. brosme* (Cuvier) Oken, 1817 407
Genus *Gaidropsarus* 407
G. pacificus (Temminck & Schlegel, 1842) 408
G. novaezealandiae (Hector, 1874) 408
G. capensis Barnard, 1925 408
G. mediterraneus (Linné, 1758) 408
Genus *Onogadus* 408
O. argentatus (Reinhardt, 1837) 408
O. ensis De Buen, 1934 —
Genus *Enchelyopus* 408
E. cimbrius Bloch & Schneider, 1801 408
Genus *Ciliata* 408
C. mustela (Linné, 1758) 408
C. septentrionalis (Collett, 1875) 408
Genus *Raniceps* 408
R. raninus (Cuvier) Oken, 1817 408
Genus *Phycis* 409
Forkbeard, *P. phycis* (Linné, 1766) —
P. blennoides (Brünnich, 1768) 409
Genus *Urophycis* 409
Codling, *U. tenuis* (Mitchill, 1815) 409
Mud Hake, *U. chuss* (Walbaum, 1792) 409
U. brasiliensis (Kaup, 1858) 409
Genus *Lota* 410
Burbot, *L. lota* (Linné, 1758) 410
L. lota lota (Linné, 1758) 410
Alekey Trout, *L. lota maculosa* (Le Sueur, 1817) 410
L. lota leptura Hubbs & Schultz, 1941 410
Genus *Molva* 410
Ling, *M. molva* (Linné, 1758) 410
M. dipterygia (Pennant, 1784) 413
Blue Ling, *M. elongata* (Otto, 1821) 413
Genus *Gadus* 413
Atlantic Cod, *G. morhua* Linné, 1758 413

Greenland Cod, *G. ogac* Richardson, 1836 415
Pacific Cod, *G. macrocephalus* Tilesius, 1810 415
Genus *Boreogadus* 415
Polar Cod, *B. saida* (Lepechin, 1774) 415
Genus *Arctogadus* 415
East Siberian Cod, *A. borisovi* Dryagin, 1932 416
Arctic Greenland Cod, *A. glacialis* (Peters, 1874) 416
Genus *Eleginus* 415
Wachna Cod, *E. navaga* (Pallas, 1811) 416
Far Eastern Navaga, *E. gracilis* (Tilesius, 1810) 416
Genus *Microgadus* 415
Atlantic Tomcod, *M. tomcod* (Walbaum, 1792) 416
Pacific Tomcod, *M. proximus* (Girard, 1854) 416
Genus *Theragra* 415
Alaska Pollack, *T. chalcogramma* (Pallas, 1811) 416
Genus *Melanogrammus* 419
Haddock, *M. aeglefinus* (Linné, 1758) 419
Genus *Pollachius* 419
Pollack, *P. pollachius* (Linné, 1758) 420
Pollock, *P. virens* (Linné, 1758) 419
Genus *Merlangius* 420
Whiting, *M. merlangus* (Linné, 1758) 420
Genus *Micromesistius* 420
Poutassou, *M. poutassou* (Risso, 1826) 420
M. australis Norman, 1937 420
Genus *Trisopterus* 423
T. minutus (O. Müller, 1776) 423
T. esmarkii (Nilsson, 1855) 423
T. luscus (Linné, 1758) 423
Genus Silvery Pouts *(Gadiculus)* 423
G. thori Schmidt, 1914 423
G. argenteus Guichenot, 1850 423

Family Hakes (Merlucciidae) 423
Genus *Merluccius* 423
Hake, *M. merluccius* (Linné, 1758) 423
American Hake, *M. bilinearus* (Mitchill, 1814) 423
Stockfish, *M. capensis* Castelnau, 1861 423
M. hubbsi Marini, 1932 423
M. gayi (Guichenot, 1848) 423
M. productus (Ayres, 1855) 423
M. australis (Hutton, 1872) 423

Suborder Cusk Eels (Ophidioidei)

Family Cusk Eels (Ophidiidae) 424
Genus Ophidium 424
O. barbatum Linné, 1758 424
Genus *Genypterus* 424
G. capensis A. Smith, 1849 —
Genus *Lepophidium* 424
L. cervinum (Goode & Bean, 1885) 424

Genus *Brotula* 424
Genus *Pluto* 396
Cusk Eel, *P. infernalis* Hubbs, 1938 396
Genus *Stygicola* 424
S. dentatus Poey, 1861 424
Genus *Lucifuga* 424
L. subterranea Poey, 1861 424

Family Pearlfishes (Carapidae) 424
  Genus *Carapus* 432

C. acus (Brünnich, 1768) 432

Suborder Viviparous Blennies (Zoarcoidei)

Family Viviparous Blennies (Zoarcidae) 424
  Genus *Zoarces* 424
    Viviparous Blenny, *Z. viviparus* Linné, 1758 424
  Genus *Macrozoarces* 425
    Mutton Fish, *M. americanus* (Bloch & Schneider, 1801) 425

Genus *Lycodes* 425
  Eel-Pount, *L. esmarki* Collett, 1875 —
Genus *Lycenchelys* 425
Genus *Gymnelis* 425
  G. viridis (Fabricius, 1780) 425

Suborder Grenadiers (Macrouroidei)

Family Grenadiers (Macrouridae) 425
  Genus *Macrourus* 425
    Smooth-Spined Rat-Tail, *M. berglax* Lacépède, 1802 425

Genus *Coelorhynchus* 425
  Soldier Fish, *C. carminatus* (Goode, 1880) 425
Genus *Coryphaenoides* 425
  Rock Grenadier, *C. rupestris* Gunnerus, 1765 425

Order Silversides (Atheriniformes)

Suborder Flyingfishes (Exocoetoidei)

Family Flying Fishes (Exocoetidae) 426
  Genus *Exocoetus* 426
    Flying Fish, *E. volitans* Linné, 1758 428
    E. obtusirostris Günther, 1866 428
  Genus *Fodiator* 428
    F. acutus (Cuvier & Valenciennes, 1846) —
  Genus *Parexocoetus* 428
  Genus *Cypselurus* 428
    C. californicus (Cooper, 1863) 428
    C. heterurus (Rafinesque, 1810) 428
    C. furcatus (Cuvier & Valenciennes, 1846) 449
  Genus Flying Halfbeaks *(Oxyporhamphus)* 428
  Genera Halfbeaks (the following five genera):
  Genus *Euleptorhamphus* 428
    E. viridis (van Hasselt, 1824) 428
  Genus *Hyporhamphus* 428
    H. unifasciatus (Ranzani 1842) 428
  Genus *Hemirhamphus* 428

H. brasiliensis (Linné, 1758) 428
Genus *Dermogenys* 428
  D. pusillus van Hasselt, 1823 429
Genus *Nomorhamphus* 428

Family Needlefishes (Belonidae) 428
  Genus *Belone* 429
    Garfish, *B. bellone* (Linné, 1758) 429
  Genus *Potamorrhaphis* 429
    P. guianensis (Schomburgk, 1843) 429
  Genus *Strongylura* 429
    S. crocodila (Le Sueur, 1821) 429
    S. marina (Walbaum, 1792) 429

Family Sauris (Scomberesocidae) 430
  Genus *Scomberesox* 430
    Atlantic Saury, *S. saurus* (Walbaum, 1792) 430

Suborder Toothed Carps (Cyprinodontoidei)

Family Empetrichthyidae 430

Family Orestiidae 430
  Genus *Orestias* 430
    O. pentlandii Valenciennes, 1839 71
    ♀ *O. cuvieri* Valenciennes 1846 433

Family Killifishes (Cyprinodontidae) 433
  Genus *Aphanius* 433
    A. sophiae (Heckel, 1846) 433
    A. chantrei (Gailland, 1895) 433
    A. anatoliae (Leidenfrost, 1912) 433

A. mento (Heckel, 1893) 433
A. apodus (Gervais, 1853) 433
A. dispar Rüppel, 1826 434
A. fasciatus (Valenciennes, 1821) 434
A. iberus (Cuvier & Valenciennes, 1846) 434
Genus *Cyprinodon* 434
  C. variegatus Lacépède, 1803 434
  C. nevadensis Eigenmann, 1889 434
Genus *Jordanella* 434
  Flag Fish, *J. floridae* Goode & Bean, 1879 434
Genus *Aphyosemion* 435
  A. arnoldi (Boulenger, 1908) 436

A. nigerianum Clausen, 1963   450

Golden Pheasant Killy, *A. sjostedti*
(Lönnberg, 1895)   436

Lyretail, *A. australe* Rachow, 1921   436

A. exiguum (Boulenger, 1911)   454

A. fallax E. Ahl, 1935   454

Two-Striped Killi, *A. bivittatum* (Lönnberg,
1859)   436

Red-Speckled Killy, *A. cognatum* Meinken,
1915   436

A. calliurum (Boulenger, 1911)   436

A. filamentosum (Meinken, 1933)   450

Subgenus (or Genus?) *Roloffia*   454

A. (Roloffia) occidentalis toddi Clausen, 1965   454

Genus *Notobranchius*   436

N. guentheri (Pfeffer, 1893)   436

Fire Killy, *N. rachovi* (Ahl, 1929)   436

N. neumanni (Hilgendorf, 1905)   450

N. melanostilus (Pfeffer, 1896)   455

Genus Pearl Fishes *(Cynolebias)*   436

C. elongatus Steindachner, 1881   454

C. nigripinnis Regan, 1912   436

Argentine Pearl Fish, *C. belotti* Steindachner,
1876   436

C. ladigesi Foersch, 1958   436

Genus Longfins *(Pterolebias)*   436

P. longipinnis Garman, 1895   436

Peruvian Longfin, *P. peruensis* Myers, 1954   436

Genus Rivulus *(Rivulus)*   436

Rivulus, *R. milesi* Fowler, 1941   455

R. cylindraceus Poey, 1861   450

Genus *Austrofundulus*   454

A. transilis Myers, 1932   454

Genus *Aplocheilichthys*   437

A. myersi Poll, 1952   437

Genus *Fundulus*   450

Goldear Killy, *F. chrysotus* Holbrook, 1866   450

Genus *Epiplatys*   437

E. annulatus (Boulenger, 1915)   450

E. dageti Poll, 1953   437

E. dageti sheljuzkoi Poll, 1953   450

E. dageti monroviae Daget & Arnoult, 1964   450

Genus *Pachypanchax*   450

Rough-Back Killy, *P. playfairi* (Günther,
1866)   450

Genus *Aplocheilus*   437

Malabar Killy, *A. lineatus* (Cuvier &
Valenciennes, 1846)   450

A. dayi (Steindachner, 1892)   455

Genus *Chriopeops*   437

Bluefin Killy, *C. goodei* (Jordan, 1879)   437

Genus *Valencia*   437

V. hispanica (Cuvier & Valenciennes, 1846)   437

Family Adrianichthyidae   437

Genus *Xenopoecilus*   437

X. poptae Weber & Beaufort, 1922   437

Family Japanese Killifishes (Oryziatidae)   437

Family Glass Carp (Horaichthyidae)   437

Genus *Horaichthys*   437

Glass Carp, *H. setnai* Kulkarni, 1940   437

Family Goodeid Topminnows (Goodeidae)   438

Family Four-Eyed Fishes (Anablepidae)   438

Genus *Anableps*   438

Four-eyed Fish, *A. anableps* Linné, 1756   438

Family Jenynsiid Topminnows (Jenynsiidae)   438

Genus *Jenynsia*   438

J. lineata (Jenyns, 1842)   438

Family Livebearers (Poeciliidae)   438

Subfamily Tomeurinae   438

Genus *Tomeurus*   439

T. gracilis Eigenmann, 1909   439

Subfamily Poeciliinae   440

Genus Livebearers *(Poecilia)*   440

Guppy, *P. reticulata* Peters, 1859   440

P. branneri Eigenmann, 1894   441

P. parae Eigenmann, 1894   441

P. melanogaster Günther, 1866   441

P. nigrofasciata (Regan, 1913)   441

P. latipinna (Le Sueur, 1821)   441

P. velifera (Regan, 1914)   441

P. sphenops Valenciennes, 1846
(Domesticated form: Black Molly)   441

Amazon Molly, *P. formosa* (Girard, 1859)   441

Genus Platyfishes and Swordtails *(Xiphophorus)*   440

Platy, *X. maculatus* (Günther, 1866)   442

Variatus, *X. variatus* (Meek, 1904)   442

X. xiphidium (Gordon, 1932)   442

Swordtail, *X. helleri* Heckel, 1848   442

Genus *Phalloceros*   440

P. caudimaculatus (Hensel, 1868)   443

Barrigudinyo, *P. caudimaculatus* forma
reticulatus Köhler, 1905   443

Genus Gambusias *(Gambusia)*   443

G. affinis (Baird & Girard, 1854)   443

G. affinis affinis (Baird & Girard, 1854)   443

G. affinis holbrooki (Girard, 1859)   452

Genus *Belonesox*   440

Lucino, *B. belizanus* Kner, 1860   443

Genus *Heterandria*   440

Mosquito Fish, *H. formosa* Agassiz, 1855   443

Genus *Poeciliopsis*   443

Genus *Phallichthys*   440

Merry Widow, *P. amates* (Miller, 1907)   443

Suborder Silversides (Atherinoidei)

Superfamily Silversides (Athcrinoidea)

Family Rainbow Fishes (Melanotaeniidae) 444
  Genus *Melanotaenia* 444
    M. maccullochi Ogilby, 1915 444
    Australian Rainbow Fish, *M. nigrans*
    (Richardson, 1843) 444

Family Silversides (Atherinidae) 444
  Genus *Atherina* 444
    A. hepsetus Linné, 1758 444
    A. mochon Cuvier, 1829 444
    A. presbyter Cuvier, 1829 444
  Genus *Pseudomugil* 453
    P. signifer Kner, 1867 453
  Genus *Telmatherina* 445
    T. ladigesi E. Ahl, 1936 445
  Genus *Atherinopsis* 445

  Jacksmelt Silverside, *A. californiensis* Girard,
    1854 445
  Genus *Leuresthes* 445
    Grunion, *L. tenius* (Ayres, 1860) 445
  Genus *Hubsiella* 446
    Sardine Silverside, *H. sardina* (Jenkins &
    Evermann, 1888) 446
  Genus *Menidia* 446
  Genus *Bedotia* 453
    B. geayi Pellegrin, 1906 453

Family Isonidae 446
  Genus *Iso* 446
  Genus *Notocheirus* 446
    N. hubbsi Clark, 1937 446

Superfamily Phallostethids (Phallostethoidea)

**Family Phallostethid Priapriumfishes
(Phallostethidae)** 446

**Family Neostethid Priapriumfishes
(Neostethidae)** 446

On the Zoological Classification and Names

For many years, zoologists and botanists have tried to classify animals and plants into a system which would be a survey of the abundance of forms in fauna and flora. Such a system, of course, may be established under very different aspects. Since Charles Darwin, his predecessors, and his successors have found that all creatures have evolved out of common ancestors, species of animals and plants have been classified according to their natural relationships. Our knowledge about the phylogeny, and thus the relationship of each living being to the other, is augmented every year by new discoveries and insights. Old ideas are replaced with more recent and more appropriate ones. Therefore, the natural classification of the animal kingdom (and the plant kingdom) is subject to changes. Furthermore, the opinions of zoologists, who are working on the classification of animals into the various groups, are anything but uniform. These differences and changes are usually insignificant. The classification of vertebrates into the classes of fish, amphibia, reptiles, birds, and mammals has been fixed for many decades. Only the Cyclostomata were recently separated from the fish and all other classes of vertebrates as the "jawless" Agnatha (comp. Vol. IV).

The animal kingdom has been split into several sub-kingdoms and these were again divided into further sections, subsections, and so on. The scale of the most important systematic categories follows in a descending rank order:

Kingdom
Sub-kingdom
Phylum
Subphylum
Class
Subclass
Superorder
Order
Suborder
Infraorder
Family
Subfamily
Tribe
Genus
Subgenus
Species
Subspecies

The scientific names of the animals and their spelling follow the international rules for the zoological nomenclature as agreed upon by the XV International Congress for Zoology and are obligatory for all zoological publications. The name of the genus, which is a Latin or Latinized noun, is singular and capitalized. After the name of the genus follows the name of the species and of the subspecies. The names of the species and subspecies may be nouns or adjectives, and they are spelled in the lower case. The name of a subgenus, which is formed in the same manner as a genus, may be added in brackets following the name of the genus. The names of the tribes, subfamilies, families, and superfamilies are plural capitalized nouns. They are formed from the name of a given genus by adding to the principal word the endings -ini for the tribe, -inae for the subfamily, -idae for the family, and -oidea for the superfamily. The names of the authors who were the first to describe and to name a species, subspecies, or group of animals should be cited with the year of this naming at least once in each scientific publication. The name of the author and year are not enclosed in brackets when the species or subspecies is classified as belonging to the same genus with which the author had originally classified it. They are in brackets when another genus name is used in the present publication. The scientific names of the genus, subgenus, species, and subspecies are supposed to be printed with different letters, usually italics.

ANIMAL DICTIONARY

1. English—German—French—Russian

For scientific names of species see the German-English-French-Russian section of this dictionary or the index.

N. A. after English names means that the name is used exclusively in North America.

ENGLISH NAME	GERMAN NAME	FRENCH NAME	RUSSIAN NAME
Alaska black-fish	Fächerfisch		Даллия
– cod	Pazifik-Kabeljau	Morue du Pacifique	Тихоокеанская треска
Allice shad	Alse	Grande alose	Алоса
American eel	Amerikanischer Aal	Anguille d'Amérique	Американский речной угорь
– flag fish	Florida-Kärpfling	Jordanelle de Floride	
– paddlefish	Löffelstör		Американский веслонос
– shad	Amerikanische Alse	Alose d'Amérique	Шэд
Anchovies	Sardellen		Анчоусовые, Анчоусы
Anchovy	Europäische Sardelle	Anchois	Европейский анчоус
Angel fish	Gemeiner Meerengel	Ange de mer	Европейский морской ангел
– shark	Gemeiner Meerengel	Ange de mer	Европейский морской ангел
– sharks	Engelhaie		Морские ангелы
Angler-fish	Atlantischer Seeteufel		Обыкновенный морской черт
Angler-fishes	Seeteufel i. e. S.	Baudroies	Морские черти
Angular rough shark	Meersau	Centrine	Центрина
Arapaima	Arapaima		Арапайма
Arctic grayling	Sibirische Äsche	Ombre arctique	Сибирский хариус
Armored cat-fish	Gefleckter Panzerwels	Corydoras à casque	
– cat-fishes	Dornwelse		Иглистые сомы
Atlantic herring	Atlantischer Hering	Hareng	Атлантическая сельдь
– salmon	– Lachs	Saumon atlantique	Обыкновенный лосось
– sturgeon	– Stör		Атлантический осетр
Baloos	Halbschnäbler	Hémiramphes	Полурылы
Barbel	Barbe	Barbeau commun	Обыкновенный усач
Barbels	Echte Barben	Barbeaux	Настоящие усачи
Basking shark	Riesenhai	Squale pélerin	Гигантская акула
– sharks	Riesenhaie		Гигантские акулы
Belonesox	Hechtkärpfling	Brochet vivipare	
Bitterling	Europäischer Bitterling	Bouvière amère	Европейский горчак
Black-fishes	Fächerfische		Даллии
Black-mouthed dogfish	Fleckhai	Chien espagnol	Пилохвост
Black-nosed dace	Schwarznase	Cyprin à nez noir	
Bleak	Ukelei	Ablette	Обыкновенная уклея
Blue ling	Mittelmeerleng		Средиземноморская мольва
– shark	Blauhai	Requin bleue	Голубая акула
– sting-ray	Violetter Stechrochen	Pastenague violette	Фиолетовый хвостокол
Bonefish	Grätenfisch	Albule commun	
Bony fishes	Knochenfische, Echte Knochen-fische	Poissons osseux	Костные рыбы, Костистые рыбы
Bottle fish	Schlinger	Saccopharynx	
Bowfin	Amerikanischer Schlammfisch	Amie	Ильная рыба
Bowfins	Schlammfische	Amies	Амии
Bramble shark	Nagelhai	Chenille	
Bream	Brachsen	Brème commune	Лещ
Breams	Brassen	Brèmes	Лещи
Brook lamprey	Bachneunauge	Lamproie de Planeri	Европейская ручьевая минога
Brook trout	Bachsaibling	Saumon de fontaine	Американская ручьевая палия
Brown shark	Grauhai	Requin griset	
– trout	Europäische Forelle	Truite de mer	Лосось-таймень
Bull ray	Afrikanischer Adlerrochen	Mourine vachette	Африканский орляк
Burbot	Aalquappe	Lote	Налим
Burbots	Aalquappen	Lotes	Налимы
Butterfly-fish	Schmetterlingsfisch	Pantodon	Рыба-бабочка
Butterfly-fishes	Schmetterlingsfische		Пантодоновые
Butterfly ray	Schmetterlingsrochen	Mourine bâtarde	Гимнура

ENGLISH NAME	GERMAN NAME	FRENCH NAME	RUSSIAN NAME
California anchovy	Nordamerikanische Sardelle	Anchois de Californie	Североамериканский анчоус
— sardine	Pazifische Sardine		Тихоокеанская сардина
Cameronensis	Kap Lopez	Cap Lopez	
Capelin	Lodde	Capelan	Мойва
Caribes	Pirayas	Caribes	Пираньи
Carp	Karpfen	Carpe ordinaire	Сазан
Carplike fish	Japanische Bitterlinge		Японские горчаки
Carps	Karpfen	Carpes	Сазаны
Cartilaginous fishes	Knorpelfische		Хрящевые рыбы
Caspian sand smelt	Kleiner Ährenfisch	Prêtre	Малая атеринка
Cat-fishes	Echte Welse	Siluridés	Сомовые
Cave fishes	Blindfische		Живородковые
Central mud minnow	Östlicher Hundsfisch	Poisson de vase	Восточноамериканская евдошка
Chain pickerel	Kettenhecht	Brochet maillé	
Characins	Salmler, Salmler i. e. S.		Харациновидные, Харациновые
Charr	Wandersaibling	Omble-chevalier	Обыкновенный голец
Charrs	Saiblinge		Гольцы
Chimaera	Gewöhnliche Langnasenchimäre		Обыкновенная длинно-рылая химера
Chimaeras	Seedrachen		Цельноголовые
Chimaeroids	Seedrachen		Цельноголовые
Chinese sturgeon	Schwertstör		Китайский меченос
Chinook salmon	Quinnat	Saumon de Californie	Чавыча
Chub	Döbel	Chevain	Голавль
Chum salmon	Keta-Lachs	Saumon-chien	Кета
Coalfish (N. A.)	Steinköhler	Lieu jaune	
Cods	Eigentliche Dorsche	Morues	Треска
Coho salmon	Kisutchs-Lachs	Saumon coho	Кижуч
Comb-tooth sharks	Grauhaie		Гребнезубые акулы
Comb-toothed shark	Grauhai	Requin griset	
Common Atlantic sturgeon	Baltischer Stör	Esturgeon commun	Балтийский осетр
— cod	Kabeljau	Morue fraîche	Атлантическая треска
— eel	Europäischer Flußaal	Anguille	Обыкновенный речной угорь
— carps	Karauschen	Carassins	Караси
— hagfish	Inger	Myxine	Обыкновенная миксина
— hammerhead shark	Glatter Hammerhai	Requin marteau	Обыкновенная молот-рыба
— lake charr	Amerikanischer Seesaibling	Truite grise	Американская озерная палия
— spiny-fish	Dornhai	Aiguillat tacheté	Нокотница
— sting-ray	Gewöhnlicher Stechrochen	Pastenague	Морской кот
Conger eel	Meeraal	Congre	Морской угорь
Couch's whiting	Blauer Wittling	Poutassou	
Cow shark (N. A.)	Grauhai	Requin griset	
Cowsharks	Grauhaie		Гребнезубые акулы
Cusk	Lumb	Brosme	Менек
Cyprinodonts	Eierlegende Zahnkärpflinge		Яицекладущие карпо-зубые
Dace	Hasel	Vandoise vraie	Обыкновенный елец
Darkie charlie	Schokoladenhai	Liche	
Deep-sea angler	Laternenangler		Гигантский удильщик
Devil-fish	Meeresteufel	Diable de mer	Морской дьявол
Devil-fishes	Teufelsrochen		Морские дьяволы
Dog salmon	Keta-Lachs	Saumon-chien	Кета
Dogfish	Europäischer Hundsfisch	Poisson-chien	Европейская евдошка
Eagle rays	Adlerrochen		Орляки
Eel-like mormyrid	Großer Nilhecht	Gymarche du Nil	Гимнарх
Eel pout	Aalmutter	Loquette	Обыкновенная бельдюга
Eels	Echte Aale	Anguilles	Речные угри
Electric eel	Zitteraal	Anguille tremblante	Электрический угорь
— eels	Zitter- und Messeraale		Электрические угри
— rays	Zitterrochen	Torpilles	Электрические скаты
Elephant shark	Riesenhai	Squale pélerin	Гигантская акула
European bitterling	Bitterlinge	Bouvières	Обыкновенные горчаки
— cat-fish	Flußwels	Silure glane	Обыкновенный сом
— moray	Mittelmeer-Muräne	Murène commune	Средиземноморская мурена
— roach	Plötze	Gardon blanc	Плотва
— smelt	Stint	Éperlan d'Europe	Обыкновенная корюшка
Eyed electric ray	Gefleckter Zitterrochen	Torpille tachetée	Пятнистый электрический скат
Featherbacks	Eigentliche Messerfische	Notoptéridés	

ENGLISH NAME	GERMAN NAME	FRENCH NAME	RUSSIAN NAME
Fierce shark	Schildzahnhai	Odontaspide féroce	
Finescale dace	Kleinschuppenrötling	Goujon à fines écailles	
Fishes	Fische	Poissons	Рыбы
Flapper skate	Glattrochen	Pocheteau blanc	Гладкий скат
Flying-fishes	Atlantische Flugfische, Fliegende Fische	Poissons volants	Долгоперы, Летучие рыбы
Four-eyed fish	Vierauge		Четырехглазая рыба
Four-eyed fishes	Vieraugen		Четырехглазые рыбы
Fresh-water eel	Europäischer Flußaal	Anguille	Обыкновенный речной угорь
— lamprey	Flußneunauge	Lamproie de rivière	Европейская речная минога
Frilled shark	Krausenhai	Requin à tunique	Плащеносная акула
— sharks	Krausenhaie		Плащеносные акулы
Frog-fish	Sargassofisch		Морская мышь
Frog-fishes	Fühlerfische i. e. S.		Морские клоуны
Gadoid fishes	Dorsche i. e. S.	Gades	Тресковые
Galaxiids	Hechtlinge i. e. S.		Галаксиевые
Gambia eel	Großer Nilhecht	Gymnarche du Nil	Гимнарх
Gar-fish	Europäischer Hornhecht	Aiguille	Европейский сарган
Gar-fishes	Hornhechte	Aiguilles	Сарганы, Саргановые
Gar-pikes	Knochenhechte		Панцырные щуки
Gars	Knochenhechte		Каймановые рыбы
Giant Danio	Malabarbärbling	Danio géant	
Gilt sardine	Ohrensardine	Allache	
Goldfish	Goldfisch, Silberkarausche	Carassin doré, Poisson rouge	Золотая рыбка, Серебряный карась
Goose-fish	Atlantischer Seeteufel		Обыкновенный морской черт
Gorbusha	Buckellachs	Saumon à bosse	Горбуша
Grass pickerel	Grass-Hecht	Brochet vermiculé	
Grayfish (N. A.)	Dornhai	Aiguillat tacheté	Нокотница
Grayling	Europäische Äsche	Ombre de rivière	Обыкновенный хариус
Graylings	Äsche		Хариусы
Great white shark	Weißhai	Requin	
Green Pollack	Steinköhler	Lieu jaune	
Greenland halibut	Grönlandkabeljau	Morue de roche	Гренландская фиордовая треска
— shark	Grönlandhai	Requin des glaces	
Gudgeon	Gewöhnlicher Gründling	Goujon	Обыкновенный пескарь
Gudgeons	Gründlinge i. e. S.	Goujons	Обыкновенные пескари
Guitar fish	Gemeiner Geigenrochen, Citarrenfisch	Guitare	Обыкновенная рохля, Гитара-рыба
Gymnotids	Echte Messeraale	Gymnotidés	
Haddock	Schellfisch	Morue noire	Пикша
Hagfishes	Inger		Миксинообразные
Hake	Seehecht	Merlu	Обыкновенный хэк
Half-beaks	Halbschnäbler	Hémiramphes	Полурылы
Hammerhead sharks	Hammerhaie		Молот-рыбы
Head-and-tail-light	Leuchtfleckensalmler	Feu de position	
Herrings	Heringe i. e. S., Heringe	Harengs, Clupéidés	Океанские сельди, Сельдевые
Hornsharks	Hornhaie		Рогатые акулы
Humpback salmon	Buckellachs	Saumon à bosse	Горбуша
Id	Aland	Ide mélanote	Язь
Inconnu	Weißlachs	Saumon du Mackenzie	Белорыбица
Indian cat-fish	Sackkiemer		Мешкожаберный сом
Kamloops trout	Regenbogenforelle	Truite de Kamloops	Радужная форель
King salmon	Quinnat	Saumon de Californie	Чавыча
Kisutch	Kisutch-Lachs	— coho	Кижуч
Kneriids	Ohrenfische		Кнерии
Ladyfish	Grätenfisch	Albule commun	
Ladyfishes	Eigentliche Grätenfische	Albules	
Lake herring (N. A.)	Amerikanische Kleine Maräne		Американская ряпушка
— sturgeon	Roter Stör		Американский пресноводный осетр
— trout	Seeforelle	Truite de lac	Озерная форель
Lampern	Flußneunauge	Lamproie de rivière	Европейская речная минога
Lampreys	Neunaugen, Rundmäuler, Australische Neunaugen	Lamproies	Миногообразные, Круглоротые, Австралийские миноги
— and hagfishes	Kieferlose		Бесчелюстные
Large-finned shark	Atlantischer Braunhai	Petit chien bleu	
Large-spotted dogfish	Großgefleckter Katzenhai	Grande roussette	Большой морской кот

ENGLISH NAME	GERMAN NAME	FRENCH NAME	RUSSIAN NAME
Latern shark	Schwarzer Dornhai	Sagre	Черная колючая акула
Lesser lamprey	Bachneunauge	Lamproie de Planeri	Европейская ручьевая минога
Ling	Leng	Lingue	Обыкновенная мольва
Long-nosed gar-pike	Schlanker Knochenhecht	Lépisostée osseux	Костяной клювонос
Loach	Europäischer Steinbeißer	Loche de rivière	Обыкновенная щиповка
Loaches	Schmerlen		Вьюновые
Lyretail	Kap Lopez	Cap Lopez	
Mackerel shark	Mako	Lamie à nez pointu	Сельдевидные акулы,
— sharks	Makrelenhaie, Heringshaie		Сельдевые акулы
Mako shark	Mako	Lamie à nez pointu	Пещерная живородка
Mammoth cave blindfish	Mammuthöhlen-Blindfisch	Amblyopsis	Гигантская манта
Manta ray	Riesenmanta		Мраморный электри-
Marbled electric ray	Marmorzitterrochen	Torpille marbrée	ческий скат
Mediterranean shark	Spitzkopfsechskiemer	Griset	Речной гольян
Minnow	Elritze	Vairon	Карповые
Minnows	Weißfische	Cyprinidés	Карпообразные
— Suckers and Loaches	Karpfenfische		
Mooneye	Mondauge		Луноглаз
Mooneyes	Mondaugen		Луноглазы
Morays	Muränen, Riffmuränen	Murènes	Мурены
Mormyrids	Nilhechte		Длиннорылы
Morry eels	Riffmuränen	Murènes	
Mosquito fish	Koboldkärpfling	Poisson-Léopard	Гамбузия
— fishes	Gambusen	Gambusies	Гамбузии
Mud pickerel	Grass-Hecht	Brochet vermiculé	
Mudfish	Amerikanischer Schlammfisch	Amie	Ильная рыба
Muskellunge	Muskellunge	Maskinongé	
Northern anchovy	Nordamerikanische Sardelle	Anchois de Californie	Североамериканский анчоус
Northern pike (N. A.)	Amurhecht		Амурская щука
Osteoglossid	Afrikanischer Knochenzüngler		Африканский косте- язычник
Osteoglossids	Knochenzüngler		Костеязычные
Pacific anchovy	Nordamerikanische Sardelle	Anchois de Californie	Североамериканский анчоус
— cod	Pazifik-Kabeljau	Morue du Pacifique	Тихоокеанская треска
— herring	Pazifischer Hering	Hareng du Pacifique	Тихоокеанская сельдь
— salmons	Pazifische Lachse	Saumons	Тихоокеанские лососи
Paddlefishes	Löffelstöre		Веслоносы
Pearl Danio	Schillerbärbling	Danio rosé	
— fish	Fierasfer	Aurin	Фиерасфер
Pelican-fish	Schlinger	Saccopharynx	
Pencil fish	Längsbandsalmler	Nannostome	
Pickerel	Hecht	Brochet commun	Обыкновенная щука
Pickerels	Hechte, Hechte i. e. S.	Brochets	Щуковые, Щуки
Pike	Hecht	Brochet commun	Обыкновенная щука
Pilchard	Pazifische Sardine, Pilchard	Sardine	Тихоокеанская сардина, Атлантическая сардина
Pink salmon	Buckellachs	Saumon à bosse	Горбуша
Pirarucu	Arapaima		Арапайма
Piraya	Piraya	Piraya	Пиранья
Pirayas	Pirayas	Caribes	Пираньи
Poeciliids	Lebendgebärende Zahnkärpflinge		Живородящие карпозубые
Pollack	Köhler	Lieu noir	Сайда
Polypterus	Flösselhechte		Многоперы
Pond loach	Schlammpeitzger	Loche d'étang	Вьюн-пескарь
Porbeagle	Heringshai	Lamie	Сельдевая акула
Powan	Große Schwebrenke	Corégone lavaret	Обыкновенный сиг
Pride	Bachneunauge	Lamproie de Planeri	Европейская ручьевая минога
Pygmy top minnow	Zwergkärpfling	Poisson moustique	
Rabbit fish	Seeratte	Rat de mer	Обыкновенная химера
— fishes	Seeratten	Chimères	
Rainbow trout	Regenbogenforelle	Truite de Kamloops	Радужная форель
Red from Rio	Roter von Rio	Tétra rouge	
— Rasbora	Keilfleckbarbe	Rasbora	
— salmon	Blaurückenlachs		Нерка
Red-eye	Rotfeder	Rotengle	Красноперка
River trout	Bachforelle	Truite de rivière	Ручьевая форель
Rudd	Rotfeder	Rotengle	Красноперка
Salmon trout	Europäische Forelle	Truite de mer	Лосось-таймень
Salmons	Lachsähnliche i. e. S.	Saumons	Лососевые

ENGLISH NAME	GERMAN NAME	FRENCH NAME	RUSSIAN NAME
Sand sharks	Blauhaie		Голубые акулы
— smelt	Großer Ährenfisch	Siouclet	Большая атеринка
tiger shark	Sandtiger	Odontaspide taureau	
Sandpiper	Bachneunauge	Lamproie de Planeri	Европейская ручьевая минога
Sardines	Echte Sardinen	Sardines	Сардины
Saurie	Atlantischer Makrelenhecht	Balaou	Атлантическая макреле-щука
Sawfish	Westlicher Sägefisch		Западная пила-рыба
Sawfishes	Sägerochen, Sägefische	Scies	Пилы-рыбы
Sea cat-fishes	Kreuzwelse		Зубастые сомы
— gars	Hornhechte		Саргановые
— lamprey	Meerneunauge	Lamproie marine	Морская минога
— trout	Europäische Forelle	Truite de mer	Лосось-таймень
Shad	Alse	Grande alose	Алоса
Shads	Alsen	Aloses	Атлантические проходные сельди
Sharks	Haie	Squales	Акулообразные
— and rays	Plattenkiemer		Пластиножаберные
Sheat-fish	Flußwels	Silure glane	Обыкновенный сом
Shovel-nosed sturgeon	Gemeiner Schaufelstör		Обыкновенный лопатонос
— sturgeons	Schaufelnasenstöre		Лопатоносы
Silver bream	Güster	Brème bordelière	Густера
— salmon	Kisutch-Lachs	Saumon coho	Кижуч
Silverside	Großer Ährenfisch	Siouclet	Большая атеринка
Silversides	Ährenfische i. e. S.		Атеринки
Silvery pouts	Silberdorsche	Merlan argenté	
Six-gilled shark	Grauhai	Requin griset	
Skates	Echte Rochen	Raies	Настоящие скаты
— and rays	Rochen		Скатовые
Small-spotted dog-fish	Kleingefleckter Katzenhai	Petite roussette	Морской кот
Smooth-hound	Südlicher Glatthai	Emissole lisse	Южная кунья акула
Sockeye salmon	Blaurückenlachs		Нерка
South-red-bellied dace	Rötling	Vairon à gorge rouge	
Speckled trout	Bachsaibling	Saumon de fontaine	Американская ручьевая палия
Spiny dog-fish	Dornhai, Schwarzer Dornhai	Aiguillat tacheté, Sagre	Нокотница, Черная колючая акула
— dog-fishes	Dornhaie	Squales	Колючие акулы
— shark	Nagelhai	Chenille	
Spoonbills	Löffelstöre		Веслоносы
Spotted Danio	Tüpfelbärbling	Danio moucheté	
— dog-fishes	Katzenhaie		Кошачьи акулы
— eagle ray	Gewöhnlicher Adlerrochen	Aigle de mer	Обыкновенный орляк
Squaloids	Stachelhaie	Squaloïdes	Колючеперые акулы
Squatinoids	Engelhaie		Морские ангелы
Stellate smooth-hound	Nördlicher Glatthai	Emissole tachetée	Северная кунья акула
Sterlet	Sterlet	Sterlet	Стерлядь
Sting rays	Stachelrochen		Хвостоколы
Sturgeons	Störe	Esturgeons	Осетры, Осетровые
Suckers	Sauger		Чукучановые
Sumatran barb	Viergürtelbarbe	Barbeau de Sumatra	
Swordtail	Schwertträger	Porte-Glaive	
Tarpon	Atlantischer Tarpun		Атлантический тарпун
Tarpons	Tarpune		Тарпуны
Tench	Schleie	Tanche commune	Линь
Tenches	Schleie	Tanches	Лини
Thornback ray	Nagelrochen	Raie bouclée	Шиповатый скат
Thresher shark	Fuchshai	Requin renard	Морская лисица
Tooth carps	Zahnkärpflinge		Карпозубовые
Tope	Hundshai	Milandre	
Topes	Fleckhaie		Пилохвосты
Torpedoes	Zitterrochen	Torpilles	Электрические скаты
Trouts	Lachse und Forellen	Saumons	Благородные лососи
True smelts	Stinte		Корюшковые
Tschavitscha	Quinnat	Saumon de Californie	Чавыча
Twaite shad	Finte	Alose finte	Финта
Vertebrates	Wirbeltiere	Vertébrés	Черепные
Viviparous blenny	Aalmutter	Loquette	Обыкновенная бельдюга
Wall-eye pollack	Alaska-Pollack		Минтай
Whale shark	Walhai		Малозубая акула
White bream	Güster	Brème bordelière	Густера
— sturgeon	Weißer Stör		Восточнотихоокеанский осетр
Whitefishes	Renken	Corégones	Сиги
Zebra fish	Zebrabärblinge	Petit Danio	

II German—English—French—Russian

N. A. bei englischen Namen bedeutet, daß dieser Name nur in Nordamerika gebräuchlich ist.

GERMAN NAME	ENGLISH NAME	FRENCH NAME	RUSSIAN NAME
Aalmutter	Viviparous blenny	Loquette	Обыкновенная бельдюга
Aalquappe	Burbot	Lote	Налим
Aalquappen	Burbots	Lotes	Налимы
Abramis	Breams	Brèmes	Лещи
— *brama*	Bream	Brème commune	Лещ
Acheilognathus	Carplike fish		Японские горчаки
Acipenser	Sturgeons	Esturgeons	Осетры
— *fulvescens*	Lake sturgeon		Американский пресно- водный осетр
— *oxyrhynchus*	Atlantic sturgeon		Атлантический осетр
— *ruthenus*	Sterlet	Sterlet	Стерлядь
— *sturio*	Common Atlantic sturgeon	Esturgeon commun	Балтийский осетр
— *transmontanus*	White sturgeon		Восточнотихоокеанский осетр
Acipenseridae	Sturgeons	Esturgeons	Осетровые
Adlerrochen	Eagle rays		Орляки
Afrikanischer Adlerrochen	Bull ray	Mourine vachette	Африканский орляк
— Knochenzüngler	Osteoglossid		Африканский косте- язычник
Agnatha	Lampreys and hagfishes		Бесчелюстные
Ährenfische i. e. S.	Silversides		Атеринки
Aitel	Chub	Chevain	Голавль
Aland	Id	Ide mélanote	Язь
Alaska-Pollack	Wall-eye pollack		Минтай
Albula	Ladyfishes	Albules	
— *vulpes*	Bonefish	Albule commun	
Alburnus		Ablettes	Уклейки
— *alburnus*	Bleak	Ablette	Обыкновенная уклея
Alopias vulpinus	Thresher shark	Requin renard	Морская лисица
Alosa	Shads	Aloses	Атлантические проходные сельди
— *alosa*	Shad	Grande alose	Алоса
— *fallax*	Twaite shad	Alose finte	Финта
— *sapidissima*	American shad	— d'Amérique	Шэд
Als	Chub	Chevain	Голавль
Alse	Shad	Grande alose	Алоса
Alsen	Shads	Aloses	Атлантические проходные сельди
Alwe	Bleak	Ablette	Обыкновенная уклея
Amblyopsidae	Cave fishes		Живородковые
Amblyopsis spelaeus	Mammoth cave blindfish	Amblyopsis	Пещерная живородка
Amerikanische Alse	American shad	Alose d'Amérique	Шэд
— Kleine Maräne	Lake herring (N. A.)		Американская ряпушка
Amerikanischer Aal	American eel	Anguille d'Amérique	Американский речной угорь
— Kahlhecht	Bowfin	Amie	Ильная рыба
— Schlammfisch	Bowfin	Amie	Ильная рыба
— Seesaibling	Common lake charr	Truite grise	Американская озерная палия
Amia	Bowfins	Amies	Амии
— *calva*	Bowfin	Amie	Ильная рыба
Amiidae	Bowfins		Амии
Amurhecht	Northern pike (N. A.)		Амурская щука
Anablepidae	Four-eyed fishes		Четырехглазые рыбы
Anableps tetrophthalmus	— fish		Четырехглазая рыба
Anchovis	Anchovy	Anchois	Европейский анчоус
Anguilla anguilla	Fresh-water eel	Anguille	Обыкновенный речной угорь
— *rostrata*	American eel	— d'Amérique	Американский речной угорь
Anguillidae	Eels	Anguilles	Речные угри
Antennariidae	Frog-fishes		Морские клоуны
Aphyosemion australe	Lyretail	Cap Lopez	
Arapaima	Arapaima		Арапайма
Arapaima gigas	Arapaima		Арапайма
Ariidae	Sea cat-fishes		Зубастые сомы

GERMAN NAME	ENGLISH NAME	FRENCH NAME	RUSSIAN NAME
Äsche	Graylings		Хариусы
Aspius		Aspes	Жерехи
– aspius		Aspe	Обыкновенный жерех
Atherina hepsetus	Silverside	Siouclet	Большая атеринка
– mochon	Caspian sand smelt	Prêtre	Малая атеринка
Atherinidae	Silversides		Атеринки
Atlantische Flugfische	Flying-fishes	Poissons volants	Долгоперы
Atlantischer Braunhai	Large-finned shark	Petit chien bleu	
– Hering	Atlantic herring	Hareng	Атлантическая сельдь
– Lachs	– salmon	Saumon atlantique	Обыкновенный лосось
– Makrelenhecht	Saurie	Balaou	Атлантическая макреле-щука
– Seeteufel	Angler-fish		Обыкновенный морской черт
– Stör	Atlantic sturgeon		Атлантический осетр
– Tarpun	Tarpon		Атлантический тарпун
Australische Neunaugen	Lampreys		Австралийские миноги
Bachforelle	River trout	Truite de rivière	Ручьевая форель
Bachneunauge	Brook lamprey	Lamproie de Planeri	Европейская ручьевая минога
Bachsaibling	– trout	Saumon de fontaine	Американская ручьевая палия
Baltischer Stör	Common Atlantic sturgeon	Esturgeon commun	Балтийский осетр
Barbe	Barbel	Barbeau commun	Обыкновенный усач
Barbus	Barbels	Barbeaux	Настоящие усачи
– barbus	Barbel	Barbeau commun	Обыкновенный усач
– meridionalis petenyi		– canin	Крапчатый усач
Belone	Gar-fishes	Aiguilles	Сарганы
– bellone	Gar-fish	Aiguille	Европейский сарган
Belonidae	Gar-fishes		Саргановые
Belonesox belizanus	Belonesox	Brochet vivipare	
Bitterlinge	European bitterling	Bouvières	Обыкновенные горчаки
Blauer Wittling	Couch's whiting	Poutassou	
Blauhai	Blue shark	Requin bleue	Голубая акула
Blauhaie	Sand sharks		Голубые акулы
Blaurückenlachs	Sockeye salmon		Нерка
Blei	Bream	Brème commune	Лещ
Blicca bjoerkna	Silver bream	– bordelière	Густера
Blikke	Silver bream	– bordelière	Густера
Blindfische	Cave fishes		Живородковые
Brachsen	Bream	– commune	Лещ
Brachydanio albolineatus	Pearl Danio	Danio rosé	
– nigrofasciatus	Spotted Danio	– moucheté	
– rerio	Zebra fish	Petit Danio	
Brasse	Bream	Brème commune	Лещ
Brassen	Breams	Brèmes	Лещи
Breiting	Bream	Brème commune	Лещ
Brosme brosme	Cusk	Brosme	Менек
Brutt	Minnow	Vairon	Речной гольян
Buckellachs	Pink salmon	Saumon à bosse	Горбуша
Callichthyidae	Callichthyidés		Панцырные сомы
Carapus acus	Pearl fish	Aurin	Фиерасфер
Carassius	Common carps	Carassins	Караси
– auratus	Goldfish	Poisson rouge	Серебряный карась
– carassius	Crucian carp	Carassin vulgaire	Обыкновенный карась
Carcharhinidae	Sand sharks		Голубые акулы
Carcharhinus plumbeus	Large-finned shark	Petit chien bleu	
Carcharias ferox	Fierce shark	Odontaspide féroce	
– taurus	Sand tiger shark	– taureau	
Carcharodon carcharias	Great white shark	Requin	
Catostomidae	Suckers		Чукучановые
Cetorhinidae	Basking sharks		Гигантские акулы
Cetorhinus maximus	– shark	Squale pélerin	Гигантская акула
Characidae	Characins		Харациновые
Characoidei	Characins		Харациновидные
Chimaera	Rabbit fishes	Chimères	Обыкновенная химера
– monstrosa	– fish	Rat de mer	Химеры
Chimaeridae		Chiméroïdes	Цельноголовые
Chimären	Chimaeras		Плащеносные акулы
Chlamydoselachidae	Frilled sharks		Плащеносная акула
Chlamydoselachus anguineus	– shark	Requin à tunique	Хрящевые рыбы
Chondrichthyes	Cartilaginous fishes		Подусты
Chondrostoma		Chondrostomes	Обыкновенный подуст
– nasus		Chondrostome nez	
Chrosomus erythrogaster	South-red-bellied dace	Vairon à gorge rouge	

GERMAN NAME	ENGLISH NAME	FRENCH NAME	RUSSIAN NAME
– neogaeus	Finescale dace	Goujon à fines écailles	Океанические сельди
Clupea	Herrings	Harengs	
– harengus	Atlantic herring	Hareng	Атлантическая сельдь
– pallasii	Pacific herring	– du Pacifique	Тихоокеанская сельдь
Clupeidae	Herrings	Clupes	Сельдевые
Clupisudis niloticus	Osteoglossid		Африканский косте-язычник
Cobitidae	Loaches		Вьюновые
Cobitis		Loches	Щиповки
– taenia	Loach	Loche de rivière	Обыкновенная щиповка
Conger conger	Conger eel	Congre	Морской угорь
Coregonus	Whitefishes	Corégones	Сиги
– albula		Corégone blanc	Ряпушка
– artedi	Lake herring (N. A.)		Американская ряпушка
– fera		– féra	Придонный сиг
– lavaretus	Powan	– lavaret	Обыкновенный сиг
– oxyrhynchus		Outil	Морской сиг
Corydoras paleatus	Armored cat-fish	Corydoras à casque	
Cyclostomata	Lampreys	Lamproies	Круглоротые
Cyprinidae	Minnows	Cyprinidés	Карповые
Cypriniformes	Minnows, Suckers and Loaches		Карпообразные
Cyprinodontidae	Cyprinodonts		Яицекладущие карпо-зубые
Cyprinodontoidei	Tooth carps		Карпозубовые
Cyprinus	Carps	Carpes	Сазаны
– carpio	Carp	Carpe ordinaire	Сазан
Dalatias licha	Darkie charlie	Liche	
Dallia	Black-fishes		Даллии
– pectoralis	Alaska black-fish		Даллия
Danio malabaricus	Giant Danio	Danio géant	
Dasyatidae	Sting-rays		Хвостоколы
Dasyatis		Pastenagues	Хвостоколы
– pastinaca	Common sting-ray	Pastenague	Морской кот
– violacea	Blue sting-ray	– violette	Фиолетовый хвостокол
Dibel	Chub	Chevain	Голавль
Dickkopf	Chub	Chevain	Голавль
Döbel	Chub	Chevain	Голавль
Doradidae	Armored cat-fishes		Иглистые сомы
Dornhai	Spiny dog-fish	Aiguillat tacheté	Нокотница
Dornhaie	Spiny dog-fishes	Squales	Колючие акулы
Dornwelse	Armored cat-fishes		Иглистые сомы
Dorsche i. e. S.	Gadoid fishes	Gades	Тресковые
Drescher	Thresher shark	Requin renard	Морская лисица
Echinorhinus brucus	Bramble shark	Chenille	
Echte Aale	Eels	Anguilles	Речные угри
– Barben	Barbels	Barbeaux	Настоящие усачи
– Knochenfische	Bony fishes	Poissons osseux	Костистые рыбы
– Messeraale	Gymnotids	Gymnotidés	
– Rochen	Skates	Raies	Настоящие скаты
– Sardinen	Sardines	Sardines	Сардины
– Störe	Sturgeons	Esturgeons	Осетровые
– Welse	Cat-fishes	Siluridés, Silures	Сомовые, Обыкновенные сомы
Echter Sandhai	Sand tiger shark	Odontaspide taureau	
Eierlegende Zahnkärpflinge	Cyprinodonts		Яицекладущие карпозубые
Eigentliche Dorsche	Cods	Morues	Треска
– Grätenfische	Ladyfishes	Albules	
– Messerfische	Featherbacks	Notoptéridés	
Eishai	Greenland shark	Requin des glaces	
Elasmobranchii	Sharks and rays		Пластиножаберные
Elderitz	Minnow	Vairon	Речной гольян
Electrophorus electricus	Electric eel	Anguille tremblante	Электрический угорь
Elektrischer Wels		Poisson-chat électrique	Электрический сом
Elritze	Minnow	Vairon	Речной гольян
Elze		Chondrostome nez	Обыкновенный подуст
Engelhaie	Squatinoids, Angel sharks		Морские ангелы
Engraulidae	Anchovies		Анчоусовые
Engraulis	Anchovies		Анчоусы
– encrasicholus	Anchovy	Anchois	Европейский анчоус
– mordax	Northern anchovy	– de Californie	Североамериканский анчоус
Erling	Minnow	Vairon	Речной гольян
Esocidae	Pickerels		Щуковые

GERMAN NAME	ENGLISH NAME	FRENCH NAME	RUSSIAN NAME
Esox	Pickerels	Brochets	Щуки
– lucius	Pike	Brochet commun	Обыкновенная щука
– masquinongyi	Muskellunge	Maskinongé	
– niger	Chain pickerel	Brochet maillé	
– reicherti	Northern pike (N. A.)		Амурская щука
– vermiculatus	Grass pickerel	Brochet vermiculé	
Etmopterus spinax	Latern shark	Sagre	Черная колючая акула
Exocoetidae	Flying-fishes		Летучие рыбы
Exocoetus	Flying-fishes	Poissons volants	Долгоперы
Europäische Äsche	Grayling	Ombre de rivière	Обыкновенный хариус
– Forelle	Brown trout	Truite de mer	Лосось-таймень
– Sardelle	Anchovy	Anchois	Европейский анчоус
Europäischer Bitterling	Bitterling	Bouvière amère	Европейский горчак
– Flußaal	Fresh-water eel	Anguille	Обыкновенный речной угорь
– Hausen		Grand esturgeon	Белуга
– Hornhecht	Gar-fish	Aiguille	Европейский сарган
– Hundsfisch	Dogfish	Poisson-chien	Европейская евдошка
– Steinbeißer	Loach	Loche de rivière	Обыкновенная щиповка
Fächerfisch	Alaska black-fish		Даллия
Fächerfische	Black-fishes		Даллии
Felsenstör	Lake sturgeon		Американский пресно-водный осетр
Fierasfer	Pearl fish	Aurin	Фиерасфер
Finte	Twaite shad	Alose finte	Финта
Fische	Fishes	Poissons	Рыбы
Fleckhai	Black-mouthed dogfish	Chien espagnol	Пилохвост
Fleckhaie	Topes		Пилохвосты
Fliegende Fische	Flying-fishes		Летучие рыбы
Florida-Kärpfling	American flag fish	Jordanelle de Floride	
Flösselhechte	Polypterus		Многоперы
Flußaale	Eels	Anguilles	Речные угри
Flußneunauge	Fresh water lamprey	Lamproie de rivière	Европейская речная минога
Flußwels	Sheat-fish	Silure glane	Обыкновенный сом
Fuchshai	Thresher shark	Requin renard	Морская лисица
Fühlerfische i. e. S.	Frog-fishes		Морские клоуны
Gadiculus	Silvery pouts	Merlan argenté	
Gadidae	Gadoid fishes	Gades	Тресковые
Gadus	Cods	Morues	Треска
– macrocephalus	Pacific cod	Morue du Pacifique	Тихоокеанская треска
– morhua	Cod common	– fraîche	Атлантическая треска
– ogac	Greenland halibut	– de roche	Гренландская фиордовая треска
Galaxiidae	Galaxiids		Галаксиевые
Galeorhinus galeus	Tope	Milandre	
Galeus	Topes		Пилохвосты
– melanostomus	Black-mouthed dogfish	Chien espagnol	Пилохвост
Gambusen	Mosquito fishes	Gambusies	Гамбузии
Gambusia	Mosquito fishes	Gambusies	Гамбузии
– affinis	– fish	Poisson-Léopard	Гамбузия
Gängling	Id	Ide mélanote	Язь
Gefleckter Panzerwels	Armored cat-fish	Corydoras à casque	
– Zitterrochen	Eyed electric ray	Torpille tachetée	Пятнистый электрический скат
Gemeiner Geigenrochen	Guitar fish	Guitare	Обыкновенная рохля
– Meerengel	Angel fish	Ange de mer	Европейский морской ангел
– Schaufelstör	Shovel-nosed sturgeon		Обыкновенный лопатонос
– Stör	Common Atlantic sturgeon	Esturgeon commun	Балтийский осетр
Geotria	Lampreys		Австралийские миноги
Gewöhnliche Karausche	Crucian carp	Carassin vulgaire	Обыкновенный карась
– Langnasenchimäre	Chimaera		Обыкновенная длинно-рылая химера
Gewöhnlicher Adlerrochen	Spotted eagle ray	Aigle de mer	Обыкновенный орляк
– Gründling	Gudgeon	Goujon	Обыкновенный пескарь
– Stechrochen	Common sting-ray	Pastenague	Морской кот
Gitarrenfisch	Guitar fish		Гитара-рыба
Glatter Hammerhai	Common hammerhead shark	Requin marteau	Обыкновенная молот-рыба
Glatthai des Aristoteles	Stellate smooth-hound	Emissole tachetée	Северная кунья акула
Glattrochen	Flapper skate	Pocheteau blanc	Гладкий скат
Gobio	Gudgeons	Goujons	Обыкновенные пескари
– gobio	Gudgeon	Goujon	Обыкновенный пескарь
Goldfisch	Goldfish	Carassin doré	Золотая рыбка

GERMAN NAME	ENGLISH NAME	FRENCH NAME	RUSSIAN NAME
Goldnerfling		Orfe	Орф
Goldorfe		Orfe	Орф
Grass-Hecht	Grass pickerel	Brochet vermiculé	
Grätenfisch	Bonefish	Albule commun	
Grauhai	Comb-toothed shark	Requin griset	
Grauhaie	Comb-tooth sharks, Sand sharks		Гребнезубые акулы, Голубые акулы
Grieslauge		Blageon	Рислинг
Grönlandhai	Greenland shark	Requin des glaces	
Grönlandkabeljau	– halibut	Morue de roche	Гренландская фиордовая треска
Große Bodenrenke		Corégone féra	Придонный сиг
Große Schwebrenke	Powan	Corégone lavaret	Обыкновенный сиг
Großer Ährenfisch	Silverside	Siouclet	Большая атеринка
– Nilhecht	Eel-like mormyrid	Gymnarche du Nil	Гимнарх
Großgefleckter Katzenhai	Large-spotted dogfish	Grande roussette	Большой морской кот
Grümpel	Minnow	Vairon	Речной гольян
Grundhai	Smooth-hound	Emissole lisse	Южная кунья акула
Gründlinge i. e. S.	Gudgeons	Goujons	Обыкновенные пескари
Güster	Silver bream	Brème bordelière	Густера
Gymnarchus niloticus	Eel-like mormyrid	Gymnarche du Nil	Гимнарх
Gymnothorax	Morays	Murènes	
Gymnotidae	Gymnotids	Gymnotidés	
Gymnotoidei	Electric eels		Электрические угри
Gymnura altavela	Butterfly ray	Mourine bâtarde	Гимнура
Haie	Sharks	Squales	Акулообразные
Halbschnäbler	Half-beaks	Hémiramphes	Полурылы
Hammerhaie	Hammerhead sharks		Молот-рыбы
Harriotta raleighana	Chimaera		Обыкновенная длинно-рылая химера
Hasel	Dace	Vandoise vraie	Обыкновенный елец
Hässling	Dace	Vandoise vraie	Обыкновенный елец
Hecht	Pike	Brochet commun	Обыкновенная щука
Hechte	Pickerels		Щуковые
Hechte i. e. S.	Pickerels	Brochets	Щуки
Hechtkärpfling	Belonesox	Brochet vivipare	
Hechtlinge i. e. S.	Galaxiids		Галаксиевые
Hemigrammus ocellifer	Head-and-tail-light	Feu de position	
Hemirhamphus	Half-beaks	Hémiramphes	Полурылы
Heptranchias perlo	Mediterranean shark	Griset	
Heringe	Herrings	Clupéidés	Сельдевые
Heringe i. e. S.	Herrings	Harengs	Океанические сельди
Heringshai	Porbeagle	Lamie	Сельдевая акула
Heringshaie	Mackarel sharks		Сельдевые акулы
Heterandria formosa	Pygmy top minnow	Poisson moustique	
Heterodontidae	Hornsharks		Рогатые акулы
Heteropneustes fossilis	Indian cat-fish		Мешкожаберный сом
Hexanchidae	Comb-tooth sharks		Гребнезубые акулы
Hexanchus griseus	Comb-toothed shark	Requin griset	
Hiodon	Mooneyes		Луноглазы
– tergisus	Mooneye		Луноглаз
Histrio histrio	Frog-fish		Морская мышь
Holocephali	Chimaeras		Цельноголовые
Hornhaie	Hornsharks		Рогатые акулы
Hornhechte	Gar-fishes	Aiguilles	Саргановые, Сарганы
Hundshai	Smooth-hound, Tope	Emissole lisse, Milandre	Южная кунья акула
Europäischer Hausen		Grand esturgeon	Белуга
Hyphessobrycon flammeus	Red from Rio	Tétra rouge	
Inger	Common hagfish, Hagfishes	Myxine	Обыкновенная миксина, Миксинообразные
Isuridae	Mackerel sharks		Сельдевидные акулы
Isurus oxyrhynchus	– shark	Lamie à nez pointu	
Japanische Bitterlinge	Carplike fish		Японские горчаки
Jordanella floridae	American flag fish	Jordanelle de Floride	
Kabeljau	Common cod	Morue fraîche	Атлантическая треска
Kaimanfische	Gar-pikes		Панцырные щуки
Kap Lopez	Lyretail	Cap Lopez	
Karauschen	Common carps	Carassins	Караси
Karibenfisch	Piraya		Пиранья
Karpfen	Carp, Carps	Carpe ordinaire, Carpes	Сазан, Сазаны
Karpfenfische	Minnows, Suckers and Loaches		Карпообразные
Karpfenfische i. e. S.	Minnows		Карповые
Katzenhaie	Spotted dog-fishes		Кошачьи акулы
Keilfleckbarbe	Red Rasbora	Rasbora	
Keta-Lachs	Dog salmon	Saumon-chien	Кета

GERMAN NAME	ENGLISH NAME	FRENCH NAME	RUSSIAN NAME
Kettenhecht	Chain pickerel	Brochet maillé	
Kieferlose	Lampreys and hagfishes		Бесчелюстные
Kisutch-Lachs	Silver salmon	Saumon coho	Кижуч
Kleine Maräne		Corégone blanc	Ряпушка
– Schwebrenke		Outil	Морской сиг
Kleiner Ährenfisch	Caspian sand smelt	Prêtre	Малая атеринка
Kleingefleckter Katzenhai	Small-spotted dog-fish	Petite roussette	Морской кот
Kleinschuppenrötling	Finescale dace	Goujon à fines écailles	
Kneriidae	Kneriids		Кнерии
Knochenfische	Bony fishes		Костные рыбы
Knochenhechte	Gar-pikes, Gars	Poissons osseux	Панцырные щуки,
			Каймановые рыбы
Knochenzüngler	Osteoglossids		Костеязычные
Knorpelfische	Cartilaginous fishes		Хрящевые рыбы
Koboldkärpflinge	Mosquito fish	Poisson-Léopard	Гамбузия
Köhler	Pollack	Lieu noir	Сайда
Krausenhai	Frilled shark		Плащеносная акула
Krausenhaie	– sharks	Requin à tunique	Плащеносные акулы
Kreuzwelse	Sea cat-fishes		Зубастые сомы
Lachsähnliche i. e. S.	Salmons	Saumons	Лососевые
Lachse und Forellen	Trouts	Saumons	Благородные лососи
Lamna	Mackarel sharks		Сельдевые акулы
– nasus	Porbeagle	Lamie	Сельдевая акула
Lampetra fluviatilis	Fresh water lamprey	Lamproie de rivière	Европейская речная
			минога
– planeri	Brook lamprey	– de Planeri	Европейская ручьевая
			минога
Langnasen-Knochenhecht	Long-nosed gar-pike	Lépisostée osseux	Костяной клювонос
Längsbandsalmler	Pencil fish	Nannostome	
Lasch	Bream	Brème commune	Лец
Laternenangler	Deep-sea angler		Гигантский удильщик
Lauel	Bleak	Ablette	Обыкновенная уклея
Lebendgebärende Zahnkärpflinge	Poeciliids		Живородящие карпозубые
Leng	Ling	Lingue	Обыкновенная мольва
Lepisosteidae	Gar-pikes		Панцырные щуки
Lepisosteiformes	Gar-pikes		Панцырные щуки
Lepisosteus	Gars		Каймановые рыбы
– osseus	Long-nosed gar-pike	Lépisostée osseux	Костяной клювонос
Leucaspius delineatus		Able de stymphale	Обыкновенная верховка
Leuchtfleckensalmler	Head-and-tail-light	Feu de position	
Leuciscus		Chevaines	Ельцы
– (Idus) idus	Id	Ide mélanote	Язь
– – – forma orfus		Orfe	Орф
– (Leuciscus) leuciscus	Dace	Vandoise vraie	Обыкновенный елец
– (Squalinus) cephalus	Chub	Chevain	Голавль
– (Telestes) souffia agassizi		Blageon	Рислинг
Linophryne arborifer	Deep-sea angler		Гигантский удильщик
Lodde	Capelin	Capelan	Мойва
Löffelstör	American paddlefish		Американский веслонос
Löffelstöre	Spoonbills, Paddlefishes		Веслоносы
Lophiidae	Angler-fishes	Baudroies	Морские черти
Lophius piscatorius	Angler-fish		Обыкновенный морской
			черт
Lota	Burbots	Lotes	Налимы
– lota	Burbot	Lote	Налим
Lumb	Cusk	Brosme	Менек
Lycenchelis, Lycodes		Lycodes	Ликоды
Maifisch	Shad	Grande alose	Алоса
Mako	Mackerel shark	Lamie à nez pointu	
Makrelenhaie	– sharks		Сельдевидные акулы
Malabarbärbling	Giant Danio	Danio géant	
Malapterurus electricus		Poisson-chat électrique	Электрический сом
Mallotus villosus	Capelin	Capelan	Мойва
Mammuthöhlen-Blindfisch	Mammoth cave blindfish	Amblyopsis	Пещерная живородка
Manta birostris	Manta ray		Гигантская манта
Mantarochen	Devil-fishes		Морские дьяволы
Marmorrochen		Raie brunette	Мраморный скат
Marmorzitterrochen	Marbled electric ray	Torpille marbrée	Мраморный электри-
			ческий скат
Mäusebeißer		Aspe	Обыкновенный жерех
Meeraal	Conger eel	Congre	Морской угорь
Meeresteufel	Devil-fish	Diable de mer	Морской дьявол
Meerneunauge	Sea lamprey	Lamproie marine	Морская минога
Meersau	Angular rough shark	Centrine	Центрина
Megalopidae	Tarpons		Тарпуны

GERMAN NAME	ENGLISH NAME	FRENCH NAME	RUSSIAN NAME
Megalops atlanticus	Tarpon		Атлантический тарпун
Melanogrammus aeglefinus	Haddock	Morue noire	Пикша
Menschenfresser-Haie	Mackerel sharks		Сельдевидные акулы
Menschenhai	Great white shark	Requin	
Merlangius merlangus		Merlan	Атлантический мерланг
Merluccius merluccius	Hake	Merlu	Обыкновенный хэк
Messerfisch		Rasoir	Чехонь
Micromesistius poutassou	Couch's whiting	Poutassou	
Misgurnus		Loches d'étangs	Вьюны
– *fossilis*	Pond loach	Loche d'étang	Вьюн-пескарь
Mittelmeer-Glatthai	Smooth-hound	Emissole lisse	Южная кунья акула
Mittelmeerleng	Blue ling		Средиземноморская мольва
Mittelmeer-Muräne	European moray	Murène commune	Средиземноморская мурена
Mobula mobular	Devil-fish	Diable de mer	Морской дьявол
Mobulidae	Devil-fishes		Морские дьяволы
Moderlieschen		Able de stymphale	Обыкновенная верховка
Molva elongata	Blue ling		Средиземноморская мольва
Molva molva	Ling	Lingue	Обыкновенная мольва
Mondauge	Mooneye		Луноглаз
Mondaugen	Mooneyes		Луноглазы
Mormyridae	Mormyrids		Длиннорылы
Muraena helena	European moray	Murène commune	Средиземноморская мурена
Muraenidae	Morays	Murènes	Мурены
Muränen	Morays	Murènes	Мурены
Muskellunge	Muskellunge	Maskinongé	
Mustelus asterias	Stellate smooth-hound	Emissole tachetée	Северная кунья акула
– *mustelus*	Smooth-hound	– lisse	Южная кунья акула
Myliobatidae	Eagle rays		Орляки
Myliobatis aquila	Spotted eagle ray	Aigle de mer	Обыкновенный орляк
Myxine glutinosa	Common hagfish	Myxine	Обыкновенная миксина
Myxini	Hagfishes		Миксинообразные
Nagelhai	Bramble shark	Chenille	
Nagelrochen	Thornback ray	Raie bouclée	Шиповатый скат
Nannostomus beckfordi anomalus	Pencil fish	Nannostome	
Nase		Chondrostome nez	Обыкновенный подуст
Nasen		Chondrostomes	Подусты
Näsling		Chondrostome nez	Обыкновенный подуст
Nesling	Dace	Vandoise vraie	Обыкновенный елец
Neunaugen	Lampreys	Lamproies	Миногообразные
Neunaugenkönig	Sea lamprey	Lamproie marine	Морская минога
Nilhechte	Mormyrids		Длиннорылы
Nordamerikanische Sardelle	Northern anchovy	Anchois de Californie	Североамериканский анчоус
Nördlicher Glatthai	Stellate smooth-hound	Emissole tachetée	Северная кунья акула
Notopteridae	Featherbacks	Notoptéridés	
Ohrenfische	Kneriids		Кнерии
Ohrensardine	Gilt sardine	Allache	
Oncorhynchus	Pacific salmons	Saumons	Тихоокеанские лососи
– *gorbuscha*	Pink salmon	Saumon à bosse	Горбуша
– *keta*	Dog salmon	Saumon-chien	Кета
– *kisutch*	Silver salmon	Saumon coho	Кижуч
– *nerka*	Sockeye salmon		Нерка
– *tschawytscha*	Chinook salmon	– de Californie	Чавыча
Orfe		Orfe	Орф
Osmeridae	True smelts		Корюшковые
Osmerus eperlanus	European smelt	Éperlan d'Europe	Обыкновенная корюшка
Osteichthyes	Bony fishes	Poissons osseux	Костные рыбы
Osteoglossidae	Osteoglossids		Костеязычные
Östlicher Hundsfisch	Central mud minnow	Poisson de vase	Восточноамериканская евдошка
Oxynotus centrina	Angular rough shark	Centrine	Центрина
Pantodon buchholzi	Butterfly-fish	Pantodon	Рыба-бабочка
Pantodontidae	Butterfly-fishes		Пантодоновые
Panzerwelse		Callichthyidés	Панцирные сомы
Pazifik-Kabeljau	Pacific cod	Morue du Pacifique	Тихоокеанская треска
Pazifische Lachse	– salmons	Saumons	Тихоокеанские лососи
– Sardine	Pilchard		Тихоокеанская сардина
Pazifischer Hering	Pacific herring	Hareng du Pacifique	Тихоокеанская сельдь
Pelecus cultratus		Rasoir	Чехонь
Perille	Minnow	Vairon	Речной гольян
Petromyzon marinus	Sea lamprey	Lamproie marine	Морская минога

GERMAN NAME	ENGLISH NAME	FRENCH NAME	RUSSIAN NAME
Petromyzones	Lampreys	Lamproies	Миногообразные
Phoxinus phoxinus	Minnow	Vairon	Речной гольян
Pierling	Minnow	Vairon	Речной гольян
Pilchard	Pilchard	Sardine	Атлантическая сардина
Piraya	Piraya	Piraya	Пиранья
Pirayas	Caribes	Caribes	Пираньи
Pisces	Fishes	Poissons	Рыбы
Plattenkiemer	Sharks and rays		Пластиножаберные
Pliete	Silver bream	Brème bordelière	Густера
Plötze	European roach	Gardon blanc	Плотва
Poeciliidae	Poeciliids		Живородящие карпо-зубые
Pollachius pollachius	Green Pollack	Lieu jaune	
– *virens*	Pollack	– noir	Сайда
Polyodon	Paddlefishes		Веслоносы
– *spathula*	American paddlefish		Американский веслонос
Polyodontidae	Spoonbills		Веслоносы
Polypterus	Polypterus		Многоперы
Pricke	Fresh water lamprey	Lamproie de rivière	Европейская речная минога
Prionace glauca	Blue shark	Requin bleue	Голубая акула
Pristidae	Sawfishes	Scies	Пилы-рыбы
Pristioidei	Sawfishes	Scies	Пилы-рыбы
Pristis	Sawfishes	Scies	Пилы-рыбы
Pristis pectinatus	Sawfish		Западная пила-рыба
Psephurus gladius	Chinese sturgeon		Китайский меченос
Pteromylaeus bovinus	Bull ray	Mourine vachette	Африканский орляк
Puntius tetrazona tetrazona	Sumatran barb	Barbeau de Sumatra	
Quermaul		Chendrostome nez	Обыкновенный подуст
Quinnat	Chinook salmon	Saumon de Californie	Чавыча
Raja	Skates	Raies	Настоящие скаты
– *batis*	Flapper skate	Pocheteau blanc	Гладкий скат
– *clavata*	Thornback ray	Raie bouclée	Шиповатый скат
– *undulata*		– brunette	Мраморный скат
Rajiformes	Skates and rays		Скатовые
Rapfen		Aspe, Aspes	Обыкновенный жерех, Жерехи
Rasbora heteromorpha	Red Rasbora	Rasbora	
Rautenschmelzschupper	Gar-pikes		Панцырные щуки
Regenbogenforelle	Rainbow trout	Truite de Kamloops	Радужная форель
Renken	Whitefishes	Corégones	Сиги
Rhincodon typus	Whale shark		Малозубая акула
Rhinichthys atronasus	Black-nosed dace	Cyprin à nez noir	
Rhinobatos cemiculus	Guitar fish		Гитара-рыба
– *rhinobatos*	Guitar fish	Guitare	Обыкновенная рохля
Rhodeus	European bitterling	Bouvières	Обыкновенные горчаки
– *sericeus amarus*	Bitterling	Bouvière amère	Европейский горчак
Riesenhai	Basking shark	Squale pélerin	Гигантская акула
Riesenhaie	– sharks		Гигантские акулы
Riesenmanta	Manta ray		Гигантская манта
Riessling		Blageon	Рислинг
Riffmuränen	Morays	Murènes	
Rochen	Skates and rays		Скатовые
Rotauge	European roach	Gardon blanc	Плотва
Rötel	Rudd	Rotengle	Красноперка
Roter Stör	Lake sturgeon		Американский пресно-водный осетр
– von Rio	Red from Rio	Tétra rouge	
Rotfeder	Rudd	Rotengle	Красноперка
Rotflosser	Rudd	Rotengle	Красноперка
Rotkarpfen	Rudd	Rotengle	Красноперка
Rötling	South-red-bellied dace	Vairon à gorge rouge	
Rotschiedel		Aspe	Обыкновенный жерех
Rotten	Id	Ide mélanote	Язь
Rümpchen	Minnow	Vairon	Речной гольян
Rundmäuler	Lampreys	Lamproies	Круглоротые
Rüsselstöre	Sturgeons	Esturgeons	Осетровые
Rüßling	Dace	Vandoise vraie	Обыкновенный елец
Rutilus rutilus	European roach	Gardon blanc	Плотва
Säbelfisch		Rasoir	Чехонь
Saccopharynx ampullaceus	Pelican-fish	Saccopharynx	
Sackkiemer	Indian cat-fish		Мешкожаберный сом
Sägefische	Sawfishes	Scies	Пилы-рыбы
Sägerochen	Sawfishes	Scies	Пилы-рыбы
Saiblinge	Charrs		Гольцы

GERMAN NAME	ENGLISH NAME	FRENCH NAME	RUSSIAN NAME
Salmler	Characins		Харациновидные
Salmler i. e. S.	Characins		Харациновые
Salmo	Trouts	Saumons	Благородные лососи
– *gairdneri*	Rainbow trout	Truite de Kamloops	Радужная форель
– *salar*	Atlantic salmon	Saumon atlantique	Обыкновенный лосось
– *trutta*	Brown trout	Truite de mer	Лосось-таймень
– *trutta fario*	River trout	– de rivière	Ручьевая форель
– – *lacustris*	Lake trout	– de lac	Озерная форель
Salmonidae	Salmons	Saumons	Лососевые
Salvelinus	Charrs		Гольцы
– *alpinus*	Charr	Omble-chevalier	Обыкновенный голец
– *fontinalis*	Brook trout	Saumon de fontaine	Американская ручьевая палия
– *namaycush*	Common lake charr	Truite grise	Американская озерная палия
Sandtiger	Sand tiger shark	Odontaspide taureau	
Sardellen	Anchovies		Анчоусовые, Анчоусы
Sardina	Sardines	Sardines	Сардины
– *pilchardus*	Pilchard	Sardine	Анчоусовые, Анчоусы
Sardinella aurita	Gilt sardine	Allache	
Sardinops caerulea	Pilchard		Тихоокеанская сардина
Sargassofisch	Frog-fish		Морская мышь
Sauger	Suckers		Чукучановые
Scaphirhynchinae	Shovel-nosed sturgeons		Лопатоносы
Scaphirhynchus platorhynchus	– sturgeon		Обыкновенный лопатонос
Scardinius erythrophthalmus	Rudd	Rotengle	Красноперка
Schaufelnasenstöre	Shovel-nosed sturgeons		Лопатоносы
Schaufelrüßler	American paddlefish		Американский веслонос
Schellfisch	Haddock	Morue noire	Пикша
Scherben	Silver bream	Brème bordelière	Густера
Schied	Aspe	Aspe	Обыкновенный жерех
Schildzahnhai	Fierce shark	Odontaspide féroce	
Schillerbärbling	Pearl Danio	Danio rosé	
Schlammfische	Bowfins	Amies	Амии
Schlammpeitzger	Pond loach	Loche d'étang	Вьюн-пескарь
Schlanker Knochenhecht	Long-nosed gar-pike	Lépisostée osseux	Костяной клювонос
Schleie	Tenches, Tench	Tanches, Tanche commune	Лини, Линь
Schleimaal	Common hagfish	Myxine	Обыкновенная миксина
Schlinger	Pelican-fish	Saccopharynx	
Schmerlen	Loaches		Вьюновые
Schmetterlingsfisch	Butterfly-fish	Pantodon	Рыба-бабочка
Schmetterlingsfische	Butterfly-fishes		Пантодоновые
Schmetterlingsrochen	Butterfly ray	Mourine bâtarde	Гимнура
Schneider	Bleak	Ablette	Обыкновенная уклея
Schnutt	Dace	Vandoise vraie	Обыкновенный елец
Schokoladenhai	Darkie charlie	Liche	
Schuppenfisch	Bleak	Ablette	Обыкновенная уклея
Schupper	Bleak	Ablette	Обыкновенная уклея
Schutt	Aspe	Aspe	Обыкновенный жерех
Schwarzbauch		Chondrostome nez	Обыкновенный подуст
Schwarzer Dornhai	Latern shark	Sagre	Черная колючая акула
Schwarznase	Black-nosed dace	Cyprin à nez noir	
Schwarznerfling	Id	Ide mélanote	Язь
Schwertfisch		Rasoir	Чехонь
Schwertrüßler	Chinese sturgeon		Китайский меченос
Schwertstör	Chinese sturgeon		Китайский меченос
Schwertträger	Swordtail	Porte-Glaive	
Scomberesox saurus	Saurie	Balaou	Атлантическая макреле-щука
Scyliorhinidae	Spotted dog-fishes		Кошачьи акулы
Scyliorhinus caniculus	Small-spotted dog-fish	Petite roussette	Морской кот
– *stellaris*	Large-spotted dogfish	Grande roussette	Большой морской кот
Seedrachen	Chimaeras		Цельноголовые
Seeforelle	Lake trout	Truite de lac	Озерная форель
Seehecht	Hake	Merlu	Обыкновенный хэк
Seekatzen		Chiméroïdes	Химеры
Seelamprete	Sea lamprey	Lamproie marine	Морская минога
Seeratte	Rabbit fish	Rat de mer	Обыкновенная химера
Seeratten	– fishes, Chimaeras	Chimères	Цельноголовые
Seestör	Lake sturgeon		Американский пресно-водный осетр
Seeteufel i. e. S.	Angler-fishes	Baudroies	Морские черти
Selachii	Sharks	Squales	Акулообразные
Semling		Barbeau canin	Крапчатый усач
Serrasalmus	Caribes	Caribes	Пираньи
– *piraya*	Piraya	Piraya	Пиранья

GERMAN NAME	ENGLISH NAME	FRENCH NAME	RUSSIAN NAME
Sibirische Äsche	Arctic grayling	Ombre arctique	Сибирский хариус
Sichling		Rasoir	Чехонь
Silberdorsche	Silvery pouts	Merlan argenté	
Silberkarausche	Goldfish	Poisson rouge	Серебряный карась
Siluridae	Cat-fishes	Siluridés	Сомовые
Silurus		Silures	Обыкновенные сомы
– glanis	Sheat-fish	Silure glane	Обыкновенный сом
Somniosus microcephalus	Greenland shark	Requin des glaces	
Sonnenfischchen		Able de stymphale	Обыкновенная верховка
Speier		Chondrostome nez	Обыкновенный подуст
Sphyrna zygaena	Common hammerhead shark	Requin marteau	Обыкновенная молот-рыба
Sphyrnidae	Hammerhead sharks		Молот-рыбы
Spitzkopfsechskiemer	Mediterranean shark	Griset	
Spitzlaube	Bleak	Ablette	Обыкновенная уклея
Spöke	Rabbit fish	Rat de mer	Обыкновенная химера
Squalidae	Spiny dog-fishes	Squales	Колючие акулы
Squaloidei	Squaloids	Squaloïdes	Колючеперые акулы
Squalus acanthias	Spiny dog-fish	Aiguillat tacheté	Нокотница
Squatina squatina	Angel fish	Ange de mer	Европейский морской ангел
Squatinidae	– sharks		Морские ангелы
Squatinoidei	Squatinoids		Морские ангелы
Stachelhaie	Squaloids	Squaloïdes	Колючеперые акулы
Stachelrochen	Sting rays	Pastenagues	Хвостоколы
Steinbeißer i. e. S.		Loches	Щиповки
Steinköhler	Green Pollack	Lieu jaune	
Stenodus leucichthys	Inconnu	Saumon du Mackenzie	Белорыбица
Sterlet	Sterlet	Sterlet	Стерлядь
Stint	European smelt	Éperlan d'Europe	Обыкновенная корюшка
Stinte	True smelts		Корюшковые
Störe	Sturgeons	Esturgeons	Осетры
Strömer		Blageon	Рислинг
Südlicher Glatthai	Smooth-hound	Emissole lisse	Южная кунья акула
Tarpune	Tarpons		Тарпуны
Teleostei	Bony fishes	Poissons osseux	Костистые рыбы
Teufelsrochen	Devil-fishes		Морские дьяволы
Theragra chalcogramma	Wall-eye pollack		Минтай
Thymallus	Graylings		Хариусы
– arcticus	Arctic grayling	Ombre arctique	Сибирский хариус
– thymallus	Grayling	– de rivière	Обыкновенный хариус
Tinca	Tenches	Tanches	Лини
– tinca	Tench	Tanche commune	Линь
Torpedinidae	Electric rays	Torpilles	Электрические скаты
Torpedo marmorata	Marbled electric ray	Torpille marbrée	Мраморный электрический скат
– torpedo	Eyed electric ray	– tachetée	Пятнистый электрический скат
Tübling	Chub	Chevain	Голавль
Tüpfelbärbling	Spotted Danio	Danio moucheté	
Ukelei	Bleak	Ablette, Ablettes	Обыкновенная уклея, Уклейки
Umbra krameri	Dogfish	Poisson-chien	Европейская евдошка
– limi	Central mud minnow	Poisson de vase	Восточноамериканская евдошка
Urophycis		Morues	Западноатлантические морские налимы
Vertebrata	Vertebrates	Vertébrés	Черепные
Vielzähner	Spoonbills		Веслоносы
Vierauge	Four-eyed fish		Четырехглазая рыба
Vieraugen	Four-eyed fishes		Четырехглазые рыбы
Viergürtelbarbe	Sumatran barb	Barbeau de Sumatra	
Violetter Stechrochen	Blue sting-ray	Pastenague violette	Фиолетовый хвостокол
Walhai	Whale shark		Малозубая акула
Wandersaibling	Charr	Omble-chevalier	Обыкновенный голец
Weißer Stör	White sturgeon		Восточнотихоокеанский осетр
Weißfische	Minnows	Cyprinidés	Карповые
Weißfische i. e. S.		Chevaines	Ельцы
Weißhai	Great white shark	Requin	
Weißlachs	Inconnu	Saumon du Mackenzie	Белорыбица
Westatlantische Gabeldorsche		Mórues	Западноатлантические морские налимы
Westlicher Sägefisch	Sawfish		Западная пила-рыба
Wettling	Minnow	Vairon	Речной гольян
Wibling	Minnow	Vairon	Речной гольян
Wilder Haifisch	Fierce shark	Odontaspide féroce	

GERMAN NAME	ENGLISH NAME	FRENCH NAME	RUSSIAN NAME
Wirbeltiere	Vertebrates	Vertébrés	Черепные
Wittling		Merlan	Атлантический мерланг
Wolfsfische		Lycodes	Ликоды
Xiphophorus helleri	Swordtail	Porte-Glaive	
Zahnkärpflinge	Tooth carps		Карпозубовые
Zebrabärbling	Zebra fish	Petit Danio	
Ziege		Rasoir	Чехонь
Zitteraal	Electric eel	Anguille tremblante	Электрический угорь
Zitterrochen	Electric rays	Torpilles	Электрические скаты
Zitter- und Messeraale	− eels		Электрические угри
Zoarces viviparus	Viviparous blenny	Loquette	Обыкновенная бельдюга
Zwergkärpfling	Pygmy top minnow	Poisson moustique	
Zwerglaube		Able de stymphale	Обыкновенная верховка
Zwergpricke	Brook lamprey	Lamproie de Planeri	Европейская ручьевая минога

III. French—German—English—Russian

L'abréviation N. A., mise entre parenthèses, indique que les noms respectifs ne sont utilisés qu'en Amérique du Nord.

FRENCH NAME	GERMAN NAME	ENGLISH NAME	RUSSIAN NAME
Able	Ziege		Чехонь
− de stymphale	Moderlieschen		Обыкновенная верховка
Ablé	Ukelei	Bleak	Обыкновенная уклея
Ablette	Ukelei	Bleak	Обыкновенная уклея
Ablettes	Ukelei		Уклейки
Acipe	Baltischer Stör	Common Atlantic sturgeon	Балтийский осетр
Aigle de mer	Gewöhnlicher Adlerrochen	Spotted eagle ray	Обыкновенный орляк
Aiglefin	Schellfisch	Haddock	Пикша
Aiguillat tacheté	Dornhai	Spiny dog-fish	Нокотница
Aiguille	Europäischer Hornhecht	Gar-fish	Европейский сарган
Aiguilles	Hornhechte	Gar-fishes	Сарганы
Aiguillon	Hecht	Pike	Обыкновенная щука
Albule commun	Grätenfisch	Bonefish	
Albules	Eigentliche Grätenfische	Ladyfishes	
Allache	Ohrensardine	Gilt sardine	
Alose d'Amérique	Amerikanische Alse	American shad	Шэд
− finte	Finte	Twaite shad	Финта
− vraie	Alse	Shad	Алоса
Aloses	Alsen	Shads	Атлантические проходные сельди
Amblyopsis	Mammuthöhlen-Blindfisch	Mammoth cave blindfish	Пещерная живородка
Amie	Amerikanischer Schlammfisch	Bowfin	Ильная рыба
Amies	Schlammfische	Bowfins	Амии
Anchois	Europäische Sardelle	Anchovy	Европейский анчоус
− de Californie	Nordamerikanische Sardelle	Northern anchovy	Североамериканский анчоус
Ange de mer	Gemeiner Meerengel	Angel fish	Европейский морской ангел
Anguille	Europäischer Flußaal	Fresh-water eel	Обыкновенный речной угорь
− d'Amérique	Amerikanischer Aal	American eel	Американский речной угорь
− de mer	Meeraal	Conger eel	Морской угорь
− tremblante	Zitteraal	Electric eel	Электрический угорь
Anguilles	Echte Aale	Eels	Речные угри
Appocalle	Grönlandhai	Greenland shark	
Arlequin	Elritze	Minnow	Речной гольян
Aspe	Rapfen		Обыкновенный жерех
Aspes	Rapfen		Жерехи
Assée	Hasel	Dace	Обыкновенный елец
Aurin	Fierasfer	Pearl fish	Фиерасфер
Balaou	Atlantischer Makrelenhecht	Saurie	Атлантическая макреле-щука
Barbeau canin	Semling		Крапчатый усач
− commun	Barbe	Barbel	Обыкновенный усач
− de Sumatra	Viergürtelbarbe	Sumatran barb	
− fluviatile	Barbe	Barbel	Обыкновеный усач
Barbeaux	Echte Barben	Barbels	Настоящие усачи

FRENCH NAME	GERMAN NAME	ENGLISH NAME	RUSSIAN NAME
Barbillon	Barbe	Barbel	Обыкновенный усач
– truité	Semling		Крапчатый усач
Barbotte	Aalquappe	Burbot	Налим
Baudroies	Seeteufel i. e. S.	Angler-fishes	Морские черти
Beurotte	Schleie	Tench	Линь
Blageon	Strömer		Рислинг
Blavin	Elritze, Strömer	Minnow	Речной гольян, Рислинг
Blizon	Stint	European smelt	Обыкновенная корюшка
Bouvière amère	Europäischer Bitterling	Bitterling	Европейский горчак
Bouvières	Bitterlinge	European bitterling	Обыкновенные горчаки
Brème bordelière	Güster	Silver bream	Густера
– commune	Brachsen	Bream	Лещ
Brèmes	Brassen	Breams	Лещи
Brochet commun	Hecht	Pike	Обыкновенная щука
– maillé	Kettenhecht	Chain pickerel	
– vermiculé	Grass-Hecht	Grass pickerel	
– vivipare	Hechtkärpfling	Belonesox	
Brochets	Hechte i. e. S.	Pickerels	Щуки
Brosme	Lumb	Cusk	Менек
Cabillaud	Kabeljau	Common cod	Атлантическая треска
Cabot	Döbel	Chub	Голавль
Callichthyidés	Panzerwelse		Панцырные сомы
Cap Lopez	Kap Lopez	Lyretail	
Capelan	Lodde	Capelin	Мойва
Carassin doré	Goldfisch	Goldfish	Золотая рыбка
Carassin vulgaire	Gewöhnliche Karausche	Crucian carp	Обыкновенный карась
Carassins	Karauschen	Common carps	Караси
Caribes	Pirayas	Caribes	Пираньи
Carpe à la lune	Gewöhnliche Karausche	Crucian carp	Обыкновенный карась
– ordinaire	Karpfen	Carp	Сазан
Carpes	Karpfen	Carps	Сазаны
Carreau	Gewöhnliche Karausche	Crucian carp	Обыкновенный карась
Centrine	Meersau	Angular rough shark	Центрина
Charin	Rotfeder	Rudd	Красноперка
Chatouille	Bachneunauge	Brook lamprey	Европейская ручьевая минога
Chenille	Nagelhai	Bramble shark	
Chevain	Döbel	Chub	Голавль
Chevaines	Weißfische i. e. S.		Ельцы
Chevesne	Döbel	Chub	Голавль
Chien de mer	Dornhai	Spiny dog-fish	Нокотница
– espagnol	Fleckhai	Black-mouthed dogfish	Пилохвост
– noir	Schwarzer Dornhai	Latern shark	Черная колючая акула
Chimère	Seeratte	Rabbit fish	Обыкновенная химера
Chimères	Seeratten	– fishes	
Chiméroïdes	Seekatzen		Химеры
Chondrostome nez	Nase		Обыкновенный подуст
Chondrostomes	Nasen		Подусты
Clupes	Heringe	Herrings	Сельдевые
Congre	Meeraal	Conger eel	Морской угорь
Corégone blanc	Kleine Maräne		Ряпушка
– féra	Große Bodenrenke		Придонный сиг
– lavaret	– Schwebrenke	Powan	Обыкновенный сиг
Corégones	Renken	Whitefishes	Сиги
Cormontant	Brachsen	Bream	Лещ
Corneau	Finte	Twaite shad	Финта
Corydoras à casque	Gefleckter Panzerwels	Armored cat-fish	
Cyprin à nez noir	Schwarznase	Black-nosed dace	
Cyprinidés	Weißfische	Minnows	Карповые
Dard	Hasel	Dace	Обыкновенный елец
Danio géant	Malabarbärbling	Giant Danio	
– moucheté	Tüpfelbärbling	Spotted Danio	
– rosé	Schillerbärbling	Pearl Danio	
– zebré	Zebrabärbling	Zebra fish	
Diable de mer	Meeresteufel	Devil-fish	Морской дьявол
Dorade de Chine	Goldfisch	Goldfish	Золотая рыбка
Durgan	Semling		Крапчатый усач
Écrivain	Nase		Обыкновенный подуст
Emissole lisse	Südlicher Glatthai	Smooth-hound	Южная кунья акула
– tachetée	Nördlicher Glatthai	Stellate smooth-hound	Северная кунья акула
Éperlan d'Europe	Stint	European smelt	Обыкновенная корюшка
Esturgeon commun	Baltischer Stör	Common Atlantic sturgeon	Балтийский осетр
Esturgeons	Störe, Echte Störe	Sturgeons	Осетры, Осетровые
Féra	Große Bodenrenke		Придонный сиг
Feu de position	Leuchtfleckensalmler	Head-and-tail-light	

FRENCH NAME	GERMAN NAME	ENGLISH NAME	RUSSIAN NAME
Fifre	Flußneunauge	Fresh water lamprey	Европейская речная миного
Gades	Dorsche i. e. S.	Gadoid fishes	Тресковые
Gambusie	Koboldkärpfling	Mosquito fish	Гамбузия
Gambusies	Gambusen	– fishes	Гамбузии
Gardon blanc	Plötze	European roach	Плотва
– rouge	Rotfeder	Rudd	Красноперка
Gendarme	Elritze	Minnow	Речной гольян
Gibèle	Gewöhnliche Karausche	Crucian carp	Обыкновенный карась
Goiffon	Gewöhnlicher Gründling	Gudgeon	Обыкновенный пескарь
Goujon	– Gründling	Gudgeon	Обыкновенный пескарь
– à fines écailles	Kleinschuppenrötling	Finescale dace	
Goujons	Gründlinge i. e. S.	Gudgeons	Обыкновенные пескари
Grand esturgeon	Europäischer Hausen		Белуга
Grande alose	Alse	Shad	Алоса
– lamproie	Meerneunauge	Sea lamprey	Морская минога
– roussette	Großgefleckter Katzenhai	Large-spotted dogfish	Большой морской кот
Griset	Spitzkopfsechskiemer	Mediterranean shark	
Grisette	Elritze	Minnow	Речной гольян
Grougnau	Gewöhnlicher Gründling	Gudgeon	Обыкновенный пескарь
Guitare	Gemeiner Geigenrochen	Guitar fish	Обыкновенная рохля
Gymnarche du Nil	Großer Nilhecht	Eel-like mormyrid	Гимнарх
Gymnote électrique	Zitteraal	Electric eel	Электрический угорь
Gymnotidés	Echte Messeraale	Gymnotids	
Hâ	Hundshai	Tope	
Hareng	Atlantischer Hering	Atlantic herring	Атлантическая сельдь
– du Pacifique	Pazifischer Hering	Pacific herring	Тихоокеанская сельдь
Harengs	Heringe i. e. S.	Herrings	Океанические сельди
Hémiramphes	Halbschnäbler	Half-beaks	Полурылы
Hotu	Nase		Обыкновенный подуст
Ide mélanote	Aland	Id	Язь
Inconnu	Weißlachs	Inconnu	Белорыбица
Jacquine	Finte	Twaite shad	Финта
Jordanelle de Floride	Florida-Kärpfling	American flag fish	
Lamie	Heringshai	Porbeagle	Сельдевая акула
– à nez pointu	Mako	Mackerel shark	
Lampresse	Meerneunauge	Sea lamprey	Морская минога
Lamprillon	Flußneunauge	Fresh water lamprey	Европейская речная миного
Lamproie de Planeri	Bachneunauge	Brook lamprey	Европейская ручьевая миного
– de rivière	Flußneunauge	Fresh water lamprey	Европейская речная миного
– marbrée	Meerneunauge	Sea lamprey	Морская минога
– marine	Meerneunauge	– –	Морская минога
Lamproies	Rundmäuler, Neunaugen	Lampreys	Круглоротые, Миногообразные
Lavaret	Große Schwebrenke	Powan	Обыкновенный сиг
Lépisostée osseux	Schlanker Knochenhecht	Long-nosed gar-pike	Костяной клювонос
Liche	Schokoladenhai	Darkie charlie	
Lieu jaune	Steinköhler	Green Pollack	
– noir	Köhler	Pollack	Сайда
Lingue	Leng	Ling	Обыкновенная мольва
Loche de rivière	Europäischer Steinbeißer	Loach	Обыкновенная щиповка
– d'étang	Schlammpeitzger	Pond loach	Вьюн-пескарь
– épineuse	Europäischer Steinbeißer	Loach	Обыкновенная щиповка
Loches	Steinbeißer i. e. S.		Щиповки
– d'étangs	Schlammpeitzger		Вьюны
Loquette	Aalmutter	Viviparous blenny	Обыкновенная бельдюга
Lote	Aalquappe	Burbot	Налим
Lotes	Aalquappen	Burbots	Налимы
Lycodes	Wolfsfische		Ликоды
Mangeur d'homme (N. A.)	Weißhai	Great white shark	
Marotte	Rotfeder	Rudd	Красноперка
Maskinongé	Muskellunge	Muskellunge	
Mélanote	Aland	Id	Язь
Merlan	Wittling		Атлантический мерланг
Merlans argentés	Silberdorsche	Silvery pouts	
Merlu	Seehecht	Hake	Обыкновенный хэк
Meule	Gewöhnliche Karausche	Crucian carp	Обыкновенный карась
Meunier	Döbel	Chub	Голавль
Milandre	Hundshai	Tope	
Mirandelle	Ukelei	Bleak	Обыкновенная уклея
Morue commune	Kabeljau	Common cod	Атлантическая треска
Morue d'Alaska	Pazifik-Kabeljau	Pacific cod	Тихоокеанская треска
– de roche	Grönlandkabeljau	Greenland halibut	Гренландская фиордовая

FRENCH NAME	GERMAN NAME	ENGLISH NAME	RUSSIAN NAME
			треска
– du Groenland	Grönlandkabeljau	– halibut	Гренландская фиордовая треска
– –Pacifique	Pazifik-Kabeljau	Pacific cod	Тихоокеанская треска
– fraîche	Kabeljau	Common cod	Атлантическая треска
– noire	Schellfisch	Haddock	Пикша
Morues	Eigentliche Dorsche, Westatlantische Gabeldorsche	Cods	Треска, Западноатлантические морские налимы
Motelle	Aalquappe	Burbot	Налим
Mourine bâtarde	Schmetterlingsrochen	Butterfly ray	Гимнура
– vachette	Afrikanischer Adlerrochen	Bull ray	Африканский орляк
Murène commune	Mittelmeer-Muräne	European moray	Средиземноморская мурена
Murènes	Muränen, Riffmuränen	Morays	Мурены
Myxine	Inger	Common hagfish	Обыкновенная миксина
Nannostome	Längsbandsalmler	Pencil fish	
Nase	Nase		Обыкновенный подуст
Notoptéridés	Eigentliche Messerfische	Featherbacks	
Odontaspide féroce	Schildzahnhai	Fierce shark	
– taureau	Sandtiger	Sand tiger shark	
Omble de fontaine	Bachsaibling	Brook trout	Американская ручьевая палия
– – ruisseau	Bachsaibling	– –	Американская ручьевая палия
– gris	Amerikanischer Seesaibling	Common lake charr	Американская озерная палия
Omble-chevalier	Wandersaibling	Charr	Обыкновенный голец
Ombre à écailles	Europäische Äsche	Grayling	Обыкновенный хариус
– arctique	Sibirische Äsche	Arctic grayling	Сибирский хариус
– boréal	– –	– –	Сибирский хариус
– commun(e)	Europäische Äsche	Grayling	Обыкновенный хариус
– de rivière	– –	Grayling	Обыкновенный хариус
Ombrette	Europäische Äsche	Grayling	Обыкновенный хариус
Orfe	Goldorfe		Орф
Orphie	Atlantischer Makrelenhecht, Europäischer Hornhecht	Saurie, Gar-fish	Атлантическая макрелещука, Европейский сарган
Orphies	Hornhechte	Gar-fishes	Сарганы
Outil	Kleine Schwebrenke		Морской сиг
Palée	Große Bodenrenke		Придонный сиг
Pantodon	Schmetterlingsfisch	Butterfly-fish	Рыба-бабочка
Pastenague	Gewöhnlicher Stechrochen	Common sting-ray	Морской кот
– violette	Violetter Stechrochen	Blue sting-ray	Фиолетовый хвостокол
Pastenagues	Stachelrochen		Хвостоколы
Peau bleue	Blauhai	– shark	Голубая акула
Pelletet	Europäischer Bitterling	Bitterling	Европейский горчак
Perce-pierre	Meerneunauge	Sea lamprey	Морская минога
Péteuse	Europäischer Bitterling	Bitterling	Европейский горчак
Petit brochet	Kettenhecht	Chain pickerel	
– chien bleu	Atlantischer Braunhai	Large-finned shark	
– Danio	Zebrabärblinge	Zebra fish	
Petite lamproie	Bachneunauge	Brook lamprey	Европейская ручьевая минога
– roussette	Kleingefleckter Katzenhai	Small-spotted dog-fish	Морской кот
Pimperneau	Europäischer Flußaal	Fresh-water eel	Обыкновенный речной угорь
Piraya	Piraya	Piraya	Пиранья
Pirayas	Pirayas	Caribes	Пираньи
Platton	Brachsen	Bream	Лещ
Pocheteau blanc	Glattrochen	Flapper skate	Гладкий скат
Poisson armé	Schlanker Knochenhecht	Long-nosed gar-pike	Костяной клювонос
– de mai	Alse	Shad	Алоса
– – vase	Östlicher Hundsfisch	Central mud minnow	Восточноамериканская евдошка
– moustique	Zwergkärpfling	Pygmy top minnow	
– papillon	Schmetterlingsfisch	Butterfly-fish	Рыба-бабочка
– rouge	Silberkarausche	Goldfish	Серебряный карась
Poisson-chat électrique	Elektrischer Wels		Электрический сом
Poisson-chien	Europäischer Hundsfisch	Dogfish	Европейская евдошка
Poisson-flamme	Roter von Rio	Red from Rio	
Poisson-loup	Hecht	Pike	Обыкновенная щука
Poisson-Léopard	Koboldkärpfling	Mosquito fish	Гамбузия
Poissons	Fische	Fishes	Рыбы
– osseux	Echte Knochenfische, Knochenfische	Bony fishes	Костистые рыбы, Костные рыбы
– volants	Atlantische Flugfische	Flying-fishes	Долгоперы

FRENCH NAME	GERMAN NAME	ENGLISH NAME	RUSSIAN NAME
Poissons-ciseaux	Pirayas	Caribes	Пираньи
Porte-épée	Schwertträger	Swordtail	
Porte-Glaive	Schwertträger	Swordtail	
Poutassou	Blauer Wittling	Couch's whiting	
Prêtre	Kleiner Ährenfisch	Caspian sand smelt	Малая атеринка
Queue-de-voile	Goldfisch	Goldfish	Золотая рыбка
Raie bouclée	Nagelrochen	Thornback ray	Шиповатый скат
– brunette	Marmorrochen		Мраморный скат
Raies	Echte Rochen	Skates	Настоящие скаты
Rasbora	Keilfleckbarbe	Red Rasbora	
Rasoir	Ziege		Чехонь
Rat de mer	Seeratte	Rabbit fish	Обыкновенная химера
Requin	Weißhai	Great white shark	
– à tunique	Krausenhai	Frilled shark	Плащеносная акула
– bleue	Blauhai	Blue shark	Голубая акула
– des glaces	Grönlandhai	Greenland shark	
– griset	Grauhai	Comb-toothed shark	
– marteau	Glatter Hammerhai	Common hammerhead shark	Обыкновенная молот-рыба
– renard	Fuchshai	Thresher shark	Морская лисица
Requins	Haie	Sharks	Акулообразные
Roche	Plötze	European roach	Плотва
Ronzon	Hasel	Dace	Обыкновенный елец
Rosière	Europäischer Bitterling	Bitterling	Европейский горчак
Rotengle	Rotfeder	Rudd	Красноперка
Saccopharynx	Schlinger	Pelican-fish	
Sagre	Schwarzer Dornhai	Latern shark	Черная колючая акула
Salogne	Rotfeder	Rudd	Красноперка
Salut	Flußwels	Sheat-fish	Обыкновенный сом
Sardine	Pilchard	Pilchard	Атлантическая сардина
Sardines	Echte Sardinen	Sardines	Сардины
Saumon à bosse	Buckellachs	Pink salmon	Горбуша
– argenté	Kisutch-Lachs	Silver salmon	Кижуч
– atlantique	Atlantischer Lachs	Atlantic salmon	Обыкновенный лосось
– coho	Kisutch-Lachs	Silver salmon	Кижуч
– de Californie	Quinnat	Chinook salmon	Чавыча
Saumon de fontaine	Bachsaibling	Brook trout	Американская ручьевая палия
– du Mackenzie	Weißlachs	Inconnu	Белорыбица
– Quinnat	Quinnat	Chinook salmon	Чавыча
Saumon-chien	Keta-Lachs	Dog salmon	Кета
Saumons	Lachsähnliche i. e. S., Lachse und Forellen, Pazifische Lachse	Salmons, Trouts, Pacific salmons	Лососевые, Благородные лососи, Тихоокеанские лососи
Scies	Sägerochen, Sägefische	Sawfishes	Пилы-рыбы
Sept-œils	Flußneunauge	Fresh water lamprey	Европейская речная минога
Silure glane	Flußwels	Sheat-fish	Обыкновенный сом
– hyène	Flußwels	Sheat-fish	Обыкновенный сом
Silures	Echte Welse		Обыкновенные сомы
Siluridés	Echte Welse	Cat-fishes	Сомовые
Siouclet	Großer Ährenfisch	Silverside	Большая атеринка
Soufie	Strömer		Рислинг
Squale pélerin	Riesenhai	Basking shark	Гигантская акула
Squales	Haie, Dornhaie	Sharks, Spiny dog-fishes	Акулообразные, Колючие акулы
Squaloïdes	Stachelhaie	Squaloids	Колючеперые акулы
Sterlet	Sterlet	Sterlet	Стерлядь
Suce-pierre	Bachneunauge, Meerneunauge	Brook lamprey, Sea lamprey	Европейская ручьевая минога, Морская минога
Tanche commune	Schleie	Tench	Линь
Tanches	Schleie	Tenches	Лини
Taupe de mer	Heringshai	Porbeagle	Сельдевая акула
Tenque	Schleie	Tench	Линь
Tétra de Rio	Roter von Rio	Red from Frio	
– rouge	– – –	– – –	
Touille	Heringshai	Porbeagle	Сельдевая акула
Touladi	Amerikanischer Seesaibling	Common lake charr	Американская озерная палия
Torpille marbrée	Marmorzitterrochen	Marbled electric ray	Мраморный электрический скат
– tachetée	Gefleckter Zitterrochen	Eyed electric ray	Пятнистый электрический скат
Torpilles	Zitterrochen	Electric rays	Электрические скаты
Tregan	Gewöhnlicher Gründling	Gudgeon	Обыкновенный пескарь
Troque	Stint	European smelt	Обыкновенная корюшка

FRENCH NAME	GERMAN NAME	ENGLISH NAME	RUSSIAN NAME
Truite argentée	Seeforelle	Lake trout	Озерная форель
– bigarée	Bachforelle	River trout	Ручьевая форель
– de Kamloops	Regenbogenforelle	Rainbow trout	Радужная форель
– – lac	Seeforelle	Lake trout	Озерная форель
– – mer	Europäische Forelle	Brown trout	Лосось-таймень
– – rivière	Bachforelle	River trout	Ручьевая форель
– – ruisseau	Bachforelle	– –	Ручьевая форель
– grise	Amerikanischer Seesaibling	Common lake charr	Американская озерная палия
– mouchetée	Bachsaibling	Brook trout	Американская ручьевая палия
Vairon	Elritze	Minnow	Речной гольян
– à gorge rouge	Rötling	South-red-bellied dace	
Vandoise vraie	Hasel	Dace	Обыкновенный елец
Vermiaux	Europäischer Flußaal	Fresh-water eel	Обыкновенный речной угорь
Vertébrés	Wirbeltiere	Vertebrates	Черепные
Vilain	Döbel	Chub	Головль
Xiphophore port-épée	Schwertträger	Swordtail	

IV. Russian–German–English–French

N. A. при английских названиях означает, что эти названия употребляются только в Северной Америке.

RUSSIAN NAME	GERMAN NAME	ENGLISH NAME	FRENCH NAME
Австралийские миноги	Australische Neunaugen	Lampreys	
Акулообразные	Haie	Sharks	Squales
Алоса	Alse	Shad	Grande alose
Американская озерная палия	Amerikanischer Seesaibling	Common lake charr	Truite grise
Американская ручьевая палия	Bachsaibling	Brook trout	Saumon de fontaine
Американская ряпушка	Amerikanische Kleine Maräne	Lake herring (N. A.)	
Американский веслонос	Löffelstör	American paddlefish	
Американский пресноводный осетр	Roter Stör	Lake sturgeon	
Американский речной угорь	Amerikanischer Aal	American eel	Anguille d'Amérique
Амии	Schlammfische	Bowfins	Amies
Амия	Amerikanischer Schlammfisch	Bowfin	Amie
Амурская щука	Amurhecht	Northern pike (N. A.)	
Анчоусовые	Sardellen	Anchovies	
Анчоусы	Sardellen	–	
Арапайма	Arapaima	Arapaima	
Атеринки	Ährenfische i. e. S.	Silversides	
Атлантическая макреле-щука	Atlantischer Makrelenhecht	Saurie	Balaou
Атлантическая сардина	Pilchard	Pilchard	Sardine
Атлантическая сельдь	Atlantischer Hering	Atlantic herring	Hareng
Атлантическая треска	Kabeljau	Common cod	Morue fraîche
Атлантические проходные сельди	Alsen	Shads	Aloses
Атлантический мерланг	Wittling		Merlan
Атлантический осетр	Atlantischer Stör	Atlantic sturgeon	
Атлантический тарпун	– Tarpun	Tarpon	
Африканский костеязычник	Afrikanischer Knochenzüngler	Osteoglossid	
Африканский орляк	– Adlerrochen	Bull ray	Mourine vachette
Балтийский осетр	Baltischer Stör	Common Atlantic sturgeon	Esturgeon commun
Белорыбица	Weißlachs	Inconnu	Saumon du Mackenzie
Белуга	Europäischer Hausen		Grand esturgeon
Бесчелюстные	Kieferlose	Lampreys and hagfishes	
Благородные лососи	Lachse und Forellen	Trouts	Saumons
Большая атеринка	Großer Ährenfisch	Silverside	Siouclet
Большой морской кот	Großgefleckter Katzenhai	Large-spotted dogfish	Grande roussette
Веслоносы	Löffelstöre	Paddlefishes, Spoonbills	
Восточноамериканская евдошка	Östlicher Hundsfisch	Central mud minnow	Poisson de vase
Восточнотихоокеанский осетр	Weißer Stör	White sturgeon	

RUSSIAN NAME	GERMAN NAME	ENGLISH NAME	FRENCH NAME
Вьюновые	Schmerlen	Loaches	
Вьюн-пескарь	Schlammpeitzger	Pond loach	Loche d'étang
Вьюны	Schlammpeitzger		Loches d'étangs
Галаксиевые	Hechtlinge i. e. S.	Galaxiids	
Гамбузии	Gambusen	Mosquito fishes	Gambusies
Гамбузия	Koboldkärpfling	– fish	Poisson-Léopard
Гигантская акула	Riesenhai	Basking shark	Squale pélerin
Гигантская манта	Riesenmanta	Manta ray	
Гигантские акулы	Riesenhaie	Basking sharks	
Гигантский удильщик	Laternenangler	Deep-sea angler	
Гимнарх	Großer Nilhecht	Eel-like mormyrid	Gymnarche du Nil
Гимнура	Schmetterlingsrochen	Butterfly ray	Mourine bâtarde
Гитара-рыба	Gitarrenfisch	Guitar fish	
Гладкий скат	Glattrochen	Flapper skate	Pocheteau blanc
Голавль	Döbel	Chub	Chevain
Голубая акула	Blauhai	Blue shark	Requin bleue
Голубые акулы	Blauhaie	Sand sharks	
Гольцы	Saiblinge	Charrs	
Горбуша	Buckellachs	Pink salmon	Saumon à bosse
Гребнезубые акулы	Grauhaie	Comb-tooth sharks	
Гренландская фиордовая треска	Grönlandkabeljau	Greenland halibut	Morue de roche
Густера	Güster	Silver bream	Brème bordelière
Даллии	Fächerfische	Black-fishes	
Даллия	Fächerfisch	Alaska black-fish	
Длиннорылы	Nilhechte	Mormyrids	
Долгоперы	Atlantische Flugfische	Flying-fishes	Poissons volants
Европейская евдошка	Europäischer Hundsfisch	Dogfish	Poisson-chien
Европейская речная минога	Flußneunauge	Fresh water lamprey	Lamproie de rivière
Европейская ручьевая минога	Bachneunauge	Brook lamprey	– – Planeri
Европейский анчоус	Europäische Sardelle	Anchovy	Anchois
Европейский горчак	Europäischer Bitterling	Bitterling	Bouvière amère
Европейский морской ангел	Gemeiner Meerengel	Angel fish	Ange de mer
Европейский сарган	Europäischer Hornhecht	Gar-fish	Aiguille
Ельцы	Weißfische i. e. S.		Chevaines
Жерехи	Rapfen		Aspes
Живородковые	Blindfische	Cave fishes	
Живородящие карпозубые	Lebendgebärende Zahnkärpflinge	Poeciliids	
Западная пила-рыба	Westlicher Sägefisch	Sawfish	
Западноатлантические морские налимы	Westatlantische Gabeldorsche		Morues
Золотая рыбка	Goldfisch	Goldfish	Carassin doré
Зубастые сомы	Kreuzwelse	Sea cat-fishes	
Иглистые сомы	Dornwelse	Armored cat-fishes	
Ильная рыба	Amerikanischer Schlammfisch	Bowfin	Amie
Каймановые рыбы	Knochenhechte	Gars	
Караси	Karauschen	Common carps	Carassins
Карп	Karpfen	Carp	Carpe ordinaire
Карповые	Weißfische	Minnows	Cyprinidés
Карпозубовые	Zahnkärpflinge	Tooth carps	
Карпообразные	Karpfenfische	Minnows, Suckers and Loaches	
Катран	Dornhai	Spiny dog-fish	
Кета	Keta-Lachs	Dog salmon	
Кижуч	Kisutch-Lachs	Silver salmon	Saumon-chien
Килец	Kleine Maräne		– coho
Китайский меченос	Schwertstör	Chinese sturgeon	Corégone blanc
Клювоносы	Knochenhechte	Gars	
Кнерии	Ohrenfische	Kneriids	
Колючеперые акулы	Stachelhaie	Squaloids	Squaloïdes
Колючие акулы	Dornhaie	Spiny dog-fishes	Squales
Корюшковые	Stinte	True smelts	
Костеязычные	Knochenzüngler	Osteoglossids	
Костистые рыбы	Echte Knochenfische	Bony fishes	Poissons osseux
Костные рыбы	Knochenfische	– –	– –
Костяной клювонос	Schlanker Knochenhecht	Long-nosed gar-pike	Lépisostée osseux
Кошачьи акулы	Katzenhaie	Spotted dog-fishes	
Крапчатый усач	Semling		Barbeau canin
Красавка	Elritze	Minnow	Vairon
Красная	Blaurückenlachs	Sockeys salmon	
Красноперка	Rotfeder	Rudd	Rotengle
Круглоротые	Rundmäuler	Lampreys	Lamproies
Кумжа	Europäische Forelle	Brown trout	Truite de mer
Летучие рыбы	Fliegende Fische	Flying-fishes	

RUSSIAN NAME	GERMAN NAME	ENGLISH NAME	FRENCH NAME
Лещ	Brachsen	Bream	Brème commune
Лещи	Brassen	Breams	Brèmes
Ликоды	Wolfsfische		Lycodes
Лини	Schleie	Tenches	Tanches
Линь	Schleie	Tench	Tanche commune
Лопатоносы	Schaufelnasenstöre	Shovel-nosed sturgeons	
Лососевые	Lachsähnliche i. e. S.	Salmons	Saumons
Лосось-таймень	Europäische Forelle	Brown trout	Truite de mer
Луноглаз	Mondauge	Mooneye	
Луноглазы	Mondaugen	Mooneyes	
Малая атеринка	Kleiner Ährenfisch	Caspian sand smelt	Prêtre
Малозубая акула	Walhai	Whale shark	
Менек	Lumb	Cusk	Brosme
Мешкожаберный сом	Sackkiemer	Indian cat-fish	
Миксинообразные	Inger	Hagfishes	
Миногообразные	Neunaugen	Lampreys	Lamproies
Минтай	Alaska-Pollack	Wall-eye pollack	
Мирон	Barbe		Barbeau
Многоперы	Flösselhechte	Polypterus	
Мойва	Lodde	Capelin	Capelan
Молот-рыбы	Hammerhaie	Hammerhead sharks	
Морская лисица	Fuchshai	Thresher shark	Requin renard
Морская минога	Meerneunauge	Sea lamprey	Lamproie marine
Морская мышь	Sargassofisch	Frog-fish	
Морская собака	Kleingefleckter Katzenhai	Small-spotted dog-fish	Petite roussette
Морская щука	Leng	Ling	Lingue
Морские ангелы	Engelhaie	Angel sharks, Squatinoids	
Морские дьяволы	Teufelsrochen	Devil-fishes	
Морские клоуны	Fühlerfische i. e. S.	Frog-fishes	
Морские черти	Seeteufel i. e. S.	Angler-fishes	Baudroies
Морской дьявол	Meeresteufel	Devil-fish	Diable de mer
Морской кот	Gewöhnlicher Stechrochen	Common sting-ray	Pastenague
Морской кот	Kleingefleckter Katzenhai	Small-spotted dog-fish	Petite roussette
Морской пес	Kleine Schwebrenke		Outil
Морской сиг	Kleingefleckter Katzenhai	Small-spotted dog-fish	Petite roussette
Морской угорь	Meeraal	Conger eel	Congre
Мраморный скат	Marmorrochen		Raie brunette
Мраморный электри-ческий скат	Marmorzitterrochen	Marbled electric ray	Torpille marbrée
Мурены	Muränen	Morays	Murènes
Налим	Aalquappe	Burbot	Lote
Налимы	Aalquappen	Burbots	Lotes
Настоящие скаты	Echte Rochen	Skates	Raies
Настоящие усачи	Echte Barben	Barbels	Barbeaux
Нерка	Blaurückenlachs	Sockeye salmon	
Нокотница	Dornhai	Spiny dog-fish	Aiguillat tacheté
Обыкновенная бельдюга	Aalmutter	Viviparous blenny	Loquette
Обыкновенная верховка	Moderlieschen		Able de stymphale
Обыкновенная длиннорылая химера	Gewöhnliche Langnasenchimäre	Chimaera	
Обыкновенная корюшка	Stint	European smelt	Éperlan d'Europe
Обыкновенная миксина	Inger	Common hagfish	Myxine
Обыкновенная молот-рыба	Glatter Hammerhai	– hammerhead shark	Requin marteau
Обыкновенная мольва	Leng	Ling	Lingue
Обыкновенная рохля	Gemeiner Geigenrochen	Guitar fish	Guitare
Обыкновенная уклея	Ukelei	Bleak	Ablette
Обыкновенная химера	Seeratte	Rabbit fish	Rat de mer
Обыкновенная щиповка	Europäischer Steinbeißer	Loach	Loche de rivière
Обыкновенная щука	Hecht	Pike	Brochet commun
Обыкновенные горчаки	Bitterlinge	European bitterling	Bouvières
Обыкновенные пескари	Gründlinge i. e. S.	Gudgeons	Goujons
Обыкновенные сомы	Echte Welse		Silures
Обыкновенный голец	Wandersaibling	Charr	Omble-chevalier
Обыкновенный елец	Hasel	Dace	Vandoise vraie
Обыкновенный жерех	Rapfen		Aspe
Обыкновенный карась	Gewöhnliche Karausche	Crucian carp	Carassin vulgaire
Обыкновенный лопатонос	Gemeiner Schaufelstör	Shovel-nosed sturgeon	
Обыкновенный лосось	Atlantischer Lachs	Atlantic salmon	Saumon atlantique
Обыкновенный морской черт	– Seeteufel	Angler-fish	
Обыкновенный орляк	Gewöhnlicher Adlerrochen	Spotted eagle ray	Aigle de mer
Обыкновенный пескарь	– Gründling	Gudgeon	Goujon
Обыкновенный подуст	Nase		Chondrostome nez
Обыкновенный речной угорь	Europäischer Flußaal	Fresh-water eel	Anguille
Обыкновенный сиг	Große Schwebrenke	Powan	Corégone lavaret

RUSSIAN NAME	GERMAN NAME	ENGLISH NAME	FRENCH NAME
Обыкновенный сом	Flußwels	Sheat-fish	Silure glane
Обыкновенный усач	Barbe	Barbel	Barbeau commun
Обыкновенный хариус	Europäische Äsche	Grayling	Ombre de rivière
Обыкновенный хэк	Seehecht	Hake	Merlu
Озерная форель	Seeforelle	Lake trout	Truite de lac
Океанические сельди	Heringe i. e. S.	Herrings	Harengs
Орляки	Adlerrochen	Eagle rays	
Орф	Goldorfe		Orfe
Осетровые	Echte Störe	Sturgeons	Esturgeons
Осетры	Störe	–	–
Палии	Saiblinge	Charrs	
Пантодон	Schmetterlingsfisch	Butterfly-fish	Pantodon
Пантодоновые	Schmetterlingsfische	Butterfly-fishes	
Панцырные щуки	Knochenhechte	Gar-pikes	
Панцырные сомы	Panzerwelse		Callichthyidés
Пещерная живородка	Mammuthöhlen-Blindfisch	Mammoth cave blindfish	Amblyopsis
Пикша	Schellfisch	Haddock	Morue noire
Пилохвост	Fleckhai	Black-mouthed dogfish	Chien espagnol
Пилохвосты	Fleckhaie	Topes	
Пилы-рыбы	Sägefische	Sawfishes	Scies
Пираньи	Pirayas	Caribes	Caribes
Пиранья	Piraya	Piraya	Piraya
Пластиножаберные	Plattenkiemer	Sharks and rays	
Плащеносная акула	Krausenhai	Frilled shark	Requin à tunique
Плащеносные акулы	Krausenhaie	– –	
Плотва	Plötze	European roach	Gardon blanc
Подусты	Nasen		Chondrostomes
Позвоночные	Wirbeltiere	Vertebrates	Vertébrés
Полурылы	Halbschnäbler	Half-beaks	Hémiramphes
Придонный сиг	Große Bodenrenke		Corégone féra
Пятнистый электрический скат	Gefleckter Zitterrochen	Eyed electric ray	Torpille tachetée
Радужная форель	Regenbogenforelle	Rainbow trout	Truite de Kamloops
Речной гольян	Elritze	Minnow	Vairon
Речные угри	Echte Aale	Eels	Anguilles
Рипус	Kleine Maräne		Corégone blanc
Рислинг	Strömer		Blageon
Рогатые акулы	Hornhaie	Hornsharks	
Рогачи	Teufelsrochen	Devil-fishes	
Ручьевая форель	Bachforelle	River trout	Truite de rivière
Рыба-бабочка	Schmetterlingsfisch	Butterfly-fish	Pantodon
Рыбы	Fische	Fishes	Poissons
Ряпушка	Kleine Maräne		Corégone blanc
Сазан	Karpfen	Carp	Carpe ordinaire
Сазаны	–	Carps	Carpes
Сайда	Köhler	Pollack	Lieu noir
Саргановые	Hornhechte	Gar-fishes	
Сарганы	–	–	Aiguilles
Сардины	Echte Sardinen	Sardines	Sardines
Северная кунья акула	Nördlicher Glatthai	Stellate smooth-hound	Emissole tachetée
Североамериканский анчоус	Nordamerikanische Sardelle	Northern anchovy	Anchois de Californie
Сельдевая акула	Heringshai	Porbeagle	Lamie
Сельдевидные акулы	Makrelenhaie	Mackerel sharks	
Сельдевые	Heringe	Herrings	Clupes
Сельдевые акулы	Heringshaie	Mackarel sharks	
Семга	Atlantischer Lachs	Atlantic salmon	Saumon atlantique
Серебряный карась	Silberkarausche	Goldfish	Poisson rouge
Сибирский хариус	Sibirische Äsche	Arctic grayling	Ombre arctique
Сиги	Renken	Whitefishes	Corégones
Скатовые	Rochen	Skates and rays	
Снеток	Stint	Smelt	Éperlan d'Europe
Собачья акула	Kleingefleckter Katzenhai	Small-spotted dog-fish	Petite roussette
Сомовые	Echte Welse	Cat-fishes	Siluridés
Средиземноморская мольва	Mittelmeerleng	Blue ling	
Средиземноморская мурена	Mittelmeer-Muräne	European moray	Murène commune
Стерлядь	Sterlet	Sterlet	Sterlet
Тарпуны	Tarpune	Tarpons	
Тихоокеанская сардина	Pazifische Sardine	Pilchard	
Тихоокеанская сельдь	Pazifischer Hering	Pacific herring	Hareng du Pacifique
Тихоокеанская треска	Pazifik-Kabeljau	– cod	Morue du Pacifique
Тихоокеанские лососи	Pazifische Lachse	– salmons	Saumons
Треска	Eigentliche Dorsche	Cods	Morues

RUSSIAN NAME	GERMAN NAME	ENGLISH NAME	FRENCH NAME
Тресковые	Dorsche i. e. S.	Gadoid fishes	Gades
Уклейки	Ukelei		Ablettes
Фиерасфер	Fierasfer	Pearl fish	Aurin
Финта	Finte	Twaite shad	Alose finte
Фиолетовый хвостокол	Violetter Stechrochen	Blue sting-ray	Pastenague violette
Хамса	Europäische Sardelle	Anchovy	Anchois
Харациновидные	Salmler	Characins	
Харациновые	– i. e. S.	–	
Хариусы	Äsche	Graylings	
Хвостоколы	Stachelrochen	Sting rays	Pastenagues
Химеры	Seekatzen		Chiméroïdes
Хрящевые рыбы	Knorpelfische	Cartilaginous fishes	
Цельноголовые	Seedrachen	Chimaeras	
Центрина	Meersau	Angular rough shark	Centrine
Чавыча	Quinnat	Chinook salmon	Saumon de Californie
Черепные	Wirbeltiere	Vertebrates	Vertébrés
Черная колючая акула	Schwarzer Dornhai	Latern shark	Sagre
Черная рыба	Fächerfisch	Alaska black-fish	
Четырехглазая рыба	Vierauge	Four-eyed fish	
Четырехглазые рыбы	Vieraugen	– fishes	
Чехонь	Ziege		Rasoir
Чукучановые	Sauger	Suckers	
Шереспер	Rapfen		Aspe
Шиповатый скат	Nagelrochen	Thornback ray	Raie bouclée
Шэд	Amerikanische Alse	American shad	Alose d'Amérique
Шиповки	Steinbeißer i. e. S.		Loches
Щуки	Hechte i. e. S.	Pickerels	Brochets
Щуковые	Hechte	–	
Электрические скаты	Zitterrochen	Electric rays	Torpilles
Электрические угри	Zitter- und Messeraale	– eels	
Электрический сом	Elektrischer Wels		Poisson-chat électrique
Электрический угорь	Zitteraal	– eel	
Южная кунья акула	Südlicher Glatthai	Smooth-hound	Emissole lisse
Язь	Aland	Id	Ide mélanote
Яицекладущие карпозу-бые	Eierlegende Zahnkärpflinge	Cyprinodonts	
Японские горчаки	Japanische Bitterlinge	Carplike fish	

Conversion Tables of Metric to U.S. and British Systems

U.S. Customary to Metric   Metric to U.S. Customary

—— Length ——

To convert	Multiply by	To convert	Multiply by
in. to mm.	25.4	mm. to in.	0.039
in. to cm.	2.54	cm. to in.	0.394
ft. to m.	0.305	m. to ft.	3.281
yd. to m.	0.914	m. to yd.	1.094
mi. to km.	1.609	km. to mi.	0.621

—— Area ——

sq. in. to sq. cm.	6.452	sq. cm. to sq. in.	0.155
sq. ft. to sq. mi.	0.093	sq. m. to sq. ft.	10.764
sq. yd. to sq. m.	0.836	sq. m. to sq. yd.	1.196
sq. mi. to ha.	258.999	ha. to sq. mi.	0.004

—— Volume ——

cu. in. to cc.	16.387	cc. to cu. in.	0.061
cu. ft. to cu. m.	0.028	cu. m. to cu. ft.	35.315
cu. yd. to cu. m.	0.765	cu. m. to cu. yd.	1.308

—— Capacity (liquid) ——

fl. oz. to liter	0.03	liter to fl. oz.	33.815
qt. to liter	0.946	liter to qt.	1.057
gal. to liter	3.785	liter to gal.	0.264

—— Mass (weight) ——

oz. avdp. to g.	28.35	g. to oz. avdp.	0.035
lb. avdp. to kg.	0.454	kg. to lb. avdp.	2.205
ton to t.	0.907	t. to ton	1.102
l. t. to t.	1.016	t. to l. t.	0.984

Abbreviations

U.S. Customary   Metric

U.S. Customary	Metric
avdp.—avoirdupois	cc.—cubic centimeter(s)
ft.—foot, feet	cm.—centimeter(s)
gal.—gallon(s)	cu.—cubic
in.—inch(es)	g.—gram(s)
lb.—pound(s)	ha.—hectare(s)
l. t.—long ton(s)	kg.—kilogram(s)
mi.—mile(s)	m.—meter(s)
oz.—ounce(s)	mm.—millimeter(s)
qt.—quart(s)	t.—metric ton(s)
sq.—square	
yd.—yard(s)	

By kind permission of Walker: Mammals of the World
©1968 Johns Hopkins Press, Baltimore, Md., U.S.A.

TEMPERATURE

CENTIGRADE   FAHRENHEIT

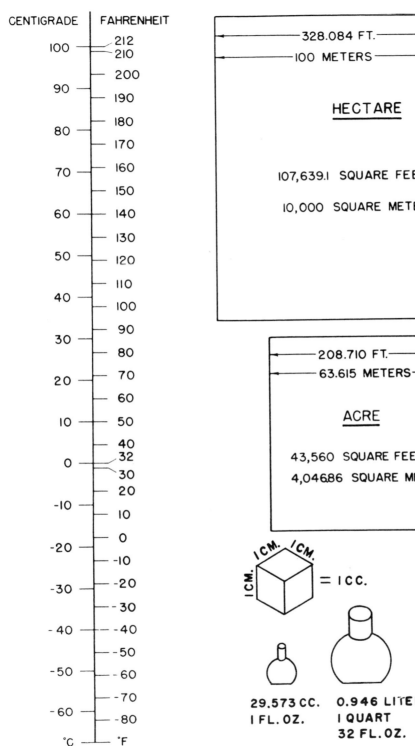

CENTIGRADE	FAHRENHEIT
100	212
	210
	200
90	190
	180
80	170
70	160
	150
60	140
	130
50	120
	110
40	100
	90
30	80
	70
20	60
10	50
	40
0	32
	30
	20
-10	10
	0
-20	-10
-30	-20
	-30
-40	-40
	-50
-50	-60
	-70
-60	-80
°C	°F

AREA

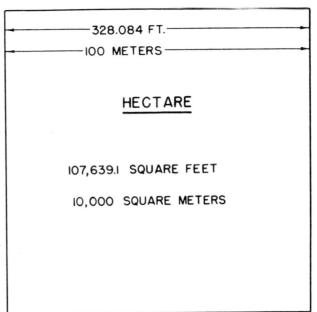

— 328.084 FT. —
— 100 METERS —

HECTARE

107,639.1 SQUARE FEET

10,000 SQUARE METERS

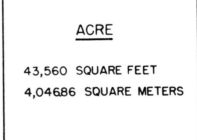

— 208.710 FT. —
— 63.615 METERS —

ACRE

43,560 SQUARE FEET

4,046.86 SQUARE METERS

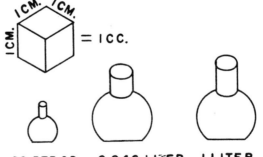

I CM.   I CM.   I CM.   I CM.   = I CC.

29.573 CC.	0.946 LITER	I LITER
I FL. OZ.	I QUART	1,000 CC.
	32 FL. OZ.	1.057 QT.

WEIGHT

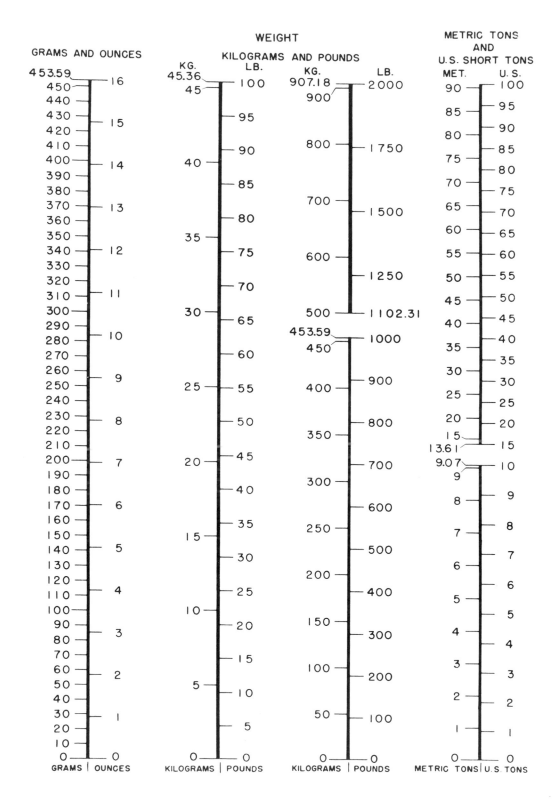

GRAMS AND OUNCES

453.59	
450	16
440	
430	15
420	
410	
400	14
390	
380	
370	13
360	
350	
340	12
330	
320	
310	11
300	
290	
280	10
270	
260	
250	9
240	
230	8
220	
210	
200	7
190	
180	
170	6
160	
150	
140	5
130	
120	
110	4
100	
90	3
80	
70	
60	2
50	
40	
30	1
20	
10	
0	0

GRAMS | OUNCES

KILOGRAMS AND POUNDS

KG. — LB.

45.36	
45	100
	95
40	90
	85
35	80
	75
	70
30	65
	60
25	55
	50
20	45
	40
	35
15	30
	25
10	20
	15
5	10
	5
0	0

KILOGRAMS | POUNDS

KG. — LB.

907.18	2000
900	
800	1750
700	1500
600	1250
500	1102.31
453.59	1000
450	
400	900
350	800
300	700
250	600
	500
200	400
150	300
100	200
50	100
0	0

KILOGRAMS | POUNDS

METRIC TONS AND U.S. SHORT TONS

MET. — U.S.

90	100
85	95
80	90
75	85
70	80
65	75
60	70
55	65
50	60
45	55
40	50
35	45
30	40
25	35
20	30
15	25
13.61	20
	15
9.07	10
9	
8	9
7	8
6	7
5	6
4	5
3	4
2	3
1	2
0	1
	0

METRIC TONS | U.S. TONS

LENGTH: MILLIMETERS AND INCHES

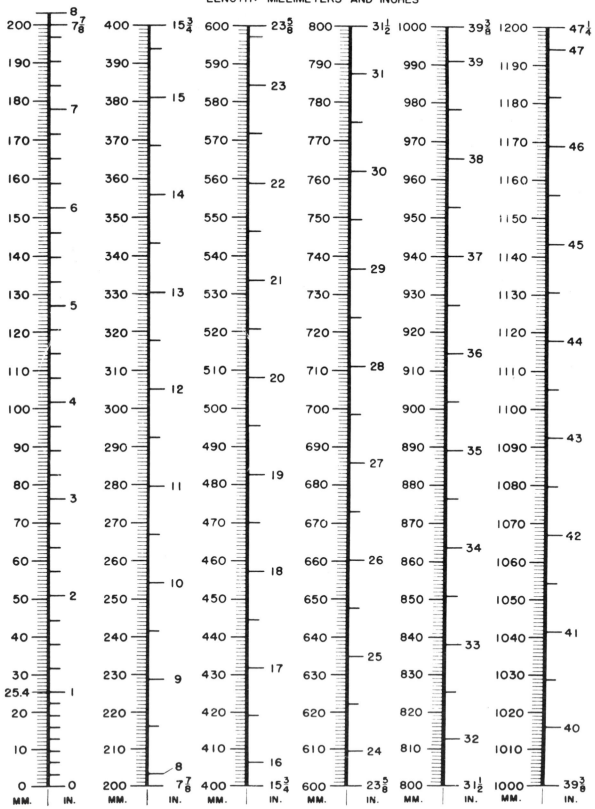

LENGTH

METERS AND FEET

METERS	FEET
	25
	24
7	23
	22
	21
6	20
	19
	18
	17
5	16
	15
	14
4	13
	12
	11
3	10
	9
	8
	7
2	6
	5
	4
1	3
	2
	1
0	0

METERS	FEET
30	100
29	95
28	
27	90
26	85
25	
24	80
23	75
22	
21	70
20	65
19	
18	60
17	55
16	
15	50
14	45
13	
12	40
11	35
10	
9	30
8	25
7	
6	20
5	15
4	
3	10
2	5
1	
0	0

KILOMETERS AND MILES

METERS	FEET
7500	25,000
7000	22,500
6500	
6000	20,000
5500	17,500
5000	
4500	15,000
4000	12,500
3500	
3000	10,000

METERS	FEET
	10,000
3000	9842
2700	9000
2400	8000
2100	7000
1800	6000
1500	5000
1200	4000
900	3000
600	2000
300	1000
0	0

KILOMETERS	MILES
160	100
155	
150	95
145	90
140	
135	85
130	80
125	
120	75
115	70
110	
105	65
100	60
95	
90	55
85	50
80	
75	45
70	40
65	
60	35
55	30
50	
45	25
40	20
35	
30	15
25	10
20	
15	5
10	
5	
0	0

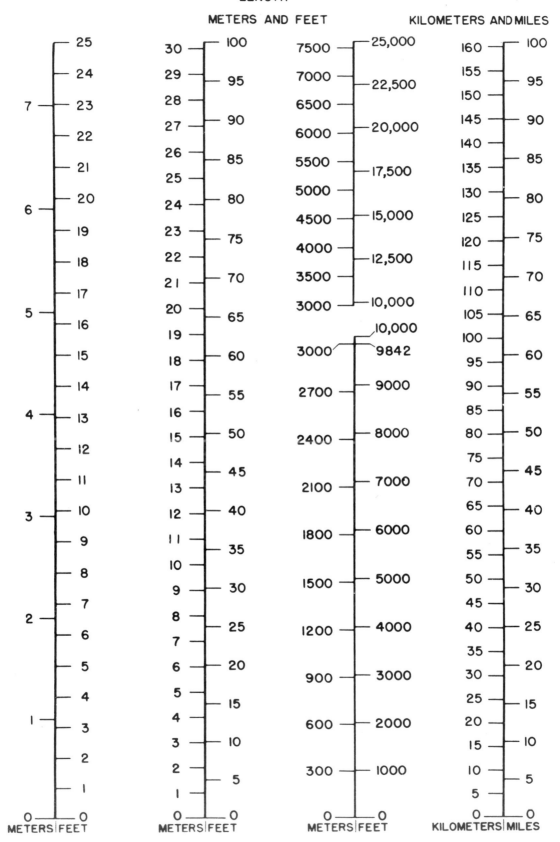

Supplementary Readings

Books and Monographs

American Fisheries Society. 1960. *A List of Common and Scientific Names of Fishes from the United States and Canada.* American Fisheries Society Special Publication 2.

Assem, J. van den. 1967. *Territory in the Three-spined Stickleback Gasterosteus aculeatus.* An Experimental Study in Intraspecific Competition. Behavior Supplement 16, E. J. Brill, Leiden.

Axelrod, H. R. and L. P. Shultz. 1955. *Handbook of Tropical Aquarium Fishes.* McGraw-Hill Book Co., New York.

Baerends, G. and J. M. Baerends-van Roon. 1950. *An Introduction to the Study of the Ethology of Cichlid Fishes.* Behavior Supplement 1, E. J. Brill, Leiden.

Bailey, R. M. and others. 1970. *A List of Common and Scientific Names of Fishes from the United States and Canada.* American Fisheries Society, Washington.

Berg, L. S. 1947. *Classification of Fishes, Both Recent and Fossil.* Edwards Bros., Inc., Ann Arbor, Michigan.

Bigelow, H. B., I. P. Parfante and W. C. Schroeder. 1948, 1953. *Fishes of the Western North Atlantic,* (2 vols). Sears Foundation for Marine Research, Yale University, New Haven, Connecticutt.

Boardman, E. T. 1944. *Guide to Higher Aquarium Animals.* Cranbrook Institute of Science, Bloomfield Hills, Michigan.

Böhlke, J. E. and C. C. G. Chaplin. 1968. *Fishes of the Bahamas and Adjacent Tropical Waters.* The Academy of Natural Sciences of Philadelphia and Livingston Publishing Co., Wynnewood, Pennsylvania.

Boulenger, G. A. 1904. *Fishes.* In the Cambridge Natural History, Vol. VII. The Macmillan Company, New York. (Reprinted, 1958, by Hafner Publishing Co., New York).

Breder, C. M. 1948. *Field Book of Marine Fishes of the Atlantic Coast from Labrador to Texas.* Putnam, New York.

Brown, A. L. 1971. *Ecology of Fresh Water.* Harvard University Press, Cambridge, Massachusetts.

Brown, M. E. (Ed.). 1957. *The Physiology of Fishes,* Vol. 1, Metabolism, Vol. 2, Behavior. Academic Press, Inc., New York.

Budker, P. 1971. *The Life of Sharks.* Columbia University Press, New York.

Carlander, K. D. 1969. *Handbook of Freshwater Fishery Biology,* Vol. 1. The Iowa State University Press, Ames, Iowa.

Coker, R. E. 1947. *This Great and Wide Sea.* University of North Carolina Press, Chapel Hill, North Carolina.

Colman, J. S. 1950. *The Sea and Its Mysteries.* G. Bell and Sons, London.

Cooper, A. 1971. *Fishes of the World.* Grosset and Dunlap, New York.

Cropp, B. 1971. *Shark Hunters.* Macmillan, New York.

Curtis, B. 1949, *The Life Story of the Fish.* Constable & Co. Ltd., London. (Reprinted by Dover)

Cushing, D. H. 1968. *Fisheries Biology.* A Study in Population Dynamics. The University of Wisconsin Press, Madison.

Daniel, J. F. 1934. *The Elasmobranch Fishes.* The University of California Press, Berkeley, California.

Eddy, S. 1957. *How to Know the Fresh-water Fishes.* Wm. C. Brown Co., Dubuque, Iowa.

— , and T. Surber. 1947. *Northern Fishes with Special Reference to the Upper Mississippi Valley,* Rev. Ed. University of Minnesota Press, Minneapolis.

Ekman, S. 1953. *Zoogeography of the Sea.* Sidgwick and Jackson, London.

Evans, H. M. 1940. *Brain and Body of Fish.* McGraw-Hill Book Company, New York.

Fish, M. P. and W. H. Mowbray. 1970. *Sounds of Western North Atlantic Fishes:* A Reference File of Biological Underwater Sounds. John Hopkins Press, Baltimore.

Gilbert, P. W. (Ed.). 1963. *Sharks and Survival.* D. C. Heath & Co., Boston.

Goadby, P. 1959. *Sharks and other Predatory Fish of Australia.* Jacaranda Press, Brisbane.

Gosline, W. A. 1971. *Functional Morphology and Classification of Teleostean Fishes.* University of Hawaii Press, Honolulu.

— , and V. E. Brock. 1960. *Handbook of Hawaiian Fishes.* University of Hawaii Press, Honolulu, Hawaii.

Graham, D. H. 1953. *A Treasury of New Zealand Fishes.* A. H. & A. W. Reed, Wellington.

Groot, C. 1965. *On the Orientation of Young Sockeye Salmon (Oncorynchus nerka) during their Seaward Migration out of Lakes.* Behaviour, Supplement No. 14, E. J. Brill, Leiden.

Guenther, K. and K. Deckert. 1956. *Creatures of the Deep Sea.* Charles Scribner's Sons, New York.

Günther, A. C. L. G. 1880. *An Introduction to the Study of Fishes.* A. and C. Black, Edinburgh.

Halstead, B. W. 1959. *Dangerous Marine Animals.* Cornell Maritime Press, Cambridge, Maryland.

Harden Jones, F. R. 1968. *Fish Migration.* St. Martin's Press, New York.

Hardisty, M. W. and I. C. Potter, Eds. 1971. *The Biology of Lampreys,* Vol. 1. Academic Press, New York.

Hasler, A. D. 1966. *Underwater Guide Posts,* Homing of Salmon. The University of Wisconsin Press, Madison.

Hay, J. 1965. *The Run.* Doubleday, New York.

Hensen, A. C. 1972. *The Cod.* Crowell, New York.

Herald, E. S. 1961. *Living Fishes of the World.* Doubleday, Garden City, New York.

Hoar, W. S. and D. J. Randall. 1969-1971. *Fish Physiology* (4 vols.). Academic Press, New York & London.

Hubbs, C. L. and K. F. Lagler. 1964. *Fishes of the Great Lakes Region.* University of Michigan Press, Ann Arbor.

Idyll, C. D. 1971. *Abyss,* The Deep Sea and the Creatures That Live in it. Crowell, New York.

Ingle, D. (Ed.) 1968. *The Central Nervous System and Fish Behavior.* The University of Chicago Press, Chicago.

Innes, W. T. 1935. *Exotic Aquarium Fishes:* A Work of General Reference, 19th Ed. Innes Publishing Co., Philadelphia.

Ivlev, V. S. 1961. *Experimental Ecology of the Feeding of Fishes.* Yale University Press, New Haven, Connecticutt.

Jones, F. R. H. 1968. *Fish Migration.* St. Martin's Press, New York.

Jordan, D. S. 1905. *A Guide to the Study of Fishes.* 2 vols. Henry Holt and Company, New York.

— . 1923. *A Classification of Fishes Including Families and Genera as far as Known.* Stanford University Publications in Biological Science 3(2): 79-243.

— . 1963. *The Genera of Fishes and a Classification of Fishes.* U.S. National Museum, Stanford University Press, Stanford, California.

— , and B. W. Evermann. 1896-1900. *The Fishes of North America.* A Descriptive Catalogue of the Species of Fishlike Vertebrates Found in the Waters of North America, North of the Isthmus of Panama. U.S. National Museum Bulletin, 47, 4 Vols.

— , and H. W. Clark. 1955. *Checklist of the Fishes and Fish-like Vertebrates of North and Middle America.* Reprint of Appendix X to the Report of the United States Commissioner of Fisheries for the Fiscal Year 1928.

Kleerekoper, H. 1969. *Olfaction in Fishes.* Indiana University Press, Bloomington & London.

Kuenan, P. H. 1955. *Realms of Water.* John Wiley and Sons, New York.

Kyle, H. M. 1926. *The Biology of Fishes.* Sidgwick and Jackson, London; The Macmillan Company, New York.

Lagler, K. F. 1956. *Freshwater Fishery Biology.* Wm. C. Brown Co., Dubuque, Iowa.

— , J. E. Bardach, and R. R. Miller, 1963. *Ichthyology,* The Study of Fishes. John Wiley & Sons, Inc., New York.

La Monte, F. 1945. *North American Game Fishes.* Doubleday, Doran and Co., Inc., New York.

— . 1952. *Marine Game Fishes of the World.* Doubleday, Garden City, New York.

Lanham, U. N. 1962. *The Fishes.* Columbia University Press, New York.

Lineaweaver, T. H. III, and R. H. Backus. 1970. *The Natural History of Sharks.* J. B. Lippincott Company, Philadelphia & New York.

MacGinity, G. E., and N. McGinity. 1949. *Natural History of Marine Animals.* McGraw-Hill Book Company, New York.

Marshall, N. B. 1954. *Aspects of Deep-Sea Biology.* Philosophical Library, Inc., New York.

—. 1960. Swimbladder Structure of Deep-Sea Fishes in Relation to their Systematics and Biology. *Discovery Reports* 31: 1–112.

—. 1970. *The Life of Fishes.* Universe Books, New York.

—. 1971. *Explorations in the Life of Fishes.* Harvard University Press, Cambridge, Massachusetts.

Matsubara, K. 1955. *Fish Morphology and Hierarchy.* Ishazaki Shoten, Tokyo.

Moore, H. B. 1958. *Marine Ecology.* John Wiley & Sons, New York.

Moy-Thomas, J. A. 1971. *Palaeozoic Fishes.* Saunders, Philadelphia.

National Geographic Society. 1965. *Book of Fishes.* National Geographic Society, Washington, D.C.

Needham, P. R. 1940. *Trout Streams,* Conditions That Determine Their Productivity and Suggestions for Stream and Lake Management. Comstock Publishing Co., Ithaca, New York.

Nieuwenhuizen, A. van den. 1964. *Tropical Aquarium Fish, Their Habits and Breeding Behavior.* Van Nostrand, Princeton, N.J.

Norman, J. R. 1931. *A History of Fishes.* Ernest Benn, London. (Reprinted. 1951. by A. A. Wyn, New York.

—, and F. C. Fraser. 1937. *Giant Fishes, Whales and Dolphins.* Putnam & Co., Ltd., London.

—, and F. C. Fraser. 1949. *Field Book of Giant Fishes.* G. P. Putnam's Sons, New York.

Ommanney, F. D., and the editors of *Life.* 1963. *The Fishes.* Time, Inc., New York.

Perlmutter, A. 1961. *Guide to Marine Fishes.* New York University Press, New York.

Pettersson, H. 1954. *The Ocean Floor.* Yale University Press, New Haven, Connecticut.

Phillips, C. 1964. *The Captive Sea.* Chilton Books, Philadelphia and New York.

Pincher, C. 1948. *A Study of Fish.* Duell, Sloan and Pearce, New York.

Radcliff, W. 1921. *Fishing From the Earliest Times.* John Murray, London.

Rasa, O. A. E. 1971. *Appetence for Aggression in Juvenile Damsel Fish.* Paul Parey, Berlin and Hamburg.

Reid, G. K. 1961. *Ecology of Inland Waters and Estuaries.* Reinhold Publishing Corp., New York.

Roule, L. 1933. *Fishes, Their Journeys and Migrations.* W. W. Norton & Co., Inc., New York.

—. 1935. *Fishes and Their Ways of Life.* W. W. Norton & Company, New York.

Rounsefell, G. A. and W. H. Everhart. 1953. *Fishery Science,* Its Methods and Applications. John Wiley & Sons, New York.

Royce, W. F. 1972. *Introduction to the Fishery Sciences.* Academic Press, Inc., New York and London.

Russell, F. S. 1928. *The Seas.* F. Warne and Co., London.

Schrenkeisen, R. M. 1963. *Field Book of Fresh-water Fishes of North America North of Mexico.* G. P. Putnam's Sons, New York.

Schultz, L. P. and E. M. Stern. 1948. *The Ways of Fishes.* D. Van Nostrand Co., Princeton, New Jersey.

Sears Foundation for Marine Research, Yale University, New Haven. Memoir I. *Fishes of the Western North Atlantic.*

—. 1948. Part 1. Lancelots, Cyclostomes, Sharks.

—. 1953. Part 2. Sawfishes, Guitar Fishes, Skates, Rays, Chimaeroids.

—. 1963. Part 3. Salmons, Trouts, Tarpons, Ladyfish, etc.

Shaw, E. 1970. Schooling in Fishes: Critique and Review in *Development and Evolution of Behavior,* Ed. by L. R. Aronson and others, pp. 452–480. W. H. Freeman and Company, San Francisco.

Simkatis, H. 1958. *Salt-water Fishes for the Home Aquarium.* J. B. Lippincott Co., Philadelphia.

Smith, F. G. W. 1954. *The Ocean.* Charles Scribner's Sons, New York.

Smith, J. L. B. 1949. *The Sea Fishes of South Africa.* Central News Agency, Capetown.

—. 1956. *The Search Beneath the Sea:* The Story of Coelacanth. Henry Holt & Co., New York.

Sterba, G. 1962. *Freshwater Fishes of the World.* Vista Books, Longacre Press, Ltd., London.

Sverdrup, H. V., M. W. Johnson & R. H. Fleming. 1942. *The Oceans.* Prentice-Hall, Englewood Cliffs, N.J.

Van Iersel, J. J. A. 1953. *An Analysis of the Parental Behavior of the Male Three-spined Stickleback (Gasteosteus aculeatus L.).* Behaviour Supplement No. 3, E. J. Brill, Leiden.

Vesey-Fitzgerald, B., and F. LaMonte. (Eds.). 1950. *Game Fish of the World.* Harper & Bros., New York.

Weatherley, A. H. 1972. *Growth and Ecology of Fish Populations.* Academic Press, New York.

Webster, D. K. 1963. *Myth and Maneater, the Story of the Shark.* Norton, New York.

Wheeler, A. 1969. *The Fishes of the British Isles and North-west Europe.* Michigan State University Press, East Lansing.

Whitley, G. P. 1940. *The Fishes of Australia.* Part 1. The Sharks, Rays, Devilfish and Other Primitive Fishes of Australia and New Zealand. Australian Zoological Handbook. Royal Zoological Society, New South Wales, Sydney.

Whitney, L. F. and P. Hahnel. 1955. *All About Guppies.* Practical Science Publishing Co., Garden City, New York.

Wickler, W. 1966. *Breeding Aquarium Fish.* D. Van Nostrand Company, Inc., Princeton, N.J. and New York.

Zim, H. S. and H. H. Shoemaker. 1957. *Fishes.* Golden Press.

SCIENTIFIC JOURNALS

Black, E. C. 1951. Respiration in Fishes. *University of Toronto Studies in Biology Series* 59: 91–111.

Black, V. S. 1951. Osmotic Regulation in Teleost Fish. *University of Toronto Studies in Biology Series* 59: 53–89.

Briggs, J. C. 1960. Fishes of Worldwide (Circumtropical) Distribution. *Copeia* (3): 171–180.

Brooks, J. L. 1950. Speciation in Ancient Lakes. *Quarterly Review of Biology* 25: 30–60 and 131–176.

Carey, F. C. 1973. Fishes with Warm Bodies. *Scientific American* 229(2): 36–44.

Gary, J. 1957. How Fishes Swim. *Scientific American* 197(2): 48–54.

Hasler, A. D. 1954. Odour Perception and Orientation in Fishes. *Journal of the Fisheries Research Board of Canada* 11(2): 107–129.

Hoagland, H. 1933. Electric Responses From the Lateral Line Nerves of Catfishes. *Journal of General Physiology* 16: 695–731.

Jones, F. R. H. and N. B. Marshall. 1953. The Structure and Function of the Teleostean Swimbladder. *Biology Review Cambridge Philosophical Society* 28: 16–83.

Kleerekoper, H. and E. C. Chagnon. 1954. Hearing in Fish with Special Reference to *Semotilus atromaculatus atromaculatus* (Mitchill). *Journal of the Fisheries Research Board of Canada* 11(2): 130–152.

Lissmann, H. W. 1946. The Neurological Basis of the Locomotory Rhythm in the Spiny Dogfish *(Scillium canicula, Acanthias vulgaris)* I. Reflex Behavior, II. The Effect of De-afferentation. *Journal of Experimental Biology* 23: 143–176.

Nursall, J. R. 1956. The Lateral Musculature and the Swimming of Fish. *Proceedings of the Zoological Society of London* 126(1): 127–143.

Ricker, W. F. 1946. Production and Utilization of Fish Populations. *Ecology Monographs* 16: 373–391.

Romer, A. S. 1947. The Early Evolution of Fishes. *Quarterly Review of Biology* 21(1): 33–69.

Schultz, R. J. 1961. Reproductive Mechanisms of Unisexual and Bisexual Strains of the Viviparous Fish *Poeciliopsis. Evolution* 15: 302–325.

GERMAN BOOKS
AND SCIENTIFIC JOURNALS

Anwand, K. 1965. *Die Schleie.* Neue Brehm-Bücherei, Ziemsen, Wittenberg Lutherstadt.

Bauch, G. 1955. *Die einheimischen Süsswasserfische.* 3. Aufl. Neumann, Radebeul/Berlin.

Beebe, W. 1935. *923 Meter unter dem Meeresspiegel.* Brockhaus, Leipzig.

Berg, L. S. 1932. *Übersicht der Verbreitung der Süsswasserfische Europas.* Zoogeographie 1.

— . 1958. *System der rezenten und fossilen Fischartigen und Fische.* VEB Deutscher Verlag der Wissenschaften.

Coker, R. E. 1966. *Das Meer–der grösste Lebenstraum.* Parey, Hamburg/Berlin.

Dzwillo, M. 1961. *Lebendgebärende Zahnkarpfen.* Kernen, Stuttgart.

Eibl-Eibesfeldt, I. 1965. *Haie, Angriff, Abwehr, Arten.* Kosmos, Franckh, Stuttgart.

— . 1967. *Grundriss der vergleichenden Verhaltensforschung.* Piper, München.

Frey, H. 1957. *Das Aquarium von A-Z.* Neumann, Radebeul.

Gerlach, R. 1950. *Die Fische.* Claassen, Hamburg.

Günther, K., und K. Deckert. 1950. *Wunderwelt der Tiefsee.* Herbig, Berlin.

Hegemann, M. 1964. *Der Hecht.* Neue Brehm-Bücherei, Ziemsen, Wittenberg Lutherstadt.

Herald, E. S., und D. Vogt. 1964. *Fische.* Knaurs Tierreich in Farben, Droemer-Knaur, München/Zürich.

Ladiges, W. 1951. *Der Fisch in der Landschaft.* Wenzel und Sohn, Braunschweig.

— . 1960. *Fische der Nordmark.* De Gruyter, Hamburg.

— . 1962. *Barben.* Kernen, Stuttgart.

— . 1963. *Bärblinge.* Kernen, Stuttgart.

— , und D. Vogt. 1965. *Die Süsswasserfische Europas* bis zum Ural und Kaspischen Meer. Parey, Hamburg/Berlin.

Luther, W., und K. Fiedler. 1967. *Die Unterwasserfauna der Mittelmeerküsten.* 2., neubearb. Aufl., Parey, Hamburg/Berlin.

Marshall, N. B. 1957. *Tiefseebiologie.* VEB Fischer, Jena.

Mohr, E. 1952. *Der Stör.* Neue Brehm-Bücherei, Ziemsen, Wittenberg Lutherstadt.

— . 1954. *Fliegende Fische.* Neue Brehm-Bücherei, Ziemsen, Wittenberg Lutherstadt.

Muus, B. J., und P. Dahlström. 1965. *Meeresfische in Farben.* BLV, München.

— . 1968. *Süsswasserfische Europas.* BLV, München.

Nikolski, N. 1957. *Spezielle Fischkunde.* VEB Deutscher Verlag der Wissenschaften, Berlin.

Norman, J. R. 1966. *Die Fische.* Parey, Hamburg/Berlin.

— , und F. C. Fraser. 1963. *Riesenfische, Wale und Delphine.* Parey, Hamburg/Berlin.

Petzold, H. G. 1968. *Der Guppy.* Neue Brehm-Bücherei, Ziemsen, Wittenberg Lutherstadt.

Pinter, H. 1966. *Aquarienfischzucht.* Kernen, Stuttgart.

Riedel, D. 1968. *Die Europäische Sardine.* Neue Brehm-Bücherei, Wittenberg Lutherstadt.

Riedl, R. 1963. *Fauna und Flora der Adria.* Parey, Hamburg.

— . 1966. *Biologie der Meereshöhlen.* Parey, Hamburg/Berlin.

Scheurig, L. 1929–1930. *Die Wanderungen der Fische.* Ergebnisse der Biologie, Bd. VI.

Schindler, O. 1953. *Unsere Süsswasserfische.* Kosmos Verlag, Stuttgart.

Steffens, W. 1969. *Der Karpfen.* Neue Brehm-Bücherei, Ziemsen, Wittenberg Lutherstadt.

Sterba, G. 1952. *Die Neunaugen.* Neue Brehm-Bücherei, Ziemsen, Wittenberg Lutherstadt.

— . 1959. *Süsswasserfische aus aller Welt.* Zimmer und Herzog, Berchtesgaden.

Thienemann, A. 1926. *Die Süsswasserfische Deutschlands.* Eine tiergeographische Skizze. Handbuch der Binnenfischerei Mitteleuropas, 3.

Villwock, W. 1960. *Eierlegende Zahnkarpfen.* Kernen, Stuttgart.

Vogt, D. 1956–1957. *Taschenbuch der tropischen Zierfische.* Bd. I und II, Kosmos, Stuttgart.

— . 1959. *Salmler I, II.* Kernen, Stuttgart.

Wickler, W. 1962. *Das Meeresaquarium.* Kosmos, Franckh, Stuttgart.

— . 1968. *Mimikry.* Kindler, München.

Wundsch, H. H. 1962. *Barsch und Zander.* Neue Brehm-Bücherei, Ziemsen, Wittenberg Lutherstadt.

Picture Credits

Index

Aasen, 183
Abbottina rivularis, **331**
Abramis (Breams)
— *ballerus,* **320**
— *brama* (Bream), **319**f, 320m, 341*
— *sapa,* **320**
Abramites, **293**
Acanthodii (Spiny sharks), **46**, 46*
Acanthodorus spinosissimus (Talking catfish), **377**
Acanthophthalmus, 357, **359**
— *kuhli kuhli,* **360**, 372*
— *myersi,* **360**
— *semicinctus,* **360**, 372*
— *shelfordi,* **360**
Acanthopsis choiorhynchus, 372*
Acanthopterygii (Spiny-rayed fishes), 131, **152**f
Acanthorhodeus, 329
Acestrocephalus, 278
Acestrorhynchus, **278**, 289
Acheilognathinae (Bitterlings), **329**, 329m
Acipenser (Sturgeons), 135
— *baeri* (Siberian sturgeon), **137**f
— *brevirostrus,* **138**
— *dabryanus,* **138**
— *fulvescens* (Lake sturgeon), 137*, **138**f
— *gueldenstaedti* ("Russian sturgeon"), **137**
— — *colchicus,* **137**
— — *gueldenstaedti,* 137
— — *persicus,* 137
— *kikuchii,* **138**
— *medirostris,* **138**
— *multiskutatus,* **138**
— *naccari* (Adriatic sturgeon), **137**
— *nudiventris,* **137**
— *oxyrhynchus* (Atlantic sturgeon), **138**
— *ruthenus* (Sterlet), **136**, 145*
— *schrencki* (Amur sturgeon), **137**f
— *sinensis,* **138**
— *stellatus,* **137**, 139*
— *sturio* (Common Atlantic sturgeon), **136**, 145*
— *transmontanus* (White sturgeon), **138**
Acipenseridae (Sturgeons), **134**f
Acipenseriformes, 134
Actinopterygii (Higher bony fishes), 45
Adaptation, 70
Adrianichthyidae, **437**
Adriatic sturgeon *(Acipenser naccari),* **137**
Aeschynichthyidae, see Diceratiidae
African bonytongue *(Clupisudis niloticus),* 202, **204**
Agassiz, Louis, 40, 45
Agnatha (Jawless fishes), 30ff, 40
Agoniates, **277**f, 278*
Agoniatinae, **277**f
Ahlstrom, 254
Alabama Shad *(Alosa alabamae),* **195**
Alaska blackfish *(Dallia pectoralis),* **252**, 255*
Alaska pollack *(Theragra chalcogramma),* 69, **416**f, 419*
Albula vulpes (Bonefish), 152, **156**
Albulidae (Bonefish family), 156, 156*
Albuloidei (Bonefishes), 152, **155**f
Alburnoides bipunctatus, **319**
Alburnus alburnus (Bleak), 312, **316**f, 316m, 341*
Alekey trout or eel-pout *(Lota lota*

maculosa), **410**
Alepisauridae (Lancet fishes), **263**, 263*
Alepocephaloidei (Deepsea slickheads), 213, **258**
Alestes longipinnis, **289**, 308*
Alestidae, see Hydrocynidae
Alestinae, **289**
Alestopetersius caudalis, 308*
Allen, 202
Alligator gar *(Lepisosteus spatula),* **147**, 157*
Allmouth or Angler *(Lophius piscatorius),* **401**f
Alopias vulpinus (Thresher shark), **97**f, 103*
Alopiidae (Thresher sharks), 97
Alosa (Shads), 193
— *alabamae* (Alabama shad), **195**
— *alosa* (Shad), **193**, 193m, 198*
— *fallax* (Twaite shad), 193m, **194**, 198*
— *sapidissima* (American shad), **194**f
Alosinae, 177
Amazon molly *(Poecilia formosa),* **441**
Amblyceps mangois, **373**
Amblycipitidae, **373**
Amblydoras hancocki, **378**
Amblyopsidae, **395**
Amblyopsis, **396**
— *spelaeus* (Mammoth cave blindfish), **396**, 417*
Amblyopsoidei (Cavefishes), **395**
American brook lamprey *(Ichthyomyzon fossor),* **35**
American eel *(Anguilla rostrata),* 160m, **163**
American hake *(Merluccius bilinearis),* **423**, 423m, 431*
American paddlefish *(Polydon spathula),* **142**f, 143*, 146*
American sea lamprey *(Petromyzon marinus dorsatus),* **38**
American shad *(Alosa sapidissima),* **194**f
American smelt *(Osmerus eperlanus mordax),* **247**
Amia calva (Bowfin), **149**ff, 150*, 151*, 157*
Amiiformes, **149**, 149m, 149*
Animocoetes brachialis, see also Petromyzonidae
Amur pike *(Esox reicherti),* **251**
Amur sturgeon *(Acipenser schrencki),* **137**f
Anablepidae, **438**
Anableps anableps (Four-eyed fish), 436*, **438**, 456/457*
Anachantobatidae, 122
Anaspida, 40
Anatomy of fish, **52**
Anchoa, **200**
Anchovia, **200**
Anchoveta (Engraulis ringens), 68, **200**
Anchoviella, **200**
Anchovies *(Engraulis),* 177, **196**, 199m
Anchovy *(Engraulis encrasicholus),* **196**f
Ancistrus, 388*, **390**
— *bufonius,* **390**
— *dolichopterus* (Blue antenna catfish), **391**, 412*
Angel shark *(Squatina squatina),* **119**f, 119*
Anglerfishes (Lophiiformes), **398**f
Anguilla (Eels)

— *anguilla* (European fresh-water eel), **160**ff, 160m, 167*
— *japonica* (Japanese eel), **163**
— *rostrata* (American eel), 160m, **163**
Anguillidae (Freshwater eels), **160**
Anguilliformes (Eels), **159**ff
Anguilloidei, 159, **160**f
Angular rough sharks (Oxynotidae), **115**
Anodinae, 294
Anodonta, (Mussel), 327/328*, 330
Anomalops, **56**
Anoptichthys jordani (Blind characin), **280**, 396, 417*
Anostomidae, **293**
Anostomus, 293
— *anostomus,* 291*
Anotopterus pharao, **262**
Antennariidae, **402**
Antennarioidei (Frogfishes), **402**
Antennarius, 402, 402*
— *moluccensis,* 402, 422*
— *scaber,* **402**
Antiarchi, **47**
Antimora rostrata (Blue hake), **406**, 406*
— *viola,* **406**
Aphanius
— *anatoliae,* **433**
— *apodus,* **433**
— *chantrei,* **433**
— *dispar,* **434**
— *fasciatus,* **434**
— *iberus,* **434**
— *mento,* **433**, 450*
— *sophiae,* **433**
Aphredoderoidei, **396**
Aphredoderus sayanus (Pirate perch), 256*, 396f
Aphyocharacinae, **279**
Aphyocharax rubripinnis, 286*
Aphyocypris pooni, **311**
Aphyosemion (Killies), 282, 435f
— *arnoldi,* **436**
— *australe* (Lyretail), **436**, 450*, 454*
— *bivittatum* (Two-striped killy), **436**
— *calliurum,* **436**
— *cognatum* (Red-speckled killy), **436**
— *exiguum,* 454*
— *fallax,* 454*
— *filamentosum,* 450*, 454*
— *nigerianum,* 450*, 454*
— *occidentalis toddi,* 454*
— *sjostedti* (Golden pheasant killy), **436**, 450*
Aplocheilichthys myersi, **437**, 450*
Aplocheilus, **437**
— *dayi,* 455*
— *lineatus* (Malabar killy), 450*
Aplochiton, **250**
Aplochitonidae, **249**
Applegate, 35, 38
Apteronotidae, 296, **303**
Apteronotus albifrons, 298*, 298, 303, 317*
Arapaima (Arapaimas)
— *gigas* (Giant or Brazilian arapaima), 49, **202**f, 207*
Arctic Greenland cod *(Arctogadus glacialis),* **416**
Arctogadus, **415**
— *borisovi* (East Siberian cod), **416**, 416*
— *glacialis* (Arctic Greenland cod), **416**
Argentine pearl fish *(Cynolebias*

belotti), **436**, 455*
Argentines (Argentinoidei), 213, **252**, 252*
Argentinoidei (Argentines), 213, **252**, 252*
Argyropelecus, 57, **254**
— *affinis* (Deepsea hatchet fish), 266*
Arriidae (Sea catfishes), **270**
Ariosoma balearica, **166**
Aristotle, 45
Arius proops (Crucifix catfish), **370**, 386*
Armbrust, W., 133
Armored catfishes, 378, 381, 389
Aronson, 441
Arowana *(Osteoglossum bicirrhosum),* **203**f, 207*, 223*
Arthrodira (Arthrodira), **46**f
Arynchobatidae, 122
Asiatic shovel-nosed sturgeons *(Pseudoscaphirhynchus),* 141
Asp *(Aspius aspius),* 312, **321**, 342*
Asphredoderus sayanus (Pirate perch), 256*
Aspius aspius (Asp), 312, **321**, 342*
Astronesthidae (Deepsea snaggletooths), **253**, 253*
Astroscopus, 296
Astrospis, 40
Astyanax, 280
— *fasciatus mexicanus* (Mexican tetra), **280**f, 291*, 396
Ateleopodidae (Deepsea atelopid), 264, 264*
Ateleopodoidei (Deepsea atelopids), 264
Atherina
— *hepsetus,* **444**
— *mochon* (Caspian sand smelt), **444**
— *presbyter,* **444**
Atherinidae (Silversides), 444 ·
Atheriniformes (Silversides), **426**ff
Atherinoidei, 444
Atherinopsis californiensis (Jacksmelt silverside), **445**
Atlantic cod *(Gadus morhua),* 41*, **413**ff, 414m, 431*
Atlantic falsecat shark *(Pseudotriakis microdon),* **102**
Atlantic gray shark *(Carcharhinus plumbeus),* **106**
Atlantic herring *(Clupea harengus),* **178**ff, 178m, 179m, 197*
Atlantic midshipman *(Porichthys porosissimus),* **397**, 418*
Atlantic salmon *(Salmo salar),* **214**ff, 214m, 215*, 217*, 219*, 224*
Atlantic saury *(Scomberesox saurus),* **430**
Atlantic sturgeon *(Acipenser oxyrhynchus),* **138**
Atlantic tarpon *(Megalops atlanticus),* **153**, 153m
Atlantic tom cod or frostfish *(Microgadus tomcod),* **416**, 419*
Atlanto-scandian herring, 178, 180f, 181*, 183f, 186
Auchenoglanis occidentalis, **366**
Auditory system, **23**f
Aulopodidae, **260**
Aulopus
— *filamentosus,* **260**
— *japonicus,* **260**
— *purpurrissatus* (Sergeant Baker), **260**
Australian bonytongue *(Sceropages leichhardti),* **204**
Australian rainbow fish *(Melanotaenia nigrans),* **444**, 453*,

Heavy type indicates the main entry, an asterisk ★ indicates an illustration, and m indicates a distribution map

456/457*
Australian sardine (Sardinops neopilchardus), 191
Austrofundulus transilis, 454*
Ayu (Plecoglossus altivelus), 245f, 245m

Babcock, 154f
Backhouse, Dieter, 284
Back-swimming Congo catfish (Synodontis nigriventris) 377f
Bagarius bagarius, 373
Bagrichths hypselopterus, 366
Bagridae (Bagrid catfishes), 365
Bahr, 36f
Barb or barbel (Barbus barbus), 337f, 339m, 341*
Barbinae (Barbins), 332f, 338m
Barbins (Barbinae), 332f, 338m
Barbourisiidae, 264, 267
Barbus (Barbs)
— barbus (Barb or barbel), 337f, 339m, 341*
— brachycephalus caspius, 338
— capito, 338
— meridionalis petenyi, 323, 338
— mosal (Indian barb), 55
Bariliinae, a.k.a. Danioninae, 306
Barilius chrystyi, 309, 352*
"Barracudina," see Paralepididae, 262, 262*
Barreleye (Opisthoproctus grimaldi), 252, 252*
Barrigudinyo (Phalloceros caudimaculatus), 443, 451*
Basking shark (Cetorhinus maximus), 49, 93*, 96f
Basking sharks (Cetorhinidae), 96
Batavia putyekanipa (Muraenesox cinerus), 165
Batfishes (Ogcocephalidae), 402f
Bathophilus, 253
Bathybentonic fishes (Deepsea bottom-dwellers), 252
Bathycongrus mystax, 166, 169*
Bathylaconoidei (Deepsea smelts), 213, 258
Bathylagidae (Deepsea smelts), 252
Bathypelagic fishes (Deepsea fishes), 253
Bathypteroidae, 261
Bathypterois, 261f
— ventralis, 266*
Bathysauridae, 260
Bathysaurus, (see also Harpodontidae), 260f
Bathrachocephalus, 369
Batrachoidiformes (Toadfishes), 397
Bdellostoma burgeri (Japanese hapfish), 33
Bdellostomatinae, 31
Bedotia geayi, 453*, 456/457*
Beebe, 259f
Belone bellone (Garfish), 429
Belonesox, 440
— belizanus (Lucino), 443, 452*
Belonidae (Needlefishes or garfishes), 429
Belonophago, 295
Belthophilus, 49
Berg, 166, 248
Bichir (Polypterus bichir), 132
Bichirs (Polypteri), 48, 131
Bitterlings (Acheilognathinae), 329, 329m
Bivibranchiinae, 293
Biwa Lake gudgeon (Gnathopogon coerulescens), 331
Black electric ray (Torpedo nobiliana), 122
Black molly (Poecilia sphenops), 51*, 441

Black telescoped fog-tail goldfish, 334*
Black-mouthed dogfish (Gaeus melanostomus), 100f
Blacknosed dace (Rhinichthys stronasus), 323
Black-tip reef shark (Carcharhinus melanopterus), 106
Black-winged hatchet fish (Carnegiella marthae), 291*
Blaxter, 185
Bleak, (Alburnus alburnus), 312, 316f, 316m, 341*
Blicca bjoerkna (Silver bream), 320
Blind characin (Anoptichthys jordani), 280, 396, 417*
Blind fish, 280f
Blister-back, see Pollock (Pollachius vireus)
Blood cells, 26*
Blood circulation, 25, 61
Blue antenna catfish (Ancistrus dolochopterus), 391, 412*
Blue hake (Antimora rostrata), 406, 406*
Blue ling (Molva elongata), 413, 413m
Blue stingray (Dasyatis violacea), 124
Bluefin killy (Chriopeops goodei), 437
Böhlke, J., 166, 171, 282
Bolster, 182
Bombay duck (Harpodon), 261, 261*
Bonefish (Albula vulpes), 152, 156
Bonefish family (Albulidae), 156, 156*
Bonefishes (Albuloidei), 152, 155f
Bonytongues (Osteoglossiformes), 202ff, 204m
Bony fishes (Osteichthys), 45, 130f
Boreogadus saida (Polar cod), 415f, 416m, 416*
Borodin, 252
Botia, 357
— berdmorei, 358
— horae, 358, 372*
— hymenophysa, 358, 372*
— lohachata, 358
— macracanthus, 358, 372*
— sidthimunki, 358, 372*
Botiinae, 356, 357f, 358m
Bottlenose dolphin (Tursiops truncatus), 108/109*
Bottom-dwelling whitefish (Coregonus), 235f, 239*
Boulengerella cuvieri, 307*
Boulengeromyrus knoepffleri, 211, 211*
Bouquet-head goldfish, 361*
Bowfin (Amia calva), 149ff, 150*, 151*, 157*
Bowfins and gars (Holostei), 48, 131, 144ff
Bowline catfish (Corydoras arcuatus), 384, 411*
Brachydanio
— albolineatus (Pearl danio), 310, 318*
— frankei, 309f
— kerri, 309, 318*
— nigrofasciatus, (Spotted danio), 309, 318*
— rerio (Zebra fish), 309, 318*
Brachygalaxias bullocki, 249
Brachymystax, 214
— lenok (Lenok), 233f
Brachyonichthyidae, 402
Brain, 22f, 23*
Bramble shark (Echinorhinus brucus), 116f
Bramble sharks (Echinorhinidae), 116f
Bream (Abramis brama), 319f, 320m, 341*
Bregmaceros, 407, 407*

Bregmacerotidae (Bregmacerotid codfishes), 406f
Bridger, 182
Bristlemouth (Cyclothone signata), 266*
Britton, 282
Brochis, 382
— coeruleus, 411*
Bronze catfish (Corydoras aeneus), 384
Brook lamprey (Lampetra planeri), 35, 37f
Brook trout (Salvelinus fontinalis), 224*, 232
Brosme brosme (Cusk), 407f, 408m, 431*
Brotula, 424
Brown bullhead (Ictalurus nebulosus), 364
Brown trout (Salmo trutta), 218, 218m, 218*, 219*, 224*
Brycon, 279
Bryconinae, 277, 279
Bryssetaeres pinniger, 398
Bubyr, 49
Buchholz, 205
Bückmann, 180
Bugeye goldfish, 361*
Bull ray (Pteromylaeus bovinus), 124
Bull-head (Cottus gobio), 237/238*
Bunocephalidae ("Frying pan catfishes"), 379
Bunocephalus bicolor, 379
Burbot (Lota lota), 256*, 410, 458/459*
Butterfly fish (Pantodon buchholzi), 204f, 285*, 427
Butterfly fishes (Pantodontidae), 202, 204
Butterfly rays (Gymnuridae), 124

Caecobarbus, 337
— geertsi (Congo blind barb), 417*
Caenotropus, 294
Calamoichthys (Reed fishes), 132f
— calabaricus (Reed fish), 132, 140f
Calandruccio, 161
Callichthyidae (Callychtyid armored catfishes), 47, 381f, 382*
Callichthys, 382
— callichthys, 383, 411*
Callorhynchidae, 126
Callorhynchus capensis, 129
Campylomormyrus, 210ff, 212*
"Candiru" (Vandellia), 381
Cannibalism, pre-birth, 88
Capelin (Mallotus villosus), 247f, 248*
Capoeta
— capoeta, 340
— heratensis natio steindachneri, 340
Carapidae (Pearlfishes), 424, 424*
Carapus acus (Pearl fish), 432*
Carassius (Common carps)
— auratus (Goldfish or Johnny carp), 335*, 345ff
— gibelio (Gibel), 64, 345m, 350
— carassius (Crucian carp), 327/328*, 345f, 345m
— humilis (See Carrassius carassius)
Carcharhinidae (Gray sharks), 102
Carcharhinus (Gray sharks or requiem sharks), 106
— melanopterus (Black-tip reef shark), 106
— menisorrah (Reef shark), 107*
— plumbeus (Atlantic gray shark), 106
Carcharias
— ferox (Fierce shark), 91
— taurus (Sand tiger shark), 88, 91f, 112*
Carchariidae (Sand sharks), 91

Carcharodon carcharias (Great white shark), 93*, 95*, 95
Carnegiella
— marthae (Black-winged hatchet fish), 291*
— strigata, 291*
Carp (Cyprinidae), 305f
Carp (Cypriniformes), 276
Carp (Cyprinus carpio), 78, 333*, 344f, 344m
Carp breeding, 78, 78*, 79*
Carp, i.n.s. (Cyprinoidei), 277
Carpet sharks (Orectolobidae), 98f
Cartilaginous fishes (Chondrichthyes), 45, 86ff
Caspialosa, 196
Caspian sand smelt (Atherina mochon), 444
Catfishes (Siluriformes), 363ff
Catla catla (Indian carp), 343
Catlocarpio siamensis, 306, 343
Catoprion mento, 287
Catoprioninae, 287
Catostomidae (Suckers), 353f, 354m
Catostomus catostomus (Longnose sutker), 256*, 354
Caularchus maeandricus (Northern clingfish), 398
Caulophryne, 403
Cavefishes (Amblyopsoidei), 395
Central mudminnow (Umbra limi), 252
Centropristis (Black bass), 64
Cephalaspidomorphi, 40
Cephalaspis, 40
Ceratias, 63, 403
— hollbolli, 404, 421*
Ceratiidae (Ceratiid anglers), 403
Ceratioidei (Deepsea angler fishes), 403
Cercomitus flagellifer, 169
Cetengraulis, 200
Cetomimidae, 264f, 267
Cetomimiformes, 264ff
Cetomimoidei ("Whale-headed" fishes), 264, 267*
Cetorhinidae (Basking sharks), 96
Cetorhinus maximus (Basking shark), 49, 93*, 96f
Chaca chaca, 375f, 393*
Chacidae, 375
Chain pickerel (Esox niger), 251, 265*
Chalcalburnus chalcoides mento, 319
Chalceus, 279
Chanidae (Milkfishes), 269ff
Channel catfish (Ictalurus punctatus), 256*, 364
Chanoidei, 269
Chanos chanos (Milkfish), 158*, 269ff
Chapman, A. B., 138
Characidae (Characins, i.n.s.), 278, 278*
Characidiidae, 289*, 290
Characinae, 277f
Characins (Characoidei), 277, 277m
Characoidei (Characins), 277, 277m
Charax, 278
Charrs (Salvelinus), 214, 230f, 231*
Chauliodontidae (Deepsea viperfishes), 254
Chauliodus sloani (Deepsea viperfish), 223*, 254f, 268
Chaunacidae, 402
Cheirodon axelrodi (Red neon), 282f, 300*
Cheirodontinae, 277, 282
Cheirolepis, 48, 48*
Chela
— cuchius, 325
— caeruleostigmata, 318*

— *dadyburjori*, 324
— *laubuca*, 325
Chelaethiops, 177
Chilodinae, 294
Chilodus punctatus (Headstander), 291*, **294**
Chimaera (Rabbit fishes)
— *monstrosa* (Rabbit fish), 104*, **126**f
Chimaeras (Holocephali), 45, 87, **126**
Chimaeridae, 126
Chinese bitterling *(Rhodeus sericeus sericeus)*, **330**
Chinese sturgeon *(Psephurus gladius)*, **143**
Chinook salmon *(Onchorhynchus tschawytscha)*, 227ff
Chirocentridae (Wolf herrings), 177, **200**f
Chirocentrodon, 200
Chirocentrus dorab (Wolf herring), 158*, **200**f
Chlamydoselachus, 47
— *anguineus* (Frilled shark), **90**, 104*
Chlamydoselachoidei (Frilled sharks), 90
Chlorophthalmidae, 261, 261*
Chologaster, 396
— *cornutus*, 396
Chondrichthyes (Cartilaginous fishes), 45, **86**ff
Chondrostei (Sturgeons and paddlefishes), 48, 131, **134**f
Chondrostoma, 55
— *nasus*, **321**, 321m
Chorisochismus dentex, **398**
Chriopeops goodei (Bluefin killy), **437**
Chrosomus
— *erythrogaster* (South red-bellied dace), **322**
— *neogaeus*, **322**
Chub *(Leuciscus cephalus)*, 314, 341*
Cichlasoma urophthalmus, 396
Ciliata
— *mustela*, **408**
— *septentrionalis*, **408**
Circulation, 25
Cirrhina molitorella, 73
Citharidium, **294**
Citharinidae, **295**
Citharininae, **295**
Citharinus, 294f
— *distichodoides*, **295**
Cladoselache, 47*
Cladoselachii, 47
Clarias, 374
— *lazera*, 399*
Clariidae (Labyrinthic catfishes), 374
Clark, 441
Clausen, 177
Clingfishes (Gobiesocidae), 398
Clupea (Herrings), 177*, 179m, 180*, 181*, 193m
— *fuegensis*, 74
— *harengus* (Atlantic herring), **178**ff, 178m, 179m, 197*
— — *membras*, 72
— *pallasii* (Pacific herring), **178**, 178m, 188f
Clupeacharacinae, 277, **279**
Clupeacharax anchoveoides, **279**
Clupeichthys, 195
Clupeidae (Herring), **177**f
Clupeiformes, 172
Clupeinae, 177
Clupeoidei, 172, **177**
Clupeoides, 195
Clupeonella, 191, 191m
Clupisudis (Osteoglossids or bony-tongues), 202f
— *niloticus* (African bonytongue),

202, **204**
Coates, 297
Cobitidae (Loaches), 355f, 357m
Cobitinae, 356, **359**
Cobitis, 360m
— *elongata* (European loach), **356**
— — *bilseli*, **356**
— *taenia*, 237/238*, **359**f, 371*
Cobitophis, 359
Codfish (Gadidae), **407**, 407m
Codfishes (Gadiformes), **405**ff
Codfishes (Gadoidei), **406**
Codling *(Urophycis tenuis)*, 409
Cod-liver oil, 89
Coelorhynchus carminatus (Soldier fish), **425**
Cold-blooded *(Poikilothermic)*, 26
Colossoma
— *bidens*, **287**
— *oculum*, **287**
Columbia transmontana, 397
Comb-toothed shark *(Hexanchus griseus)*, 90
Comb-toothed sharks (Notidan-oidei), 89f
Commercial fishing, 67ff
Common Atlantic sturgeon *(Acipenser sturio)*, **136**, 145*
Common codfishes *(Gadus)*, 413
Common gudgeon *(Gobio gobio)*, 237/238*, **331**f, 331m, 341*
Common hammerhead *(Sphyrna zygaena)*, 103*, **115**
Common stingray *(Dasyatis pastinaca)*, 117*, **123**
Conger conger (Conger eel), 165, 167*
Conger eel *(Conger conger)*, 165, 167*
Conger eels (Congridae), 160, **165**
Congo blind barb *(Caecobarbus geertsi)*, 417*
Congridae (Conger eels), 160, **165**
Copeina, 290
— *guttata*, 291*
Copella arnoldi, 290, 291*
Coregoninae (Whitefishes), 234
Corydoras, 382ff
— *aeneus* (Bronze catfish), **384**
— *arcuatus* (Bowline catfish), **384**, 411*
— *barbatus*, 411*
— *caudimaculatus*, 411*
— *hastatus* (Micro catfish), **383**f, 411*
— — *julii* (Leopard corydoras), **384**
— *macropterus*, 411*
— *melanistius* (Guiana catfish), **384**, 411*
— *metae*, **384**
— *paleatus*, **384**, 411*
— *punctatus*, **384**
— *pygmaeus*, **383**ff
— *reticulatus*, 411*
— *schultzei*, **384**
— *schwartzi*, **384**
Coregonus (Whitefishes), **235**, 235m, 235*
— *aeronius*, 242f
— *albula* (European whitefish), **240**ff, 255*
— — *artedi* (See C. artedi)
— — *sardinella*, 241
— *artedi* (Shallow water cisco, a.k.a. Lake herring), 240, **241**
— *fera*, **243**
— *lavaretus* (Houting), 241f
— *oxyrhynchus*, 240, **243**, 255*
— *nasus* (Tschirr), 239f, 243
— *pidschian* (See C. acronius)
— *tugun* (Tugun), **239**
Corica, 195
Corynopoma riisei (Swordtail characin), 279, 280*, 286*
Coryphaenoides rupestris (Rock

grenadier), **425**, 432*
Cottus gobio (Bull-head), 237/238*
Couch's whiting or poutassou *(Micromesistius poutassou)*, **420**, 423*
Couciero, 297
Count, 254
Cow sharks (Hexanchidae), 90
Cox, 297
Craig, 183
Crenuchidae, **294**, 294m
Crenuchus spilurus, 50, **294**
Cromeria, 273f
Cromeriidae, 269, **273**f, 273*
Crossopterygii (Lobefin fish), 20*
Crucian carp *(Carassius carassius)*, 327/328*, **345**f, 345m
Crucifix catfish *(Arius proops)*, 370, 386*
Cryptosaras conesi, **404**
Ctenoluciidae, **289**f, 289m
Ctenopharyngodon, 313m, 353
— *idella*, 73, **313**
Ctenothrissiformes, 268
Culter alburnus, 324
Cultrinae (Cultrine carp), **324**, 324m
Curimatidae, **294**
Curimatinae, **294**
Cusk *(Brosme brosme)*, **407**f, 408m, 431*
Cusk eels (Ophidioidei), 424
Cutthroat trout *(Salmo clarki)*, 226
Cyclostomata (Cystostomes), 30f
Cyclothone, 254
— *signata* (Bristlemouth), 266*
Cyema atrum (Deepwater eel), 169*, 170, 170*
Cyemidae, 159, **170**
Cynolebias (Pearl fishes), 282, 436
— *belotti* (Argentine pearl fish), **436**, 455*
— *elongatus*, 454*
— *ladigesi*, **436**, 450*
— *nigripinnis*, **436**, 455*
Cynopotamos, 278
Cyprinidae (Carp), **305**f
Cypriniformes (Carp), 276
Cyprininae, **344**
Cyprinodon
— *nevadensis*, **434**
— *variegatus*, **434**
Cyprinodontidae (Killifishes), 433
Cyprinodontoidei (Toothed carps), 430
Cyprinoidei (Carp, i.n.s.), 277, **305**ff
Cyprinus (Carps)
— *carpio* (Caro), 78, 333*, **344**f, 344m
Cypselurus
— *californicus*, **428**
— *furcatus*, 449*
— *heterurus*, **428**
Cystostomes (Cyclostomata), 30f

Dace *(Leuciscus leuciscus)*, **314**
Daget, 295
Dalatiidae (Spineless dogfishes), 116
Dalatius licha (Darkie Charlie), 116
Dallia pectoralis (Alaska blackfish), **252**, 255*
Dalliidae, 250
Danio
— *devario*, **309**
— *malabaricus* (Giant danio), **309**, 318*
— *regina*, **309**
Danioninae (Danois), 306, 306m
Danois (Danoininae), 306, 306m
Danube salmon *(Hucho hucho)*, 233, 233*, 233m
Dapedius, 48*
Darkie Charlie *(Dalatius)*, **116**

Darwin, Charles, 30
Dasyatidae (Stingrays), 123
Dasyatis
— *centroura* (Stingray), 123*
— *pastinaca* (Common stingray), 117*, **123**
— *sabina*, 124*
— *violacea* (Blue stingray), **124**
Dean, 33
De Beaufort, 169
Deepsea angler *(Linophryne arborifer)*, **403**, 421*
Deepsea anglerfishes (Cerati-oidei), **403**
Deepsea ateleopids (Ateleopo-doidei, Ateleopodidae), 264, 264*
Deepsea bonefishes (Ptero-thrissus), **156**
Deepsea bristlemouths (Gonosto-matidae), **253**f
Deepsea catfishes (Moridae), **406**
Deepsea hatchet fish *(Argyropelecus affinis)*, 266*
Deepsea hatchet fishes (Sternop-tychidae), **254**
Deepsea herring *(Dolichopteryx)*, **258**
Deepsea loosejaw *(Ultimostomias mirabilis)*, **253**
Deepsea scaleless dragon fishes (Melanostomiatidae), **253**
Deepsea scaly dragonfishes (Sto-miatidae), **253**
Deepsea slickheads (Alepocepha-loidei), 213, **258**
Deepsea smelts (Bathylaconoidei), 213, **258**
Deepsea smelts (Bathylagidae), **252**
Deepsea snaggletooths (Astron-esthidae), **253**, 253*
Deepsea stalkeye fish *(Idiacanthus fasciola)*, 266*
Deepsea stalkeye fishes (Idia-canthidae), **253**
Deepsea viperfish *(Chauliodus sloani)*, 223*, **254**f, 268
Deepsea viperfishes (Chauliodon-tidae), **254**
Deepwater eel *(Cyema atrum)*, 169*, **170**, 170*
Delage, Yves, 160
Dentriceps clupeoides (Denticipitoid herring), **172**
Denticipitoid herring *(Denticeps clupeoides)*, **172**
Denticipitoidei, 172
Dermogenys, 428
— *pusillus*, **429**, 451*
Devold, 184
Devilfish *(Mobula mobular)*, **125**
Dianema longibarbis, 411*
Diaphus elucens, 266*
Diceratiidae, 403
Digestive tract, 62
Dinichthys, 46*
Dipterus, 47*
Distichodontinae, **295**
Distichodus sexfaciatus, **295**, 308*
Dixonina nemoptera, 156
Dog salmon *(Oncorhynchus keta)*, 227f, 256*
Dogfishes (Scyliorhinidae), 99f, 112*
Dolichopteryx (Deepsea herring), **258**
Doradidae (Doradid armored cat-fishes), **378**
Doras costatus, **378**
Dorosoma (Gizzard shad), 195
Dorosomatinae, 177
Dorypterus, 48
Downsherring, or Eastern channel

herring, 179, 185
Drepanaspis, 40, 42*
Dussumieria, 196
Dussumieriinae, 177
Dwarf catfishes *(Mystus)*, 365

Eagle rays (Myliobatoidei), 123
East Siberian cod *(Arctogadus borisovi)*, 416, 416*
Echeneis naucrates (Remora), 103*, 113*
Echidna
— *nebulosa*, 168*
— *zebra* (Zebra moray), 164, 174*
Echinorhinidae (Bramble sharks), 116f
Echinorhinus brucus (Bramble shark), 116f
Echo dispersal region, 74, 74*
Echolocation, 68, 69*
Eel cods (Muraenolepioidei), 406, 406*
"Eel mothers" *(Zoarces)*, 424
Eels (Anguilliformes), 159ff
Egg cells, 28, 28*
Eggs, 27f
Eibl-Eibesfeldt, Irenäus, 166
Eigenmann, 249, 395f
Eigenmannia, 298, 304
— *virescens*, 304, 317*
Elachocharax, 290
Elasmobranchii, 45, 87, 89ff
Electric catfish *(Malapterurus electricus)*, 296, 376
Electric eel *(Electrophorus electricus)*, 296, 303, 317*
Electric eels (Gymnotoidei), 277, 295f, 297m
Electric field, 206*
Electric organ, 206, 296
Electric rays (Torpedinoidei), 121
Electric shocks, 56
Electrophoridae, 296, 303
Electrophorus electricus (Electric eel), 296, 303, 317*
Eleginus, 415
— *gracilis* (Far Eastern navaga), 416, 416*
— *navaga* (Wachna cod or navaga), 416
Ellis, 298
Elopichthys bambusa (Sheltostshek), 311
Elopidae, 152, 153m
Elopiformes (Tarpons), 152
Elopoidei, 152f
Elops saurus (Lady fish), 152f, 153*, 158*
Embryonic development, 64f
Empetrichthyidae, 430
Enchelyopus cimbrius, 408
Engraulicypris, 306f
Engraulidae (Anchovies), 177, 196
Engraulis (Anchovies), 196, 199m
— *anchoveta*, 74
— — *capensis* (South African anchovy), 200
— *encrasicholus* (Anchovy), 196f
— *japonicus* (Japanese anchovy), 69, 200
— *mordax* (North American anchovy), 78, 200
— *ringens* (Anchoveta), 68, 200
Epalzeorhynchus
— *kallopterus*, 362*
— *siamensis*, 362*
Epinephelus, 55
Epiplatys
— *annulatus*, 450*, 455*
— *dageti*, 437
— — *monroviae*, 450*
— — *sheljuzkoi*, 450*
Eremichthys, 324
Erythrinidae, 289

Erythrinus erythrinus, 289
Erythroculter
— *illishaeformis*, 324f
— *mongolicus*, 325
Esocidae (Pikes), 250, 250m
Esocoidei, 250
Esomus
— *danrica* (Flying barb), 310, 318*
— *malayensis*, 310, 318*
Esox (Pikes)
— *americanus*, 251
— *lucius* (Northern pike), 223*, 250f, 265*, 327/328*
— *masquinongyi* (Muskellunge), 251, 265*
— *niger* (Chain pickerel), 251, 265*
— *reicherti* (Amur pike), 251
— *vermiculatus* (Grass pickerel), 251, 265*
Ethmalosa, 195
Ethmopterus spinax (Latern shark), 116
Etremeus, 196
Eugnathichthys, 295
Euleptorhamphus viridis, 428
Eupharyngidae (Gulpers), 170, 170*, 267
Eupharynx
— *pelecanoides*, 171, 176*
— *richardi*, 171
Eurasian catfishes (Siluridae), 366
European bitterling *(Rhodeus sericeus amarus)*, 301*, 327/328*, 329f, 371*
European charr *(Salvelinus alpinus)*, 231f
European freshwater eel *(Anguilla anguilla)*, 160ff, 160m, 167*
European grayling *(Thymallus thymallus)*, 237/238*, 244f, 255*
European loach *(Cobitis elongata)*, 356
European moray eel *(Muraena helena)*, 165, 168*, 174*
European mudminnow *(Umbra krameri)*, 251f
European roach *(Rutilus rutilus)*, 312, 312m
European smelt *(Osmerus eperlanus eperlanus)*, 246f
European sturgeon *(Huso huso)*, 135
European wels *(Silurus glanis)*, 366ff, 394*
European whitefish *(Coregonus albula)*, 240ff, 255*
Eutaeniophoridae, 264
Eutheria (Higher mammals), 20*
Eutropiella debauvi, 368f
Evermannellidae (Sabertooth fishes), 263
Evolution, 19
Excretory organs, 62
Exocoetidae (Flying fishes), 288, 426, 449*
Exocoetoidei (Flying fishes), 213, 426
Exocoetus, 426f
— *obtusirostris*, 428
— *volitans*, 428, 449*
Exodon, 278
Eyed electric ray *(Torpedo torpedo)*, 122
Eyes, 23, 24*, 59

False cat sharks (Pseudotriakidae), 102
False tetra *(Hyphessobrycon simulans)*, 283
Far Eastern navaga *(Eleginus gracilis)*, 416, 416*
Faraday, 296
Farlowella, 391
— *acus*, 412*
Featherback family (Notop-

teridae), 205
Feathertail tetra *(Phenocogrammus interruptus)*, 289, 308*
Fertilization, 64
Fessard, 297
Fielder, Arkady, 378
Fierasferidae, see Carapidae
Fierce shark *(Carcharias ferox)*, 91
Fins, 58
Fire killy *(Nothobranchius rachovi)*, 436, 450*, 455*
Fischer, Wolfgang, 203, 377
Fish colorations, 55
Fish forms, 50
Fish markings, 75f, 75*
Fisher, 249
Fisheries, 65f
Fishes (Pisces), 45ff
Fishing devices, 74
Fishing yields, 65f, 65*
Fjord cod, see *Gadus ogac*
Flag fish *(Jordanella floridae)*, 434, 450*
Flapper skate *(Raja batis)*, 123
Flying barb *(Esomus danrica)*, 310, 318*
Flying halfbeaks *(Oxyporhamphus)*, 428
Flying fishes (Exocoetidae), 288, 426, 449*
Fodiator, 428, 449*
Food chain, 66, 66*
Forkbeards *(Phycis)*, 409, 409m, 409*
Forkbeards *(Urophycis)*, 409, 409m, 409*
Foster, 444
Four-eyed fish *(Anableps anableps)*, 436*, 438, 456/457*
Free-swimming whitefish *(Coregonus)*, 235f, 239*
Freshwater eels (Anguillidae), 160
Freshwater lamprey *(Lampetra fluviatilis)*, 36, 38, 41*, 237/238*
Fridriksson, 183
Frilled shark *(Chlamydoselachus anguineus)*, 90, 104*
Frilled sharks (Chlamydoselachoidei), 90
Frisch, Karl von, 273, 315
"Frog codfish" *(Raniceps raninus)*, 408f, 432*
Frogfishes (Antennarioidei), 402
Frostfish, see Atlantic tomcod *(Microgadus tomcod)*
"Frying pan catfishes" (Bunocephalidae), 379
Fuhrmann, 204
Fundulidae, 433, 435
Fundulus chrysotus (Goldear killy), 450*

Gadiculus (Silvery pouts), 423, 423*
— *argenteus*, 423
— *thori*, 423
Gadidae (Codfish), 407, 407m
Gadiformes (Codfishes), 405ff
Gadoidei (Codfishes), 406
Gadus (Common codfishes), 413
— *macrocephalus*, (Pacific cod), 414m, 415
— *morhua* (Atlantic cod), 41*, 413ff, 414m, 431*
— *ogac* (Greenland cod or fjord cod), 414m, 415, 415*
Gaidropsarus, 407m, 408*
— *capensis*, 408
— *mediterraneus*, 408
— *novaezealandiae*, 408
— *pacificus*, 408
Galatheathauma, 56
Galaxias
— *alepidotus*, 249
— *attenuatus*, 249

— *maculatus*, 249
Galaxiidae (Galaxiids), 248f, 249*
Galaxiids (Galaxiodei), 91, 213, 248, 249m
Galaxioidei (Galaxiids), 91, 213, 248, 249m
Galeocerdo cuvieri (Tiger shark), 105
Galeoidei, 91
Galeorhinus galeus (Tope), 94*, 105f
Galeus melanostomus (Blackmouthed dogfish), 100f
Gambusia (Gambusias), 51*, 440, 443
— *affinis affinis*, 443
— — *holbrooki*, 452*
Garden eels (Heterocongridae), 160, 166, 166*
Gar fish *(Belone bellone)*, 429
Garra (Sucking barbs), 337, 340
— *taeniata*, 362*
Gars (Lepisosteidae), 144f, 147*
Gasteropelecidae (Hatchet fishes), 288, 288m, 288*
Gasteropelecus sternicla, 307*
Gasterosteus aculeatus (Three-spined stickleback), 327/328*
Gastromyzon (Suckerbelly loaches), 355
— *borneensis*, 362*
Geisler, R., 282
Gemminger bones, 206
Genital organs, 63
Genyomyrus, 211, 212*
— *donnyi*, 211
Genypterus, 424, 424*
Geotria (Lampreys), 34
Gerlach, Richard, 284
Gesner, Conrad, 241
Giant catfish *(Pangasianodon gigas)*, 369
Giant danio *(Danio malabaricus)*, 309, 318*
Giant or Brazilian arapaima *(Arapaima gigas)*, 49, 202f, 207*
Gibel *(Carassius auratus gibelio)*, 64, 345m, 350
Gigantactis, 403*
Gigantura vorax, 268
Giganturidae, 268, 268*
Giganturoidei (Giganturids), 264
Gilbertolus, 278
Gill, 160
Gill breathing, 61
Gizzard shad *(Dorosoma)*, 195
Glandulocaudinae, 277, 279
Glass carp *(Horaichthys setnai)*, 436*, 437f
Glyptothorax trilineatus, 373
Gmelin, 160
Gnathocharax, 278, 278*
Gnatholemus, 293
Gnathonemus, 211
— *petersi*, 208*, 212
— *tamandua*, 223*
Gnathopogon coerulescens (Biwa Lake gudgeon), 331
Gobiesocidae (Clingfishes), 398
Gobiesociformes, 398
Gobio (Gudgeons)
— *albipinnatus*, 332
— *gobio* (Common gudgeon), 237/238*, 331f, 331m, 341*
— *kessleri*, 332
— *uranoscopus*, 237/238*, 332
Gobiobotia, 306, 330ff
Gobioninae (Gudgeons), 330
Goblin sharks (Scapanorhynchidae), 92
Goldear killy *(Fundulus chrysotus)*, 450*
Golden pheasant killy *(Aphyosemion sjostedti)*, 436, 450*
Goldfish breeds, 334*, 361*
Goldfish or Johnny carp *(Carassius*

auratus), 335*, **345ff**
Gonichthys couvi, 259
Gonopodium, **439***
Gonorynchidae, 274
Gonorynchiformes, 269
Gonorynchoidei, 269, **274**
Gonorynchus gonorynchus, 158*, **274**f
Gonostomatidae (Deepsea bristle-mouths), 253f
Goodeidae (Goodeid top-minnows), 436m, **438**
Goosefishes (Lophiidae), **401**
Gordon, M., 441, 443
Gorgasia maculata, 173*
Gosline, W. A., 163, 166
Grass pickerel *(Esox vermiculatus),* 251, 265*
Grassé, 274
Grasseichthyidae, 274
Grasseichthys gabonensis, 274
Grassi, 161
Gray sharks (Carcharinidae), 102
Gray sharks or requiem sharks *(Carcharhinus),* 106
Graylings (Thymallinae, *Thymallus),* 244, 244m
Gray's cutthroat eel *(Synaphobranchus pinnatus),* 169
Great blue shark *(Prionace glauca),* 102f, 103*, 108/109*
Great crested newt *(Triturus cristatus),* 327/328*
Great hammerhead *(Sphyrna mokkaran),* **106**f
Great silver water-beetle *(Hydrous piceus),* 327/328*
Great white shark *(Charcharodon carcharias),* 93*, 95*, **95**
Greenland cod or fjord cod *(Gadus ogac),* 414m, **415**, 415*
Greenland shark or large sleeper *(Somniosus microcephalus),* **116**
Greenwood, 159, 166
Grenadiers (Macrouroidei), 425, 425*
Grevé, C., 141
Grunion *(Leuresthes tenuis),* 445f, 445*
Gudgeons (Gobioninae), 330
Guiana catfish *(Corydoras melanistius),* 384, 411*
Guitar fish *(Rhinobatos cemiculus),* 121
Guitar fishes (Rhinobatoidei), 120f
Gulpers (Eupharyngidae), **170**, 170*, 267
Guppy *(Poecilia reticulata),* 51*, **440**f, 451*, 460*
Gymnallabes, 374f
— *typus,* **375**
Gymnarchid *(Gymnarchus niloticus),* 206, 206*, 208*, **212**, 212*
Gymnarchidae, **212**
Gymnarchus niloticus (Gymnarchid), 206, 206*, 208*, **212**, 212*
Gymnelis viridis, **425**
Gymnocharacinus, 282
Gymnorhamphichthys hypostomus, 304
Gymnotidae, **303**
Gymnotoidei (Electric eels), 277, 295f, 297m
Gymnotus carapo, 296, 298, **303**
Gymnura altavela, 124
Gymnuridae (Butterfly rays), 124
Gynogenesis, 442
Gyrinocheilidae, **354**, 354m
Gyrinocheilus aymonieri (Siamese gyrinocheilid), 354f, 362*
Gyrodus, 48

Haddock *(Melanogrammus aeglefinus),* **419**, 419m, 431*
Haemopis sanguisuga (Leech), 327/328*

Hagfishes (Myxini), 20, 31, 35*
— (Myxinidae), 31, 31m
Hake *(Merluccius merluccius),* **423**, 423m
Hakes (Merlucciidae), **423**
Halichoeres centriquadrus (Wrass), 128*
Halieutaea retifera, **403**
Holosauropsis macrochir, 176*
Hammerhead sharks (Sphyrnidae), 106
Harengula, 190m, 191
— *zunasi,* **191**
Harms, 30
Harpodon (Bombay duck), **261**, 261*
Harpodontidae (Bombay ducks), 261
Harriotta raleighana, **129**
Hartt, A., 83
Hass, 193
Hatchet fishes (Gasteropelecidae), **288**, 288m, 288*
Head-and-tail-light *(Hemigrammus ocellifer),* 280, 286*
Headstander *(Chilodus punctatus),* 291*, **294**
Hediger, 343
Heincke, 178
Hemibrycon, 280
Hemiculter leucisculus, **324**
Hemicyclaspis, 42*
Hemigrammus, 280
— *ocellifer* (Head-and-tail-light), 280, 286*
— *pulcher,* 286*
— *rhodostomus,* 286*
Hemiodontidae, **293**
Hemiodontinae, 293
Hemiodus semitaeniatus, 291*
Hemirhamphus brasiliensis, **428**
Hemistichodus, **295**
Hems, 349
Hepsetidae, **289**
Hepsetus odoe, 278, **289**
Heptotranchias perlo, (Narrow-headed seven-gilled shark), **90**
Herald, 171, 254, 260, 262f, 401, 403
Herre, Albert W. C. T., 447
Herrings *(Clupea),* 177*, 179m, 180*, 181*, 193m
Hervey, 349
Hetevandria, 440
— *formosa* (Mosquito fish), **443**, 452*
Heterocongridae (Garden eels), 160, 166, 166*
Heterodontidae (Horn sharks), 90f
Heterodontus (Horn sharks), 47, **91**
— *japonicus* (Japanese horn shark), 91, 91*
Heteropneustes fossilis (Indian catfish or stinging catfish), 375, 393*
Hexanchidae (Cow sharks), 90
Hexanchus, 47
— *corinus* (Six-gilled shark), **90**
— *griseus* (Comb-toothed shark), 90
Hildebrand, S. F., 152, 155f
Hildebrandichthys, 200
Hillstream loaches (Homalopteridae), 354m, **355**
Hilsa, 195
Himantolophidae, 404
Himantolophus groenlandicus, **404**
Hiodon (Mooneyes)
— *alosoides,* 205
— *salenops,* 205
— *tergisus* (Mooneye), **205**
Hiodontidae (Mooneyes), 202, **205**
Hippopotamyrus, 211
Histrio histrio, 402
Hjort, 258

Hokkaido-Sakhalin-Herring, 188
Holliday, 185
Holocephali (Chimaeras), 45, 87, **126**
Holostei (Bowfins and gars), 48, 131, **144**ff
Homalopteridae (Hillstream loaches), 354m, **355**
Homaloptera, **355**
Homeiothermic (Warm-blooded), 26
Hoplerythrinus unitaeniatus, **289**
Hoplias malabaricus, **289**, 307*
Hoplo catfish *(Hoplosternum thoracatum),* **383**, 411*
Hoplocharax goethei, **278**
Hoplosternum, 382
— *littorale,* **383**
— *thoracatum* (Hoplo catfish), **383**, 411*
Horaichthyidae, 437
Horaichthys setnai (Glass carp), 436*, **437**f
Hormones, 26
Hornsharks *(Heterodontus),* 90f
Houting *(Coregonus lavaretus),* **241**f
Hubbs, C. L., 396
Hubbsiella sardina (Sardine silverside), **446**
Hucho, 214
— *hucho* (Danube salmon), **233**, 233*, 233m
— *taimen,* **233**
Humantin *(Oxynotus centrina),* **115**
Humboldt, Alexander von, 283, 296
Huso
— *dauricus* (Kaluga), **136**
— *huso* (European sturgeon), **135**
Hybodontoidea, 47
Hybopsis, 323
Hydrocininae, 289
Hydrocinus
— *goliath,* **289**
— *lineatus,* **289**
Hydrocynidae, 288, 289*
Hydrolagus colliei, 128*
Hydrolycus, 278
Hydrous piceus (Great silver water-beetle), 327/328*
Hymenophysa, see *Botia berdmorei, B. horae, B. hymenophysa,* and *B. sidthimunki*
Hyperlophus, 195
Hyperopisus bebe, 208*, **211**
Hyphessobrycon, 280
— *callistus,* 286*
— *erythrostigma* (Tetra Perez), **282**
— *flammeus* (Red tetra), 280, 286*
— *heterorhabdus,* 286*
— *ornatus,* **280**
— *rubrostigma,* 286*
— *simulans* (False tetra), **283**
Hypomesus, 246
Hypophthalmichthyinae, **353**, 353m
Hypophthalmichthys
— *molitrix,* **353**
— *nobilis,* 73
Hypopomus, 303*, **304**
Hyporhamphus unifasciatus, **428**
Hyrcanogobius bergi (Goby fish), 49
Hyrth, 212

Icefishes (Salangidae), 49, **248**, 248*
Ichthyoborinae, 295, 295*
Ichthyoborus, **295**
Ichthyomyzon fossor (American brook lamprey), 35
Ictaluridae (North American fresh-water catfishes), 364
Ictalurus
— *furcatus,* **364**

— *nebulosus* (Brown bullhead), **364**
— *punctatus* (Channel catfish), 256*, **364**
Ide or Orfe *(Leuciscus idus),* **314**f, 341*
Idiacanthidae (Deepsea stalkeye fishes), 253
Idiacanthus, 257
— *fasciola* (Deepsea stalkeye fish), 266*
Idus, subgenus of *Leuciscus*
— *idus,* 314
Iguanodectinae, 279
Ilisha, 195f
Iguanodectinae, 279
Ilyophidae, **169**
Ilyophis brummeri, **169**
Inconnu *(Stenodus leucichthys),* 234f, 255*
Indian barb *(Barbus mosal),* 55
Indian carp *(Catla catla),* 343
Indian catfish or stinging catfish *(Heteropneustes fossilis),* 375, 393*
Indian putykanipa *(Muranesox talabon),* **165**
Indian sardine, see *Sardinella*
Ipnopidae, 262
Iranocypris, 337
Isichthys, **211**, 211*
Iso, 446
Isonidae, **446**
Isuridae (Mackerel sharks), 92f
Isurus (Mako sharks), 95
— *oxyrhynchus* (Mako shark), **96**, 96*

Jacksmelt silverside *(Atherinopsis californiensis),* **445**
Jacobson's organ or vomeronasal organ, 23
Jamoytius, 40
— *kerwoodi,* 43
Japanese anchovy *(Engraulis japonicus),* 78, **200**
Japanese eel *(Anguilla japonica),* 163
Japanese goblin shark *(Scapanorhynchus owstoni),* 92, **92**
Japanese hagfish *(Bdellostoma burgeri),* 33
Japanese horn shark *(Heterodontus japonicus),* 91, 91*
Japanese killifishes (Oryziatidae), **437**
Japanese sardine *(Sardinops melanosticta),* 191f, 192*
Jarvik, E., 43
Jawless fishes (Agnatha), 30ff
Jenkinsia, 196
Jenynsia lineata, **438**
Jenysiidae (Jenynsiid top-minnows), **438**
Johnson, 185
Johnson's black anglerfish *(Melanocetus johnsoni),* **404**, 421*
Jordanella floridae (Flag fish), **434**, 450*

Kaluga (Huso dauricus), **136**
Kasidoridae, 264
Katsuwonus (Oceanic bonito), 54
Kaup, 161
Kidney system, 25f, 27*
Killies *(Aphyosemion),* 282, 435f
Killifishes (Cyprinodontidae), **433**
Klausewitz, 166, 379
Klunzinger, 201
Knaak, 382
Kneria, **271**
— *polli,* 285*
Kneriidae (Kneriids), 269, **271**ff, 272*
Kneriids (Kneriidae), 269, **271**ff, 272*
Kobbengrund herring, 179

Merry widow (Phallichthys amates), **443**f, **451***

Metamorphosis in lamprey larvae, 34ff, **36***

Metynnis, 287
— roosevelti, **307***
— schreitmuelleri, **307***

Mexican tetra (Astyanax fasciatus mexicanus), 280f, **291***

Micro catfish (Corydoras hastatus), **383**f, **411***

Microdon, 48

Microgadus, 415
— tomcod (Atlantic tomcod or frostfish), **416**, **419***
— proximus (Pacific tomcod), **416**

Microglanis parahybae, **380**

Micromesistius
— australis, **420**
— poutassou (Couch's whiting or poutassou), **420**, **423***

Microphysogobio, 330

Midshipmen (Porichthys), 56, 397

Milkfish (Chanos chanos), 158*, **269**f

Milkfishes (Chanidae), 269ff

Mimagoniates microlepis, **281***

Minch or Scottish West Coast herring, 179f

Minnow (Phoxinus phoxinus), **315**f, 315m, **371***

Minnows (Leuciscinae), 311ff, 311m

Mirapinnatoidei ("Wonder finned fishes"), 264

Mirapinnidae, 264

Misgurnus, 356, 359, 359m
— erikssoni, 356
— fossilis (Pond loach), 327/328*, **359**, **371***

Mistichthys luzonensis (Goby), 49

Moapa, 323f

Mobula mobular (Devil fish), **125**

Mobulidae (Mantas), 125

Mochocidae (Upside-down catfishes), **376**f

Moenkhausia, 280

Mohr, 183

Molliensia, (See Poecilia)

Molo, (See Merlangius merlangus), 420

Molva (Ling)
— elongata (Blue ling), **413**, 413m
— dipterygia, **413**, 413m
— molva (Ling), **410**, 413*, 413m

Mondini, 160

Monk fishes (Squatinidae), 119

Monognathidae, **171**

Mooneye (Hiodon tergisus), **205**

Mooneyes (Hiodontidae), 202, **205**

Moonfish or platy (Xiphophorus maculatus), **439***, **442**, **452***, **454***

Mora, 406

Morays (Muraenidae), 160, **164**

Mordacinae, 34

Moridae (Deepsea codfishes), **406**

Moringua
— bicolor, **163**
— javanica, **163**f
— macrochir, **163**

Moringuidae, 160, **163**f

Mormyridae, **210**

Mormyridiformes (Mormyrids), 202, **205**, 205m

Mormyrids (Mormyridiformes), 202, **205**, 205m

Mormyromast, 209

Mormyrops, 209, **211***
— boulengeri, **211**

Mormyrus, **211**, 211*
— kannume, **209**

Morris, W., 160

Mosquito fish (Heterandria formosa), **443**, **452***

425*

Macrourus berglax (Smooth-spined rat-tail), **425**

Macrozoarces americanus (Mutton fish), **425**

Mako shark (Isurus oxyrhynchus), 96, 96*

Mako sharks (Isurus), 95

Malabar killy (Aplocheilus lineatus), **450***

Malacosteidae, 253

Malapteruridae (Electric catfishes), **376**

Malapterurus electricus (Electric catfish), 296, **376**

Malayan bonytongue (Sceropages formosus), **204**

Mallotus, 246
— villosus (Capelin), **247**f, **248***

Mammoth cave blindfish (Amblyopsis spelaeus), **396**, **417***

Mangold, 241

Manta birostris (Manta ray), 49, 118*, **125**, 127*

Manta ray (Manta birostris), 49, 118*, **125**, 127*

Mantas (Mobulidae), 125

Marbled electric ray (Torpedo marmorata), 117*, **112**

Marcgraf, Georg, 296

Marcoy, 203

Marcusenius (Parrot mormyrid), **210**, 210*
— isidori, **210**
— monteiri, **210**
— montede, **210**
— sphecodes, **210**
— stanleyanus, **210**

Marshall, 254, 257f, 404

Masu (Oncorhynchus masou), 227

Matle, Paul, 349

Matlhes, 209, 295

Maurolicus, 254
— muelleri, 266*

Meda, 323f

Mediterranean cod (See Merlangius merlangus), 470

Megalamphodus, 282

Megalobrama, 324

Megalopidae (Tarpons), 153f

Megalops (Tarpons), 153f, 154*
— atlanticus (Atlantic tarpon), **153**, 153m
— cyprinoides (Ox-eye herring), 153, 153m, 155, 158*

Melanocetidae, 404

Melanocetus johnsoni (Johnson's black anglerfish), **404**, 421*

Melanogrammus aeglefinus (Haddock), **419**, 419m, 431*

Melanostomiatidae (Deepsea scaleless dragon-fishes), 253

Melanotaenia
— maccullochi, **444**, 453*
— nigrans (Australian rainbow fish), **444**, 453*, 456/457*

Melanotaeniidae (Rainbow fishes), 444

Melostoma temmincki, 81

Menidia, 446

Merlangius merlangus (Whiting), 69, **420**, 420*

Merlucciidae (Hakes), **423**

Merluccius
— australis, 423m, **423**
— bilinearis (American hake), **423**, 423m, 431*
— capensis (Stockfish), 69, **423**, 423m
— gayi, **423**, 423m
— hubbsi, 74, 423m, **423**
— merluccius (Hake), **423**, 423m
— productus, 423m, **423**

— fasciatus, **291***

Leptobotia, 356, **358**

Leptocephalus, **153**, 153*
— brevirostris (See European eel), **161**
— morrisii (See Conger eel), 160

Leptolepsis, 48

Lestidium, 262

Leucaspius dilineatus, **319**, **341***

Leuciscinae (Minnows), 311ff, 311m

Leuciscus, 313f
— cephalus (Chub), **314**, 341*
— Idus (Ide or Orfe), **314**f, 341*
— forma orfus, **315**
— leuciscus (Dace), **314**
— souffia agassizi, **314**

Leuresthes tenius (Grunion), **445**f, **445***

Lile, 195

Limbs, 20f, 20*, **21***

Limnaca (Snail), 327/328*

Linea lateralis (Lateral line), 60

Ling (Molva molva), **410**, 413*, 413m

Linné, Carl, 45

Linophryne arborifer (Deepsea angler), **403**, 421*

Linophrynidae (Linophrynid anglers), 403

Lion's head goldfish, 334*, 348, 361*

Lissmann, 209, 297, 376

Little Red Riding Hood goldfish, 361*

Live bearers (Poecilia), 440

Lizard fishes (Synodontidae), 260

Loaches (Cobitidae), 355f, 357m

Locomotion, 20, 20*

Lohnisky, 37

Longfin (Pterolebias longipinnis), **436**, **450***

Longnose gar (Lepisosteus osseus), **147**, 157*

Longnose sucker (Castostomus catostomus), 256*, **354**

Lophiidae (Goosefishes), **401**

Lophiiformes (Angler fishes), 398f

Lophioidei, 401

Lophius, 58, **401***
— piscatorius (Allmouth or Angler), **401**f

Loricaria (Whiptail loricaria), **390**f
— parval, **391**, 412*

Loricariidae (Loricarrid armored catfishes), **389**, **390***

Lota (Burbots)
— lota (Burbot), 256*, **410**, 458/459*
— leptura, 410
— lota, 410
— maculosa (Alekey trout or eel-pout), 410

Lotella, 406

Lucifuga subterranea, **424**

Lucino (Belonesox belizanus), **443**, 452*

Luecken, 433

Lüling, 202f, 289

Luminescence, 56

Lycenchelys, 425

Lycodes esmarki, 425

Lycondontis javanicus, 175*

Lyretail (Aphyosemion australe), **436**, 450*, 454*

MacCulloch, 249

Machin, 376

Mackenzie, 249

Mackerel sharks (Isuridae), 92f

Macristium chavesi, 268f, 269*

Macrouroidei (Grenadiers), 425,

Kofoed, 143

Kosswig, 280, 433

Kosswig, C., 443

Kowala, 195

Kyrptopterus bichirrhis, 368, 388*, 393*

Labeo, **340**f
— bicolor, **343**
— erythrura, **343**
— velifer, **343**
— wecksi, 352*

Labyrinth (inner ear), 23, 25*

Labyrinthic catfishes (Clariidae), 374

Ladiges, Werner, 36, 287f, 366, 370, 374

Lady fish (Elops saurus), 152f, 153*, 158*

Laemonema, 406

Laevoceratiidae, see Diceratiidae

Lake charr (Salvilinus namaycush), **232**f

Lake herring, see Shallowwater cisco

Lake sturgeon (Acipenser fulvescens), 137*, **138**f

Lake trout (Salmo trutta lacustris), 218ff, 224*

Lamna (Porbeagle sharks), 88
— nasus (Probeagle shark), **95**f, 104*

Lamnidae, see Isuridae

Lampanyctus
— erocodilus, 404
— leucopsarus, **259**

Lamprey (Lamprey), 19*

Lampetra (Lamprey)
— fluviatilis (Freshwater lamprey), 36, 38, 41*, 237/238*
— planeri (Brook lamprey), 35, 37f

Lampreys (Petromyzones), 20, 31
— Petromyzonidae, 31, **34**f, 34m, 35*, 36*, 37*

Lancetfishes (Alepisauridae), 263, 263*

Lanternfishes (Myctophoidei), 213, **258**

Lanternfishes, i.n.s. (Myctophidae), 258f

Larger putyekanipa (Muraenesox tababonoides), **165**

Larger spotted dogfish (Scyliorhinus stellaris), **100**, 104*

Lantern shark (Etmopterus spinax), **116**

Latrunculus, 49

Lebiasina, 290

Lebiasinidae, **290**

Lebour, 185

Leech (Haemopis sanguisuga), 327/328*

Leiocassis, 365
— siamensis, **366**, 393*

Lenok (Brachymystax lenok), **233**f

Lepadogaster bimaculatus, 398

Lepidion, **406**

Lepidocephalus, 356
— guntea, **360**
— thermalis, 372*

Lepidotus, 48

Lepisosteidae (Gars), 144f, 147*

Lepisosteiformes, 144, 147m

Lepisosteus (Gar pikes)
— oculatus, **147**
— osseus (Longnose gar), **147**, 157*
— platostomus, **147**
— platyrhincus, **147**
— productus, **147**
— spatula (Aliigator gar), **147**, 157*
— tristoechus, **147**
— tropicus, **147**

Leporinus, **293**

Mouth-breeding catfishes (*Tachysuridae*), **369**
Mud hake (*Urophycis chuss*), **409**
Mudminnows (Umbridae), 250, **251**f
Müller, Johannes, 45
Muraena helena (European Moray), 165, **168***, **174***
— *paradalis*, **164**f, **164***
Muraenesocidae (Pike eels), 160, **165**
Muraenesox
— *cinereus* (Batavia putyekanipa), **165**
— *talabon* (Larger putyekanipa), **165**
— *talabonoides* (Indian putyekanipa), **165**
Muraenidae (Morays), 160, **164**
Muraenolepidae, **406**
Muraenolepioidei (Eel cods), **406**
Muraenolepis (Eel cod), **406**, **406***
Musculature, 58
Muskellunge (*Esox masquinongyi*), **251**, **265***
Mussehenbroek, van, 296
Mussel (*Anodonta*), **327**/**328***
Mustelus
— *asterias* (Stellate smooth hound), **101**
— *mustelus* (Smooth hound), **101**
Mutton fish (*Macrozoares americanus*), **425**
Myctophidae (Lantern fishes, i.n.s.), **258**f
Myctophoidei (Lantern fishes), 213, **258**
Myctophum
— *affine*, **259**f
— *phengodes*, 266*
— *punctatum*, **259**
Myleinae, 287
Myleus pacu, **287**
Myliobatidae, 124
Myliobutis, 125*
— *aquila* (Spotted eagle ray), 118*, **124**
Myliobatoidei (Eagle rays), 123
Mylopharyngodon piceus, 313
Myloplus schultzei, 307*
Mylossoma, 287
Myomyrus, 211
Myroconger compressus, **164**
Myrocongridae, 160, **164**
Mystus (Dwarf catfishes), 365
— *nemurus*, 365
— *tengara*, 365
— *vittatus* (Striped dwarf catfish), 365, 393*
Myxine glutinosa (North Atlantic or Common hagfish), **31**ff, 32*, 33*, 41*
Myxini (Hagfishes), 20, 31, 35*
Myxinidae (Hagfishes), 31, 31m
Myxininae, 31

Nannaethiops
— *tritaeniatus*, 308*
— *unitaeniatus*, 308*
Nannocharax, **290**, 295
Nannostominae (Pencil fishes) **290**f
Nannostomus, 293
— *anomalus*, 286*
— *beckfordi*
— — *beckfordi*
— — *anomalus*
— — *aripirangensis*, 286*
Narkidae, 121
Narrow-headed seven-gilled shark (*Heptranchias perlo*), 90
Naucrates ductor (Pilor fish), 118*

Navaga, see *Eleginus navaga*
Needle fishes or Gar fishes (Belonidae), **250**
Nelson, K., 279
Nematobrycon palmeri, 50, 286*
Nematolosa, 195
Nemichthyidae (Snipe eels), 169
Nemichthys scolopaceus (Snipe eel), 169f, 169*, 176*
Neoceratias, 404*
Neochanna apoda, **249**
Neoeucirrichthys maydelli, 356, 359
Neolebias
— *ansorgei*, 308*
— *landgrafi*, 308*
Neon tetra (*Paracheirodon innesi*), **283**, **300***
Neoscopelidae, 260
Neoscopelus macrolepidotus, 260
Neostethidae, 446
Nervous system, 59
New England trout (Retropinnidae), **248**
Nichols, 154
Nikolski, 160f, 404
Niwaella, 356
— *delicatta*, 357
Noemacheilinae, 356, **358**
Noemacheilus, **358**f, 358m
— *angorae bureschi*, 359
— *barbatulus* (Stone loach), 237/238*, **359**, 371*
— *fasciatus*
— — *kuiperi*, 372*
Nomorhamphus, 478f
North American fresh-water catfishes (Ictaluridae), **364**
North Atlantic or Common hagfish (*Myxine glutinosa*), 31ff, 32*, 33*, 41*
North Irish herring or herring of the Isle of Man, 180
North Sea bank herring, 179, 185
Northern clingfish (*Caularchus maeandricus*), 398
Northern pike (*Esox lucius*), 223*, **250**f, 265*, **327**/**328***
Notacanthidae (Spiny eels), 171
Notacanthiformes, 171
Nothobranchius
— *guentheri*, **436**, 455*
— *melanostilus*, 455*
— *neumanni*, 455*
— *rachovi* (Fire killy), **436**, 450*, 455*
Notidanoidei (Comb-toothed sharks), 89f
Notocheirus hubbsi, **446**
Notopteridae (Featherback family), 205
Notopteroidea, 205
Notopterus chitala, **205**
Notropis (Shiners), 322
— *hypselopterus*, 322
— *lutrensis*, 322f
— *welaka*, 322
Noturus, 365
Novumbra hubbsi (Olympic mudminnow), 252
Novumbridae, see Umbridae, 250
Nybelin, 156

Oceanic trout (*Salmo trutta trutta*), 218f, 224*
Odontosternarchus, 303
Odontostilbe, 282f
Oedemognatus, 303
Ogcocephalidae (Batfishes), 402f
Ogcocephalus
— *nasutus*, **403**
— *vespertilio*, **403**, 421*
Oil sardines, See *Sardinella*
Olah, 135

Olfactory organ, 22f
Olfactory sense, 59f
Oligosarchus, 278
Olympic mudminnow (*Novumbra hubbsi*), 252
Omosudidae, 263
Omosudis lowii, 266*
Oncorhynchus (Pacific salmon), 73, 213, 226, **227**, 227m
— *gorbuscha* (Pink salmon), 227f, 228m, 228*
— *keta* (Dog salmon), 227f, 256*
— *kisutch* (Silver salmon), 227f
— *masou* (Masu), 227
— *nerka* (Sockeye salmon), 227f, 228m, 256*
— *tschawytscha* (Chinook salmon), 227ff
Oneirodidae, 404
Onogadus
— *argentatus*, 408
— *ensis*, 408
Opal fish (Goldfish), 361*
Opichthyidae (Snake eels), 160, 166f
Opichthys
— *gomesii*, 169
— *ophis*, 169
Ophidiidae, 424
Ophidioidei (Cusk eels), 424
Ophidium barbatum, 424
Opisthonema, 195
Opisthoproctidae (Barreleyes), 252
Opisthoproctus grimaldii (Barreleye), 252, 252*
Opsanus tau (Oyster toadfish), 398, 418*
Orchid dragoneye goldfish, 361*
Orectolobidae (Carpet sharks), 98f
Orectolobus maculatus (Wobbegong), 99, 103*, 114*
Oreglanis siamensis, 374
Orestias, 430
— *cuvieri*, 71, **433**
— *pentlandii*, 71
Orestiidae, 430
Orfe, see Ide
Oryziatidae (Japanese killifishes), **437**
Osmeridae (Smelts), 246
Osmerus
— *eperlanus*
— — *eperlanus* (European smelt), 246f
— — *mordax* (American smelt), 247
Osphronemus goramy, 81
Ostariophysi group, 276, 305
Osteichthys (Bony fishes), 45, 130f
Osteogenoiosus militaris, 369f
Osteoglossidae, 202
Osteoglossiformes (Osteoglossids or bonytongues), 202ff, 204m
Osteoglossids or bonytongues (*Clupisidis*), 202f
Osteoglossoidei, 202
Osteoglossum
— *bicirrhosum* (Arowana), 203f, 207*, 223*
— *ferrairai*, 204
Osteogtraci, 40
Ostracodermata (Ostracoderms), 39, 46
Ostracoderms (Ostracodermata), 39, 46
Otocinclus, 390, **392**
— *vittatus*, 412*
Ovipary, 88
Ovovivipary, 88
Owen, Richard, 45
Ox-eye herring (*Megalops cyprinoides*), 153, 153m, 155, 158*
Oxynotidae (Angular rough

sharks), 115
Oxynotus centrina (Humantin), 115
Oxyprohamphus (Flying halfbeaks), **428**
Oyster toadfish (*Opsanius tau*), **398**, 418*

Pachypanchax playfairi (Rough-back killy), 450*
Pacific cod (*Gadus macrocephalus*), 414m, **415**
Pacific herring (*Clupea pallasii*), **178**, 178m, 188f
Pacific salmon (*Oncorhynchus*), 73, 213, 226, **227**, 227m
Pacific sardine (*Sardinops caerulea*), 191f
Pacific tomcod (*Microgadus proximus*), **416**
Paddle fishes (Polyodontidae), 134, **142**
Paez, 287
Palaeodus, 40
Palaeoniscus, 48
Pangasianodon gigas (Giant catfish), 369
Pangasiidae (Giant catfishes), 369
Pantodon, 288
— *buchholzi* (Butterfly fish), 204f, 285*, 427
Pantodontidae (Butterfly fishes), 202, **204**
Parabramis, 324
Paracanthopterygii (Primitive spinny-finned fishes), 395
Paracheirodon innesi (Neon tetra), **283**, 300*
Paradontinae, 293
Paragoniatinae, 279
Parakneria, 271ff
Paralepididae, 262, 262*
Paralepis
— *atlantica*
— *barysoma*, 262
Parasitic catfishes (Trichomycteridae), 380f
Pareutropius mandevillei, 399*
Parexocoetus, **428**
Parr, 258
Parrot mormyrid (*Marcusenius*), **210**, 210*
Parrish, 182
Parthenogenesis, 64
Pearl danio (*Brachydanio albolineatus*), **310**, 318*
Pearl fishes (*Cynolebias*), 282, 436
Pearleyes (Scopelarchidae), 262f
Pearlfish (*Carapus acus*), 432*
Pearlfishes (Carapidae), 424, 424*
Pearl-scaled goldfish, 348, 361*
Pelecus cultratus, **325**
Pelican fish (*Saccopharynx ampullaceus*), 170
Pellegrin, 166
Pellonula, 195
Pellonulinae, 177
Pencil fishes (Nannostominae), **290**f
Percopsidae, 397
Percopsiformes, 395ff
Percopsis omiscomaycus, **397**
Percopsoidei (Trout-perches), 395, 397
Peruvian longfin (*Pterolebias peruensis*), **436**
Petchora herring, 189
Peters, 435f
Peters, N., 280
Petersinae, 289
Petrocephalus, **210**, 210*
Petromyzon
— *marinus* (Sea lamprey), 35, **37**f, 38m, 41*

— —*dorsatus* (American sea lamprey), 38
Petromyzonidae (Lampreys), 31, **34f**, 34m, 35*, 36*, 37*
Petromyzoninae, 34
Pfeiffer, 133
Phago, **295**
Phagoborus, **295**
Phallichthys, 440
— *amates* (Merry widow), **443f**, 451*
Phalloceros, 440
— *caudimaculatus* (Barrigudinyo), **443**, 451*
— — *reticulatus*, 443, 451*
Phallostethidae (Phallostethid priaprium fishes), 446f, 447*
Phallostethoidea, 446
Phenacogrammus interruptus, (Feather-tail tetra), **289**, 308*
Photocorynidae (Photocorynid anglers), 403
Phoxinus phoxinus (Minnow), **315f**, 315m, 371*
Phraetolaemiidae, 269, **274**
Phractolaemus ansorgei, **274**, 285*
Phreatichthys, **337**
Phycis (Forkbeards), 409, 409m, 409*
— *blennoides*, **409**
— *phycis*
Physailia pellucida, **369**
Pike (Exocoetoidei), 213
Pike eels (Muraenesocidae), 160, **165**
Piked dogfish or spur dog (*Squalus acanthias*), 104*, **115f**
Pikes (Esocidae), **250**, 250m
Pilchard (*Sardina pilchardus*), **192f**
Pimelodidae (Pimelodid catfish), **380**
Pink salmon (*Oncorhynchus gorbuscha*), 227f, 228m, **228**
Pinter, 377
Piranha (*Serrasalmus piraya*), **284**, 292*
Piranhas (Serrasalmidae), **283f**, 283m, 284*
Pirate perch (*Aphredoderus sayanus*), 256*, **396f**
Pisces, **45ff**
Pisoodonophis cruentifer, **169**
Placodermi (Placoderms), 39, **46**
Placoderms (Placodermi), 39, **46**
Placoid scales, 86
Planorbis, 327/328*
Platy, see *Xiphophorus maculatus*
Platyfishes and swordtails (*Xiphophorus*), 440
Platyproctidae, **258**
Platysomus, 48
Plecoglossidae, 245
Plecoglossus altivelus (Ayu), **245f**, 245m
Plecostomus, **390**
— *punctatus*, 412*
Pleuracanthodii, 47
Pleuronectes microcephalus (Lemon dab), 55
Plotosidae (Plotosid sea catfishes), **379f**
Plotosus
— *canius*, **379**
— *lineatus*, 379, 385*
Pluto infernalis, 396
Poecilia (Livebearers), 440
— *branneri*, **441**, 451*
— *formosa* (Amazon molly), **441**
— *latipinna*, **441**, 452*
— *melanogaster*, **441**, 451*
— *nigrofasciata*, **441**, 451*
— *parae*, **441**, 451*
— *reticulata*, (Guppy), 51*, **440f**, 451*, 460*

— *sphenops* (Black molly), 51*, **441**
— *velifera*, **441**, 452*
Poeciliidae (Livebearers), **438f**
Poeciliinae, 440
Poeciliopsis, 440, **443**
Poecilobrycon
— *eques*, 286*, **293**
— *espei*, 291*
— *ocellatus*, 286*
— *unifasciatus*, **293**
Poecilocharax
— *bovallii*, **294**
— *weitzmani*, **294**
Poecilothrissa, 195
Poikilothermic (Cold-blooded), 26
Polar cod (*Boreogadus saida*), **415f**, 416m, 416*
Poll, 289
Pollachius
— *pollachius* (Pollack), **420**, 420m
— *virenas* (Pollock, saithe or blister-back), **419**, 420m, 431*
Pollack (*Pollachius virens*), **420**, 420m
Pollock, saithe or blister-back (*Pollachius virens*), **419**, 420m, 431*
Polster, 182
Polyodon spathula (American paddlefish), 142f, 143*, 146*
Polyodontidae (Paddlefishes), 134, **142**
Polypteri (Bichirs), 48, 131
Polypteridae, 132
Polypterids (Polypteriformes), 132ff, 132m, 133*
Polypteriformes (Polypterids), 132ff, 132m, 133*
Polypterus (*Polypterus*), 132f
Polypterus (*Polypterus*), 132f
— *bichir* (Bichir), **132**
— *ornatipinnis*, 132f, 139*
— *senegalus*, 132, 140*
Pomolobus, 196
Pond loach (*Misgurnus fossilis*), 327/328*, **359**, 371*
Porbeagle shark (*Lamna nasus*), 95f, 104*
Porbeagle sharks (*Lamna*), 88
Porichthys (Midshipmen), 56, 397
— *notatus*, **397**
— *porosissimus* (Atlantic midshipman), 397, 418*
Porotergus, **303**
Portheus molossus, 48
Potamalosa, 195
Potamorrhaphis guianensis, **429**
Poutassou, see Couch's whiting (*Micromesistius poutassou*), **420**, 423*
Priestly, 183
Primitive spiny-finned fishes (Paracanthopterygii), **395**
Prionace glauca (Great blue shark), 102f, 103*, 108/109*
Pristella riddlei, 282, 286*
Pristigasterinae, 177
Pristioidei (Sawfish), 120
Pristiophorus japonicus, 119*
Pristiophoroidei (Saw sharks), 119
Pristis
— *pectinatus* (Sawfish), 118*, **120**
— *pristis*, **120**
Probolodus, 287
Prochilodinae, **294**
Prochilodus
— *insignis*, 291*
— *reticulatus magdalenae*, **294**
Promyllantor latedorsalis, **166**
Protosphyraena, 48
Prototroctes, **250**
Psephurus gladius, (Chinese sturgeon), **143**
Pseudocorynopoma doriae, 291*
Pseudomugil signifer, 453*

Pseudoplatystoma fasciatum, **380**, 387*
Pseudorasbora parva (Stone moroko), **331**
Pseudoscaphirhynchus (Asiatic shovel-nosed sturgeons), 141
— *fedtschenkoi*, **141**
— *hermanni*, **141**
— *kaufmanni*, **141**
Pseudotriakidae (False cat sharks), 102
Pseudotriakis microdon (Atlantic false cat shark), **102**
Psilorhynchidae, 354m, **355**
Psilorhynchus, **355**
Pteraspidomorphi, 40
Pteraspis, 40, 42*
Pterengraulis, 200
Pterichthys, 46*
Pterolebias
— *longipinnis* (Longfin), **436**, 450*
— *peruensis*, (Peruvian longfin), **436**
Pteromylaeus bovinus (Bull ray), 124
Pterothrissidae, 156
Pterothrissus (Deepsea bonefishes), **156**
— *gissu*, 156
Pterygoblichthys, **390**
Pterygolepis, 40, 42*
Ptychocheilus (Squawfish), 312
Ptyctodontida, 47
Puntius, 337
— *arulius*, 351*
— *conchonius*, **339**, 351*
— *cumingii*, 351*
— *eugrammus*, 351*
— *everetti*, **339**
— *fasciolatus*, **339**
— *filamentosus*, 351*
— *gelius*, 351*
— *goniotus*, 81
— *holotaenia*, 339
— *hulstaerti*, **339**, 352*
— *javanicus*, 338
— *lateristriga*, **339**
— *nigrofasciatus*, **339**, 351*
— *oligolepis*, 351*
— *orphoides*, 81
— *pentazona*, **339**
— — *hexazona*
— — *pentazona*, **339**, 351*
— *schwanefeldi*, 300*, 351*
— *semifasciolatus*, 351*
— *tetrazona*
— — *partipentazona*, **339**
— — *tetrazona* (Sumatran barb), **339**, 351*
— *titteya*, **339**, 351*
— *vittatus*, 351*
— *viviparus*, 339
Putyekanipa, see *Muraenesox*
Pyrrhulina, 290
— *vittata*, 291*
Pyrrhulininae, 290

Rabbit fish (*Chimaera monstrosa*), 104*, **126f**
Rafinesque, 142
Rainbow fishes (Melanotaeniidae), **444**
Rainbow trout (*Salmo gairdneri*), 78, 221*, 222*, 224*, **225ff**, 237/238*
Raja (Skates), 88*
— *batis* (Flapper skate), **123**
— *clavata* (Thornback ray), 117*, **123**
— *undulata* (Skate), 128*
Rajidae (True rays), 120, 122f
Rajiformes (Rays and skates), 19*, 47, 87, **120**
Rajoidei (Rays), 122
Rana temporaria, 327/328*
Raniceps raninus ("Frog codfish"),

408, 432*
Rasbora (*Rasbora heteromorpha*), 302*, **310**, 310m
Rasbora
— *daniconius*, **310**
— *heteromorpha*, (*Rasbora*), 302*, **310**, 310m
— *lateristriata elegans*, **310**
— *maculata*, 305, **311**
— *trilineata* (Scissors-tail rasbora), **310**
— *vaterifloris*, 310f
Rasborichthys altior, **325**
Rasborinae, a.k.a., Danioninae, 306
Rasborinus lineatus (a.k.a. *Rasborichthys altior*), **325**
Rays (Rajoidei), 122
Rasso, Prince of Bavaria, 366
Rays and skates (Rajiformes), 19*, 47, 87, **120**
Red dragoneye goldfish, 361*
Red neon (*Cheirodon axelrodi*), **282f**, 300*
Red tetra (*Hyphessobrycon flammeus*), **280**, 286*
Redfieldius, 48
Red-speckled killy (*Aphyosemion cognatum*), **436**
Reedfish (*Calamoichthys calabaricus*), **132**, 140*
Reedfishes (*Calamoichthys*), 132f
Reef shark (*Carcharhinus menisorrah*), 107*
Regan, 269
Reighard, 150
Remora (*Echeneis naucrates*), 103*, 113*
Respiration, 61
Retropinnidae (New England trout), **248**
Rhabdalichops, **304**
Rhamdia
— *guatemalensis*, **396**
— *sebae*, **380**
Rhamphichthyidae, 296, **304**
Rhamphichthyinae, **304**
Rhamphichthys rostratus, 304
Rhaphiodon gibbus, **278**
Rhaphiodontinae, 277
Rhenanida, 47
Rhincodon typus (Whale shark), 93*, **98**, 110/111*
Rhincodontidae (Whale sharks), 98
Rhinichthys stronasus (Blacknose dace), **323**
Rhinobatidae, 121
Rhinobatoidei (Guitar fishes), 120f
Rhinobatos
— *cemiculus*, **121**
— *productus* (Guitar fish), 117*
— *rhinobatos*, **121**
Rhinochimaeridae, 126, 129*
Rhinomuraena, **164**, 164*
Rhinosardina, 195
Rhipidistia lungfishes, 48
Rhoadsinae, 277, **283**
Rhodeus, 329
— *sericeus*
— *amarus* (European bitterling), 301*, 327/328*, **329f**, 371*
— — *sericeus* (Chinese bitterling), **330**
Rhynchobatidae, 121
Richter, 296
Risso, A., 91
River trout (*Salmo trutta fario*), 218, 220f, 224*, 237/238*
Rivulus, 436f
— *cylindraceus*, 450*
— *milesi*, 455*
Roach (*Rutilus rutilus*), (see European roach)

Roccus saxatilis (Striped bass), 256*
Rock grenadier *(Coryphaenoides rupestris)*, 425, 432*
Roeboexodon, 278
Roeboides quatemalensis, 278, 291*
Rohtee, 332
Roloffia, 436
Rondeletia, 267
Rondeletiidae, 264, 267, 267*
Rosauridae, 268
Rosen, D. E., 441, 444, 446
Rössel, 379
Rough-back killy *(Pachypanchax playfairi)*, 450*
Roule, 169
Rudd *(Scardinius erythrophthalmus)*, 313, 313m, 327/328*, 341*
Runnström, 182
"Russian sturgeon" *(Acipenser gueldenstaedti)*, 137
Rutilus
— *frisii meidingeri*, 313
— *pigus virgo*, 313
— *rutilus* (European roach), 312, 312m
— — *caspicus* (Vobla), 312
— — *heckeli* (Taran), 312

Sabanejewia, 356
— *aurata*, 360
— — *bulgarica*, 357
Sabertooth fishes (Evermannellidae), 263
Saccopharyngidae (Swallowers), 170, 170*
Saccopharyngoidei, 159, 170
Saccopharynx ampullaceus (Pelican-fish), 170
Sachs, Carl, 297
Sachs organ, 297f
Saithe, See Pollock *(Pollachius vireus)*
Salangidae (Icefishes), 49, 248, 248*
Salminus, 279
Salmo, 213
— *carpio*
— *clarki* (Cutthroat trout), 226
— *gairdneri* (Rainbow trout), 78, 221*, 222*, 224*, 225ff, 237/238*
— *salar* (Atlantic salmon), 214ff, 214m, 215*, 217*, 219*, 224*
— *trutta* (Brown trout), 218, 218m, 218*, 219*, 224*
— — *fario* (River trout), 218, 220f, 224*, 237/238*
— — *lacustris* (Lake trout), 218ff, 224*
— — *trutta* (Oceanic trout), 218f, 224*
Salmon (Salmonoidei), 213ff
Salmon catch, 83ff
Salmon migration, 214ff, 214*, 228ff
Salmonidae (Salmons), 213
Salmoniformes, 213f
Salmoninae, 213
Salmonoidei (Salmon), 213ff
Salvelinus (Charrs), 214, 230f, 231*
— *alpinus* (European charr), 231f
— *fontinalis* (Brook trout), 224*, 232
— *namaycushi* (Lake charr), 232f
Salviani, J., 97
Sanchez, 203
Sand sharks (Carcharhinidae), 91
Sand tiger shark *(Carcharias taurus)*, 88, 91f, 112*
Sarcocheilichthys sinensis, 329, 331
Sarcopterygii (Lungfishes and lobefins), 45, 48, 131
Sardina (True sardines), 192f, 193m, 193*
— *pilchardus* (Pilchard), 192f

Sardine silverside *(Hubbsiella sardina)*, 446
Sardinella (Sardines), 191, 191m
— *aurita*, 191
Sardines *(Sardinella)*, 191, 191m
Sardinops, 191, 191m
— *caerulea* (Pacific sardine), 191f
— *melanosticta* (Japanese sardine), 191f, 192*
— *neopilchardus* (Australian sardine), 191
— *ocellata* (South African sardine), 69, 191f
Satan eurystomus, 364f
Sator respectus, 293, 293*
Saurida, 261
Saury (Scomberesocidae), 430
Saw sharks (Pristiphoroidei), 119
Sawfish *(Pristis pectinatus)*, 118*, 120
Saurogobio, 331
— *dabryi*, 332
Scales, 54, 75*
Scaly dragonfishes (Stomiatoidei), 56, 213
Scanorhynchidae (Goblin sharks), 92
Scapanorhynchus owstoni (Japanese goblin shark), 92, 92
Scaphirhynchinae (Shovel-nosed sturgeons), 135, 141
Scaphirhynchus
— *albus*, 141
— *mexicanus*, 141
— *platorhynchus*, 141, 146*
Scardinius erythrophthalmus (Rudd), 313, 313, 313m, 327/328*, 341*
Sceropages
— *formosus* (Malayan bonytongue), 204
— *leichhardti* (Australian bonytongue), 204
Schilbeidae (Schilbeid catfishes), 368
Schindleria, 49
Schizodon, 293
Schizothoracinae, 343, 343m
Schizothorax, 343
Schmidt, Johannes, 161, 440
Schomburgk, 202
Schrank, 135
Schultz, Harald, 283
Scissors-tail rasbora *(Rasbora trilineata)*, 310
Scomber japonicus, 74
Scomberesocidae (Saury), 430
Scomberesox saurus (Atlantic saury), 430
Scopelarchidae (Pearleyes), 262f
Scopelosauridae, 263
Scyliorhinidae (Dogfishes), 99f, 112*
Scyliorhinus
— *caniculus* (Small-spotted dogfish), 100, 104*, 114*
— *stellaris* (Larger-spotted dogfish), 100, 104*
Sea catfishes (Ariidae), 370
Sea lamprey *(Petromyzon marinus)*, 35, 37f, 38m, 41*
Searsiidae, 258
Selachii (Sharks), 19*, 87, 89f
Sergeant Baker *(Aulopus purpurissatus)*, 260
Serrasalmidae (Piranhas), 283f, 283m, 284*
Serrasalmus
— *nattereri*, 284, 299*
— *piraya* (Piranha), 284, 292*
Serrivomer sector, 170
Serrivomeridae, 159, 170
Sex organs, 27
Shad *(Alosa alosa)*, 193, 193m, 198*
Shads *(Alosa)*, 193

Shallowwater cisco or lake herring *(Coregonus artedi)*, 240, 241
Sharks (Selachii), 19*, 87, 89f
Shaw, Evelyn, 446
Shelf herring, 179f
Sheltostshek *(Elophichthys bambusa)*, 311
Shiners *(Notropis)*, 322
Shovel-nosed sturgeon *(Scaphirhynchus platorhynchus)*, 141, 146*
Shovel-nosed sturgeons (Scaphirhynchinae), 135, 141
Siamese gyrinocheild *(Cyrinocheilus aymonieri)*, 354f, 362*
Siberian sturgeon *(Acipenser baeri)*, 137f
Siluridae (Eurasian catfishes), 366
Siluriformes (Catfishes), 363ff
Silurus
— *aristotelis*, 368
— *glanis* (European wels), 366ff, 394*
Silver bream *(Blicca bjoerkna)*, 320
Silver salmon *(Onchorhynchus kisutch)*, 227f
Silversides (Atheriniformes), 426ff
Silvery pouts *(Gadiculus)*, 423, 423*
Simenchelyidae (Snubnose eels), 169
Simenchelys parasiticus, 169
"Singing catfishes of the Ukayali", 378
Sinibotia, 358
Sisoridae (Sucker catfishes), 373
Six-gilled shark *(Hexanchus corinus)*, 90
Skate, *(Raja undutata)*, 128*
Skates, *(Raja)*, 88*
Skeleton, 21f, 22*, 56
Skin, 54
Skygazer goldfish, 361*
Sleeper shark *(Somniosus rostratus)*, 116
Smaller spotted dogfish *(Scyliorhinus caniculus)*, 10, 100, 104*, 114*
Smelts (Osmeridae), 246
Smith, H. M., 375
Smooth dogfish *(Triaenodon obesus)*, 101f, 113*
Smooth hound *(Mustelus mustelus)*, 101
Smooth hounds or smooth dogfishes (Triakidae), 101
Smooth newt *(Triturus vulgaris)*, 327/328*
Smooth-spined rat-tail *(Macrourus berglax)*, 425
Snail *(Limnaea)*, 327/328*
Snake eels (Ophichthyidae), 160, 166ff
Snipe eel *(Nemichthys scolopaceus)*, 169f, 169*, 176*
Snipe eels (Nemichthyidae), 169
Snubnose eels (Simenchelyidae), 169
Sockeye salmon *(Oncorhynchus nerka)*, 227f, 228m, 256*
Soldier fish *(Coelorhynchus carminatus)*, 425
Somniosus
— *microcephalus* (Greenland shark or large sleeper), 116
— *rostratus* (Sleeper shark), 116
Sorubim lima, 380, 400*
South African anchovy *(Engraulis capensis)*, 200
South African sardine *(Sardinops ocellata)*, 69, 191f
South American sardine *(Sardinops sagax)*, 78, 191f
South red-bellied dace *(Chrosomus erythrogaster)*, 322
Spawning, 64

Sphyrna
— *mokkaran* (Great hammerhead), 106f
— *zygaena* (Common hammerhead), 103*, 115
Sphyrnidae (Hammerhead sharks), 106
Spineless dogfishes (Dalatiidae), 116
Spiniyomer goodei, 170
Spiny dogfish (Doradidae)
Spiny dogfishes (Squalidae), 56, 115
Spiny eels (Notacanthidae), 171
Spiny sharks, 46, 46*
Spiny-rayed fishes (Acanthopterygii), 131, 152f
Spotted danio *(Brachydanio nigrofasciatus)*, 309, 318*
Spotted dragoneye goldfish, 361*
Spotted eagle ray *(Myliobatis aquila)*, 118*, 124*
Spratelloides, 196
Sprat *(Sprattus sprattus)*, 190f
Sprats *(Sprattus)*, 189f, 190m
Sprattus (Sprats), 189f, 190m
— *fuegensis*
— *sprattus* (Sprat), 190f
Squalidae (Spiny dogfishes), 56, 115
Squalius, subgenus of *Leuciscus cephalus*, 314
Squalus acanthias (Piked dogfish, or spur dog), 104*, 115f
Squatina squatina (Angel shark), 119f, 119*
Squatinidae (Monk fishes), 119
Squatinoidei (Squatinoids), 119
Squatinoids (Squatinoidei), 119
Squawfish *(Ptychocheilus)*, 312
Steatogenys elegans, 298, 304
Steelhead trout, 226
Stegocephalia (Amphibian), 20*
Stegostoma fasciatum (Zebra shark), 99
Stellate smoothhound *(Mustelus asterias)*, 101
Stenodus leucichthys (Inconnu), 234f, 255*
Stensiö, E. A., 39f, 43
Sterba, G., 30, 35, 37
Sterlet *(Acipenser ruthenus)*, 136, 145*
Sternacho gitou, 303
Sternarchella, 303
Sternarchorhamphus, 303
Sternarchorhunchus, 298*, 303f
Sternoptychidae (Deepsea hatchet fishes), 254
Sternopyginae, 304
Sternopygus, 304
Stethaprioninae, 277, 279, 287
Stilbiscus, 163
— *edwardsi*, 164
Stinging catfish, see Indian catfish
Stingray *(Dasyatis centroura)*, 123*
Sting rays (Dasyatidae), 123
Stockel, 249
Stockfish *(Merluccius capensis)*, 69, 423, 423m
Stomatorhinus, 210, 211*
Stomiatidae (Deepsea scaly dragonfishes), 253
Stomiatoidei, 56, 213, 252
Stone loach *(Noemacheilus barbatulus)*, 237/238*, 359, 371*
Stone moroko *(Pseudorasbora parva)*, 331
Striped bass *(Roccus saxatilis)*, 256*
Striped dwarf catfish *(Mystus vittatus)*, 365, 393*
Strongylura
— *crocodila*, 429
— *marina*, 429

Sturgeons (Acipenseridae), **134f**
Sturgeons and paddle fishes (Chondrostei), 48, 131, **134f**
Stygichthys typlops, 281f, 281*
Stygicola dentata, 417*, **424**
Stylophthalmus paradoxus, see *Idia canthidae*, 253
Sucker catfishes (Sisoridae), **373**
Suckers (Catostomidae), 353f, 354m
Suckerbelly loaches (*Gastromyzon*), **355**
Sucking barbs (*Garra*), 337, **340**
Swallowers (Saccopharyngidae), **170**, 170*
Swim bladder, 60f
Swimming, 50f
Swordtail characin (*Corynopoma riisei*), 279, 280*, 286*
Swordtail or helleris (*Xiphophorus helleri*), 51*, **442**, 452*
Synaphobranchidae, **169**
Synaphobranchus pinnatus (Gray's cutthroat eel), **169**
Synaptolemus, **293**
Synaptomidei, 293
Synodontidae (Lizard fishes), **260**
Synodontis
— *angelicus*, 377, 399*
— *nigriventris* (Back-swimming Congo catfish), 377f
— *schall*, 377, 399*
Synodus
— *foetens*, 260f
— *lucioceps*, **261**
— *synodus*, **261**
Syrski, 160

Tachysuridae (Mouth-breeding catfishes), **369**
Tachysurus, 369
— *sagor*, **370**
Taeniura lymna, **124**, 128*
Taleichthys, 246
Talking catfish (*Acanthodoras spinosissimus*), **377**
Tanichthys albonudes (White cloud mountain fish), **311**, 318*
Taran (*Rutilus rutilus heckeli*), **312**
Tarpons (Megalopidae), 153f
Teeth, 24
Teleost fishes, see bony fishes
Teleostei (True bony fishes, or teleosts), 130f, **150**
Telestes, subgenus of *Leuciscus soufliaagassizi*, 314
Telmatherina ladigesi, **445**, 453*
Temeridae, 121
Tench (*Tinca tinca*), **321f**, 322m, 327/328*, 336*, 341*
Tesseraspis, 40
Tetra perez (*Hyphessobrycon erythrostigma*), **282**
Tetragonopterinae (Tetras), 277, 279f
Tetras (Tetragonopterinae), 277, 279f
Thalassophryne (Toadfishes), 397
— *maculosa*, 418*
Thalassotia, 397

Thaumaturidae, 213
Thayeria
— *boehlkei*, 286*
— *obliqua*, 286*
Thelodonti, 40
Theragra, 415
— *chaleogramma* (Alaska pollack), 69, **416f**, 419*
Thienemann, A. F., 48
Thompson, 143
Thoracopterus niederristi, 427f
Thornback ray (*Raja clavata*), 117*, **123**
Three-spined stickleback (*Gasterosteus aculeatus*), 327/328*
Thresher shark (*Alopias vulpinus*), 97f, 103*
Thresher sharks (Alopiidae), 97
Thymallinae (Graylings), 244
Thymallus (Grayling), **244**, 244m
— *arcticus* (Arctic grayling)
— *thymallus* (European grayling), 237/238*, **244f**, 255*
Thyrsoidea macrurus, 164, **165**
Tiger shark (*Galeocerdo cuvieri*), **105**
Tilapia
— *galilea*, 71
— *macrochir*, 80
— *melanopleura*, 80
— *mossambica* (Mozambique cichlid), 78, 80
— *zillii*, 71
Tinca (Tenches)
— *tinca* (Tench), **321f**, 322m, 327/328*, 336*, 341*
Titanichthys, 46
Toadfishes (Batrachoidiformes), **397**
Toadfishes (*Thalassophryne*), 397, 418*
Tomcod, see *Microgradus*, 416
Tomeurinae, 439
Tomeurus gracilis, **439f**
Tooth carps (Cyprinodontoidei), 430
Tope (*Galeorhinus galeus*), 94*, **105f**
Tor tor, **338**
Torpedinidae, 121
Torpedinoidei (Electric rays), 121
Torpedo, 121*
— *marmorata* (Marbled electric ray), 117*, **122**
— *nobiliana* (Black electric ray), **122**
— *torpedo* (Eyed electric ray), **122**
Trachinocephalus myops, **261**
Trawl, 68, 68*
Triaenodon obesus (Smooth dogfish), 101f, 113*
Triakidae (Smooth hounds or smooth dogfishes), 101
Tribolodon, 315
Trichinocephalus myops, **261**
Trichomycteridae (Parasitic catfishes), **380f**
Triportheus, **279**
Trisopterus
— *esmarkii*, **423**, 423*
— *luscus*, **423**
— *minutus*, **423**

Triturus
— *alpestris*, 327/328*
— *cristatus* (Great crested newt), 327/328*
— *vulgaris* (Smooth newt), 327/328*
Troglanis, 364
— *pattersoni*, **365**
Trout breeding, 79f, 79*, 80*
Trout-perches (Percopsoidei), 395, **397**
True bony fishes or teleosts (Teleostei), 130f, **150**
True rays (Rajidae), 120, 122f
True sardines (*Sardina*), **192f**, 193m, 193*
Tschernavin, 254
Tschirr (*Coregonus nasus*), **239f**, 243
Tugun (*Coregonus tugun*), **239**
Tursiops truncatus (Bottlenose dolphin), 108/109*
Twaite shad (*Alosa fallax*), 193m, **194**, 198*
Two-striped killy (*Aphyosemion bivittatum*), **436**
Typhlias pearsei, 396
Typhlichthys, 396
Typhlobagrus kronei, 380
Typhlogarra, 337, **340**
Tyttocharax, **279**

Ultimostomias mirabilis (Deepsea loosejaw), 253
Umbra
— *krameri* (European mudminnow), 251f
— *limi* (Central mudminnow), **252**
— *pygmaea*, **252**
Umbridae (Mudminnows), 250, 251f
Upside-down catfishes (Mochocidae), 376f
Urethra fish (*Vandellia*), 381
Urophycis (Forkbeards), 409, 409m, 409*
— *brasiliensis*, **409**
— *chuss* (Mud hake), 409
— *tenuis* (Codling or hake), **409**

Vaillantella, 358
Valencia hispanica, **437**
Vandellia (Uretha fish or "Candiru"), 381
Variatus (*Xiphophorus variatus*), **442**, 452*
Varicorhinus, **340**
Velocity of propulsion, 54
Vertebrata, 19ff
Villwock, 433
Vimba
— *elongata*, **320**
— *vimba*, **320**
Vinciguerra, 254
— *lucetia*, **254**
Viviparous blenny (*Zoarces viviparus*), 57, **424f**, 432*
Vobla (*Rutilus rutilus caspicus*), 312

Wachna cod or navaga (*Eleginus navaga*), **416**
Walbaum, 142
Wallagonia attu, **368**
Warm-blooded (homeiothermic), 26
Water pollution, effect of, 81f
Weber, 169, 276
Weberian apparatus, 276f
Weitzmann, 290
Whale shark (*Rhincodon typus*), 93*, **98**, 110/111*
Whale sharks (Rhincodontidae), 98
"Whale-headed" fishes (Cetomimoidei), **264**, 267*
Whiptail loricaria (*Loricaria*), 390f
White cloud mountain fish (*Tanichthys albonubes*), **311**, 318*
White sturgeon (*Acipenser transmontanus*), **138**
Whitefishes (*Coregonus*), 235, 235m, 235*
Whitehead, 156
Whiting (*Merlangius merlangus*), 69, **420**, 420*
Wickler, Wolfgang, 279
Winge, Ölaf, 440
Wobbefong (*Orectolobus maculatus*), 99, 103*, 114*
Wolf herring (*Chirocentrus dorab*), 158*, **200f**
Wolf herrings (Chirocentridae), 177, **200f**
"Wonder-finned fishes" (Mirapinnatoidei), 264
Wunder, 330
Wrass (*Halichoeris centriguadrus*), 128*

Xenocara, see *Ancistrus*
Xenocyprinidae, 325f
Xenocypris, 329
Xenomystus nigri, 205, 208*
Xenopoecilus poptae, **437**
Xenurobrycon, 279
Xiphorphorus (Platyfishes and swordtails), 440
— *helleri* (Swordtail or helleris), 51*, **442**, 452*
— *maculatus* (Moonfish or platy), 439*, **442**, 452*, 454*
— *variatus* (Variatus), **442**, 452*
— *xiphidium*, **442**

"Yegoro" herring, 189

Zebra fish (*Brachydanio rerio*), **309**, 318*
Zebra moray (*Echidna zebra*), **164**, 174*
Zebra shark (*Stegostoma fasciatum*), **99**
Zoarces ("Eel mothers"), 424
viviparus (Viviparous blenny), 57, **424f**, 432*
Zoarcidae, 424
Zoarcoidei (Viviparous blennies), **424**

Abbreviations and Symbols

C, °C . . . . Celsius, degrees centigrade

C.S.I.R.O. . . . Commonwealth Scientific and Industrial Res. Org. (Australia)

f . . . . following (page)

ff . . . . following (pages)

L . . . . total length (from tip of nose [bill] to end of tail)

I.R.S.A.C. . . . Institute for Scientific Res. in Central Africa, Congo

I.U.C.N. . . . . Intern. Union for Conserv. of Nature and Natural Resources

BH . . . . body height

HRL . . . . head-rump length (from nose to base of tail or end of body)

N, N- . . . . . . North, Northern, North-

NE, NE- . . . . Northeast, Northeastern, Northeast-

E, E- . . . . East, Eastern, East-

S, S- . . . . South, Southern, South-

TL . . . . tail length

SE, SE- . . . . . Southeast, Southeastern, Southeast-

SW, SW- . . . Southwest, Southwestern, Southwest-

W, W- . . . . . West, Western, West-

♂ . . . . male

♂♂ . . . . males

♀ . . . . female

♀♀ . . . . females

♂♀ . . . . pair

+ . . . . extinct

$\frac{2 \cdot 1 \cdot 2 \cdot 3}{2 \cdot 1 \cdot 2 \cdot 3}$ . . . tooth formula, explanation in Volume X

▷ . . . . following (opposite page) color plate

▷▷ . . . . Color plate or double color plate on the page following the next

▷▷▷ . . . . Third color plate or double color plate (etc.)

⬧ . . . . Endangered species and subspecies